# Solid-State NMR

Applications in biomembrane structure

# Biophysical Society-IOP Series



### About the Series

The Biophysical Society and IOP Publishing have forged a new publishing partnership in biophysics, bringing the world-leading expertise and domain knowledge of the Biophysical Society into the rapidly developing IOP ebooks program.

The program publishes textbooks, monographs, reviews, and handbooks covering all areas of biophysics research, applications, education, methods, computational tools, and techniques. Subjects of the collection will include: bioenergetics; bioengineering; biological fluorescence; biopolymers *in vivo*; cryo-electron microscopy; exocytosis and endocytosis; intrinsically disordered proteins; mechanobiology; membrane biophysics; membrane structure and assembly; molecular biophysics; motility and cytoskeleton; nanoscale biophysics; and permeation and transport.

# Solid-State NMR

Applications in biomembrane structure

**Edited by**
**Frances Separovic and Marc-Antoine Sani**
*School of Chemistry, Bio21 Institute, University of Melbourne, Melbourne, VIC 3010, Australia*

**IOP** Publishing, Bristol, UK

ISBN 978-0-7503-2532-5 (ebook)
ISBN 978-0-7503-2530-1 (print)
ISBN 978-0-7503-2533-2 (myPrint)
ISBN 978-0-7503-2531-8 (mobi)

DOI 10.1088/978-0-7503-2532-5

Version: 20201201

IOP ebooks

British Library Cataloguing-in-Publication Data: A catalogue record for this book is available from the British Library.

Published by IOP Publishing, wholly owned by The Institute of Physics, London

IOP Publishing, Temple Circus, Temple Way, Bristol, BS1 6HG, UK

US Office: IOP Publishing, Inc., 190 North Independence Mall West, Suite 601, Philadelphia, PA 19106, USA

# Contents


**Preface** xiv

**Foreword** xv

**Editor biographies** xviii

**List of contributors** xix

**1 Solid-state NMR methods for studying membrane systems** **1-1**
*Gerhard Gröbner and Philip Williamson*

1.1 Introduction 1-2
1.2 Using NMR as a reporter on membrane structure 1-3
1.3 Dynamics in biological membranes 1-8
1.4 Probing the surface charge of membranes with solid-state NMR 1-12
1.5 Lateral domain formation 1-14
1.6 Lipid/protein interactions 1-17
1.6.1 Insight into protein/lipid headgroup association by $^{31}P$ CP MAS NMR 1-17
1.6.2 Studying annular lipids bound to integral membrane proteins 1-19
1.6.3 Studying non-annular lipids bound to integral membrane proteins 1-20
1.7 Membrane remodeling 1-22
1.8 Summary 1-24
Acknowledgements 1-24
References 1-24

**2 Characterization of lipid behaviour in model biomembranes using $^{2}H$ solid-state NMR** **2-1**
*Michael R Morrow and Jenifer L Thewalt*

2.1 Introduction 2-1
2.2 The quadrupole interaction 2-2
2.3 The wideline $^{2}H$ NMR spectra of chain-deuterated phospholipids in bilayers 2-3
2.4 The quadrupole echo 2-7
2.5 Characterizing phospholipid bilayer phases using wideline $^{2}H$ NMR 2-11

2.6 Using spectral subtraction to identify boundaries of two-phase coexistence regions in bilayer temperature-composition phase diagrams 2-14

2.7 Beyond difference spectroscopy—defining liquid crystalline phase coexistence with $^2$H NMR 2-17

2.8 Summary 2-18

References 2-18

**3 Deuterium solid-state NMR of whole bacteria: sample preparation and effects of cell envelope manipulation 3-1**

*Valerie Booth*

3.1 Structure and composition of bacterial cell envelopes 3-1

3.2 Sample preparation 3-3

3.3 Static $^2$H NMR of gram(−) and gram(+) bacteria 3-4

3.4 $^2$H NMR of bacteria with cell envelope perturbation 3-7

3.5 What have we learned from $^2$H NMR of bacteria so far? 3-8

References 3-9

**4 Solid-state NMR study of microalgal membranes and cell walls 4-1**

*Alexandre Poulhazan, Alexandre A Arnold, Dror E Warschawski and Isabelle Marcotte*

4.1 Overview 4-1

4.2 Microalgal cell membrane composition and architecture 4-3

4.2.1 Microalgal cell wall and plasma membrane 4-3

4.2.2 Photosynthetic organelles (thylakoids and chloroplast) 4-4

4.2.3 lipid droplets and free fatty acids 4-5

4.3 Technical considerations 4-6

4.3.1 Observable nuclei for microalgal membranes studies 4-6

4.3.2 Solution and solid-state NMR 4-7

4.4 NMR of lipid extracts and model membranes 4-9

4.5 *In vivo*, in-cell and *in situ* NMR 4-10

4.5.1 $^1$H-NMR based approaches 4-10

4.5.2 1D $^{13}$C solid-state NMR methods 4-12

4.5.3 2D-$^{13}$C and $^{15}$N-NMR experiments 4-13

4.6 Conclusion and future prospects 4-15

Acknowledgements 4-16

References 4-16

| | | |
|---|---|---|
| **5** | **Determining the mechanism of action of host defense peptides: solid-state NMR approaches**<br>*Suzana K Straus, Allen Takayesu and Prashant Kumar* | **5-1** |
| 5.1 | Introduction | 5-1 |
| 5.2 | Model membranes | 5-3 |
| 5.3 | Whole cells | 5-4 |
| 5.4 | Conclusions | 5-8 |
| | References | 5-8 |
| **6** | **$^{19}$F NMR of biomembranes**<br>*Stephan L Grage, Sergii Afonin and Anne S Ulrich* | **6-1** |
| 6.1 | Introduction | 6-2 |
| 6.2 | $^{19}$F NMR | 6-3 |
| | 6.2.1 Advantages of $^{19}$F as NMR probe | 6-3 |
| | 6.2.2 Limitations of $^{19}$F NMR | 6-4 |
| | 6.2.3 $^{19}$F NMR bridges the gap | 6-5 |
| | 6.2.4 Hardware requirements | 6-7 |
| | 6.2.5 Typical $^{19}$F NMR chemical shifts | 6-8 |
| 6.3 | Introducing $^{19}$F into biomembranes | 6-8 |
| | 6.3.1 $^{19}$F-labels | 6-8 |
| | 6.3.2 Biosynthetic incorporation of $^{19}$F | 6-12 |
| | 6.3.3 Chemical introduction of $^{19}$F | 6-13 |
| 6.4 | Line narrowing techniques for $^{19}$F NMR of biomembranes | 6-13 |
| | 6.4.1 Multipulse solid-state $^{19}$F NMR | 6-14 |
| | 6.4.2 Oriented samples for solid-state $^{19}$F NMR | 6-15 |
| | 6.4.3 Magic angle spinning $^{19}$F NMR | 6-17 |
| | 6.4.4 Solution NMR of membrane proteins in membrane mimetics | 6-17 |
| 6.5 | Solid-state $^{19}$F NMR distance measurements in biomembranes | 6-18 |
| | 6.5.1 Distance measurements in oriented samples | 6-18 |
| | 6.5.2 Distance measurements under MAS | 6-20 |
| | 6.5.3 Multidimensional solid-state $^{19}$F NMR experiments | 6-21 |
| | 6.5.4 Ligands bound to membrane proteins | 6-21 |
| | 6.5.5 Intermolecular interactions of oligomeric membrane proteins and peptides | 6-22 |
| | 6.5.6 Distance measurements in solution | 6-23 |
| 6.6 | Solid-state $^{19}$F NMR orientation measurements in biomembranes | 6-24 |
| | 6.6.1 Orientation dependence in solid-state $^{19}$F NMR | 6-24 |
| | 6.6.2 Mobility and orientation | 6-25 |

6.6.3 Molecular orientation in membranes from $^{19}$F NMR on oriented samples 6-25
6.6.4 Peptide re-alignment transitions in membranes revealed by solid-state $^{19}$F NMR 6-29
6.6.5 Peptide mobility in membranes 6-29
6.6.6 Orientation and mobility of lipids and small lipid-soluble molecules 6-30
6.7 Probing the environment 6-31
6.7.1 Conformational changes 6-32
6.7.2 Probing solvent environment 6-32
6.8 $^{19}$F NMR in native membranes 6-34
6.9 Conclusions 6-34
Acknowledgements 6-35
References 6-35

**7 Structure, topology and dynamics of membrane-associated peptides by solid-state NMR 7-1**
*Burkhard Bechinger and Evgeniy Salnikov*

7.1 Introduction 7-1
7.2 Solid-state NMR spectroscopic approaches 7-2
7.3 The anisotropy of solid-state NMR interactions 7-4
7.4 Sample preparation 7-5
7.4.1 Peptides 7-5
7.4.2 Sample preparation: bicelles and supported lipid bilayer 7-5
7.5 Oriented solid-state NMR spectroscopy to study the conformation, topology and dynamics of peptides in membranes 7-7
7.6 Sample heterogeneity and orientational distributions 7-10
7.7 Membrane lipid composition 7-12
7.8 Taking peptide motions into consideration 7-14
7.9 Signal enhancement by DNP 7-15
Acknowledgement 7-16
References 7-16

**8 Solid-state NMR and dynamic nuclear polarization studies of molecular interactions in membranes 8-1**
*Vivien Yeh and Boyan B Bonev*

8.1 Introduction 8-1
8.2 Membranes in the presence of small organic molecules 8-4

8.3 Membrane-active proteins and peptides 8-4
8.4 Transmembrane proteins 8-5
8.5 Membrane studies by NMR 8-6
8.6 Native membranes 8-7
8.7 Wideline and oriented membranes NMR 8-8
8.7.1 Deuterium NMR 8-8
8.7.2 Lipid phase analysis by $^{31}P$ wideline NMR 8-9
8.7.3 High resolution NMR in oriented membranes – $^{15}N$ PISEMA 8-11
8.8 High resolution sample spinning solid-state NMR 8-11
8.8.1 Membrane studies by $^{13}C$ MAS NMR 8-12
8.8.2 Membrane interactions monitored by $^{31}P$ MAS NMR 8-15
8.8.3 Correlation NMR spectroscopy in membranes 8-17
8.9 Dynamic nuclear polarization 8-21
8.10 Summary 8-23
Acknowledgements 8-23
References 8-23

**9 Mapping dynamics in membrane proteins with solid-state NMR: methods and examples** **9-1**
*Kaustubh R Mote and P K Madhu*

9.1 Introduction 9-1
9.2 NMR strategies to monitor dynamics 9-3
9.3 Monitoring dynamics with CP-based methods 9-6
9.3.1 WISE 9-6
9.3.2 CPPI 9-7
9.3.3 LG-CP 9-8
9.3.4 Dynamics in oriented NMR 9-8
9.4 Monitoring dynamics with heteronuclear DD recoupling-based methods 9-10
9.4.1 DIPSHIFT 9-10
9.4.2 REDOR 9-12
9.4.3 Symmetry-based schemes 9-13
9.5 Monitoring dynamics with CSA recoupling methods 9-14
9.5.1 2DCSA and SUPER 9-15
9.5.2 Symmetry-based schemes for CSA recoupling 9-16
9.6 Examples of studying dynamics in membrane proteins 9-17
9.6.1 Cross-polarisation based filtering 9-17
9.6.2 Rotational-echo based experiments 9-17

9.6.3 Protein structures using SLF experiments 9-19

9.7 Schemes for higher MAS and stronger couplings 9-20

9.7.1 REDOR-schemes with dipole–dipole coupling scaling 9-21

9.7.2 Equivalence of DIPSHIIT and REDOR pulse schemes 9-22

9.7.3 DIPSHIFT and REDOR at higher MAS frequencies 9-24

9.8 Conclusions 9-25

Acknowledgements 9-25

References 9-26

**10 Solid-state NMR studies of peripherally membrane-associated proteins: dealing with dynamics, disorder and dilute conditions 10-1**

*Patrick C A van der Wel*

10.1 Peripheral and conditional membrane proteins 10-1

10.1.1 What are peripheral membrane proteins? 10-1

10.1.2 Structural studies of peripheral membrane proteins 10-2

10.2 Solid-state NMR studies of peripheral membrane proteins 10-3

10.3 Case studies of ssNMR on peripheral membrane proteins 10-5

10.3.1 Lipid-bound cytochrome c as a pro-apoptotic lipid peroxidase 10-5

10.3.1.1 Preparing samples for MAS ssNMR 10-8

10.3.1.2 Balancing sample size, concentration and biological relevance 10-9

10.3.1.3 Optimizing the MAS ssNMR experimental conditions 10-9

10.3.2 Myelin basic protein 10-10

10.3.3 The Pleckstrin homology domain of PLC-δ1 10-12

10.4 Lessons learned 10-14

10.5 Conclusions 10-16

Acknowledgements 10-16

References 10-16

**11 Structural dynamics of G protein-coupled receptors in lipid membranes investigated by solid-state NMR spectroscopy 11-1**

*Daniel Huster*

11.1 Introduction 11-2

11.2 Preparative options for studying GPCRs by NMR spectroscopic tools 11-3

11.2.1 Construct optimization and expression systems 11-3

11.2.2 Isotopic labeling 11-5

11.2.3 Reconstitution into membranes and membrane mimetics 11-7

11.3 Methodological options for studying dynamics of GPCRs by solid-state NMR 11-8

11.3.1 Isotropic chemical shift changes 11-10

11.3.2 Influence of motions on anisotropic chemical shift interactions 11-10

11.3.3 Molecular order parameters 11-11

11.3.4 Relaxation rates 11-12

11.4 Examples 11-12

11.4.1 Chemokine receptor CXCR1 11-13

11.4.2 Neuropeptide Y receptor type 2 11-14

11.4.3 Growth hormone secretagogue receptor 11-16

11.5 Conclusions 11-18

Acknowledgement 11-19

References 11-19

**12 Hybridizing isotropic and anisotropic solid-state NMR restraints for membrane protein structure determination 12-1**

*Daniel K Weber, Erik K Larsen, Tata Gopinath and Gianluigi Veglia*

12.1 Oriented-sample solid-state NMR (OS-ssNMR) structure determination of membrane proteins 12-2

12.2 Chemical shift anisotropy 12-4

12.3 Dipolar coupling 12-8

12.4 Scaling of orientational restraints due to conformational and topological dynamics 12-9

12.5 Hybridization of isotropic and anisotropic restraints 12-12

12.6 Summary 12-15

Acknowledgments 12-16

References 12-16

**13 Beyond structure: understanding dynamics in membrane protein complexes 13-1**

*Nikita Malik and Lynmarie Thompson*

13.1 Introduction 13-1

13.2 Membrane protein dynamics 13-3

13.3 NMR studies of dynamics of specific membrane protein systems 13-7

13.3.1 Anabaena sensory rhodopsin 13-7

13.3.2 Outer membrane proteins 13-11
13.3.3 Bacterial transmembrane chemoreceptors 13-13
13.4 Conclusions 13-16
Acknowledgement 13-17
References 13-17

**14 Solid-state NMR methods for investigations of membrane protein structure and dynamics 14-1**
*Peng Xiao and Vladimir Ladizhansky*

14.1 Introduction 14-1
14.2 Structure determination 14-2
14.2.1 Sample preparation 14-2
14.2.2 Dipolar recoupling techniques 14-4
14.2.3 Resonance assignments 14-6
14.2.4 NMR structural restraints 14-10
14.3 Proton detection in membrane proteins 14-14
14.4 Dynamics 14-17
14.4.1 Protein dynamics 14-17
14.4.2 NMR methods for studying protein dynamics 14-19
14.4.3 Opportunity for observing mobile regions in membrane proteins 14-22
14.5 Conclusions and perspective 14-23
Acknowledgements 14-23
References 14-24

**15 Proton-detected solid-state NMR and its applications to membrane proteins 15-1**
*Julia Kotschy and Rasmus Linser*

15.1 Introduction 15-1
15.2 Fast MAS and deuteration 15-2
15.3 Spectroscopic possibilities of proton-detected ssNMR 15-6
15.3.1 General trends in proton-detected ssNMR 15-6
15.3.2 Signal assignment in proton-detected ssNMR 15-9
15.3.3 H/D exchange and interactions with small molecules 15-13
15.3.4 Protein structure calculation in proton-detected ssNMR 15-13
15.3.5 Relaxation and elucidation of protein dynamics 15-17

15.4 Proton detection for membrane proteins 15-19
15.4.1 General considerations 15-19
15.4.2 Preparative issues 15-20
15.4.2.1 Deuterated expression and H/D exchange 15-20
15.4.2.2 Solubilization, reconstitution and crystallization of membrane proteins 15-21
15.4.3 Example studies using ssNMR for membrane proteins 15-23
15.4.4 Perspectives 15-27
References 15-28

# Preface

Nuclear magnetic resonance (NMR) is crucial for providing structural and dynamic information of biomolecules at an atomic level. Increasingly, solid-state NMR is growing in importance in biology as a result of new developments in technology that have led to significant gains in sensitivity. As an essentially non-invasive spectroscopy, solid-state NMR allows investigation of large supramolecular complexes, from cell membranes to megadalton membrane-bound proteins, at atomic resolution. Solid-state NMR is fast becoming the benchmark for studies of the interplay between biomolecules and the role of dynamics in a biological function and cellular mechanisms. Novel approaches, techniques and analyses are being rapidly developed to overcome limitations and move to another level of biological questions.

This eBook delivers a range of exciting contributions at the forefront of biomembrane research that push the boundaries of solid-state NMR and also provide refined and detailed practical pointers for tackling complex biological questions. The chapters are authored by leaders in the field and are designed to cater for both beginners and experts. The book includes studies of the orientation, conformation, dynamics and interaction of lipids, peptides and proteins in membrane systems; novel solid-state NMR techniques, including developments in resolution and sensitivity enhancement as well as computational tools and dynamic analyses of biological molecules; and recent advances in high-resolution structure determination of membrane-associated proteins and peptides by multidimensional solid-state NMR spectroscopy.

The chapters cover four main areas of biological solid-state NMR: (i) applications to biological membranes and whole cells, (ii) membrane active peptides, (iii) dynamics of membrane proteins, and (iv) high-resolution NMR of membrane proteins. The journey of biological solid-state NMR and its application to biomembranes is stimulating and resonates with enthusiasm as the authors share their latest discoveries and provide a roadmap for solving key biological questions.

Finally, we must thank all the authors for their magnificent efforts, cooperation and patience while the chapters were being reviewed. We also thank Professor Mibel Aguilar for inviting us to edit this book as part of the BPS–IOP eBooks program series as well as the editorial team at IOP Publishing for their encouragement and assistance in putting the book together: Sarah Armstrong, Jessica Friccione and Michael Slaughter.

Marc-Antoine Sani and Frances Separovic
University of Melbourne, Australia

# Foreword

**Biomembranes: the Holy Grail**

Life would not exist without membranes. From the most primitive prebiotic envelopes to the double membranes surrounding human nuclei, all cells or cell compartments need to have a specific 'skin' that provides them with existence and function.

The quest for this strange skin started with the isolation of fatty acids and phospholipids two centuries ago by Chevreul and Gobley. Langmuir further showed that lipids could form monolayers at air–water interfaces and the existence of a lipid bilayer in water was postulated by Gorter and Grendel in 1925. Intrinsic and attached proteins were added later to the model by Danielli and Davson and a 'unified' picture was proposed by Bothorel and Belle in 1968 and by Singer and Nicolson in 1972 where membrane proteins would appear rather static in a dynamic flat lipid bilayer. Nowadays, the model has been enriched by considering the dynamic nature of proteins and the fact that membranes are heterogenous in their composition and dynamics: specific lipids and embedded proteins may form domains of very different dynamics, called 'rafts', that convey essential biological functions.

As a result, the initial two-dimensional image of a lipid–protein bilayer has evolved. Even though this is the most abundant topology, monolayers in pulmonary alveola have been identified and non-bilayer structures may be found in Golgi and Endoplasmic Reticulum or during contacts and fusions between cells or cell organelles.

This rich topology landscape is not a surprise because, from a very general point of view, these lipid–protein assemblages in water belong to the class of thermotropic and lyotropic liquid-crystals, i.e., to say their phases, in a Gibbs thermodynamic sense, can vary as a function of temperature and composition (dilution). Non-lamellar phases of hexagonal or cubic topology have been found. Under specific conditions, membrane fragments can even be isolated: they consist of flat sub-micronic bilayers capable of harboring proteins and are called bicelles or nanodiscs.

Biomembranes are thus halfway between liquids and crystals: they keep the order of orientation but have lost their order of position because their components diffuse in the plane of the membrane. This particular property is ideal for choosing nuclear magnetic resonance (NMR) as the technique of choice to study them. In contrast to x-ray crystallography, electron microscopy and vibrational spectroscopy, NMR provides both an atomic and overall structural description of the membrane components and, most importantly, a dynamic view of the movements over more than ten time decades, from picosecond to seconds, accurately illustrating the isomerization of the bonds or the membrane fluctuations and cell fusion at the micron-to-millimeter scale. It should also be mentioned that NMR can work on whole systems in their native state, i.e., performing NMR experiments on living cells using the isotope labeling strategy and examining the response of selected isotopes in a jungle of other atoms.

Professor Frances Separovic and Dr Marc-Antoine Sani have gathered in this book many illustrations of NMR applications, primarily solid-state NMR, to study the structure and dynamics of membranes. Approximately 35 specialists worldwide have contributed to 15 chapters providing "l'état de l'art" of NMR applied to biomembranes. Although NMR relaxation ($T_{1Z}$, $T_{2E}$, $T_{1D}$, $T_{1Q}$, $T_{2CP}$, etc) can describe many motional modes in membranes it was chosen not to be presented here but to focus mainly on structure and dynamics of membrane proteins as obtained from mono- or multidimensional spectra.

Different facets of solid-state NMR are presented here: they are divided between static (wide line spectroscopy) and magic angle sample spinning (MAS) experiments. The technology of spinning the rotor at the magic angle has made tremendous progress in recent years and a proton-resolved 2D-experiement has been recently obtained on a membrane protein in the solid state by Pintacuda and Loquet groups. They succeeded in spinning the rotor at 111 kHz in a 1 GHz NMR machine, thus allowing one to average the static dipolar interaction and obtain proton lines of a few Hz width with increased sensitivity.

In this book wonderful examples of resolving membrane protein structure using MAS solid-state NMR are presented by the groups of Madhu, Veglia, Thompson, Linser and Ladizhansky. On the other hand, valuable topological and dynamic information can be obtained from static, wide line, NMR of membranes hosting peptides or proteins. While intrinsic orientation information is lost with MAS experiments, this information can be obtained from experiments on membranes aligned in between glass plates, oriented sample (OS) solid-state NMR, or membranes that are magnetically oriented by the magnetic field, such as bicelles. The orientation and dynamics of protein helical segments with respect to the plane of the membrane can thus be very easily obtained. After the pioneering work of the Opella group, excellent examples are given in this book by Huster, Bechinger and Veglia, the latter nicely combining MAS and OS solid-state NMR.

As described above, this information can also be obtained from the naturally occurring isotopes, $^{1}H$, $^{31}P$, $^{13}C$ and $^{14}N$, and allows the extraction of lipid–protein interactions in the membrane because phospholipids are the main components. Gröbner and Williamson present a fine overview, and specific examples of the use of NMR to decipher the mechanism of action of host defense peptides and the binding of peripheral proteins to the membrane lipid domains are given by the Straus and van der Wel groups.

NMR has a low intrinsic sensitivity compared to other techniques, such as mass spectrometry or vibrational spectroscopy. However, the latter are not very quantitative, and NMR has the advantage of being isotope specific, i.e., isotope enrichment, which is readily available, allows monitoring a selected isotope in a forest of other atoms. NMR is well suited to study the behavior of a single labeled molecule in natural complex systems. This labelling strategy is well known in liquid-state NMR for the determination of the structure of soluble proteins or the identification of metabolic pathways. It was successfully applied 50 years ago by Seelig who pioneered deuterium NMR of lipid bilayers by showing that order parameters could accurately describe the microdynamics of membranes. In the 1980s, the Davis and Dufourc groups consolidated this theory by showing that lipid phase diagrams and order parameters obtained from carbon–deuterium segments could be analyzed according to the de Gennes concept of order parameters in liquid crystals and could lead to valuable information on membrane phases, bilayer thickness and membrane elasticity. In this book, the deuterium labeling strategy is illustrated by the work of Thewalt, Morrow and Booth, the latter demonstrating that deuterium isotope labelling and solid-state NMR could be extended to whole bacterial cells. Following this labeling strategy, the Ulrich group has also applied fluorine-19 NMR to biomembranes, taking advantage of the high sensitivity of this nucleus. In this monograph, a particular application of isotope labeling ($^{13}C$, $^{15}N$) to microalgae membranes is also described by Marcotte's group. NMR filtering using adapted pulse sequences allows obtaining valuable information on the structure and dynamics of membranes and identification of membrane components in this complex natural system.

As already mentioned, sensitivity is an issue in solid-state NMR experiments. However, this problem is seriously beginning to be addressed by taking advantage of DNP (dynamic nuclear polarization). The enormous polarization transfer from electrons of a stable organic radical (added to the membrane sample) in close proximity to the nuclei of interest can increase the NMR signal-to-noise ratio by several orders on magnitude. After the pioneering works of Griffin and Oschkinat, the groups of Bechinger and Bonev describe applications to peptides within oriented-membranes and combination of DNP and MAS NMR for low concentrations in membranes. Following on from improvements in NMR signal for biological systems that are difficult to obtain at high concentrations for intrinsic reasons, solid-state NMR cryoprobes are being developed by several manufacturers. The sensitivity gain is nonetheless lower than that obtained with DNP. However, it does not require addition of an external organic radical nor lowering the sample temperature below biologically relevant temperatures. The first experiments are still to be presented to the scientific community and I am sure that the two editors of this book will include these new perspectives in a future edition.

I wish you a good journey through these pages and I hope that you will get as close as possible to the Holy Grail.

Erick Dufourc, Paris

26th August 2020

# Editor biographies

## Frances Separovic

Professor Emeritus Frances Separovic AO FAA School of Chemistry, Bio21 Institute
University of Melbourne, VIC 3010, Australia
Email: fs@unimelb.edu.au Phone: +61 3 9035 7539

Professor Frances Separovic is a Biophysical Chemist and deputy director of the Bio21 Institute, University of Melbourne, Australia. Her research group studies the structure and dynamics of molecules in biological membranes, with a focus on peptide antibiotics and toxins, primarily using solid-state NMR techniques. Frances is a Fellow of the Biophysical Society, International Society of Magnetic Resonance and the Australian Academy of Science; an IUPAC Distinguished Women of Chemistry and an Officer of the Order of Australia. http://separovic.chemistry.unimelb.edu.au

## Marc-Antoine Sani

Dr Marc-Antoine Sani
School of Chemistry, Bio21 Institute
University of Melbourne, VIC 3010, Australia
Email: msani@unimelb.edu.au Phone: +61 3 9034 2402

Dr Marc-Antoine Sani is a senior research fellow at the School of Chemistry, University of Melbourne. His expertise precesses around DNP NMR and solid-state NMR of biomolecules, with a particular focus on in-cell studies. He obtained his PhD jointly from Umeå University and Bordeaux University. He is an editorial board member of ANZMAGazine and the Victorian representative for the Australian Society for Biophysics.

# List of contributors

**Sergii Afonin**
Karlsruhe Institute of Technology (KIT), Institute of Biological Interfaces (IBG-2) and Institute of Organic Chemistry (IOC), P.O. Box 3644, 76021 Karlsruhe, Germany

**Alexandre A Arnold**
Department of Chemistry, Université du Québec à Montréal, P.O. Box 8888, Montréal, QC, H3C 3P8 Canada

**Burkhard Bechinger**
Université de Strasbourg/CNRS, UMR7177, Institut de Chimie, 4, rue Blaise Pascal, 67070 Strasbourg, France

**Boyan B Bonev**
School of Life Sciences, University of Nottingham, Queen's Medical Centre, Nottingham NG7 2UH, UK

**Valerie Booth**
Department of Biochemistry and Department of Physics and Physical Oceanography, Memorial University of Newfoundland, St. John's, NL, A1B 3X9, Canada

**Erick J Dufourc**
Institute of Chemistry and Biology of Membranes and Nanoobjects, CNRS, University of Bordeaux, Bordeaux Polytechnic Institute, Bordeaux, France

**Tata Gopinath**
Department of Biochemistry, Molecular Biology, and Biophysics, University of Minnesota, Minneapolis, MN 55455, USA

**Stephan L Grage**
Karlsruhe Institute of Technology (KIT), Institute of Biological Interfaces (IBG-2) and Institute of Organic Chemistry (IOC), P.O. Box 3644, 76021 Karlsruhe, Germany

**Gerhard Gröbner**
Department of Chemistry, Umeå University, Umeå, Sweden

**Daniel Huster**
Institute of Medical Physics and Biophysics, University of Leipzig, Härtelstr. 16-18, D-04107 Leipzig, Germany

**Julia Kotschy**
Faculty of Chemistry and Chemical Biology, Technical University Dortmund, Otto-Hahn-Straße 4a, 44227 Dortmund, Germany

**Prashant Kumar**
Department of Chemistry, University of British Columbia, 2036 Main Mall, Vancouver, BC, V6T 1Z1, Canada

**Vladimir Ladizhansky**
Department of Physics and Biophysics Interdepartmental Group, University of Guelph, Guelph, ON N1G 2W1 Canada

**Erik K Larsen**
Department of Biochemistry, Molecular Biology, and Biophysics, University of Minnesota, Minneapolis, MN 55455, USA

**Rasmus Linser**
Faculty of Chemistry and Chemical Biology, Technical University Dortmund, Otto-Hahn-Straße 4a, 44227 Dortmund, Germany

**P K Madhu**
Tata Institute of Fundamental Research Hyderabad, 36/P Gopanpally Village, Ranga Reddy District, Hyderabad-500 046, India

**Nikita Malik**
Department of Chemistry, University of Massachusetts Amherst, Amherst, MA 01003, USA

**Isabelle Marcotte**
Department of Chemistry, Université du Québec à Montréal, P.O. Box 8888, Montréal, QC, H3C 3P8 Canada

**Michael R Morrow**
Department of Physics & Physical Oceanography, Memorial University of Newfoundland, St. John's, Newfoundland A1B 3X7, Canada

**Kaustubh R Mote**
Tata Institute of Fundamental Research Hyderabad, 36/P Gopanpally Village, Ranga Reddy District, Hyderabad-500 046, India

**Alexandre Poulhazan**
Department of Chemistry, Université du Québec à Montréal, P.O. Box 8888, Montréal, QC, H3C 3P8 Canada

**Evgeniy Salnikov**
Université de Strasbourg/CNRS, UMR7177, Institut de Chimie, 4, rue Blaise Pascal, 67070
Strasbourg, France

**Suzana K Straus**
Department of Chemistry, University of British Columbia, 2036 Main Mall, Vancouver, BC, V6T 1Z1, Canada

**Allen Takayesu**
Department of Chemistry, University of British Columbia, 2036 Main Mall, Vancouver, BC, V6T 1Z1, Canada

**Jennifer L Thewalt**
Department of Physics and Department of Molecular Biology & Biochemistry, Simon Fraser University, Burnaby, British Columbia V5A 1S6, Canada

**Lynmarie Thompson**
Department of Chemistry, University of Massachusetts Amherst, Amherst, MA 01003, USA

**Anne S Ulrich**
Karlsruhe Institute of Technology, Institute of Biological Interfaces IBG-2 and Institute of Organic Chemistry, P.O. Box 3644, 76021 Karlsruhe, Germany

**Patrick C A van der Wel**
Zernike Institute for Advanced Materials, University of Groningen, Nijenborgh 4, 9747 AG Groningen, The Netherlands

**Gianluigi Veglia**
Department of Biochemistry, Molecular Biology and Biophysics, Department of Chemistry, University of Minnesota, Minneapolis, MN 55455, USA

**Dror E Warschawski**
Department of Chemistry, Université du Québec à Montréal, P.O. Box 8888, Montréal, QC, H3C 3P8 Canada
Sorbonne Université, École Normale Supérieure, PSL University, CNRS, Laboratoire des biomolécules, LBM, 75005 Paris, France

**Daniel K Weber**
Department of Biochemistry, Molecular Biology, and Biophysics, University of Minnesota, Minneapolis, MN 55455, USA

**Philip Williamson**
Centre for Biological Sciences, University of Southampton, Southampton, United Kingdom

**Peng Xiao**
Department of Physics and Biophysics Interdepartmental Group, University of Guelph, Guelph, ON N1G 2W1 Canada

**Vivien Yeh**
School of Life Sciences, University of Nottingham, Queen's Medical Centre, Nottingham NG7 2UH, UK

# Chapter 1

# Solid-state NMR methods for studying membrane systems

**Gerhard Gröbner and Philip Williamson**

## Overview

Human life would not exist without biological membranes, which are fascinating lamellar bilayers composed of lipids and proteins. These incredibly thin leaflets are spontaneously formed in aqueous environment by lipids, which are amphiphilic molecules occurring in huge variations with respect to their hydrophilic headgroups and their hydrophobic fatty acid chains. Eukaryotic cells can easily obtain over a thousand different lipid variations. These lipids provide membranes with their basic physicochemical features, essential to exert their biological function in a concerted action with their inherent proteins. To understand the unique features of membranes—from the individual lipid to the cooperative structural and dynamic behavior of the entire leaflet —NMR has provided a wealth of atomic-level insight. In contrast to many other structural biology approaches, NMR can use non-perturbing reporters (naturally or synthetically introduced) to monitor the behavior of the membrane under physiological conditions, including elevated temperature, and it can follow the structural and dynamic response of membranes and their constituents upon physiological activities. This way it can elucidate the essential lipid–lipid and lipid–protein interactions that underpin the membrane's function. NMR is ideal as it can exploit the naturally occurring nuclei ($^{31}$P, $^{14}$N, $^{13}$C, $^{1}$H) within the lipids as well as synthetically incorporated deuterons as non-perturbing reporters of the structure and dynamics within the bilayer. In this manner a wide range of membrane functions have been elucidated, ranging from basic interactions with ions, small drug molecules, but also to peptides and proteins involved in numerous biological events. NMR has also contributed to the study of domain formation in membranes, where nanodomains and rafts have emerged as important factors in the regulation of several important biological processes. Some of the recent exciting developments are summarized in this chapter.

doi:10.1088/978-0-7503-2532-5ch1  

## 1.1 Introduction

Membranes are essential for human life and virtually every biological process involves these three to four nm wide bilayers composed of lipids and proteins (see for schematic picture figure 1.1). To function properly, a cell relies on spatially separating intracellular molecular interactions from the extracellular surroundings, while providing controlled molecular communication between both spaces, a task the plasma membrane fulfills ideally. Thus, membranes act as a selective barrier to enable the cell to carry out molecular reactions and processes inside without undesired leakage of biomolecules out; while enabling extracellular molecular communication to control and modulate these intracellular processes in response to the needs of the organism. Hence the plasma membrane helps a cell to maintain and control its internal wellbeing while being separated from the outside and biochemical events there.

Due to the unique features of lipid-based membranes, which provide spatial separation of molecular compounds and their specific reactions, it is not surprising that evolution generated additional specialized membranes inside eukaryotic cells. These membranes define intracellular compartments, mainly organelles, allowing internal processes to run simultaneously and exert their tasks by distributing them to organelles that keep these processes spatially and temporally separated to avoid interference (Luckey 2008, Liljas *et al* 2009, van Meer *et al* 2008).

To exert all these functions—from communication with the cellular exterior to running intracellular processes simultaneously—membranes had to adapt to these tasks by becoming highly specialized; made possible by varying the lipid composition and the properties of membrane proteins embedded within them. For this purpose, nature has generated thousands of lipid species, with phospholipids being the most common, to fine-tune their properties to physiological needs. By varying the lipid species, membranes obtain their basic physicochemical properties like thickness, fluidity, heterogeneity, surface charge and their three-dimensional dynamic and structural architecture: properties required for membrane-associated

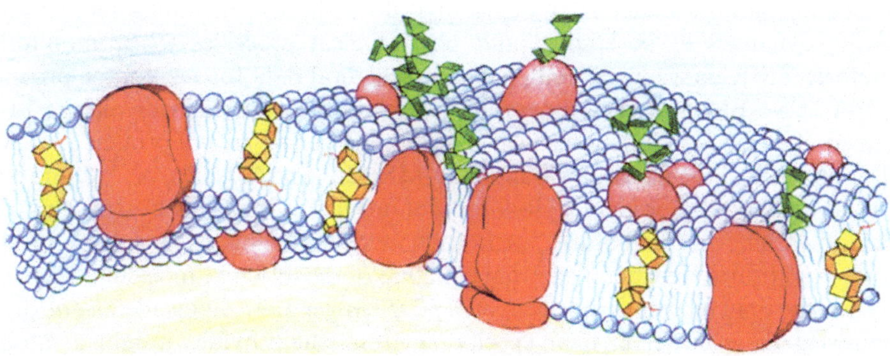

**Figure 1.1.** A biological membrane composed of lipids, sterols (yellow) and glycosylated molecules (green) with membrane proteins (red) embedded. Adapted from Liljas *et al* (2009) copyright John Wiley & Sons; courtesy of Vanessa Flegler.

proteins to function, with lipid–protein interactions often playing a crucial part in physiological events (Luckey 2008). Since phospholipids are essential constituents in nearly all biological membranes (Marsh 2013, Luckey 2008, Wahle 1983, Ardail *et al* 1990), it is not surprising that specialized membranes show a huge variation in these phospholipids with respect to fatty acid composition and headgroups (which define the class of phospholipids and the membrane surface potential). Besides these phospholipids, the sterol cholesterol plays an important role in tuning membrane properties, such as structural order and viscosity but also in modulating dynamics of lipid–lipid and lipid–protein interactions and transport of molecules across the membrane.

To understand how biological functions are spatially and temporarily organized in a two-dimensional lipid-based leaflet is a challenging task for the life sciences (Luckey 2008, Yeagle 2011, Engelman 2005, van Meer *et al* 2008). Understanding biological membranes, ranging from structural and dynamic features of the lipids to the complex structure and activity of membrane-associated proteins, has been difficult by conventional high-resolution (HR) structural methods, severely restricting our atomic-level insight into their function. Fortunately, solid-state NMR of intact membranes has emerged as a powerful tool to provide deep-level insight by studying the spatial, dynamic and functional organization of membranes under near physiological conditions (Dingeldein *et al* 2018, Fusco *et al* 2017, Kovacik *et al* 2018, Sodt *et al* 2016, Thysell *et al* 2019, Dingeldein *et al* 2017); information difficult to obtain by other means. Solid-state NMR can even describe the role of individual membrane constituents, including the proteins and lipids and their interactions.

## 1.2 Using NMR as a reporter on membrane structure

To provide an atomic-level insight into the structural and dynamic organization of a membrane at various levels of complexity, from an individual lipid molecule to supramolecular lipid/protein domains of many nm size, solid-state NMR and HR-like magic angle spinning (MAS) NMR spectroscopy have proved invaluable (Pikula *et al* 2015). Compared to other biophysical techniques, NMR permits the study of membranes under near physiological conditions, including temperature, allowing dynamics to be studied and physiological processes to be monitored. Therefore, NMR data ideally complements structural data for membrane proteins, e.g. from cryo-EM and crystallography, enabling a comprehensive picture to be obtained of membrane function as viewed from the perspective of protein and lipid.

The presence of the naturally abundant NMR active nuclei $^{31}P$, $^{14}N$, $^{13}C$ and even $^{1}H$, ensures that NMR is an ideal tool for monitoring the lipid constituents and their behavior within the membrane, with each nucleus acting as a non-perturbing reporter and providing a detailed atomic picture of the different molecular segments that span the lipid bilayer (Lindstrom *et al* 2005, Auger 1997, Dingeldein *et al* 2017). Monitoring these nuclei in their specific environment provides insight at atomic resolution into different layers of the membrane, from the hydrophilic surface formed by the lipid headgroups to the hydrophobic membrane interior formed by the fatty acid chains.

While $^{31}$P and $^{14}$N nuclei reside in the lipid headgroup and provide a local view of the headgroup region at the interface to the aqueous environment, $^{13}$C and $^{1}$H provide a more general view across the entire bilayer, with focus on the hydrophobic bilayer region (where most intensity is observed for both nuclei). In addition, specifically deuterated lipids are often used to address the behavior of specific lipid regions, useful when spectral overlap in the $^{1}$H and $^{13}$C spectra prohibit unlabeled studies. These allow changes in order, due to the presence of other lipid species or membrane-associated proteins/peptides, to be monitored (Brown and Seelig 1977, Lindstrom *et al* 2006, Overall *et al* 2019). The use of deuterated lipids is especially important in neutron-reflectometry (NR) of bilayers as both techniques complement each other (Wacklin *et al* 2016). NR provides an overall picture across the entire membrane system, including the location of molecular components and their relative abundance (Paracini *et al* 2018). In contrast, NMR gives an atomic picture from atoms and bonds in the individual molecule, including structural and dynamic characteristics. The potential to resolve individual sites is demonstrated in figure 1.2, where a $^{13}$C MAS NMR spectrum of a complex mixture of lipids is shown.

As seen here, exploiting the natural abundance of $^{13}$C is sufficient to obtain intense and well-resolved NMR spectra of complex lipid bilayers in their biologically relevant liquid-crystalline disordered state; a state characterized by fast intra- and intermolecular lipid dynamics and increased disorder in the fatty acid and headgroup region (Dufourc *et al* 1992). Such spectra enable identification of carbon sites (and lipid classes) across the entire membrane, providing a unique picture of the structural and dynamical features of the headgroup, lipid glycerol interface and the

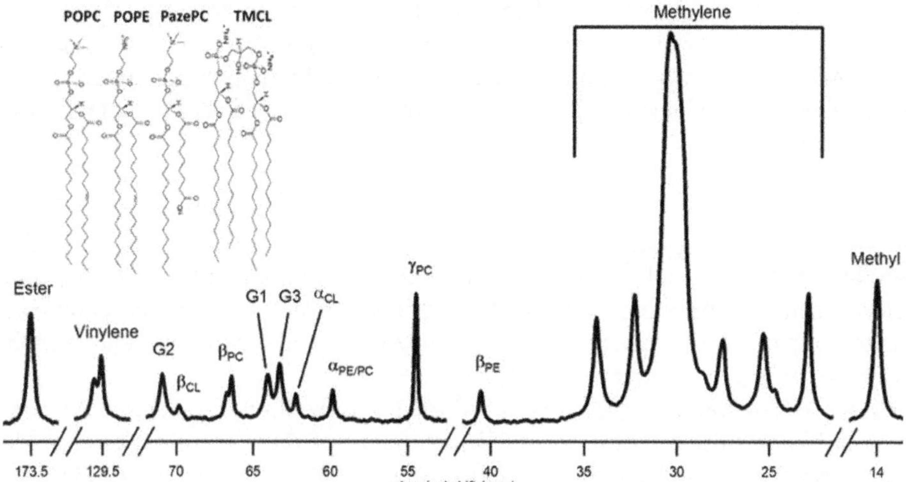


**Figure 1.2.** Direct polarization $^{13}$C MAS NMR spectrum of multilamellar vesicles (MLVs) of a lipid mixture mimicking mitochondrial outer membrane, composed of POPC/POPE/TMCL doped with an oxidized lipid species PazePC to mimic oxidative stress at a 33:36:21:10 molar ratio. Measurement was carried out on a 500 MHz NMR at 318 K with 10 kHz spinning speed. Most resonances were identified, and respective lipid locations and species are indicated. Resonance assignments are in Table I of Dingeldein *et al* (2018), reproduced with permission. Copyright 2018, Elsevier.

main hydrophobic membrane regions of the various fatty acid groups. Due to the wide range of isotropic chemical shift values for carbon nuclei in different molecular environments (ca. 200 ppm), identification of the individual lipid moieties via their specific isotropic chemical shift values is straightforward (Forbes *et al* 1988), and much less affected by spectral overlapping compared to $^{1}$H MAS NMR spectra of these systems.

Since the main classes of phospholipids are based on differences in headgroup structure, it is not surprising that the intrinsic and sensitive spin-½ $^{31}$P nucleus was exploited in early NMR studies of membranes. The 100% natural abundance of $^{31}$P ensures that even natural membranes can be studied with good sensitivity and without cumbersome labelling (Sani *et al* 2009). Although differences in local chemical environment permit the resolution of the individual lipid classes on the basis of their isotropic chemical shifts (Petzold *et al* 2009, Sani *et al* 2009), many of the early studies utilized wideline static $^{31}$P NMR to report on the phase behavior of the lipids as the resolution of the individual lipid species is not required (Cullis and De Kruijff 1979, Filippov *et al* 2015, Cullis and Hope 1978). As seen in figure 1.3, the static $^{31}$P NMR spectrum of lipids is defined largely by the chemical shielding anisotropy and the orientational distribution of the lipids relative to the magnetic field, which is determined by the lipid phase (Seelig 1978). Accordingly the $^{31}$P NMR spectra of lipids provide an excellent fingerprint of the phase adopted by a particular assembly of lipids (Seelig 1977, Cullis and Hope 1978, Seelig 1978). As with the examples seen in figure 1.3, for lipids in a liquid-crystalline bilayer the rapid rotational motion of the lipid within the bilayer coupled with the powder distribution of the sample results in an axially symmetric lineshape. Even in the case of lipid mixtures, the relatively small distribution in the isotropic chemical shifts of the naturally occurring lipid headgroups (typically $<$3–4 ppm) ensures that the spectrum is dominated by the orientational distribution of the lipids present rather than the bilayer composition. This is also seen in the wideline $^{31}$P NMR lineshapes of complex native (Sani *et al* 2009) and model membranes (figure 1.4). At lower temperatures, when many lipids enter the gel phase, there is a reduction in the $T_2$ of the $^{31}$P signals and a concomitant increase in linewidth as seen in figure 1.3 (top column, right). The spectrum exhibits asymmetric broadening of the features associated with the axially symmetric lineshape of the high-temperature $L_\alpha$ phase. Measurement of the second moment for these powder-like $^{31}$P lineshapes as a function of temperature can provide a convenient method for the characterization of the phase transition (Leonard and Dufourc 1991).

Under *in vivo* conditions most membranes are in the fluid, liquid-crystalline state which is biologically relevant. Nevertheless, many of these membranes can adopt other phases at non-physiological temperatures (Muzic *et al* 2019). The phase transition into the biological active state occurs slightly below physiological temperature where they exert their function (in humans 310 K). This melting event occurs at slightly lower temperatures in isolated native membranes, which may indicate that this transition, slightly below its physiological active state, might be important for cell function and survival. This has been demonstrated in bacteria grown under various temperature conditions, which indicate that phase transitions of natural

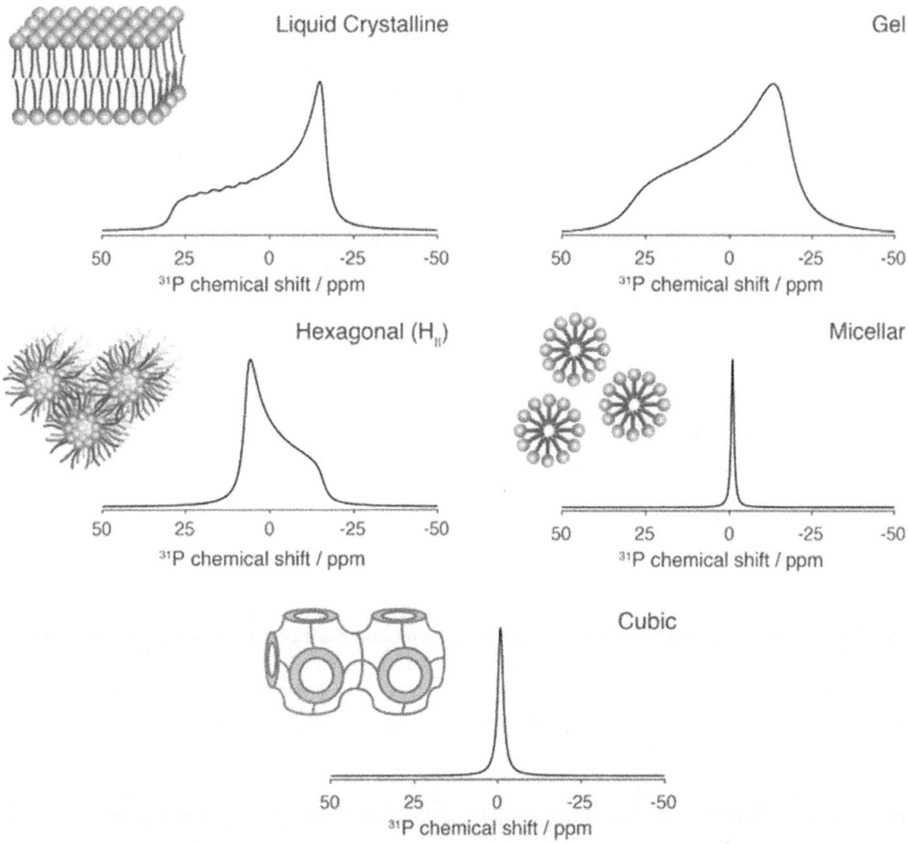


**Figure 1.3.** Phospholipid bilayers can adopt a wide range of lyotropic and thermotropic phases, dependent on headgroup and fatty acid properties of the constituting lipid species and their relative composition. Static $^{31}$P NMR lineshapes are ideal to identify different phases due to phase specific differences in orientational distribution of the lipids within the sample and the individual lipid mobilities and lateral diffusion processes (Seelig 1978, Cullis and Hope 1978).

membranes are important (Muzic *et al* 2019). Nevertheless, non-bilayer phases can also occur under *in vivo* conditions for numerous events, ranging from membrane invagination, cell division, membrane disruption by toxins, antimicrobial peptides, pore-forming proteins, etc (Lidman *et al* 2016, Shai 2002). In such cases $^{31}$P NMR can provide an invaluable tool with which to monitor formation of non-bilayer phases.

Since biological membranes are composed of lipids and proteins, extensive research has been carried out to establish if the constituent lipids alone are sufficient for the basic membrane structure and behavior. For many lipid extracts from biological membranes this is the case (Sani *et al* 2009), but many naturally occurring lipids readily adopt an alternative lipid phase if it is the only bilayer constituent (Yeagle 2011). These phases include hexagonal (mainly $H_{II}$), cubic and micellar, with each generating a distinctive $^{31}$P NMR lineshape (see figure 1.3). Many of these

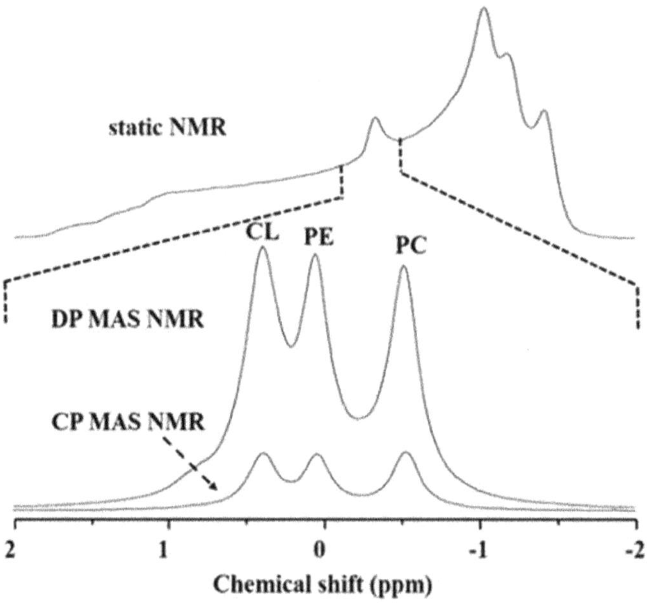


**Figure 1.4.** $^{31}$P NMR spectra of mitochondrial outer membrane-mimicking vesicles under oxidative stress conditions DOPC/DOPE/TOCL/PoxnoPC at 33:36:21:10 molar ratio; 303 K and 500 MHz NMR): (top) wideline static $^{31}$P Hahn-echo NMR spectrum; and (bottom) HR MAS NMR spectra obtained by direct polarization (DP) of all nuclei or cross polarization (CP) NMR techniques. Intensities for DP and CP spectra are plotted in absolute intensities for comparison.

non-lamellar phases can be explained by the structure of the specific phospholipid based on geometric constraints, such as the hydrophilic headgroup size and charge compared to the diameter and volume of the hydrophobic fatty acid region (Yeagle 2011). Therefore, stable lipid membranes often contain a mixture of lipid species, which only in combination form a stable bilayer as seen in Nature. As a typical example, the wideline $^{31}$P NMR lineshape of multi-component model membranes is shown single in figure 1.4. Here, the lipid species, such as cardiolipin (CL) or the oxidized lipid on its own, would not form a lamellar stable bilayer. But with this mitochondrial membrane-mimicking composition, the main lipid components are in a lamellar bilayer, as clearly seen in figure 1.4. The superposition of lamellar lineshapes, due to the differences in chemical shift anisotropies of contributing CL, phosphatidylcholine (PC) and phosphatidylethanolamine (PE) lipids, displays a convoluted $^{31}$P spectrum (figure 1.4).

While wideline $^{31}$P NMR provides a wealth of information about basic features of membranes, monitoring individual lipid components in complex membranes is cumbersome, achieved by biological or chemical labelling of the specific lipid compound of interest, often by deuteration (Davis 1983, Macdonald *et al* 1991, Molugu *et al* 2017, Weisz *et al* 1992). However, the emergence of high-resolution (HR) MAS NMR techniques allows differentiation of individual lipid species without specific labelling, but essential if studying isolated native membranes or when specific lipid components cannot be deuterated. With improvements in resolution and sensitivity brought about by the introduction of MAS, $^{1}$H, $^{13}$C and $^{31}$P NMR have proven to be a valuable capability with which to elucidate complex biological mechanisms, follow individual molecular components in biological systems, such as intact biospecimens (Dudka *et al* 2020), and monitor their response to biological activity.

By combining solid-state MAS with sensitive $^{31}$P NMR, it became possible to identify individual lipid classes according to their headgroup specific $^{31}$P chemical shift values (Meneses and Glonek 1988). As seen in figure 1.4, the individual lipid components can easily be identified by the $^{31}$P MAS NMR approach in a complex bio-membrane model composed of DOPC/DOPE/TOCL bilayers doped with PoxnoPC. Identification of individual phospholipids by $^{31}$P MAS NMR opened up the possibility, as for $^{13}$C MAS NMR, to extract the dynamic behavior of individual lipid components and their hydrophilic and hydrophobic moieties by employing dynamic filtering NMR experiments like CP and DP (Dingeldein *et al* 2018). The CP NMR signals are much weaker compared to that obtained by DP MAS NMR (see figure 1.4); a situation which reflects the degree of structural disorder and fast motional processes occurring for the different lipid species. This differential intensity for CP versus DP is much more pronounced in membranes prone to phase transitions and phase separations (Dingeldein *et al* 2018).

The MAS NMR tracing of individual lipids also enables study of their specific role and behavior during physiological processes (Carbone and Macdonald 1996, Bokvist *et al* 2004, Sani *et al* 2009); processes ranging from raft formation to binding of regulatory proteins and transport across the bilayer (Luckey 2008). In addition, lipid species and their relative compositions in membranes are of growing interest as potential markers in diseases and diagnostics, an area where often $^{31}$P and $^{1}$H HR MAS NMR on intact biopsies is employed (Dudka *et al* 2020). This diagnostic area, together with natural abundance of thousands of different lipid species necessary to fine tune membrane properties in biological systems, created a new omics research area, called 'Lipidomics' (Mouritsen 2011).

## 1.3 Dynamics in biological membranes

Solid-state NMR has proven to be a particularly useful tool in the analysis of dynamics in biological membranes, with the ability to probe motions across a broad range of timescales from nanoseconds to seconds. This has permitted the investigation of a wealth of processes, including the partitioning of drugs into bilayers, the influence of integral membrane proteins and membrane dynamics, and the interaction of antimicrobial peptides with membrane surfaces.

Much of the pioneering work in this field exploited deuterium NMR based methods (Seelig 1977, Molugu and Brown 2019). Although deuterium is not highly naturally abundant, the unambiguous labelling of particular sites within a lipid allowed site-specific studies to be undertaken; permitting the analysis of, for example, the headgroup or chain regions of the lipid. As a spin-1 nucleus, deuterium has a relatively large quadrupolar interaction (typically ~162 kHz for a static $CH_2$

group) that results in a distribution of resonances across hundreds of kHz. However, the motional processes that occur in most lipid structures ensure that the quadrupolar interaction is scaled sufficiently to permit accurate lineshapes to be acquired with a standard quadrupole echo sequence (Davis *et al* 1976), providing valuable insights into lipid structure and dynamics.

Deuterium NMR is a particularly valuable tool with which to understand the dynamics and packing of the lipid chains within the hydrophobic core of the bilayer (Weisz *et al* 1992, Molugu *et al* 2017, Mihailescu *et al* 2011). Due to the significant changes in chain dynamics observed upon transition from the liquid crystalline to the gel phase, significant broadening of the deuterium spectra occurs (Seelig 1977, Molugu *et al* 2017). This can readily be used to quantitate the phase transition behavior through analysis of the second moment of the spectrum as a function of temperature (see chapter 2).

Analysis of site-specific and chain-perdeuterated lipids by deuterium NMR spectrum has provided us with valuable insights into the dynamic gradient that runs through the bilayer, with differences in dynamics along the length of the acyl chain leading to a change in the quadrupolar splitting, from the broader spectrum for the more rigid $CH_2$ segments close to the glycerol backbone of the lipid to the narrower $^2$H splittings of the CH2 and CH3 at the ends of the acyl chains (Weisz *et al* 1992, Molugu and Brown 2019, West *et al* 2020). This allows quantitation of the order parameters through the lipid bilayer. For liquid crystalline bilayers, the lineshapes are sufficiently well resolved that much of the Pake pattern fine structure, originating from the individual deuterons along a perdeuterated lipid chain, can be identified and analyzed separately by de-Paking (Sani *et al* 2013, McCabe and Wassall 1997). This de-Paking of powder-like deuterium NMR spectra allows creation of an order parameter profile describing the dynamic gradient through the lipid bilayer. Such profiles have been used to study the interaction of proteins, nanoparticles (Russ *et al* 2016) and drugs with the lipid bilayer. Such perturbations in the order parameter profile provide insights into the localisation and mode of interaction within the lipid bilayer.

Although many of the studies have relied on deuterated synthetic lipids, with recent improvements in instrumental sensitivity and the development of techniques to introduce deuterated metabolites into organisms, $^2$H NMR is now being adopted to study native membranes in live organisms (Thurmond *et al* 1994) (see chapter 3). In such cases, deuterated fatty acids incorporated into the membranes of bacteria and algae can be studied to provide insights into the change in bilayer structure and dynamics in response to changes in environment (e.g. temperature) or challenges from antimicrobials (e.g. antimicrobial peptides) (Bouhlel *et al* 2019, Laadhari *et al* 2016, Warnet *et al* 2015, Warnet *et al* 2016, Tardy-Laporte *et al* 2013).

In the absence of the deuteration, insights into chain and headgroup dynamics can also be obtained from natural abundance $^1$H and $^{13}$C MAS NMR studies (see chapter 4). As described in section 1.3, these techniques allow unambiguous identification of chemical groups in different membrane regions to provide a unique possibility to monitor local structure (order parameter) and dynamics across the entire bilayer; and especially changes in response to perturbation of the membranes by membrane-active molecules (specific lipid species, drugs, proteins, peptides, other membrane effectors) or phase behaviour and basic physicochemical properties caused by temperature, hydration or pressure.

By applying $^{13}$C MAS NMR in combination with dynamic filtering, a dynamic profile of individual lipids and their various segments can be generated across a bilayer, ranging from the hydrophobic acyl region to the hydrophilic solvent-exposed headgroups. Extracting dynamic information via NMR is based on differences in the local dynamics inclusive of timescale and order of lipid moieties and their consequence on signal intensities. As shown previously (Nowacka *et al* 2010, Schmidt *et al* 2014), $^{13}$C INEPT (insensitive nuclei enhanced by polarization transfer) can extract information about fast motional processes while CP (cross polarization) based NMR approaches are sensitive to more restricted lipid dynamics on slower timescales with larger order parameters. By changing temperature and lipid composition, site-specific dynamic changes can be followed in different membrane regions and on respective timescales. As shown in figure 1.5, the effect of phase transition and abundance of additional lipid species is clearly visible for basic mitochondrial membrane-like lipid vesicles composed of POPC/POPE/TMCL at 43:36:23 molar ratio (Dingeldein *et al* 2019, Dingeldein *et al* 2018). By increasing temperature, the membrane 'melts' into a disordered biologically-relevant liquid-disordered state; a process characterized by a decrease in the strong CP NMR signals, most pronounced for the methyl groups of the fatty acid region, where the melting of the acyl chains increases disorder and fast dynamics (Weisz *et al* 1992). Simultaneously this increase in fast chain dynamics and disorder, which reflects an increase in membrane fluidity, causes a significant enhancement of the INEPT signal at higher temperatures. This simple behaviour is clearly seen in the hydrophobic region, where a higher chain order together with very restricted motional processes exists in the gel phase.

At low temperatures the $^{13}$C–$^{1}$H direct dipolar interactions of the methylene, methyl and vinyl segments are not dynamically averaged and, therefore, generate an efficient CP buildup and intense NMR signal, as seen in figure 1.5 (left row, bottom). This behaviour is not surprising since the acyl chain region is characterized by high chain order and very restricted intra- and intermolecular dynamics in the gel phase (Seelig 1977, Weisz *et al* 1992, Molugu *et al* 2017). At these lower temperatures, the INEPT signal is weak since its driving force is due to J-coupling, which works effectively under fast motions. Therefore, increasing temperature to gradually melt the lipids and increase their dynamics is seen directly as a gradual increase in INEPT signal and an attenuation of the CP signal intensity (see figure 1.5 left, bottom). This reflects a gradual melting of fatty acids in the hydrophobic membrane region, where above 318 K all acyl chains (including the myristoyl chains of the TMCL component) are completely in the lamellar liquid-crystalline phase and the INEPT signal is maximal while the CP signal remains relatively low. This behavior is seen for various chain segments (vinylene, methylene and methyl segments) and is summarized in figure 1.5 (right, top lane left box). However, this straightforward behavior becomes more complex for other lipid segments in the membranes, as summarized in figure 1.5 (right). Here, additional motional processes besides those

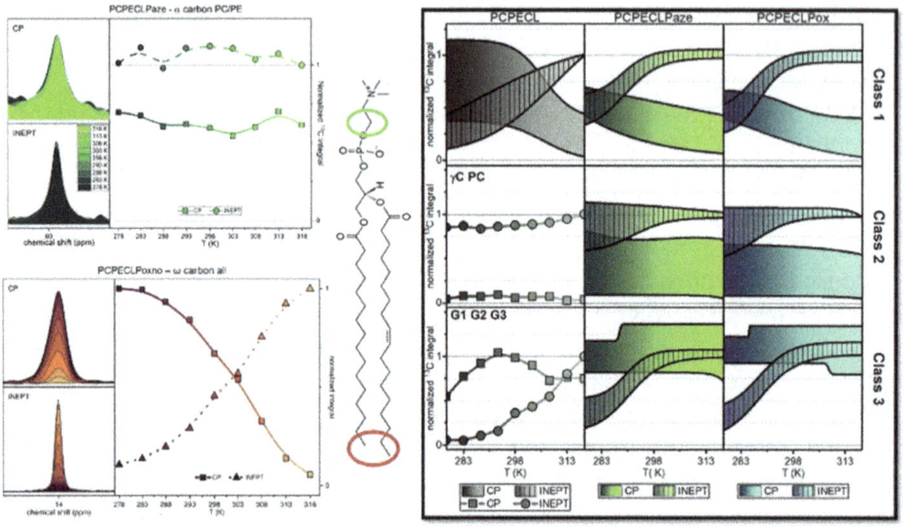


**Figure 1.5.** Principle of dynamic filtering $^{13}$C CP versus INEPT MAS NMR exploring dynamics at different locations of lipid bilayers as a function of phase behaviour and the presence of oxidized lipid species. (Left) temperature-dependent CP and INEPT NMR fingerprints for selected lipid resonances of POPC/POPE/TMCL vesicles with either PoxnoPC or PazePC as minor oxidized lipid component present: (top) α-carbons (59.9 ppm) in the headgroup region of the PC/PE lipids; and (bottom) methyl carbons at fatty acid ω−1 position (14 ppm) of all lipid components present in the vesicles. Depicted plots present the differences in the dynamics and local order of the headgroup region compared to the hydrophobic membrane core during the main phase transition. (Right) summary of dynamic behaviour of all carbon sites across the entire bilayer identified by CP versus INEPT NMR analysis of the individual NMR integrals: (i) three classes of carbon moieties with different dynamic behaviour and temperature dependence: class 1 with strong increase in INEPT signal with temperature and moderate decrease in CP intensity; class 2 with medium decrease in INEPT signal and near constant CP intensity; and class 3 with a strong increase in INEPT signal and nearly constant CP signal across the temperature region. (ii) Presence of oxidized lipids species induces major structural disorder and dynamic alterations across the entire bilayer. Reproduced and adapted with permission from Dingeldein *et al* (2018). Copyright 2018, Elsevier.

in the acyl region exist (Bjorklund *et al* 2013, Dufourc *et al* 1992, Molugu *et al* 2017). Nevertheless, in all cases the presence of oxidized lipid species, which arise under intracellular oxidative stress conditions and often trigger intrinsic apoptosis, change the general behavior of the CP and INEPT signals and their temperature dependence; a clear indication that these OxPls reduce the membrane order across the entire bilayer and increase orientational freedom and faster local dynamics.

A more quantitative $^{13}$C based approach to the analysis of dynamics within a lipid bilayer exploits the characterization of the dipolar couplings between the protons and carbon-13 in the system whose magnitude, as discussed, depends on the degree of motion present (Ferreira *et al* 2016, Ferreira *et al* 2013). Typically, when performing MAS these interactions are scaled and would lead to spectral broadening and poor resolution if measured directly in the $^{13}$C MAS spectrum. Alternative 2D approaches have been developed that recouple the $^{1}$H–$^{13}$C dipolar coupling, correlating the dipolar spectrum with the isotropic $^{13}$C chemical shifts whilst preserving the dipolar lineshape (Gross *et al* 1997). Such approaches are invaluable for complex lipid mixtures or when isotope labelling proves intractable.

## 1.4 Probing the surface charge of membranes with solid-state NMR

Many of the roles supported by membranes within the cell arise through the association of proteins with the surface of the lipid bilayer, interactions that are frequently driven through electrostatic interactions between the charged protein and anionic lipids present in many cellular membranes (Whited and Johs 2015). Examples of such interactions include the interaction of antimicrobial peptides and the anchoring of Src and MARCKS (Sani and Separovic 2016, Murray *et al* 1999). Such interactions are even thought to play an important role in the development of neurodegenerative disease such as Alzheimer's and Parkinson's diseases where such interactions may play an important role in driving amyloid formation (Bokvist *et al* 2004). Solid-state NMR has played an important role in dissecting these interactions as a number of techniques have been developed that permit the characterization of the membrane surface charge and how it responds to the binding of protein.

Many of these methods use anisotropic interactions present in the lipid headgroup to report on the headgroup orientation with respect to the lipid bilayer (figure 1.6(I)). This relies on the motional averaging of the anisotropic interaction. which is dependent upon the relative orientation to the axis of motional averaging (see figure 1.6(I)). As many headgroups possess a sizeable dipole, e.g. between the phosphate and the quaternary ammonium group, the orientation alters in response to the surface charge present.

Early studies exploited deuterated choline or glycerol headgroups (Roux *et al* 1994, Scherer and Seelig 1985, Macdonald *et al* 1991, Bechinger and Seelig 1991, Sixl and Watts 1983). These studies revealed that the quadrupolar splittings of the deuterons connected to the α, β and γ positions of the choline headgroups exhibit a strong dependence on the bilayer surface charge. Increasing concentration of anionic lipids, such as PG and PS, decrease the quadrupolar splitting of the β and γ-deuterons whilst increasing that of the α deuterons. This reflects the changes in the headgroup geometry as the dipole of the choline headgroup realigns with respect to the axis of motional averaging due to changes in electrostatic interactions.

These observations mirror those exploiting the chemical shielding anisotropy where increasing surface charge results in an increase in anisotropy (Lindstrom *et al* 2005, Scherer and Seelig 1985, Semchyschyn and Macdonald 2004). The effects are also observable in the quadrupolar spectrum of $^{14}$N in the choline headgroups. Typically, the $^{14}$N quadrupolar interaction would prohibit direct detection but, in the case of the choline headgroup, the symmetry and dynamics result in a 'small' quadrupolar interaction (tens of kHz) that allows detection using a standard quadrupolar echo. Accordingly, as the surface charge is increased, such measurements allow monitoring of the progressive reduction in the quadrupolar interaction (figure 1.6(A), (B) and (E)).

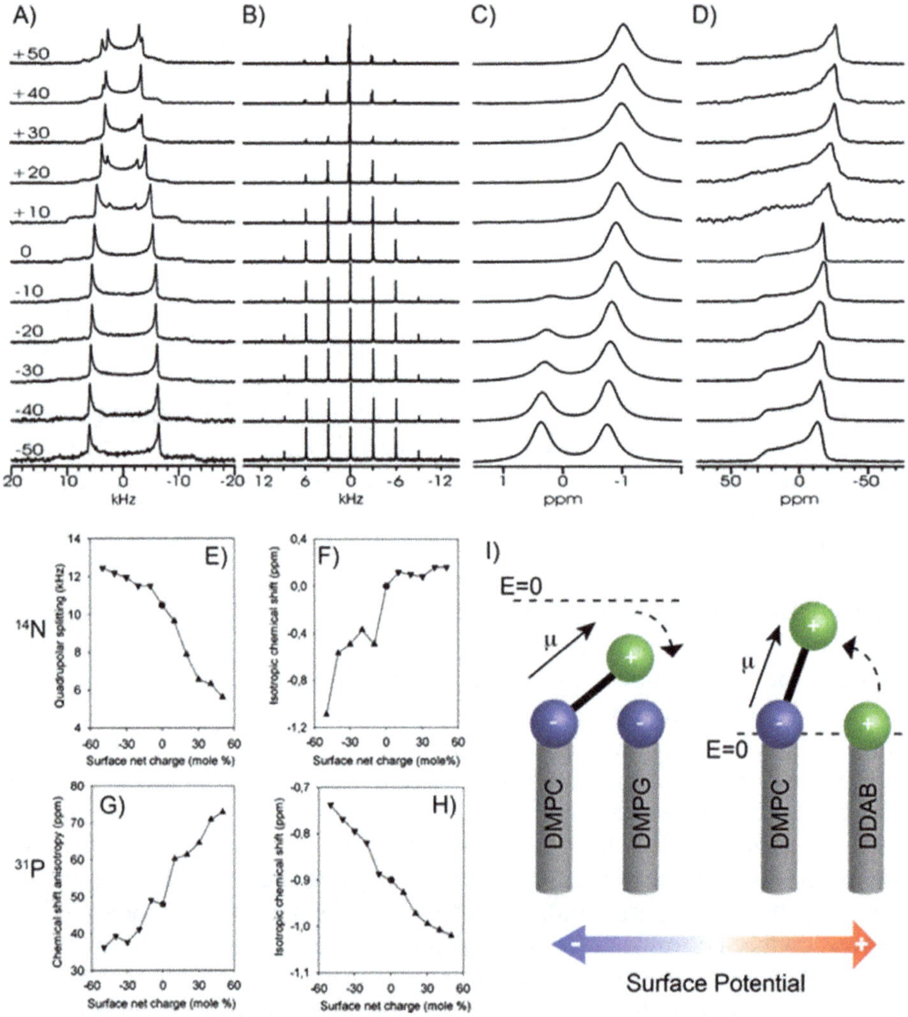


**Figure 1.6.** NMR as a molecular voltmeter: the composition of the lipid bilayer influences its surface potential. The strong dipole between the negatively charged phosphate and the positively charged quaternary ammonium group in the DMPC lipid responds to variation in the surface potential by reorientation of the lipid headgroup with respect to the lipid long axis (I). Due to the rapid axial rotation of the lipids, this results in dynamic scaling of the anisotropic interactions (as shown for $^{14}$N and $^{31}$P, A/E and D/G, respectively), which provides a readout of the membrane surface charge. The change in surface potential also distorts the electron shells around the phosphate and quaternary ammonium group and the plane of the electric field changes sign and moves with respect to the bilayer surface. This results in a change in the $^{31}$P and $^{14}$N isotropic chemical shifts as a function of surface charge (B, C, F and H) to provide an alternate readout of the surface potential present. Reproduced and adapted with permission from the Lindstrom *et al* (2005). Copyright 2005, American Chemical Society.

These measurements all rely on static NMR measurements with limited resolution of the individual lipid species and inherent insensitivity compared to MAS. This can be overcome in part through the application of low speed spinning (when the spinning frequency < size of the anisotropic interaction in kHz) where the properties of the anisotropic interactions are encoded in the sideband families which can be analyzed as described by Herzfeld and Berger (Herzfeld and Berger 1980) (see figure 1.6(B)). Alternative schemes have also been proposed where the anisotropic interactions are correlated with their isotropic counterparts. This has been used to characterize accurately $^{31}$P chemical shift anisotropies in complex lipid mixtures (Warschawski *et al* 2018).

The influence of the surface charge does not only influence the relative orientations of the headgroup with respect to the lipid bilayer, as this would leave the isotropic components of the interactions unchanged. However, careful analysis of the isotropic $^{31}$P and 14N shifts obtained from MAS spectra reveals small but systematic changes in value in the order of 1 and 2 ppm, showing that changes in surface potential deform the electron distribution around the nucleus (see figure 1.6(B),(C), (E) and (G)) (Roux *et al* 1989, Lindstrom *et al* 2005). This arises from the perturbation of the electric field at the surface of the bilayer in response to changes in surface charge (figure 1.6(H)).

## 1.5 Lateral domain formation

When the Singer–Nicholson model of the cell membrane was developed, the lipid-bilayer was viewed as a homogeneous mixture of lipids that support the activities of the protein embedded within. It is, however, becoming clear that the lateral organization of the lipids within the bilayer is more complex, with the lateral segregation of lipids driven both by the intrinsic physical properties of the lipids and through their interaction with membrane proteins. This led to the proposal by Simons and co-workers (Simons and Ikonen 1997) of lipid rafts based primarily on the presence of detergent resistant membrane fractions. Subsequently, however, a wealth of biophysical data (Raghunathan and Kenworthy 2018, van Duyl *et al* 2003, Filippov *et al* 2004), including NMR (Holland *et al* 2006, Yasuda *et al* 2015) data has emerged to support the concept of lateral segregated phases composed of liquid ordered ($L_o$) and liquid-disordered ($L_d$) states; with a growing body of evidence that such domains are important in the regulation of signaling pathways (Nakayama *et al* 2018, Horejsi and Hrdinka 2014) and the segregation of protein to particular areas of the cell membrane (Mollinedo and Gajate 2020, Lu and Fairn 2018).

Despite the common acceptance of lipid rafts, their exact nature remains the focus of much research, with debate over the exact size, number, physical properties and persistence of these domains. This is further complicated as their form is likely dependent upon a diverse array of parameters including temperature, hydration and types of lipid present. The most transparent observation of lipid rafts comes from $^2$H NMR study by Yasuda *et al* (2015) where site selective deuteration was used to study the dynamics of a widely used raft-forming mixture of dioleoyl-phosphatidylcholine (DOPC), sphingomyelin (Sph) and cholesterol. This simple labelling scheme enabled the clear identification of two pools of chain deuterated Sph as resolved by differences in the size of the quadrupolar coupling, reflecting the differences in motion experienced by the Sph in the DOPC-rich $L_d$ phase compared to the more rigid cholesterol-rich $L_o$ phase (Yasuda *et al* 2015). The order parameter profile from a series of measurements of site specifically labelled Sph molecules revealed that this property is reflected along the length of the lipid chain. The resolution of such well-defined spectra infers that there is limited exchange between the $L_d$ and $L_o$ phases on the timescale of the NMR experiment.

Earlier studies of DOPC/dipalmitoyl-phosphatidylcholine (DPPC)/cholesterol mixtures are reported by Veatch *et al* (2004) where $^2$H-NMR of chain deuterated DPPC and deuterated cholesterol were used to study the phase behavior of MLVs and compared with fluorescence microscopy of giant unilamellar vesicles. These studies identified three clear phases: the first shows a clear liquid crystalline $L_\alpha$ phase at high temperature. The second below the mixing temperature ($T_{mix}$), shows a loss of resolution, particularly in the methylene region, indicative of lipid exchange between the $L_o$ and $L_d$ phase. Given the difference in the size of the quadrupolar interactions the authors concluded an upper size domain limit of ~80 nm, assuming that the interface between the $L_o$ and $L_\alpha$ phase does not impede exchange. Interestingly, comparable fluorescence microscopy studies indicated the presence of significantly larger domains in the order of 1 μm. Below this temperature, the spectrum appears as a superimposition of the $L_\alpha$ and the $L_o$ phase. This suggests a demixing of the lipid and the formation of domains in excess of 160 nm.

These observations mirror those using static NMR $^{31}$P NMR in DOPC/Sph/Cholesterol mixtures (Koukalova *et al* 2017). In binary mixtures of DOPC/Sph (90%/10%) the $^{31}$P static NMR spectrum is a superimposition of two powder patterns with different shielding anisotropy. As the intensity of these two components is roughly equal and does not mirror the concentrations of the two species, the two spectra are taken to represent two domains: the first Sph depleted where the DOPC exists in a mobile $L_d$ state; while the second $L_o$ domain is rich in Sph, resulting in a reduction in DOPC headgroup mobility. These observations are in agreement with Monte-Carlo fluorescence energy transfer measurements (MC-FRET) which suggest that the size of these domains is in the order of ~10 nm. Similar results are obtained for ternary mixtures, with all domain-forming mixtures exhibiting powder spectra with two components. Interestingly, careful analysis of the linewidths both in the static and complementary MAS spectra provide no evidence of gel phase properties, such as enhanced $T_2$ relaxation, suggesting that within these domains the lipids exhibit a high degree of mobility more akin to the $L_\alpha$ phase.

The size and lateral diffusion of the lipids in vesicles containing lipid rafts have also been characterized using $^{31}$P or $^2$H 2D exchange NMR (Arnold *et al* 2004). As the chemical shielding anisotropy and quadrupolar interaction exhibit an orientational dependence, movement over the vesicle surface will be reflected by a change in the resonance frequency, with the assumption that the rotational correlation time of the vesicle is slow with respect to the timescale of the NMR interaction. In the context of a 2D exchange experiment, the resulting NMR spectrum effectively correlates the position prior to and that at the end of the exchange period, a process which can be modelled mathematically as a random walk, allowing the diffusion of

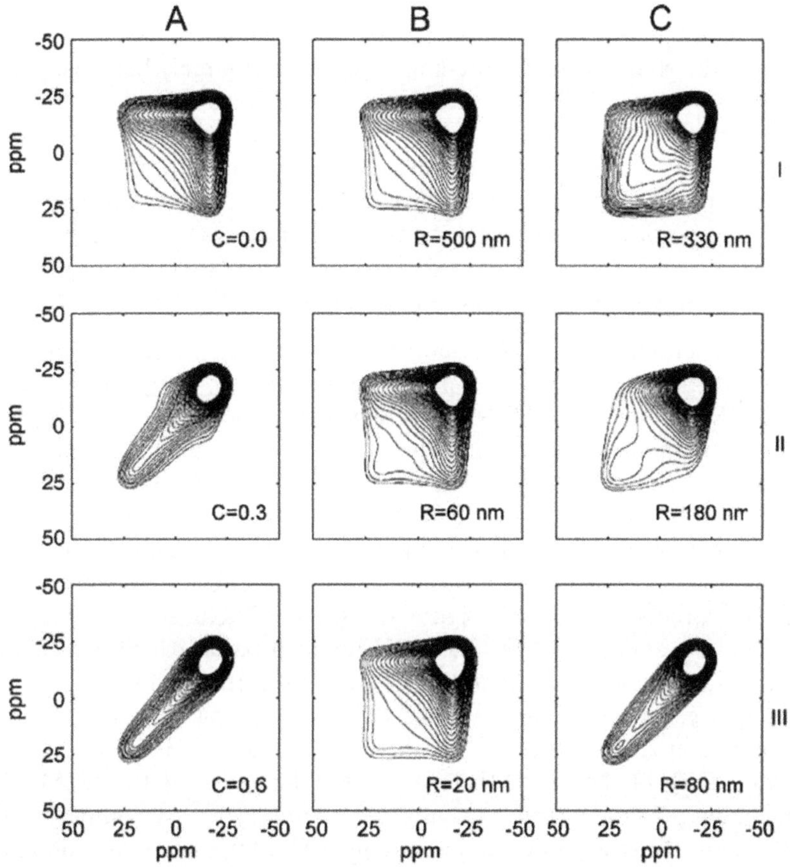


**Figure 1.7.** Diffusion in the plane of the lipid bilayer can be monitored by 2D $^{31}P$ exchange NMR where off diagonal intensity reports on the changes in lipid orientation over a given mixing time. The presence of lipid rafts or domains in the membrane restrict the diffusion of lipids, either within a particular domain or reducing the area over which they can diffuse. These studies can be used to report on (A) the area of the different concentration of obstacles, (B, the radius of the obstacles, and (C) and fluid domains. Reproduced with permission from Arnold *et al* (2004). Copyright 2004, Elsevier.

the lipids over the surface of the vesicle to be characterized. The presence of lipid domains restricts the available surface over which lipid can diffuse, either restricting it to the liquid-ordered domains or acting as barriers to diffusion in the liquid-ordered domain. Modelling of these processes allows the characterization of the lipid domains (Arnold *et al* 2005, Macdonald *et al* 2013). These techniques have been used to characterize the domains present in binary mixtures of dimyristoyl-phosphatidylcholine/distearoyl-phosphatidylcholine (figure 1.7).

More recent studies have employed center-band only detection of exchange (CODEX). These experiments benefit from enhanced sensitivity afforded by MAS as well as permitting the resolution of the different lipid species and also monitor dephasing of the magnetization during the mixing period to characterize the diffusion of the lipids over the surface of the vesicle (Saleem *et al* 2012).

## 1.6 Lipid/protein interactions

Traditionally the study of lipid/protein interactions has largely been the preserve of electron spin resonance (ESR) as the exchange of the lipids from the annular to bulk lipids is too rapid for the detection to resolve the free and the bound species by NMR. However, $^{31}$P and $^{2}$H NMR are now playing an important role in understanding how proteins interact more specifically, either through surface association with the bilayer surface or directly at annular and non-annular lipid binding sites.

### 1.6.1 Insight into protein/lipid headgroup association by $^{31}$P CP MAS NMR

To obtain a general molecular overview about membrane headgroup region and its response to the lipid–lipid and lipid–protein interactions, $^{13}$C dynamic filtering (see figure 1.4), $^{31}$P DP and CP MAS NMR (Dingeldein *et al* 2017) can be applied. This allows discrimination of membrane components and domains with high mobility from those that are more rigid and immobile. Since the identification of the individual phospholipids by $^{31}$P MAS NMR is straightforward, as seen in figure 1.4, the dynamic behavior of individual lipid components in complex membranes and changes due to bound proteins and other membrane effectors by bound biomolecules can be extracted. As seen in figure 1.4, the CP NMR signal is much weaker compared to that obtained by DP MAS NMR: a situation which reflects the large disorder and fast motional reorientations occurring in the lipid species.

To obtain a general molecular overview about the impact of PoxnoPC on these membranes at a lipid–lipid and lipid–protein level, we additionally carried out direct polarization (DP) $^{31}$P MAS NMR and cross polarization (CP) $^{31}$P MAS NMR experiments on our model systems (Dingeldein *et al* 2017). By comparing DP and CP NMR data, we could discriminate membrane components and domains that exhibit high mobility from the more rigid which exhibit very slow motional processes.

This behavior is much more pronounced in membranes prone to phase transition and phase separations. As shown by Dingeldein *et al* (2017), differences between CP and DP $^{31}$P NMR intensities are dramatic for POPC/POPE/TMCL vesicles, where the presence of oxidized lipids augments the differences, as clearly seen in figure 1.8 (middle panel). As seen in the $^{13}$C CP NMR profiles for the fatty acid methyl groups in the presence of the oxidized PoxnoPC lipid (Dingeldein *et al* 2018) (see also figure 1.5, left panel bottom), the $^{31}$P CP NMR signals of PC and PE components showed the same behavior. While at low temperatures the CP signal intensity is very pronounced, it completely vanished at 318 K. Clearly, the increase in molecular disorder and fast intra- and intermolecular dynamics results in the motional averaging of the $^{1}$H-$^{31}$P dipolar couplings required for CP in both headgroup moieties, which indicated similar general dynamic behavior for phosphate-bearing headgroups at the membrane surface (Weisz *et al* 1992).

Interestingly, despite the similar dynamics, the relative intensities in the CP signals of both lipids change with increasing temperature, indicating subtle molecular differences in the CP driving process. Presumably, the headgroup orientation changes differently for PC and PE relative to their individual rotational long axis

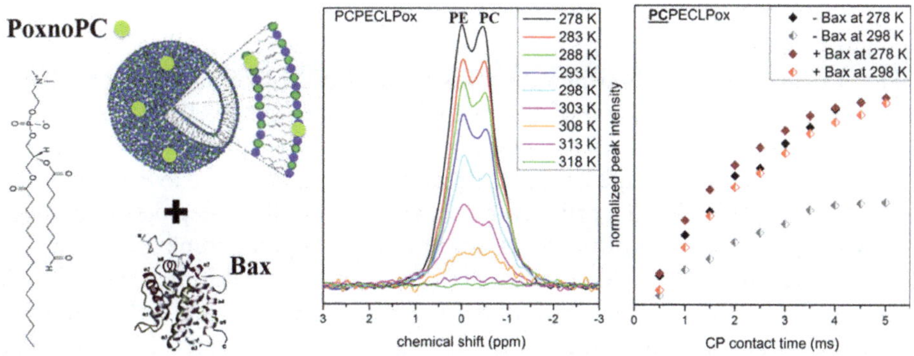


**Figure 1.8.** Association of apoptotic human Bax protein with mitochondrial outer membrane-like POPC/POPE/TMCl vesicles doped with PoxnoPC to mimic oxidative apoptotic stress. Middle panel: monitoring of lipid dynamics and disorder in headgroup region by temperature dependent $^{31}$P CP MAS NMR spectra of lipid vesicles shows loss of CP signal intensities during phase transition into a melted liquid-disordered phase (complete at 308 K). Right panel: $^{31}$P CP NMR signal buildup for PC resonance plotted as a function of contact time; comparison of vesicles prior to and upon presence of Bax protein at two temperatures with 278 K being in the gel-phase (dark symbols) and 298 K where ca. 50% of bilayer is already molten (light symbols). Reproduced and adapted with permission from Dingeldein *et al* (2017). Copyright 2017, Elsevier.

during melting, unsurprisingly since PE headgroups are known to tilt differently at low temperatures compared to PC bilayers. During melting, fast rotation about the molecular long axis on the ns-timescale cause slightly different averaging of the $^{1}$H-$^{31}$P tensor, resulting in a different dynamic scaling of the dipolar coupling (Dufourc *et al* 1992). As shown in detail by Dingeldein *et al* (2017), the ratio between CP intensities of PC versus PE lipid resonances changes gradually with temperature from around 1 at 278 K to around 1.2 at 308 K. The CP intensities were too low to be analyzed due to fast dynamics and disorder at elevated temperature. Nevertheless, this gradual change in peak intensity ratio reflects the degree of melting of coexisting phases during the broad overall phase transition, stretching from 283 K to 308 K. Upon finishing all melting events, a homogenous liquid crystalline $L_\alpha$ phase exists with a loss in $^{31}$P CP signal. This behavior is also found via $^{13}$C CP for some lipid segments (section 1.5), indicating an increase in local dynamics and disorder (Dingeldein *et al* 2018).

The binding of membrane-active proteins, such as the apoptotic Bax protein, to these membranes changes the picture quite dramatically. Bax is a pore-forming protein of the Bcl-2 (B-cell lymphoma 2) protein family involved in regulation of the mitochondrial pathway of programmed cell death or apoptosis (Kerr *et al* 1972, Cory and Adams 2002). Anti-apoptotic (cell-protecting) and apoptotic (cell-killing) members of this family meet at the mitochondrial outer membrane (MOM) system, where they arbitrate a life or death sentence. In the case of a death verdict, Bax protein will be recruited to the mitochondrion where it will partially penetrate into the membrane and cause membrane perforation. In a cell, this process is tightly regulated by Bcl-2 family whose pro- and anti-apoptotic members meet at the MOM to control its integrity and release of apoptotic factors such as cytochrome c (Lidman *et al* 2016).

The presence of partially penetrated Bax in our MOM-like bilayers here changes the organization and dynamic of the membrane quite significantly, as seen in the relevant $^{31}$P NMR results. As described in detail (Dingeldein *et al* 2017), the ratio between PC and PE CP intensities is already quite different at low temperatures (around 1.7) and plotting the ratio over temperature reveals a deep deflection over a narrow temperature area with final values around 1.4 upon complete melting. This shows a more cooperative phase behavior, as confirmed by calorimetric studies, due to an ordering effect by the partial penetration of Bax via its helices into the membranes. This ordering effect on the headgroups, even at higher temperatures, ensures an efficient CP presumably as a result of the reduction in membrane dynamics and fluidity.

These changes induced by Bax can also be seen in the intensity of CP NMR signal as a function of CP contact times; a method providing straightforward information about the size and nature of dipolar interactions and the CP modulation due to $T1\rho$ relaxation process. As seen in figure 1.8, the differences are more pronounced at low temperatures between Bax-free and Bax-containing bilayers, illustrating the changes in CP signal due to lipid dynamics and molecular order (Dingeldein *et al* 2017) .

The presence of Bax reduces the partial averaging of the dipolar interactions due to lipid dynamics and disorder. Therefore, these larger dipolar interactions drive much more effective CP mechanisms across the entire temperature region. Even at higher temperatures, where the CP signal for PC and PE lipids is not detectable in protein-free vesicles, the formation of a Bax-membrane complex with increased lipid organization and reduced dynamic freedom causes a pronounced change in CP build up and $T1\rho$ relaxation, with both effects resulting in a a visible CP NMR spectrum at high temperatures, as shown recently by Dingeldein *et al* (2017).

### 1.6.2 Studying annular lipids bound to integral membrane proteins

The influence of protein on the $^{31}$P spectrum of phospholipids is dependent on the nature of the interaction between the lipid and the protein. As eluded to above, exchange between the annular and bulk lipids typically occurs upon a timescale that is too rapid to allow the free and bound populations to be distinguished. Nevertheless, methods have been developed that permit the analysis of lipid/protein interactions.

A powerful technique for studying protein/lipid specificity was developed by Gawrisch and co-workers who developed saturation transfer-difference MAS NMR for the analysis of rhodopsin/lipid interactions (Soubias and Gawrisch 2005, Soubias *et al* 2006). This technique relies on the saturation of a particular protein resonance, with the magnetization redistributed through the protein by spin-diffusion. The magnetization is then transferred to the surrounding lipids by the strong proton–proton dipolar contacts during the long saturation time. Exchange during this period serves to enhance the effect as the reduction in signal intensity is not restricted to the annular lipids, with rapid exchange ensuring that the pool of bulk lipid

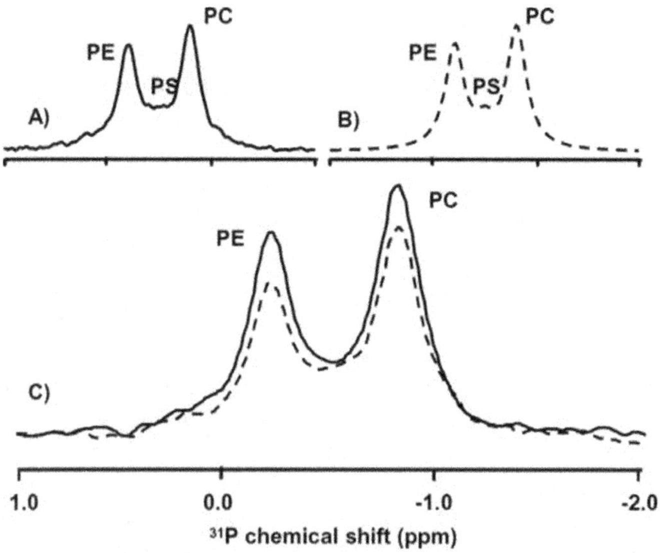


**Figure 1.9.** Saturation transfer difference (STD)-NMR is a powerful tool with which to identify the interaction of lipids with integral membrane proteins. Selective irradiation of the protein leads to the saturation of the protein resonances and any lipids which are in close proximity to the protein and hence dipolar coupled. Exchange of the lipids with the bulk pool during the extended saturation period ensures the degree to which a pool of lipids is saturated and reflects its residency and affinity for the protein. In this example, rhodopsin in rod outer segment disks is studied with the PC, PS and PE clearly resolved (A) and deconvoluted (B). Selective irradiation of the protein protons followed by CP transfer to the lipid phosphate reveals the preferential saturation of polyunsaturated PE lipids, (C) demonstrating their affinity for rhodopsin. Reproduced with permission from Soubias *et al* (2006). Copyright 2006, ABMSB Journals.

progressively becomes saturated. This leads to a suppression of signals from the protons in the $^{1}$H MAS spinning spectra, and can be utilized to identify specific interaction between classes of lipid chains and the protein (Yeliseev *et al* 2005). The relatively poor resolution of different classes of lipid classes in the proton spectra makes their identification challenging. This challenge can be overcome by the transfer of the proton magnetization from the protons within the lipid to the $^{31}$P within the lipid headgroups by CP allowing identification of particular phospholipids which bind at the lipid/protein interface. This is feasible as the magnetization transfer to the $^{31}$P in the lipid headgroup is dominated by strong dipolar couplings to the lipid headgroup and glycerol backbone (Sodt *et al* 2016). These studies have revealed that for the G-protein coupled receptor, rhodopsin, binding of polyunsaturated phosphatidylethanolamine (PE) is preferred over less saturated phosphatidylcholine (PC) species (figure 1.9).

### 1.6.3 Studying non-annular lipids bound to integral membrane proteins

Some lipids exhibit far higher affinity for particular classes of integral membrane proteins than classical annular lipids, with some acting essentially as cofactors and are a pre-requisite for integral membrane protein function (Lee 2011a, 2011b,

Powl *et al* 2008a, Marius *et al* 2008). In such cases, the absence of rapid exchange between the bulk and annular lipids limits the extent of lipid saturation during STD-NMR experiments. However, in those situations, exchange between the non-annular lipid binding sites and the annular/bulk lipid can be sufficiently slow to permit the identification of bound lipids. This has been exploited to investigate how the lipid composition influences the activity of the $K^+$ channel, KcsA (Marius *et al* 2012). Static and MAS $^{31}$P NMR studies conducted on KcsA, revealed that in all cases the presence of the protein in the bilayer led to an increase in the rate of transverse relaxation, resulting in a broadening in the spectrum. These properties are largely lipid non-specific with both the POPC and the anionic lipid POPG exhibiting similar enhancements in relaxation rates, an observation previously attributed to restricted diffusion of the lipids arising from the presence of the protein within the bilayer (figure 1.10). The spectra of KcsA in POPC and POPC/POPG reveal interesting differences, with the latter showing the presence of an additional $^{31}$P resonance at −0.14 ppm. This resonance is attributed to POPG bound to 'non-annular' lipid binding sites at the interface between the KcsA subunits, with the perturbation in chemical shift arising from the interaction with charged residues in the non-annular lipid binding site. The fact that this resonance is resolved indicates that the exchange between the bulk lipid and the 'non-annular' binding site is slow on the NMR

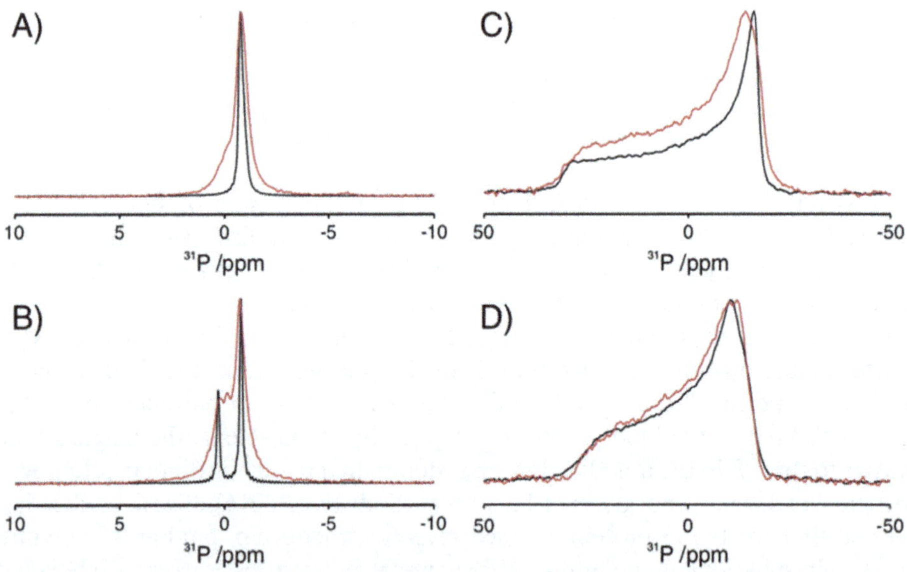


**Figure 1.10.** NMR can report on the interactions of lipids with integral membrane proteins. Here the potassium channel, KcsA, has been reconstituted into: (top) POPC, and (bottom) POPC/POPG lipid vesicles. The presence of KcsA (red) in the lipid vesicles results in a broadening of the $^{31}$P resonances in both the MAS (A and B) and static (C and D) NMR spectra. In the MAS spectra the lipid species are clearly resolved with a single resonance from the (A) POPC and (B) POPC/POPG. The reconstitution of KcsA into (B) POPC/POPG bilayers gives rise to a third resonance derived from POPG bound to a non-annular lipid binding site within KcsA, with the perturbation of the POPG chemical shift arising from electrostatic interactions at the lipid binding site. Reproduced with permission from Marius *et al* (2012). Copyright 2012, Elsevier.

timescale, and was subsequently observed in complementary ESR studies (Bolivar *et al* 2012). The specificity of this interaction was demonstrated through the mutagenesis of key arginine residues in the anionic lipid binding site, that abolished binding and furthermore altered the gating behavior of the channel. Subsequent, solid-state NMR studies of KcsA revealed the presence of an interaction network spanning from these high specificity anionic lipid binding sites across the turret region to the channel, giving rise to the observed dependence on anionic lipids for KcsA gating (Koers *et al* 2013).

## 1.7 Membrane remodeling

Traditionally the view has been that the lipid environment plays an important role in the regulation of integral membrane proteins, either acting non-specifically by providing an appropriate lipid environment (bilayer thickness, fluidity) to facilitate membrane protein function or through the binding of specific lipids at non-annular binding sites, acting as cofactors to ensure protein activity (Lee 2011a, Powl *et al* 2008a, 2008b, Marius *et al* 2008). However, integral membrane proteins can also influence the structure of the lipid bilayer, recruiting or modifying particular classes of lipids leading to the formation of domains rich in particular classes of lipid. The effect of this is to modify the local physical properties of the bilayer, influencing properties such as fluidity and rigidity. These properties have been exploited in biology to remodel cellular membranes to carry out particular functions from the budding of vesicles to the formation of viral replication complexes. As $^{2}$H and $^{31}$P NMR report on the types of structures and dynamics adopted by lipids, these have proved to be valuable tools with which to study such processes.

The Hepatitis C Virus (HCV) reproduces in small replication complexes formed from the endoplasmic reticulum (ER). These small double membraned structures host the viral replication machinery allowing replication of the viral genome to occur undetected within the cell (Gouttenoire *et al* 2014, Paul *et al* 2011, Gouttenoire *et al* 2010b, Gouttenoire *et al* 2010a, Gouttenoire *et al* 2009b, Gouttenoire *et al* 2009a). The formation of these replication centers and their persistence is thought to depend upon the processing of the polyprotein (NS3–5B), with the integral membrane protein NS4B playing a key role in remodeling the membrane. A number of solid-state NMR studies have been conducted on NS4B (Ashworth Briggs *et al* 2015, Fogeron *et al* 2015, Jirasko *et al* 2020) in conjunction with other biophysical studies (Gouttenoire *et al* 2014, Paul *et al* 2011, Gouttenoire *et al* 2010b, Gouttenoire *et al* 2010a, Gouttenoire *et al* 2009b, Gouttenoire *et al* 2009a) to understand how the protein remodels the membrane to form these replication complexes. A key component is the second amphipathic helix (AH2), which is thought to interact specifically with the ER membranes.

To investigate how AH2 may assist in the remodeling of the ER membranes static $^{31}$P NMR experiments have been conducted to investigate how AH2 interacts with both neutral and negatively charged lipid bilayers. These studies revealed no interaction with neutral POPC bilayer but upon the addition to negatively charged bilayer composed of POPC and POPG an increase in the chemical shielding anisotropy is observed at low concentrations, consistent with a reduction of mobility associated with the binding of the peptide to the bilayer surface. Interestingly however, upon further titration of the AH2, significant perturbations were seen in the $^{31}P$ lineshape, with a progressive decrease in the spectral intensity arising from lipids aligned parallel to the magnetic field. The resulting spectrum is indicative of the bulk of the lipids being aligned perpendicular to the magnetic field, similar to that observed in the spectra of aligned lipid bilayers. In this case, however, the increase in chemical shift anisotropy and progressive deformation suggest that the presence of the AH2 serves to reduce the bilayer rigidity and binds tightly to the anionic lipid present in the system, properties required to facilitate the formation of the replication centres (figure 1.11). Complementary molecular dynamics studies

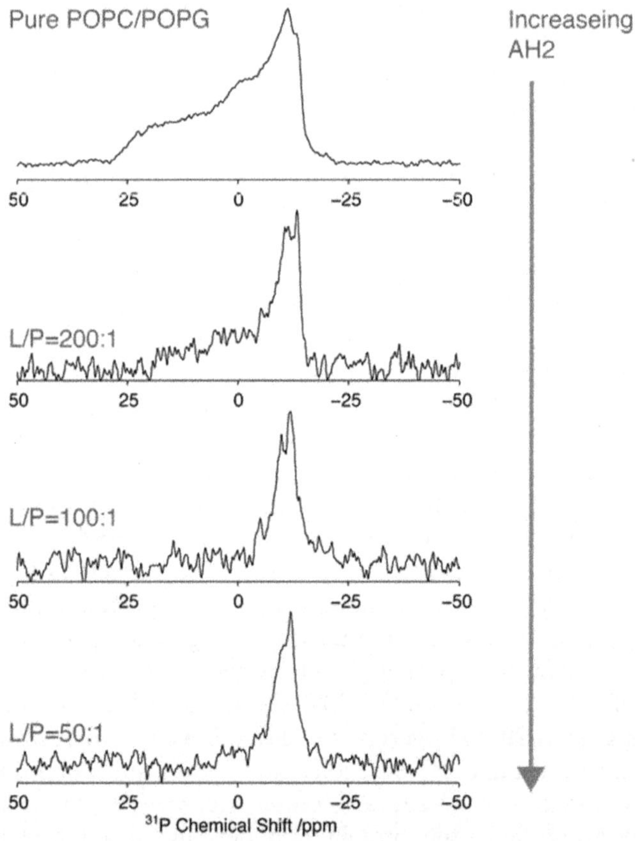


**Figure 1.11.** Static $^{31}P$ NMR provides insights into lipid distribution. Lipids possess a negative anisotropy of the diamagnetic susceptibility tensor, which favours their alignment perpendicular to the magnetic field. For a lipid vesicle or MLVs, the phospholipids have a spherical (powder) distribution. In the presence of proteins, such as AH2-NS4B, the rigidity of the lipid bilayer is reduced, resulting in the deformation of the lipid vesicles into ellipsoidal structures that favour the adoption of the energetically more stable state with more lipids aligned perpendicular to the field. Such observations allow us to determine how the presence of proteins can facilitate the formation of new bilayer structures. Reproduced with permission from Ashworth Briggs *et al* (2015). Copyright 2015, Elsevier.

revealed similar clustering of the anionic lipid phosphoinositol-4-phosphate. With the recruitment of PI4 kinase by NS5A, a pre-requisite to replication complex formation, these results suggest that the phosphorylation of the lipid headgroups leads to their recruitment by AH2 of NS4B, reducing the membrane rigidity and facilitating its deformation, potentially through the repulsive interactions between the anionic headgroups, facilitating the formation of the replication complex. This property appears far from unique with recent reports revealing similar changes in the membrane rigidity in nuclear membranes, resulting in the formation of membrane invaginations (Dazzoni *et al* 2020).

With recent improvements in membrane protein expression and improvements in NMR sensitivity, these methods can now provide us with valuable insights into how Nature harnesses the interactions between lipids and proteins to further cell processes.

## 1.8 Summary

The diversity of NMR active nuclei in lipids has made solid-state NMR an attractive tool with which to probe the structure and dynamics of biological membranes. As highlighted above, with the continuing improvements in instrumentation and methodology coupled with developments in membrane protein expression, even more challenging studies are accessible to this methodology, often without recourse to isotope labelling. As such, solid-state NMR represents an important analytical tool to help unravel the molecular properties of cell membranes and the processes they support.

## Acknowledgements

GG acknowledges financial support from the Swedish Research Council, Swedish Cancer Foundation, Kempe Foundation, Knut and Alice Wallenberg foundation ('NMR for Life' Programme), SciLifeLab, Swedish National NMR Centre and Umeå Insamlingsstiftelse for medical research. PTW acknowledges financial support from the EPSRC.

## References

Ardail D, Privat J P, Egret-Charlier M, Levrat C, Lerme F and Louisot P 1990 Mitochondrial contact sites. Lipid composition and dynamics *J. Biol. Chem.* **265** 18797–802

Arnold A, Cloutier I, Ritcey A M and Auger M 2005 Temperature and pressure dependent growth and morphology of DMPC/DSPC domains studied by Brewster angle microscopy *Chem. Phys. Lipids* **133** 165–79

Arnold A, Paris M and Auger M 2004 Anomalous diffusion in a gel-fluid lipid environment: A combined solid-state NMR and obstructed random-walk perspective *Biophys. J.* **87** 2456–69

Ashworth Briggs E L, Gomes R G, Elhussein M, Collier W, Findlow I S, Khalid S, Mccormick C J and Williamson P T 2015 Interaction between the NS4B amphipathic helix, AH2, and charged lipid headgroups alters membrane morphology and AH2 oligomeric state—Implications for the Hepatitis C virus life cycle *BBA-Biomembranes* **1848** 1671–7

Auger M 1997 Membrane structure and dynamics as viewed by solid-state NMR spectroscopy *Biophys. Chem.* **68** 233–41

Bechinger B and Seelig J 1991 Interaction of electric dipoles with phospholipid head groups—a H-2 and P-31 NMR-study of phloretin and phloretin analogs in phosphatidylcholine membranes *Biochemistry,* **30** 3923–9

Bjorklund S, Ruzgas T, Nowacka A, Dahi I, Topgaard D, Sparr E and Engblom J 2013 Skin membrane electrical impedance properties under the influence of a varying water gradient *Biophys. J.* **104** 2639–50

Bokvist M, Lindstrom F, Watts A and Grobner G 2004 Two types of Alzheimer's beta-amyloid (1-40) peptide membrane interactions: Aggregation preventing transmembrane anchoring versus accelerated surface fibril formation *J. Mol. Biol.* **335** 1039–49

Bolivar J H, East J M, Marsh D and Lee A G 2012 Effects of lipid structure on the state of aggregation of potassium Channel KcsA *Biochemistry* **51** 6010–6

Bouhlel Z, Arnold A A, Warschawski D E, Lemarchand K, Tremblay R and Marcotte I 2019 Labelling strategy and membrane characterization of marine bacteria Vibrio splendidus by *in vivo* (2)H NMR *BBA-Biomembranes* **1861** 871–8

Brown M F and Seelig J 1977 Ion-induced changes in head group conformation of lecithin bilayers *Nature* **269** 721–3

Carbone M A and Macdonald P M 1996 Cardiotoxin II segregates phosphatidylglycerol from mixtures with phosphatidylcholine: P-31 and H-2 NMR spectroscopic evidence *Biochemistry* **35** 3368–78

Cory S and Adams J M 2002 The Bcl2 family: regulators of the cellular life-or-death switch *Nat. Rev.* **2** 647–56

Cullis P R and De Kruijff B 1979 Lipid polymorphism and the functional roles of lipids in biological-membranes *BBA-Biomembranes,* **559** 399–420

Cullis P R and Hope M J 1978 Effects of fusogenic agent on membrane structure of erythrocyte-ghosts and mechanism of membrane-fusion *Nature* **271** 672–4

Davis J H 1983 The description of membrane lipid conformation, order and dynamics by H-2-NMR *BBA-Biomembranes* **737** 117–71

Davis J H, Jeffrey K R, Bloom M, Valic M I and Higgs T P 1976 Quadrupolar echo deuteron magnetic-resonance spectroscopy in ordered hydrocarbon chains *Chem. Phys. Lett.* **42** 390–4

Dazzoni R, Grelard A, Morvan E, Bouter A, Applebee C J, Loquet A, Larijani B and Dufourc E J 2020 The unprecedented membrane deformation of the human nuclear envelope, in a magnetic field, indicates formation of nuclear membrane invaginations *Sci. Rep.* **10** 5147

Dingeldein A P, Sparrman T, Aden J, Wacklin H P, Clifton L A and Grobner G 2019 Mitochondrial membrane organization under oxidative stress: insight by solid-state nmr and neutron reflectometry *Biophys. J.* **116** 508a

Dingeldein A P G, Pokorna S, Lidman M, Sparrman T, Sachl R, Hof M and Grobner G 2017 Apoptotic bax at oxidatively stressed mitochondrial membranes: lipid dynamics and permeabilization *Biophys. J.* **112** 2147–58

Dingeldein A P G, Sparrman T and Grobner G 2018 Oxidatively stressed mitochondria-mimicking membranes: a molecular insight into their organization during apoptosis *BBA-Biomembranes* **1860** 2644–54

Dudka I, Thysell E, Lundquist K, Antti H, Iglesias-Gato D, Flores-Morales A, Bergh A, Wikstrom P and Grobner G 2020 Comprehensive metabolomics analysis of prostate cancer tissue in relation to tumor aggressiveness and TMPRSS2-ERG fusion status *BMC Cancer* 20

Dufourc E J, Mayer C, Stohrer J, Althoff G and Kothe G 1992 Dynamics of phosphate head groups in biomembranes. Comprehensive analysis using phosphorus-31 nuclear magnetic resonance lineshape and relaxation time measurements *Biophys. J.* **61** 42–57

Engelman D M 2005 Membranes are more mosaic than fluid *Nature* **438** 578–80

Ferreira T M, Coreta-Gomes F, Ollila O H S, Moreno M J, Vaz W L C and Topgaard D 2013 Cholesterol and POPC segmental order parameters in lipid membranes: solid state H-1-C-13 NMR and MD simulation studies *Phys. Chem. Chem. Phys.* **15** 1976–89

Ferreira T M, Sood R, Barenwald R, Carlstrom G, Topgaard D, Saalwachter K, Kinnunen P K J and Ollila O H S 2016 Acyl chain disorder and azelaoyl orientation in lipid membranes containing oxidized lipids *Langmuir* **32** 6524–33

Filippov A, Oradd G and Lindblom G 2004 Lipid lateral diffusion in ordered and disordered phases in raft mixtures *Biophys. J.* **86** 891–6

Filippov A V, Khakimov A M and Munavirov B V 2015 P-31 NMR studies of phospholipids *Annu. Rep. NMR Spectro.* **85** 27–92

Fogeron M L *et al* 2015 Wheat germ cell-free expression: two detergents with a low critical micelle concentration allow for production of soluble HCV membrane proteins *Protein Express. Purif.* **105** 39–46

Forbes J, Husted C and Oldfield E 1988 High-field, high resolution proton magic-angle sample spinning nuclear magnetic resonance spectroscopic studies of gel and liquid crystalline lipid bilayers and the effects of cholesterol *J. Am. Chem. Soc.* **110** 1059–65

Fusco G *et al* 2017 Structural basis of membrane disruption and cellular toxicity by alpha-synuclein oligomers *Science,* **358** 1440–3

Gouttenoire J, Castet V, Montserret R, Arora N, Raussens V, Ruysschaert J M, Diesis E, Blum H E, Penin F and Moradpour D 2009a Identification of a novel determinant for membrane association in Hepatitis C virus nonstructural protein 4B *J. Virol.* **83** 6257–68

Gouttenoire J, Montserret R, Kennel A, Penin F and Moradpour D 2009b An amphipathic alpha-helix at the C terminus of hepatitis C virus nonstructural protein 4B mediates membrane association *J. Virol.* **83** 11378–84

Gouttenoire J, Montserret R, Paul D, Castillo R, Meister S, Bartenschlager R, Penin F and Moradpour D 2014 Aminoterminal amphipathic alpha-Helix AH1 of hepatitis C virus nonstructural protein 4B possesses a dual role in RNA replication and virus production *PLoS Pathog.* **10** e1004501

Gouttenoire J, Penin F and Moradpour D 2010a Hepatitis C virus nonstructural protein 4B a journey into unexplored territory *Rev. Med. Virol.* **20** 117–29

Gouttenoire J, Roingeard P, Penin F and Moradpour D 2010b Amphipathic alpha-Helix AH2 is a major determinant for the oligomerization of Hepatitis C virus nonstructural protein 4B *J. Virol.* **84** 12529–37

Gross J D, Warschawski D E and Griffin R G 1997 Dipolar recoupling in MAS NMR: A probe for segmental order in lipid bilayers *J. Am. Chem. Soc.* **119** 796–802

Herzfeld J and Berger A E 1980 Sideband intensities in NMR-spectra of samples spinning at the magic angle *J. Chem. Phys.* **73** 6021–30

Holland G P, Mcintyre S K and Alam T M 2006 Distinguishing individual lipid headgroup mobility and phase transitions in raft-forming lipid mixtures with $^{31}$P MAS NMR *Biophys. J.* **90** 4248–60

Horejsi V and Hrdinka M 2014 Membrane microdomains in immunoreceptor signaling *FEBS Lett.* **588** 2392–7

Jirasko V, Lakomek N A, Penzel S, Fogeron M L, Bartenschlager R, Meier B H and Bockmann A 2020 Proton-detected solid-state NMR of the cell-free synthesized alpha-helical transmembrane protein NS4B from Hepatitis C virus *Chem. Bio. Chem.* **21** 1453–60

Kerr J F, Wyllie A H and Currie A R 1972 Apoptosis: a basic biological phenomenon with wide-ranging implications in tissue kinetics *Br. J. Cancer* **26** 239–57

Koers E J *et al* 2013 Dynamic nuclear polarization NMR spectroscopy: revealing multiple conformations in lipid-anchored peptide vaccines *Angew. Chem. Int. Edit.* **52** 10905–8

Koukalova A, Amaro M, Aydogan G, Grobner G, Williamson P T F, Mikhalyov I, Hof M and Sachl R 2017 Lipid driven nanodomains in giant lipid vesicles are fluid and disordered *Sci. Rep.* **7** 5460

Kovacik A, Vogel A, Adler J, Pullmannova P, Vavrova K and Huster D 2018 Probing the role of ceramide hydroxylation in skin barrier lipid models by H-2 solid-state NMR spectroscopy and x-ray powder diffraction *BBA-Biomembranes* **1860** 1162–70

Laadhari M, Arnold A A, Gravel A E, Separovic F and Marcotte I 2016 Interaction of the antimicrobial peptides caerin 1.1 and aurein 1.2 with intact bacteria by (2)H solid-state NMR *Biochim. Biophys. Acta* **1858** 2959–64

Lee A G 2011a Biological membranes: the importance of molecular detail *Trends Biochem. Sci.* **36** 493–500

Lee A G 2011b Lipid-protein interactions *Biochem. Soc.* T **39** 761–6

Leonard A and Dufourc E J 1991 Interactions of cholesterol with the membrane lipid matrix. A solid state NMR approach *Biochimie* **73** 1295–302

Lidman M, Pokorna S, Dingeldein A P, Sparrman T, Wallgren M, Sachl R, Hof M and Grobner G 2016 The oxidized phospholipid PazePC promotes permeabilization of mitochondrial membranes by Bax *BBA-Biomembranes* **1858** 1288–97

Liljas A, Liljas L, Piskur J, Lindblom G R, Nissen P and Kjeldgaard M 2009 *Textbook of Structural Biology* (Singapore: World Scientific)

Lindstrom F, Thurnhofer S, Vetter W and Grobner G 2006 Impact on lipid membrane organization by free branched-chain fatty acids *Phys. Chem. Chem. Phys.* **8** 4792–7

Lindstrom F, Williamson P T F and Grobner G 2005 Molecular insight into the electrostatic membrane surface potential by N-14/P-31 MAS NMR spectroscopy: nociceptin-lipid association *J. Am. Chem. Soc.* **127** 6610–6

Lu S M and Fairn G D 2018 Mesoscale organization of domains in the plasma membrane—beyond the lipid raft *Crit. Rev. Biochem. Mol. Biol.* **53** 192–207

Luckey M 2008 *Membrane Structural Biology* (New York: Cambridge University Press)

Macdonald P M, Leisen J and Marassi F M 1991 Response of phosphatidylcholine in the gel and liquid-crystalline states to membrane-surface charges *Biochemistry* **30** 3558–66

Macdonald P M, Saleem Q, Lai A and Morales H H 2013 NMR methods for measuring lateral diffusion in membranes *Chem. Phys. Lipids* **166** 31–44

Marius P, De Planque M R and Williamson P T F 2012 Probing the interaction of lipids with the non-annular binding sites of the potassium channel KcsA by magic-angle spinning NMR *BBA-Biomembranes* **1818** 90–6

Marius P, Zagnoni M, Sandison M E, East J M, Morgan H and Lee A G 2008 Binding of anionic lipids to at least three nonannular sites on the potassium channel KcsA is required for channel opening *Biophys. J.* **94** 1689–98

Marsh D 2013 *Handbook of Lipid Bilayers* (Boca Raton, FL: CRC Press)

McCabe M A and Wassall S R 1997 Rapid deconvolution of NMR powder spectra by weighted fast Fourier transformation *Solid State Nucl. Magn. Reson.* **10** 53–61

Meneses P and Glonek T 1988 High-resolution P-31 NMR of extracted phospholipids *J. Lipid Res.* **29** 679–89

Mihailescu M, Soubias O, Worcester D, White S H and Gawrisch K 2011 Structure and dynamics of cholesterol-containing polyunsaturated lipid membranes studied by neutron diffraction and NMR *J. Membrane Biol.* **239** 63–71

Mollinedo F and Gajate C 2020 Lipid rafts as signaling hubs in cancer cell survival/death and invasion: implications in tumor progression and therapy *J. Lipid Res.* **61** 611–35

Molugu T R and Brown M F 2019 Cholesterol effects on the physical properties of lipid membranes viewed by solid-state NMR spectroscopy *Adv. Exp. Med. Biol.* **1115** 99–133

Molugu T R, Lee S and Brown M F 2017 Concepts and methods of solid-state NMR spectroscopy applied to biomembranes *Chem. Rev.* **117** 12087–132

Mouritsen O G 2011 Lipidology and lipidomics—quo vadis? A new era for the physical chemistry of lipids *Phys. Chem. Chem. Phys.* **13** 19195–205

Murray D, Arbuzova A, Hangyas-Mihalyne G, Gambhir A, Ben-Tal N, Honig B and Mclaughlin S 1999 Electrostatic properties of membranes containing acidic lipids and adsorbed basic peptides: theory and experiment *Biophys. J.* **77** 3176–88

Muzic T, Tounsi F, Madsen S B, Pollakowski D, Konrad M and Heimburg T 2019 Melting transitions in biomembranes *BBA-Biomembranes* 1861

Nakayama H, Nagafuku M, Suzuki A, Iwabuchi K and Inokuchi J I 2018 The regulatory roles of glycosphingolipid-enriched lipid rafts in immune systems *FEBS Lett.* **592** 3921–42

Nowacka A, Mohr P C, Norrman J, Martin R W and Topgaard D 2010 Polarization transfer solid-state NMR for studying surfactant phase behavior *Langmuir* **26** 16848–56

Overall S A, Zhu S Y, Hanssen E, Separovic F and Sani M A 2019 *In situ* monitoring of bacteria under antimicrobial stress using P-31 solid-state NMR *Int. J. Mol. Sci.* **20** 181

Paracini N, Clifton L A, Skoda M W A and Lakey J H 2018 Liquid crystalline bacterial outer membranes are critical for antibiotic susceptibility *Proc. Natl Acad. Sci. USA* **115** E7587–94

Paul D, Romero-Brey I, Gouttenoire J, Stoitsova S, Krijnse-Locker J, Moradpour D and Bartenschlager R 2011 NS4B self-interaction through conserved C-terminal elements is required for the establishment of functional Hepatitis C virus replication complexes *J. Virol.* **85** 6963–76

Petzold K, Olofsson A, Arnqvist A, Grobner G and Schleucher J 2009 Semiconstant-time P,H-COSY NMR: analysis of complex mixtures of phospholipids originating from *Helicobacter pylori J. Am. Chem. Soc.* **131** 14150–1

Pikula S, Bandorowicz-Pikula J and Groves P 2015 Recent advances in NMR studies of lipids *Annu. Rep. NMR Spectro.* **85** 195–246

Powl A M, East J M and Lee A G 2008a Anionic phospholipids affect the rate and extent of flux through the mechanosensitive channel of large conductance MscL *Biochemistry* **47** 4317–28

Powl A M, East J M and Lee A G 2008b Importance of direct interactions with lipids for the function of the mechanosensitive channel MscL *Biochemistry* **47** 12175–84

Raghunathan K and Kenworthy A K 2018 Dynamic pattern generation in cell membranes: Current insights into membrane organization *BBA-Biomembranes* **1860** 2018–31

Roux M, Neumann J M, Hodges R S, Devaux P F and Bloom M 1989 Conformational-changes of phospholipid headgroups induced by a cationic integral membrane peptide as seen by deuterium magnetic-resonance *Biochemistry* **28** 2313–21

Roux M, Nezil F A, Monck M and Bloom M 1994 Fragmentation of phospholipid-bilayers by myelin basic-protein *Biochemistry* **33** 307–11

Russ K A *et al* 2016 C-60 fullerene localization and membrane interactions in RAW 264.7 immortalized mouse macrophages *Nanoscale* **8** 4134–44

Saleem Q, Lai A, Morales H H and Macdonald P M 2012 Lateral diffusion of bilayer lipids measured via P-31 CODEX NMR *Chem. Phys. Lipids* **165** 721–30

Sani M A, Keech O, Gardestrom P, Dufourc E J and Grobner G 2009 Magic-angle phosphorus NMR of functional mitochondria: *in situ* monitoring of lipid response under apoptotic-like stress *FASEB J.* **23** 2872–8

Sani M A and Separovic F 2016 How membrane-active peptides get into lipid membranes *Acc. Chem. Res.* **49** 1130–8

Sani M A, Weber D K, Delaglio F, Separovic F and Gehman J D 2013 A practical implementation of de-Pake-ing via weighted Fourier transformation *PeerJ* **1** e30

Scherer P G and Seelig J 1985 Structure and dynamics of 2 phospholipid head groups in the plasma-membrane of an eukaryotic cell-line – application of H-2-NMR and P-31-NMR *Biol. Chem.* **366** 846

Schmidt P, Thomas L, Muller P, Scheidt H A and Huster D 2014 The G-protein-coupled neuropeptide Y receptor type 2 is highly dynamic in lipid membranes as revealed by solid-state NMR spectroscopy *Chemistry* **20** 4986–92

Seelig J 1977 Deuterium magnetic resonance – theory and application to lipid membranes *Q. Rev. Biophys.* **10** 353–418

Seelig J 1978 P-31 nuclear magnetic-resonance and head group structure of phospholipids in membranes *BBA-Biomembranes,* **515** 105–40

Semchyschyn D J and Macdonald P M 2004 Conformational response of the phosphatidylcholine headgroup to bilayer surface charge: torsion angle constraints from dipolar and quadrupolar couplings in bicelles *Magn. Reson. Chem.* **42** 89–104

Shai Y 2002 Mode of action of membrane active antimicrobial peptides *Biopolymers* **66** 236–48

Simons K and Ikonen E 1997 Functional rafts in cell membranes *Nature* **387** 569–72

Sixl F and Watts A 1983 Headgroup interactions in mixed phospholipid-bilayers *Proc. Natl Acad. Sci. USA* **80** 1613–5

Sodt A J, Soubias O, Gawrisch K and Pastor R W 2016 Lipid–lipid coupling to membrane curvature by simulation and NMR *Biophys. J.* **110** 243a

Soubias O and Gawrisch K 2005 Probing specific lipid-protein interaction by saturation transfer difference NMR spectroscopy *J. Am. Chem. Soc.* **127** 13110–1

Soubias O, Teague W E and Gawrisch K 2006 Evidence for specificity in lipid–rhodopsin interactions *J. Biol. Chem.* **281** 33233–41

Tardy-Laporte C, Arnold A A, Genard B, Gastineau R, Morancais M, Mouget J L, Tremblay R and Marcotte I 2013 A (2)H solid-state NMR study of the effect of antimicrobial agents on intact *Escherichia coli* without mutating *Biochim. Biophys. Acta* **1828** 614–22

Thurmond R L, Niemi A R, Lindblom G, Wieslander A and Rilfors L 1994 Membrane thickness and molecular ordering in Acholeplasma-Laidlawii strain-a studied by H-2 NMR-spectroscopy *Biochemistry* **33** 13178–88

Thysell E *et al* 2019 Gene expression profiles define molecular subtypes of prostate cancer bone metastases with different outcomes and morphology traceable back to the primary tumor *Mol. Oncol.* **13** 1763–77 van Duyl B Y, Ganchev D, Chupin V, De Kruijff B and Killian J A 2003 Sphingomyelin is much more effective than saturated phosphatidylcholine in excluding unsaturated phosphatidylcholine from domains formed with cholesterol *FEBS Lett.* **547** 101–6

van Meer G, Voelker D R and Feigenson G W 2008 Membrane lipids: where they are and how they behave *Nat. Rev. Mol. Cell Biol.* **9** 112–24

Veatch S L, Polozov I V, Gawrisch K and Keller S L 2004 Liquid domains in vesicles investigated by NMR and fluorescence microscopy *Biophys. J.* **86** 2910–22

Wacklin H P, Bremec B B, Moulin M, Rojko N, Haertlein M, Forsyth T, Anderluh G and Norton R S 2016 Neutron reflection study of the interaction of the eukaryotic pore-forming actinoporin equinatoxin II with lipid membranes reveals intermediate states in pore formation *BBA-Biomembranes* **1858** 640–52

Wahle K W J 1983 Fatty-acid modification and membrane lipids *Proc. Nutr. Soc.* **42** 273–87

Warnet X L, Arnold A A, Marcotte I and Warschawski D E 2015 In-cell solid-state NMR: an emerging technique for the study of biological membranes *Biophys. J.* **109** 2461–6

Warnet X L, Laadhari M, Arnold A A, Marcotte I and Warschawski D E 2016 A $^{2}$H magic-angle spinning solid-state NMR characterisation of lipid membranes in intact bacteria *Biochim. Biophys. Acta* **1858** 146–52

Warschawski D E, Arnold A A and Marcotte I 2018 A new method of assessing lipid mixtures by $^{31}$P magic-angle spinning NMR *Biophys. J.* **114** 1368–76

Weisz K, Grobner G, Mayer C, Stohrer J and Kothe G 1992 Deuteron nuclear magnetic resonance study of the dynamic organization of phospholipid/cholesterol bilayer membranes: molecular properties and viscoelastic behavior *Biochemistry* **31** 1100–12

West A, Zoni V, Teague W E, Leonard A N, Vanni S, Gawrisch K, Tristram-Nagle S, Sachs J N and Klauda J B 2020 How do ethanolamine plasmalogens contribute to order and structure of neurological membranes? *J. Phys. Chem.* B **124** 828–39

Whited A M and Johs A 2015 The interactions of peripheral membrane proteins with biological membranes *Chem. Phys. Lipids* **192** 51–9

Yasuda T, Tsuchikawa H, Murata M and Matsumori N 2015 Deuterium NMR of raft model membranes reveals domain-specific order profiles and compositional distribution *Biophys. J.* **108** 2502–6

Yeagle P L 2011 *The Structure of Biological Membranes* (Boca Raton, FL: CRC Press)

Yeliseev A A, Wong K K, Soubias O and Gawrisch K 2005 Expression of human peripheral cannabinoid receptor for structural studies *Protein Sci.* **14** 2638–53

# Chapter 2

# Characterization of lipid behaviour in model biomembranes using $^2$H solid-state NMR

**Michael R Morrow and Jenifer L Thewalt**

## Abstract

The interaction between the quadrupole moment of a deuteron and the electric field gradient of a carbon–deuterium bond on a lipid acyl chain perturbs the Zeeman splitting of the spin-1 deuteron in an orientation-dependent manner. Partial averaging of that perturbation by asymmetric reorientation of the acyl chains in lipid bilayers provides a powerful tool for probing and characterizing model membrane orientational order and phase behaviour. This chapter begins with a review of the quadrupole interaction as it pertains to $^2$H solid-state NMR spectroscopy in lipid bilayers. It then goes on to describe how $^2$H solid-state NMR spectroscopy is used to obtain acyl chain orientational order parameter profiles for liquid crystal phase lipid bilayers and to distinguish disordered and ordered lipid bilayer phases. The chapter continues with a discussion of how $^2$H solid-state NMR spectroscopy has been used to characterize the phase behaviour of mixed-lipid bilayers and ends with a discussion of important contributions made to the determination of sterol-lipid phase diagrams using this approach.

## 2.1 Introduction

Wideline $^2$H solid-state NMR is a valuable tool for characterizing bilayers containing deuterium–labeled lipids. $^2$H NMR spectra of chain deuterated lipids are sensitive to the amplitudes and timescales of chain motions at each labeled chain segment while replacement of hydrogen by deuterium on lipid acyl chains has little effect on bilayer organization and thermodynamics beyond slight perturbations such as the reduction of some phase transition temperatures by no more than a few degrees (Petersen *et al* 1975).

The deuteron is a spin-1 nucleus with an electric quadrupole moment. For a carbon–deuterium bond having a specific orientation with respect to an applied

doi:10.1088/978-0-7503-2532-5ch2  

magnetic field, the deuteron Zeeman energy levels are perturbed by the orientation-dependent electric quadrupole interaction as illustrated in figure 2.1. The result is a spectral doublet with a splitting that reflects the bond orientation. For chain-deuterated lipids in randomly oriented bilayers, the $^2$H NMR spectrum arising from a given deuterated chain segment reflects the extent to which the quadrupole interaction is averaged by chain segment reorientation relative to a symmetry axis (typically the bilayer normal) and on whether that reorientation is or is not axially symmetric on the $\sim 10^{-5}$ s timescale characteristic of the $^2$H NMR experiment. If, as is often the case, the lipid chain is fully deuterated (perdeuterated), the spectrum is a superposition of doublet components and contains information about how constraints on segment reorientation depend on position along the chain. It is this sensitivity to segment reorientation amplitude, motional symmetry, and timescale that makes $^2$H NMR a useful diagnostic of deuterated bilayer organization and phase behavior.

In this chapter, we will briefly review the fundamentals of the quadrupole interaction and then describe how segment reorientation is reflected by the $^2$H NMR spectra of chain-deuterated lipid bilayers. We will also discuss the quadrupole echo pulse sequence and why it is necessitated by the large quadrupole splittings encountered in lipid bilayer studies. We will then go on to describe how lipid bilayer phases are distinguished by $^2$H NMR lineshapes and how bilayer phase affects decay of the quadrupole echo. Vist and Davis (1990) introduced $^2$H NMR spectral difference spectroscopy and used it to identify regions of two-phase coexistence in the temperature-composition phase diagram for mixtures of dipalmitoyl phosphatidylcholine and cholesterol. We will discuss this technique and review some of its subsequent applications in the determination or refinement of phase diagrams for lipid bilayer systems.

## 2.2 The quadrupole interaction

The deuteron is a spin-1 nucleus. For molecules containing carbon–deuterium bonds, the deuteron Zeeman levels in an applied magnetic field can be perturbed by the interaction of their electric quadrupole moments, $eQ$, with the electric field gradient (EFG) at the nucleus. Dipole interactions are also present but are weak compared to quadrupole interactions and are typically ignored when considering $^2$H NMR spectra of deuterated lipid bilayers. The Hamiltonian describing the quadrupole interaction can be written in spherical tensor notation (Seelig 1977, Davis 1983,

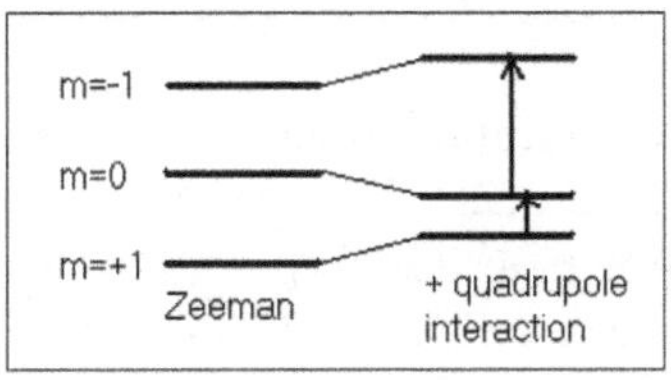


**Figure 2.1.** Perturbation of the deuteron Zeeman levels by the quadrupole interaction. The magnitude of the quadrupole interaction relative to the Zeeman interaction is exaggerated.

Molugu *et al* 2017) or Cartesian tensor notation (de Lange and Burnell 2003). In Cartesian notation, the EFG can be diagonalized by transformation to its principal axis system in which the elements become

$$V_{zz} = \frac{d^2V}{dz^2},\ V_{yy} = \frac{d^2V}{dy^2},\ V_{xx} = \frac{d^2V}{dx^2}. \tag{2.1}$$

It is usual to identify the principal component of the EFG as $eq = V_{zz}$, and an asymmetry parameter as $\eta \equiv V_{xx} - V_{yy}$. The principal axis system of the EFG tensor can be thought of as being attached to a molecule or molecular segment while spin angular momentum is quantized with respect to the applied magnetic field in the laboratory frame. Transforming the EFG tensor to the laboratory frame using the Euler angles $(\alpha, \beta, \gamma)$ allows the Hamiltonian for the quadrupole interaction to be expressed as

$$\mathcal{H}_Q = \frac{e^2qQ}{4I\left(2I-1\right)}[3I_z^2 - I(I+1)]\left[\frac{1}{2}(3\cos^2\beta - 1) + \frac{1}{2}\eta\sin^2\beta\cos 2\alpha\right]. \tag{2.2}$$

For the case of a spin-1 nucleus ($I = 1$) and axial symmetry ($\eta = V_{xx} - V_{yy} = 0$), this becomes

$$\mathcal{H}_Q = \frac{e^2qQ}{4}\left[3I_z^2 - 2\right]\left[\frac{1}{2}(3\cos^2\beta - 1)\right]. \tag{2.3}$$

The corresponding eigenvalues ($m = 0, \pm 1$) are then

$$E_m = \frac{e^2qQ}{4}[3m^2 - 2]\left[\frac{1}{2}(3\cos^2\beta - 1)\right]. \tag{2.4}$$

For CD bonds, $\frac{e^2qQ}{h} \approx 167$ kHz (Davis and Jeffrey 1977, Davis 1983).

## 2.3 The wideline $^2$H NMR spectra of chain-deuterated phospholipids in bilayers

For a $^2$H NMR experiment, the Zeeman splittings will be perturbed by the orientation-dependent quadrupole interaction, as shown in figure 2.1. The selection rule is $\Delta m = \pm 1$, so the result is a doublet split by

$$\Delta\nu_Q = \frac{3}{2}\frac{e^2qQ}{h}\left[\frac{1}{2}(3\cos^2\beta - 1)\right] = \frac{3}{4}\frac{e^2qQ}{h}(3\cos^2\beta - 1). \tag{2.5}$$

If CD bonds in a sample are immobile, then $\beta$ is, in effect, the angle between a CD bond and the applied magnetic field. The resulting spectra for static CD bonds at $\beta = 0°$ and $\beta = 90°$ are shown as the top and bottom spectra in figure 2.2 if referenced to the upper frequency scale. If the static CD bonds are randomly oriented, as in a powder, the result is Pake doublet. Derivations of the distribution of intensity in a Pake doublet can be found in many sources including Seelig (1977),

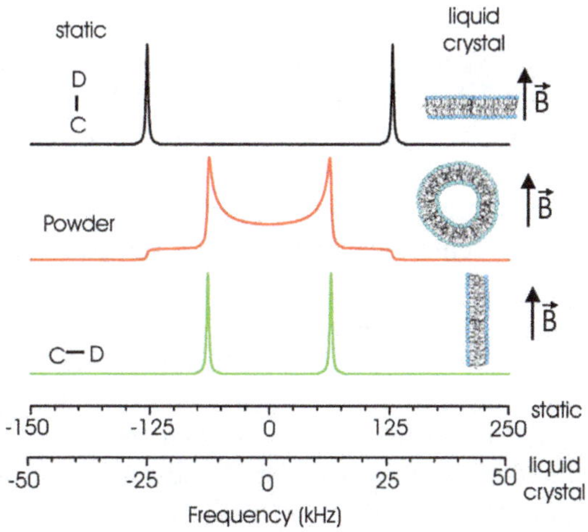


**Figure 2.2.** Simulated $^{2}$H NMR spectra. Using the upper frequency scale, the top and bottom spectra are for static CD bonds oriented parallel to ($\beta = 0°$) or perpendicular to ($\beta = 90°$) the applied magnetic field, respectively. The centre spectrum is for randomly oriented static CD bonds (a powder). The lower frequency scale is for a single deuterated acyl chain segment, with $S_{CD} = 0.2$, in liquid crystal bilayers of lipids undergoing axially symmetric reorientation. The top and bottom spectra are then for samples with bilayer normal oriented parallel to or perpendicular to the applied magnetic field, respectively, and the centre spectrum is for a sample with randomly oriented bilayer normals.

Davis (1983) and Molugu *et al* (2017). The distribution of intensity reflects the fact that there are more ways to orient spins at $\beta = 90°$ than at $\beta = 0°$.

If a CD bond is attached to a molecular segment undergoing (fast) axially symmetric reorientation about an axis oriented at an angle $\beta$ from the applied magnetic field, it is useful to first transform from the principal axis system (PAS) to the molecule-fixed system. In this case, angle $\theta$ is the angle between the EFG principal axis and the symmetry axis for reorientation and $\beta$ is the angle between the symmetry axis for reorientation and the applied magnetic field. This, in effect, introduces another factor of $P_2(\cos\theta) = \frac{3\cos^2\theta - 1}{2}$ into the splittings. If $\theta$ is modulated by conformational fluctuations of the molecule, then the factor of $P_2(\cos\theta)$ must be averaged over such internal motions. The result is a quadrupole splitting given by

$$\Delta\nu_Q = \frac{3}{4}\frac{e^2qQ}{h}(3\cos^2\beta - 1)S_{CD} \tag{2.6}$$

where $S_{CD} = \left\langle \frac{1}{2}(3\cos^2\theta - 1)\right\rangle$ is the orientational order parameter for the molecular segment to which the CD bond is attached. The orientational order parameter thus reflects the extent to which the acyl chain is ordered with respect to the symmetry axis for molecular motion. For isotropic reorientation of the CD bond, $S_{CD} = 0$.

Equation 2.6 shows that, for fast, axially symmetric reorientation, the splitting of the Pake doublet is scaled by $S_{CD}$ so that a larger range of segment reorientation results in a smaller quadrupole splitting. For lipid bilayers, the axis of symmetry for molecular reorientation is typically the bilayer normal. If the lower frequency scale is used, the spectra in figure 2.2 would apply to a deuterated acyl chain segment, with orientational order parameter $S_{CD} = 0.2$, in liquid crystal bilayers of lipids undergoing axially symmetric reorientation. The top and bottom spectra correspond to bilayer normals being oriented parallel to or perpendicular to the applied magnetic field, respectively, and the centre spectrum is for a sample with randomly oriented bilayer normals.

For bilayers of di-saturated phospholipids having a specific chain length, the main transition temperature is identified as the temperature at which a gel phase, in which acyl chains are relatively ordered and reorientation is not axially symmetric on the timescale of the $^2$H NMR experiment (~$10^{-5}$ s), can coexist with a liquid crystal (or $L_\alpha$) phase in which the chains are less ordered and reorientation about the bilayer normal is axially symmetric on the $^2$H NMR timescale. The main transition temperatures for some common phosphatidylcholines are −2 °C for dilauroyl phosphatidylcholine (DLPC or 12:0 PC), 24 °C for dimyristoyl phosphatidylcholine (DMPC or 14:0 PC), 41 °C for dipalmitoyl phosphatidylcholine (DPPC or 16:0 PC), and 55 °C for distearoyl phosphatidylcholine (DSPC or 18:0 PC). The gel to liquid crystal transition temperatures for bilayers comprising chain perdeuterated versions of these lipids are typically a few degrees lower (Petersen *et al* 1975). For a given bilayer composition, the transition reported by $^2$H NMR observations of lipid bilayers containing chain-deuterated lipids corresponds to the specific heat anomaly observed by differential scanning calorimetry. This suggests that the enthalpy change at the gel-to-liquid crystalline bilayer transition is largely associated with changes in acyl chain order.

The $^2$H NMR spectrum, for bilayers of chain-perdeuterated, di-saturated phospholipids in the liquid crystal phase, is a superposition of Pake doublets with quadrupole splittings reflecting the variation of orientational order parameter with segment position along the acyl chain. In the liquid crystal phase, motions of chain segments closer to the hydrophilic/hydrophobic interface of the bilayer are more constrained than those of chain segments closer to the bilayer centre. The result is a characteristic liquid crystal phase spectrum with prominent edges, corresponding to the overlap of doublets arising from ordered segments near the bilayer interface, and smaller splittings corresponding to the increasingly disordered segments toward the bilayer centre. Figure 2.3(a) shows the $^2$H NMR spectrum of chain-perdeuterated dilauroyl phosphatidylcholine (DLPC-$d_{46}$) bilayers in the liquid crystal phase at 25 °C. The acyl chains of DLPC-$d_{46}$ have deuterated methylene segments from position 2 to position 11. The final segment of each chain is a deuterated methyl group. The quadrupole splitting for the methyl deuterons reflects additional averaging of the quadrupole interaction due to fast reorientation about the methyl axis. The prominence of the methyl doublet reflects its smaller splitting and, if excitation is uniform across the spectrum, an area that is 3/2 that of a methylene doublet.

The orientational order parameters for each chain segment can be obtained from the corresponding quadrupole splittings. This is facilitated by transforming the

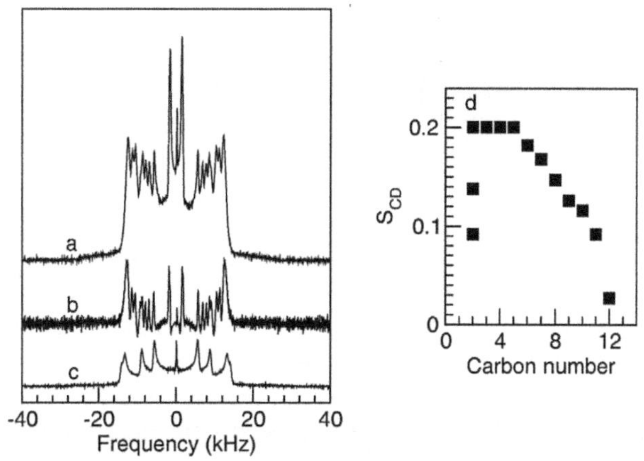


**Figure 2.3.** (a) $^2$H NMR spectrum for bilayers of DLPC-$d_{46}$ at $T$ = 25 °C. (b) Data from the spectrum in (a) transformed, using a DePakeing algorithm, to give the spectrum that would be seen for a sample of bilayers oriented with bilayer normals perpendicular to the magnetic field. (c) $^2$H NMR spectrum for bilayers of DLPC deuterated on the C2 segments of both acyl chains (DLPC-$d_4$) at $T$ = 25 °C. (d) Orientational order parameter profile for DLPC-$d_{46}$. Spectra provided by Timothy Porter.

spectrum to remove the effect of random bilayer normal orientation to obtain the spectrum that would be expected for a single bilayer orientation. This process, known as dePakeing, was introduced by Myer Bloom and co-workers (Bloom *et al* 1981, Sternin *et al* 1983, Whittall *et al* 1989). Figure 2.3(b) shows the result of dePakeing the 25 °C $^2$H NMR spectrum for DPLC-$d_{46}$ bilayers using the dePakeing algorithm of McCabe and Wassall (1995). Assigning each doublet to a chain segment can be complicated slightly by the fact that deuterons on carbon 2 of the *sn* −2 chain in di-saturated PCs typically have smaller splittings than those on the *sn*−1 chain due to geometric constraints (Engel and Cowburn 1981). This is illustrated in figure 2.3(c) which shows the $^2$H NMR spectrum of DLPC-$d_4$ which is deuterated on the alpha carbons (carbon 2) of both acyl chains. The two deuterons on carbon-2 of the *sn*−1 chain give rise to the Pake doublet with the largest splitting. The two doublets with smaller splittings are from deuterons on the corresponding segment of the *sn*−2 chain. The splittings for the alpha carbon deuterons on DLPC-$d_4$ at 25 °C are slightly higher than the corresponding splittings for DLPC-$d_{46}$. This reflects the fact, as mentioned above, that chain perdeuteration lowers the gel-liquid crystal transition temperature by 3–4 degrees relative to the case for bilayers of the same lipid with no deuteration or, as in the case of DLPC-$d_4$, little deuteration. As a result, bilayers of the more lightly deuterated lipid at 25 °C are closer to the transition temperature, and thus slightly more ordered, than the corresponding sample of chain-perdeuterated lipid.

Comparison with the DLPC-$d_4$ spectrum indicates that deuterons on carbon 2 of the *sn*−1 chain contribute to the doublet with the largest splitting while one deuteron on the *sn*−2 chain carbon-2 coincides with the carbon-11 doublet and the other nearly coincides with the carbon-8 doublet. With the carbon-2 doublets accounted for, the rest of the quadrupole splittings can be assigned and the orientational order

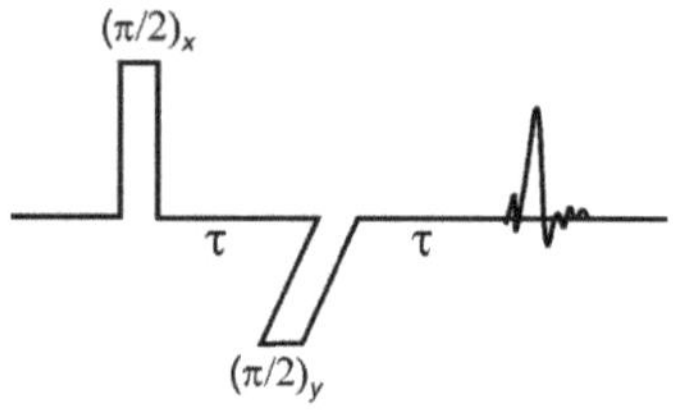


**Figure 2.4.** The quadrupole echo pulse sequence.

parameter profile shown in figure 2.3(d) can be obtained. Doublets arising from deuterons on carbons 3, 4 and 5 cannot be resolved from those of the *sn*−1 carbon 2 deuterons and are shown with a common orientational order parameter. This overlap of orientational order parameters for segments near the interface is referred to as the plateau feature in the orientational order parameter profile.

Before discussing how changes in the spectrum can be used to study lipid bilayer phase behavior, it would be appropriate to comment on a technical issue that arises due to the range of frequencies spanned by the $^2$H NMR spectrum of a deuterated phospholipid bilayer.

## 2.4 The quadrupole echo

To observe $^2$H NMR spectra from bilayers of deuterated lipids, it was first necessary to address an important practical issue. The spectra encountered typically span frequency ranges in the 10s of kHz. Because the frequency content in a radio-frequency pulse is inversely proportional to its length, exciting spins over such a range necessarily implies using very short pulses (~5 − 10 μs). To get a very short $\pi/2$ pulse, the magnitude of the radio-frequency field, expressed as a precession frequency $\omega_1$, must be large which implies a pulse amplifier with an output power of ~1 kW. At the same time, the large spectral range of intensity in a bilayer $^2$H NMR spectrum implies that the free induction decay will be short (~10 − 20 μs). This is of same order of magnitude as the typical time for a preamplifier to recover from residual effects of the necessary large RF pulse. There is thus a need to move the signal acquisition window away from the pulse.

The solution, introduced by Myer Bloom and co-workers, was to generate a quadrupole echo and analyse the free induction decay starting from the top of the echo (Davis *et al* 1976). The quadrupole echo sequence, consisting of two $\frac{\pi}{2}$ pulses differing in phase by 90° and separated by an interval $\tau$, is illustrated in figure 2.4. Myer Bloom has used the quadrupole echo sequence to illustrate the application of density matrix methods to the understanding of spin-1 system dynamics (Bloom 1988). We will discuss the quadrupole echo sequence using the notation in Bloom *et al* (2003).

For a spin-1 particle (i.e., a deuteron), spin operators are 3 × 3 matrices. The operators corresponding to observable magnetizations are

$$I_x = \frac{1}{\sqrt{2}}\begin{pmatrix} 0 & 1 & 0 \\ 1 & 0 & 1 \\ 0 & 1 & 0 \end{pmatrix},\ I_y = \frac{1}{\sqrt{2}}\begin{pmatrix} 0 & -i & 0 \\ i & 0 & -i \\ 0 & i & 0 \end{pmatrix},\ I_z = \begin{pmatrix} 1 & 0 & 0 \\ 0 & 0 & 0 \\ 0 & 0 & -1 \end{pmatrix} \tag{2.7}$$

The density matrix is also a three-dimensional matrix. It can be written as a linear combination of 8 orthogonal matrices $T_i$, plus the unit matrix, normalized so that $\mathrm{Tr}\{T_\alpha T_\beta\} = \delta_{\alpha\beta}$.

Density matrix will then have the form

$$\rho(t) = \sum_q a_q(t) T_q. \tag{2.8}$$

The $a_q(t)$ coefficients can be thought of as components of a unit magnitude vector reorienting in 9-dimensional space. The state of the system is given by the orientation of the vector in that space.

Again, following Bloom *et al* (2003), a convenient set of basis matrices for this expansion is

$$\begin{aligned} T_x &= \frac{1}{\sqrt{2}} I_x,\ T_y = \frac{1}{\sqrt{2}} I_y,\ T_z = \frac{1}{\sqrt{2}} I_z,\ T_{z^2} = \frac{1}{\sqrt{6}}\left(3I_z^2 - 2\right), \\ T_{xz} &= \frac{1}{\sqrt{2}}\left(I_x I_z + I_z I_x\right),\ T_{yz} = \frac{1}{\sqrt{2}}\left(I_y I_z + I_z I_y\right), \\ T_{x^2-y^2} &= \frac{1}{\sqrt{2}}\left(I_x^2 - I_y^2\right),\ \text{and}\ T_{xy} = \frac{1}{\sqrt{2}}\left(I_x I_y + I_y I_x\right). \end{aligned} \tag{2.9}$$

At equilibrium, the net magnetization of the system is along the $z$-axis so, using the Curie law, the only non-zero coefficient in the density matrix is $a_z(0) = \frac{\sqrt{2}\hbar\omega_0}{3kT}$. The equilibrium density matrix is thus

$$\rho(0) = \sum_q a_q(0) T_q = a_z(0) T_z. \tag{2.10}$$

The evolution of the density matrix under the influence of a given Hamiltonian is described by the Liouville equation

$$i\hbar\frac{d\rho}{dt} = [\mathcal{H}, \rho] \tag{2.11}$$

so that the evolution of the coefficients in the density matrix is given by

$$i\hbar\frac{da_p}{dt} = \sum_q a_q \mathrm{Tr}\{T_p[\mathcal{H}, T_q]\}. \tag{2.12}$$

To analyse the quadrupole echo sequence in the rotating frame, it will be necessary to have Hamiltonians for pulses along the $x$ and $y$ axes and for the quadrupole interaction. They are, respectively,

$$\mathcal{H}_1 = -\sqrt{2}\,\hbar\omega_1 \boldsymbol{T}_x,\ \mathcal{H}_2 = -\sqrt{2}\,\hbar\omega_1 \boldsymbol{T}_y,\ \mathcal{H}_Q = \sqrt{\frac{2}{3}}\,\hbar\omega_Q \boldsymbol{T}_{z^2} \tag{2.13}$$

where $\omega_1$ is the precession frequency about the stationary component of the radio frequency magnetic field in the rotating frame and $\omega_Q$ is the angular frequency associated with the quadrupole interaction.

The effect, on the density matrix, of a radio frequency pulse along the $x$-axis in the rotating frame is given by

$$i\hbar \frac{d\rho(t)}{dt} = [\mathcal{H}_1, \rho(t)]. \tag{2.14}$$

For a system starting from equilibrium, the two commutators needed to characterize the evolution of the density matrix under the influence of this Hamiltonian are

$$[\boldsymbol{T}_x, \boldsymbol{T}_z] = \frac{-i}{\sqrt{2}} \boldsymbol{T}_y \ \text{and}\ [\boldsymbol{T}_x, \boldsymbol{T}_y] = \frac{i}{\sqrt{2}} \boldsymbol{T}_z. \tag{2.15}$$

Using these plus the identity, $\mathrm{Tr}\{\boldsymbol{T}_\alpha \boldsymbol{T}_\beta\} = \delta_{\alpha\beta}$, we find that the equations of motion for the two relevant coefficients in the density matrix are

$$\frac{da_y}{dt} = \omega_1 a_z \ \text{and}\ \frac{da_z}{dt} = -\omega_1 a_y. \tag{2.16}$$

These equations describe the components of a vector precessing with angular velocity $\omega_1$ in the $\boldsymbol{T}_y \times \boldsymbol{T}_z$ plane. Evolution of the density matrix evolution can be represented visually using precession diagrams (Bloom 1988, Bloom *et al* 2003). If the pulse length is chosen to satisfy $\omega_1 t = \pi/2$, then the density matrix at the end of the pulse is $\boldsymbol{\rho} = a_z(0)\boldsymbol{T}_y$.

After the first pulse, the system can evolve for a time $\tau$ under the influence of the quadrupole Hamiltonian,

$$\mathcal{H}_Q = \sqrt{\frac{2}{3}}\,\hbar\omega_Q \boldsymbol{T}_{z^2}. \tag{2.17}$$

To characterize the evolution of the density matrix under the influence of this Hamiltonian, starting from $\boldsymbol{\rho} = a_z(0)\boldsymbol{T}_y$, the useful commutators are

$$\left[\boldsymbol{T}_{z^2}, \boldsymbol{T}_y\right] = \frac{-3i}{\sqrt{6}} \boldsymbol{T}_{xz} \ \text{and}\ [\boldsymbol{T}_{z^2}, \boldsymbol{T}_{xz}] = \frac{3i}{\sqrt{6}} \boldsymbol{T}_y. \tag{2.18}$$

The resulting equations of motion for $a_y$ and $a_{xz}$ are then

$$\frac{da_y}{dt} = \omega_Q a_{xz} \ \text{and}\ \frac{da_{xz}}{dt} = -\omega_Q a_y \tag{2.19}$$

which describe the components of a vector precessing with angular velocity $\omega_Q$ in the $\boldsymbol{T}_y \times \boldsymbol{T}_{xz}$ plane.

After evolving under the influence of the quadrupole interaction for an interval $\tau$ following the first $\pi/2$ pulse, the density matrix will be

$$\rho = a_z(0)[(\cos \omega_Q\tau)\boldsymbol{T}_y - (\sin \omega_Q\tau)\boldsymbol{T}_{xz}]. \tag{2.20}$$

For randomly oriented bilayers of deuterated phospholipids, there will be a range of quadrupole interactions and the magnitude of the net oscillating magnetization along the y-axis of the rotating frame will decrease with increasing $\tau$ as phase differences accumulate.

The second pulse of the quadrupole echo sequence is shifted in phase, by $\pi/2$ from the first so it is useful to consider the effect of a radio-frequency pulse, again of strength $\omega_1$, applied along the $y$-axis.

Under the influence of this pulse, the density matrix evolves according to

$$i\hbar\frac{d\rho}{dt} = [\mathcal{H}_2, \rho] = -\sqrt{2}\hbar\omega_1[\boldsymbol{T}_y, \rho]. \tag{2.21}$$

The commutators needed to obtain expressions for the effect of this pulse on the coefficients of the basis matrices in the density matrix are

$$[\boldsymbol{T}_y, \boldsymbol{T}_{xz}] = \frac{i}{\sqrt{2}}(\boldsymbol{T}_{x^2-y^2} - \sqrt{3}\,\boldsymbol{T}_{z^2}), \tag{2.22}$$

$$[\boldsymbol{T}_y, \boldsymbol{T}_{z^2}] = \frac{3i}{\sqrt{6}}\boldsymbol{T}_{xz}, \tag{2.23}$$

and

$$[\boldsymbol{T}_y, \boldsymbol{T}_{x^2-y^2}] = \frac{-i}{\sqrt{2}}\boldsymbol{T}_{xz}. \tag{2.24}$$

The resulting equations of motion for $a_{z^2}$, $a_{xz}$, and $a_{x^2-y^2}$ are

$$\frac{da_{z^2}}{dt} = \omega_1\sqrt{3}\,a_{xz}, \quad \frac{da_{xz}}{dt} = -\omega_1(\sqrt{3}\,a_{z^2} - a_{x^2-y^2}), \quad \frac{da_{x^2-y^2}}{dt} = -\omega_1 a_{xz}. \tag{2.25}$$

To see how this can describe precession of a vector in a plane within the space defined by the basis matrices, it is helpful to group these expressions to obtain

$$\frac{da_{xz}}{dt} = -2\omega_1\left(\frac{\sqrt{3}}{2}a_{z^2} - \frac{1}{2}a_{x^2-y^2}\right) \text{ and } \frac{d}{dt}\left(\frac{\sqrt{3}}{2}a_{z^2} - \frac{1}{2}a_{x^2-y^2}\right) = 2\omega_1 a_{xz}. \tag{2.26}$$

This describes precession, with angular frequency $2\omega_1$, in a plane spanned by $\boldsymbol{T}_{xz}$ and $\left(\frac{\sqrt{3}}{2}\boldsymbol{T}_{z^2} - \frac{1}{2}\boldsymbol{T}_{x^2-y^2}\right)$. So, if the density matrix immediately before the second pulse is

$$\rho = a_z(0)\big[(\cos \omega_Q\tau)\boldsymbol{T}_y - (\sin \omega_Q\tau)\boldsymbol{T}_{xz}\big] \tag{2.27}$$

then immediately after the pulse for which $\omega_1 t = \pi/2$, the density matrix will be

$$\rho = a_z(0)\left[(\cos \omega_Q \tau)\boldsymbol{T}_y + (\sin \omega_Q \tau)\boldsymbol{T}_{xz}\right]. \tag{2.28}$$

The effect of a $\pi/2$ pulse along the y-axis is thus to invert the coefficient of the $\boldsymbol{T}_{xz}$ basis matrix in the density matrix.

Evolution of the density matrix under the influence of $\mathcal{H}_Q = \sqrt{\frac{2}{3}}\hbar\omega_Q \boldsymbol{T}_{z^2}$ for a further interval $\tau$ will finally leave it in the state $\rho = a_z(0)$ independent of $\omega_Q$. The effect of the quadrupole echo sequence is thus an echo at time $2\tau$ following the initial $\pi/2$ pulse. If $\tau$ is chosen to be longer than the recovery time of the preamplifier, then the free induction decay, undistorted by preamplifier recovery effects, can be acquired starting from the top of the echo. As noted by Davis [1983], failure to ensure that the signal to be Fourier transformed starts from the top of the echo can result in spectral distortions. Approaches to prepare the echo in order to mitigate such problems have been described (Prosser *et al* 1991).

## 2.5 Characterizing phospholipid bilayer phases using wideline $^2$H NMR

The usefulness of $^2$H NMR spectroscopy as a tool to characterize lipid bilayer phase behavior reflects the sensitivity of spectra to characteristic changes in chain-deuterated lipid chain motions at bilayer phase transitions. Figure 2.5 shows the results of $^2$H NMR observations, from Singh *et al* (2008), for chain-perdeuterated dipalmitoyl phosphatidylcholine (DPPC-$d_{62}$) bilayers over a series of temperatures. For 37 °C and higher, the spectra are characteristic of the liquid crystalline or $L_\alpha$ phase. The spectra are superpositions of Pake doublets with quadrupole splittings corresponding to the orientational order parameters for deuterons on each segment of the deuterated acyl chains. Chain reorientation in this phase is axially symmetric on the timescale of the $^2$H NMR experiment. The prominent edges at about ±12.5 kHz arise from the overlap of deuterons with splittings near the plateau of the orientational order parameter. Between 37 °C and 36 °C, the bilayer undergoes a transition from the liquid crystal to the rippled gel or $P_{\beta'}$ phase. As noted above, this so-called main transition temperature is lower than the corresponding transition temperature for bilayers of non-deuterated DPPC by a few degrees because of the effects of chain-perdeuteration. In the gel phases, acyl chains are more ordered and reorientation is not fast enough to be axially symmetric on the $^2$H NMR timescale. Spectral components on the methylene chain segments are not distinguishable. The prominent feature between approximately ±8 kHz arises from the methyl deuterons and reflects additional fast reorientation about the methyl axis on the ordered chains. Below about 29 °C, the bilayers undergo a transition into a less-rippled gel phase, the $L_{\beta'}$ phase. Distinctions between spectra in the $P_{\beta'}$ and $L_{\beta'}$ phases are subtle but the $P_{\beta'}$ phase spectra display a flatter methyl feature and shoulders at about ±19 kHz.

As temperature is reduced through the $L_{\beta'}$ phase, the width of the methyl group feature increases slightly and there is a slight shift of intensity to larger splittings. Below 0 °C, spectral intensity begins to increase around ±63 kHz as the highly

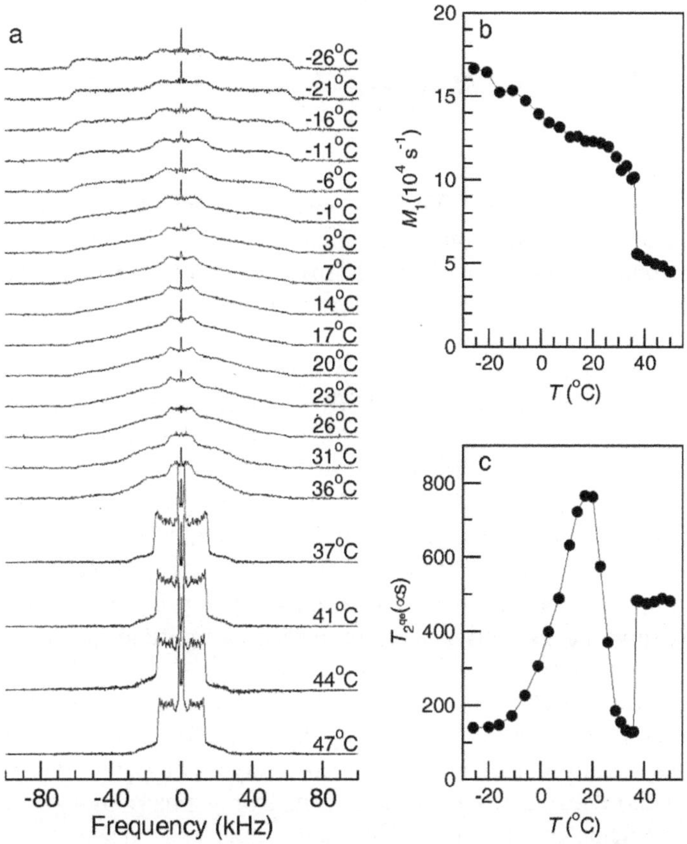


**Figure 2.5.** (a) $^2$H NMR spectra of DPPC-$d_{62}$ bilayers for a series of temperatures. (b) Temperature dependence of first spectral moments for DPPC-$d_{62}$ bilayers. (c) Temperature dependence of quadrupole echo decay times for DPPC-$d_{62}$ bilayers. Adapted by permission from Singh *et al* (2008), copyright (2008) Springer Nature.

ordered lamellar crystalline ($L_C$) or subgel phase begins to form. At equilibrium, acyl chains in this phase are effectively immobilized on the $^2$H NMR timescale. Reorganization of the acyl chains necessary to form such a fully immobilized lipid phase requires extended incubation just below the phase transition and the low temperature spectra in figure 2.5, which were acquired with equilibration times of about an hour, are not quite characteristic of completely static CD bonds. The splitting of the methyl feature in the lowest temperature spectra is about a third of the splitting for other deuterons. This reorientation about the methyl axis remains unquenched in the $L_C$ phase.

$^2$H NMR spectral moments are sensitive to changes in the acyl chain orientational order parameter profile and thus provide useful tools for characterizing the states of bilayers and the effects of various perturbing factors (Davis 1979, 1983, Morrow *et al* 1992). For a $^2$H NMR spectrum, $S(\omega)$, the first moment is

$$M_1 = \frac{\int_0^\infty \omega S(\omega) d\omega}{\int_0^\infty S(\omega) d\omega}, \tag{2.29}$$

where $\omega$ is angular frequency. The first spectral moment is proportional to the average quadrupole splitting and sharp changes in $M_1$ with temperature typically reflect bilayer phase transitions. Figure 2.5(b) shows the temperature dependence of $M_1$ for DPPC-$d_{62}$ (Singh *et al* 2008). The liquid crystal to gel transition is apparent as a discontinuity in $M_1$. The pretransition seems to correspond to a change in the slope of $M_1$ versus $T$ around 26 °C. Because of slow equilibration, the onset of the $L_C$ phase is not very apparent in figure 2.5(b). In high pressure experiments, where reorganization into the $L_C$ phase at 196 MPa occurred around 10 °C, a step in $M_1$ versus $T$ was more apparent (Singh *et al* 2008).

Refocusing of the quadrupole echo reflects the extent to which the quadrupole interaction experienced by individual deuterons remains unaffected by slow reorientation over the timescale of the experiment. The quadrupole echo decay behavior for a given sample can be obtained by systematically varying the pulse separation, $\tau$, and thus the echo time, $2\tau$, and measuring the echo amplitude $A(2\tau)$. The rate of the initial decay, in the limit of small $2\tau$, is an average, $\langle R \rangle$, over all of the deuterons contributing to the signal. The quadrupole echo decay time measured in the limit of small $2\tau$ is often reported as the echo decay time, $T_2^{qe}$, for that sample. Figure 2.5(c) shows the temperature dependence of $T_2^{qe}$ for bilayers of DPPC-$d_{62}$. It is apparent that average echo decay rate can be quite sensitive to phase changes.

For a given deuteron, labeled $i$, the quadrupole echo decay rate is a sum of contributions $R_j$ from motions, labeled $j$, that are characterized by a correlation time $\tau_{cj}$ and by the second moment, $\Delta M_{2j}$, of the part of the quadrupole interaction modulated by that motion. If fast and slow motions are distinguished by comparison of correlation times with $(\Delta M_{2j})^{-1/2}$ (Bloom and Sternin 1987), then $R_j \propto \Delta M_{2j}\tau_{cj}$ for faster motions and $R_j \propto (\tau_{cj})^{-1}$ for slower motions (Abragam 1961, Bloom and Sternin 1987).

In general, correlation times will increase with decreasing temperature. In the liquid crystal phase, quadrupole echo decay reflects contributions from fast motions like reorientation about the bilayer normal, and slower motions like bilayer undulations. In figure 2.5(c), the echo decay rate is not very sensitive to temperature suggesting that as the correlation times for motions in the liquid crystalline phase change with temperature, contributions to echo decay rate from fast and slow motion change in opposite directions. There is an abrupt decrease in quadrupole echo decay time at the liquid crystal to gel transition. Motions which are slow in the liquid crystal phase become slower and contribute less to echo decay. On the other hand, some motions that are fast in the liquid crystal phase also become slower and contribute more to echo decay. Between 36 °C and about 20 °C, the correlation times for these motions increase with decreasing temperature, the contributions to echo decay rate decrease, and echo decay time rises. The observation of a maximum in $T_2^{qe}$, and thus a minimum in average echo decay rate, near 20 °C implies that there must be a residual fast motion, possibly localized chain conformational fluctuations, that slow and begin to dominate echo decay near that temperature. These motions continue to freeze out as the sample cools into the $L_C$ phase.

## 2.6 Using spectral subtraction to identify boundaries of two-phase coexistence regions in bilayer temperature-composition phase diagrams

The use of $^2$H NMR spectral subtraction to obtain the binary temperature-composition phase diagram of a lipid mixture was introduced by Vist and Davis (1990) in their pioneering study of the DPPC/cholesterol phase diagram. A binary phase diagram can be plotted with temperature along the vertical axis and the mole fraction of one mixture component along the horizontal axis. In general, for bilayer mixtures comprising two components, there will be single phase regions in temperature/composition space, corresponding to the liquid crystal phase and one or more ordered phases, separated by regions in which two phases coexist. Precise determination of the boundaries of such coexistence regions can provide useful information about the interactions of the bilayer components and insights into the role of bilayer heterogeneity in more complex systems.

The application of $^2$H NMR difference spectroscopy to the determination of a temperature-composition phase diagram can be illustrated by the example from Morrow *et al* (1991), of DMPC-DSPC-$d_{70}$ shown in figure 2.6. For this phase diagram, composition is specified by $x_{\mathrm{DSPC}-d_{70}}$, the mole fraction of DSPC-$d_{70}$. For a temperature and overall sample composition $x$ falling within the liquid crystal/gel coexistence region of the phase diagram, the sample will separate into liquid crystal and gel domains corresponding to the endpoint compositions, $x_f$ and $x_g$, at the liquid crystal and gel boundaries, respectively, of the two-phase region for that temperature. The isotherm connecting those two points is the tie line. The fraction of the sample in a given phase is given by the lever rule which, for the more fluid phase, is

$$f = \frac{x_g - x}{x_g - x_f}. \tag{2.30}$$

If exchange of deuterated lipids between coexisting liquid crystal and gel domains is slow on the $^2$H NMR timescale, the spectrum from a sample falling within the two-phase region will be a superposition of spectra corresponding to the endpoint compositions for that temperature. The relative areas of the two spectral components will be determined by the lever rule.

For the DMPC/DSPC-$d_{70}$ example, $x_f < x_g$ for tie lines in the two-phase region. Consider two compositions, $x_A$ and $x_B$, within the two-phase region, ordered such that $x_A > x_B$. If spectra $S_A$ and $S_B$ are obtained for these compositions at a specific temperature within the two-phase region, then it should be possible to find fractions $K$ and $K'$ such that

$$S_g = S_A - KS_B \text{ and } S_f = S_B - K'S_A \tag{2.31}$$

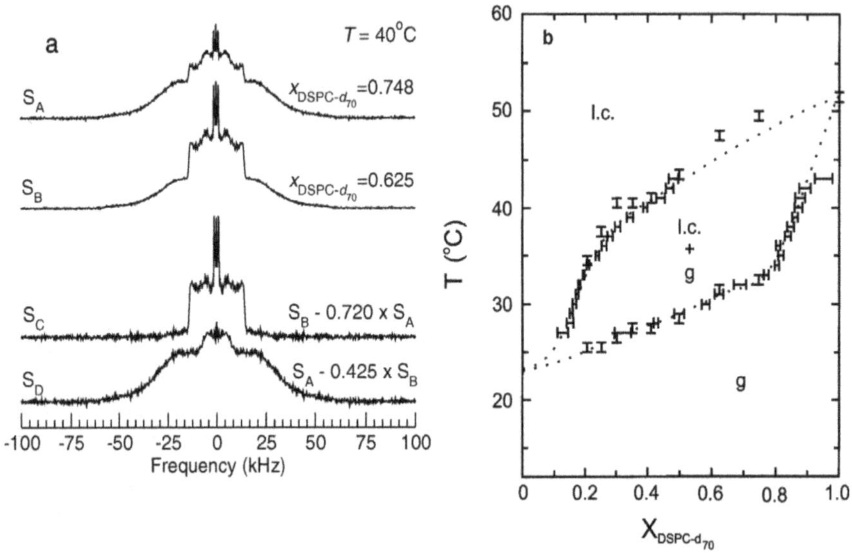


**Figure 2.6.** (a) $^2$H NMR spectra of DMPC/DPPC-$d_{70}$ bilayers at T = 40 °C. (b) Phase diagram for DMPC/DPPC-$d_{70}$ bilayer mixtures. Regions corresponding to liquid crystal and gel are indicated by l.c. and g, respectively. Vertical error bars show phase boundaries inferred from examination of spectra at fixed sample composition for a series of temperatures. Horizontal error bars show phase boundaries inferred by spectral subtraction. Reprinted from Morrow *et al* (1991). Copyright (1991) with permission from Elsevier.

show no features characteristic of the liquid crystalline or gel phases, respectively, and are, in fact, the spectra corresponding to the two endpoint compositions. The endpoint compositions are then given by

$$x_g = \frac{(1 - KC)x_A x_B}{x_B - KCx_A} \quad \text{and} \quad x_f = \frac{(1 - (K'/C))x_A x_B}{x_A - (K'/C)x_B} \tag{2.32}$$

where $C$ is an 'optional' correction factor, of order 1, that corrects for differential effects of quadrupole echo decay on the amplitudes of the liquid crystal and gel spectral components. More detail regarding the correction factor can be found in Morrow *et al* (1991).

In figure 2.6, the spectra labeled A and B are the $T = 40$ °C spectra for the compositions indicated. The spectra labeled $S_C$ and $S_D$ are the result of carrying out the indicated subtractions to obtain the liquid crystal and gel endpoint spectra, respectively. For this pair of spectra, $K' = 0.720$ and $K = 0.625$. Taking account of the correction factor $C$, this pair of spectra yield endpoint compositions $x_f = 0.420$ and $x_g = 0.866$. The endpoint compositions shown with horizontal error bars on the DMPC/DSPC-$d_{70}$ phase diagram shown in figure 2.6 were obtained from spectral subtractions, for various pairs of overall compositions, between $T = 27$ °C and $T = 42$ °C. Direct observation of transitions at specific overall compositions, shown by points with vertical error bars, was used to further define the boundary of the two-phase coexistence region. The resulting phase diagram provides information about the extent to which the DMPC/DSPC system departs from ideal mixing.

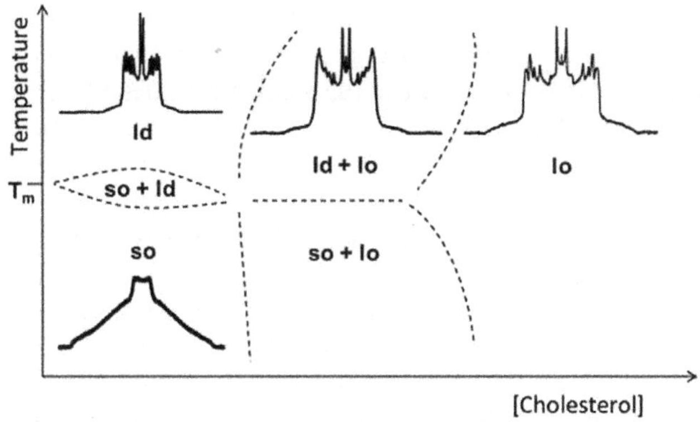


**Figure 2.7.** A schematic example of a phosphatidylcholine/cholesterol temperature/composition phase diagram. $T_m$ refers to the solid ordered (**so**) to liquid disordered (**ld**) phase transition temperature of the pure phospholipid. Sphingomyelin/cholesterol has similar phase behaviour. The spectra are characteristic for a given phase region: **so** is pure palmitoylsphingomyelin (PSM) at 30 °C; **ld** is pure PSM at 48 °C; **ld + lo** is PSM/cholesterol 80:20 at 48 °C; lo is PSM/cholesterol 60:40 at 48 °C.

Contributions involving phase diagram determination by $^2$H NMR difference spectroscopy have had the most impact in studies of phospholipid-sterol phase diagrams. As Vist and Davis (1990) discovered, the addition of 30 mol % cholesterol to DPPC-$d_{62}$ yields a new liquid crystalline phase whose properties are relatively insensitive to temperature. This phase is known as the 'liquid ordered' or **lo** phase. It is 'liquid' since its lipid molecules retain rotational and translational freedom, and 'ordered' since the conformational freedom of the phospholipid acyl chains is very much curtailed—becoming nearly kink-free. Consequently, the DPPC-$d_{62}$ $^2$H NMR $S_{CD}$ values measured in the presence of 30 mol % cholesterol are nearly double those measured in the absence of cholesterol. Below the gel to liquid crystalline phase transition temperature of DPPC-$d_{62}$, the presence of between ~10 and 25 mol % cholesterol yields $^2$H NMR spectra with gel and **lo** components that are amenable to difference spectroscopy. Thus, at a given temperature the calculation of gel and **lo** end-point cholesterol concentrations is straightforward. Phase boundaries separating single-phase gel or **lo** regions from the 'gel + **lo**' coexistence region of the phase diagram result from the calculation of these endpoints as a function of temperature. Note that in phospholipid-sterol phase diagrams, the gel phase is usually renamed the 'solid ordered' or **so** phase (Ipsen *et al* 1987), where 'solid' indicates that rotational and translational diffusion is very slow. The **so** phase encompasses both the gel and ripple ($P_{\beta'}$) phases of DPPC (figure 2.7).

The definition of 'phase' here is contextual. Ideally, gel and **lo** phase coexistence should result in phase separation of a system into two regions: one gel, one **lo**. This does not happen in phospholipid-cholesterol membranes, and the lack of observable 'macroscopic' (i.e., micron-scale) phase separation has fuelled a great deal of discussion among experimentalists that study membrane phase behaviour using a variety of imaging techniques, such as fluorescence microscopy. A $^2$H NMR gel phase spectrum will be generated by a nanoscale patch of gel phase lipids, since slow diffusion of these lipids precludes their sampling other membrane environments on the NMR timescale. There is no requirement for such gel phase nanodomains to be contiguous. Thus, $^2$H NMR defines the physical properties of the sum total of gel phase and **lo** phase lipids but does not yield direct information as to the shape or size of the phase domains. In this respect, $^2$H NMR is similar to differential scanning calorimetry.

The phosphatidylcholine-cholesterol phase diagram has been shown to be quite general. Unsaturated lipids such as 1-palmitoyl-2-oleoyl-*sn*-glycero-3-phosphocholine (POPC) and similar PCs having a double bond at different positions on the *sn*−2 chain all yield gel + **lo** phase coexistence below their main transition temperatures in the presence of ~10 to 25 mol % cholesterol (Thewalt and Bloom 1992, Thewalt *et al* 1992). Relatively few $^2$H NMR investigations into the phase behaviour of membranes containing sterols other than cholesterol have been published, but there is some consistency in DPPC-sterol phase diagrams (see, e.g., Hsueh *et al* 2005).

Recently, the generality of $^2$H NMR-derived PC-cholesterol phase diagrams was extended to sphingomyelin-cholesterol (Keyvanloo *et al* 2018). Sphingomyelin and cholesterol form the bulk of the outer leaflet of mammalian red blood cells (Murate *et al* 2015) and are major constituents of the plasma membranes of many mammalian cell types.

Difference spectroscopy is also crucial to unravelling the **so** regions of the more complicated phase diagrams associated with three-component lipid membranes (Veatch *et al* 2007, Davis *et al* 2009). In these cases, the **so** region is a triangle and the ‘lever rule’ must be generalized to utilize the triangle corners as endpoints. Note, however, that apart from some unusual biomembranes such as the layered, tightly packed lipids that are responsible for leak-proofing mammalian skin, known as the *stratum corneum* (Bouwstra and Ponec 2006), non-liquid crystalline phases have limited biological function. Another unusual natural lipid assembly is the surface-active layer at the air-water interface in lungs, which is close to a transition temperature and cycles between ordered and disordered states during breathing (Lopez-Rodriguez and Pérez-Gil 2014).

## 2.7 Beyond difference spectroscopy—defining liquid crystalline phase coexistence with $^2$H NMR

The observation of coexisting ordered and ‘disordered’ (**lo** + **ld**) liquid crystalline phases in membranes of defined lipid and sterol composition continues to inspire the cell biology community: to what extent does life capitalize on the ability of lipids to spontaneously organize? Since membrane proteins, whether lipidated or transmembrane, often have ‘preferences’ regarding their lipid environment, it is possible that non-uniform lipid distributions could lead to concentrated regions of membrane proteins that have functional consequences. Such ideas are commonly found under the rubric of the ‘raft hypothesis’ originally formulated by Simons and Ikonen (1997), who built on early PC-cholesterol phase diagram determinations.

The phenomenon of **lo** + **ld** phase coexistence has been studied most thoroughly in model membranes composed of DPPC, dioleoylPC (DOPC) and cholesterol, and the literature is replete with beautiful images of phase-separated vesicles. The phase diagram of binary mixtures of DPPC-$d_{62}$ and ~10 to 30 mol % cholesterol also features a region of **lo** + **ld** phase coexistence at temperatures above the **so** to **ld** transition. Like the domains in the **so** + **lo** phase coexistence region, the **lo** and **ld** domains cannot be easily visualized.

In figure 2.7, $^2$H NMR spectra characteristic of the **ld** and **lo** phases are shown. As already discussed, these spectra are qualitatively similar; each spectrum results from a homogeneous membrane environment and are superpositions of sharp, well defined Pake doublets. In the **ld** + **lo** coexistence region, the spectrum's superposed Pake doublets are less well resolved. This 'blurry' spectrum is a result of individual deuterated phospholipids sampling different regions on the NMR timescale. By analogy with the well understood phenomenon of chemical exchange, such spectra can be analyzed by calculating an exchange rate from the line widths of dePaked spectra. The initial applications of this technique assumed 'fast exchange' (Hsueh *et al* 2005), but this approach may be an oversimplification. Analysis of blurry spectra in terms of 'intermediate exchange', combined with known diffusion coefficients, resulted in realistic estimates of minimum domain size of ~80 nm (Keyvanloo *et al* 2018). The use of selectively deuterated phospholipids is expected to allow coexisting **ld** and **lo** domains to be characterized with even more confidence.

## 2.8 Summary

The $^2$H solid-state NMR spectrum of a deuterated lipid acyl chain within a model membrane provides information about whether or not chain reorientation is axially symmetric on characteristic timescale of the NMR experiment. For reorientation that is fast and axially symmetric, orientational order parameters of deuterated chain segments can be extracted from the resulting spectra. This sensitivity of the spectrum to acyl chain motion has made $^2$H solid-state NMR spectroscopy a valuable tool for the characterization of lipid bilayer and model membrane phase behaviour and has been particularly useful for locating two-phase coexistence boundaries in the phase diagrams of binary lipid and lipid-sterol mixed bilayers. $^2$H NMR has also been used to examine the perturbation of model and intact membranes by other amphiphilic or amphipathic molecules including antimicrobial peptides and lung surfactant proteins (Roldan *et al* 2017; Miranda *et al* 2018). In future, the spectral difference methods described here might also provide a way to obtain binary phase diagrams of such systems.

## References

Abragam A 1961 *The Principles of Nuclear Magnetism* (London: Oxford University Press)

Bloom M 1988 NMR studies of membranes and whole cells *Physics of NMR Spectroscopy in Biology and Medicine* ed B Maraviglia (Amsterdam: Elsevier)

Bloom M, Burnell E E and de Lange C A 2003 Density matrix methods in NMR *NMR of Ordered Liquids* ed E E Burnell and C de Lange (Dordrecht: Kluwer)

Bloom M, Davis J H and MacKay A L 1981 Direct determination of the oriented sample NMR spectrum from the powder spectrum for systems with local axial symmetry *Chem. Phys. Lett.* **89** 198–202

Bloom M and Sternin E 1987 Transverse nuclear spin relaxation in phospholipid bilayer membranes *Biochemistry* **26** 2101–5

Bouwstra J and Ponec M 2006 The skin barrier in healthy and diseased state *BBA-Biomembr.* **1758** 2080–95

Davis J H 1979 Deuterium magnetic resonance study of the gel and liquid crystalline phases of dipalmitoyl phosphatidylcholine *Biophys. J.* **27** 339–58

Davis J H 1983 The description of membrane lipid conformation, order and dynamics by $^2$H-NMR *Biochim. Biophys. Acta* **737** 117–71

Davis J H and Jeffrey K R 1977 The temperature dependence of chain disorder in potassium palmitate-water. A deuterium NMR study *Chem. Phys. Lipids* **20** 87–104

Davis J H, Clair J J and Juhasz J 2009 Phase equilibria in DOPC/DPPC-d(62)/cholesterol mixtures *Biophys. J.* **96** 521–39

Davis J H, Jeffrey K R, Bloom M, Valic M I and Higgs T P 1976 Quadrupolar echo deuteron magnetic resonance spectroscopy in ordered hydrocarbon chains *Chem. Phys. Letters* **42** 390–4

De Lange C A and Burnell E E 2003 Basics of NMR of molecules in uniaxial anisotropic environments *NMR of Ordered Liquids* ed E E Burnell and C de Lange (Dordrecht: Kluwer)

Engel A K and Cowburn D 1981 The origin of multiple quadrupole couplings in the deuterium NMR spectra of the 2 chain of 1,2 dipalmitoyl-*sn*-glycero-3-phosphorylcholine *FEBS Lett.* **126** 169–71

Hsueh Y-W, Gilbert K, Trandum C, Zuckermann M and Thewalt J 2005 The effect of ergosterol on dipalmitoylphosphatidylcholine bilayers: a deuterium NMR and calorimetric study *Biophys. J.* **88** 1799–808

Ipsen J H, Karlström G, Mouritsen O G, Wennerström H and Zuckermann M J 1987 Phase equilibria in the phosphatidylcholine-cholesterol system *Biochim. Biophys. Acta* **905** 162–72

Keyvanloo A, Shaghaghi M, Zuckermann M J and Thewalt J L 2018 The phase behavior and organization of sphingomyelin/cholesterol membranes: a deuterium NMR study *Biophys. J.* **114** 1344–56

Lopez-Rodriguez E and Pérez-Gil J 2014 Structure-function relationships in pulmonary surfactant membranes: from biophysics to therapy *Biochim. Biophys. Acta* **1838** 1568–85

McCabe M A and Wassall S R 1995 Fast-Fourier transform dePaking *J. Magn. Reson. Ser.* B **106** 80–2

Miranda C, Booth V K and Morrow M R 2018 Effects of amphipathic polypeptides on membrane organization inferred from studies using bicellar lipid mixtures *Langmuir* **34** 11759–71

Molugu T R, Lee S and Brown M F 2017 Concepts and methods of solid-state NMR spectroscopy applied to biomembranes *Chem. Rev.* **117** 12087–132

Morrow M R, Srinivasan R and Grandal N 1991 The phase diagram of dimyristoyl phosphatidylcholine and chain-perdeuterated distearoyl phosphatidylcholine: a deuterium NMR spectral difference study *Chem. Phys. Lipids* **58** 63–72

Morrow M R, Whitehead J P and Lu D 1992 Chain-length dependence of lipid bilayer properties near the liquid crystal to gel phase transition *Biophys. J.* **63** 18–27

Murate M, Abe M, Kasahara K, Iwabuchi K, Umeda M and Kobayashi T 2015 Transbilayer distribution of lipids at nano scale *J. Cell Sci.* **128** 1627–38

Petersen N O, Kroon P A, Kainosho M and Chan S I 1975 Thermal phase transitions in deuterated lecithin bilayers *Chem. Phys. Lipids* **14** 343–9

Prosser R S, Davis J H, Dahlquist F W and Lindorfer M A 1991 $^2$H nuclear magnetic resonance of the gramicidin A backbone in a phospholipid bilayer *Biochemistry* **30** 4687–96

Roldan N, Pérez-Gil J, Morrow M R and Garcia-Álvarez B 2017 Divide & conquer: surfactant protein SP-C and cholesterol modulate phase segregation in lung surfactant *Biophys. J.* **113** 847–59

Seelig J 1977 Deuterium magnetic resonance: theory and application to lipid membranes *Q. Rev. Biophys.* **10** 353–418

Simons K and Ikonen E 1997 Functional rafts in cell membranes *Nature* **387** 569–72

Singh H, Emberley J and Morrow M R 2008 Pressure induces interdigitation differently in DPPC and DPPG *Eur. Biophys. J.* **37** 783–92

Sternin E, Bloom M and MacKay A L 1983 De-Pake-ing of NMR spectra *J. Magn. Reson.* **55** 274–82

Thewalt J L and Bloom M 1992 Phosphatidylcholine-cholesterol phase diagrams *Biophys. J.* **63** 1176–81

Thewalt J L, Hanert C E, Linseisen F M, Farrall A J and Bloom M 1992 Lipid-sterol interactions and the physical properties of membranes *Acta Pharm.* **42** 9–23

Veatch S L, Soubias O, Keller S L and Gawrisch K 2007 Critical fluctuations in domain-forming lipid mixtures *Proc. Natl Acad. Sci. USA* **104** 17650–5

Vist M R and Davis J H 1990 Phase equilibria of cholesterol/dipalmitoyl-phosphatidylcholine mixtures: $^2$H nuclear magnetic resonance and differential scanning calorimetry *Biochemistry* **29** 451–64

Whittall K P, Sternin E, Bloom M and MacKay A L 1989 Time- and frequency-domain 'dePakeing' using inverse theory *J. Magn. Reson.* **84** 64–71

# Chapter 3

# Deuterium solid-state NMR of whole bacteria: sample preparation and effects of cell envelope manipulation

**Valerie Booth**

## Overview

We live in a time when there is an astounding array of techniques to unveil the inner workings of the molecules found in and around our cells. Many of these techniques were originally developed for probing systems with simple compositions of one or two types of biomolecule. However, there is a growing effort to extend such techniques to the much more complex milieu in which the molecules of interest normally function in living organisms. Nuclear magnetic resonance (NMR) is one powerful technique for probing the detailed structure and dynamics of biomolecules including proteins and lipids. While traditionally applied to samples of simple composition, NMR is increasingly being applied in more complex contexts including to intact, whole cells. Deuterium ($^2$H)-NMR has been of great utility in probing the structure and dynamics of $^2$H-labelled lipids in bilayers; and it has also been successfully applied to whole cells. This chapter will introduce readers to this approach as applied to bacteria, as well as highlight some of the key findings from recent years, in particular with respect to how antimicrobial peptides (AMPs) interact with membranes in bacteria.

## 3.1 Structure and composition of bacterial cell envelopes

Bacteria have traditionally been divided into two groups, Gram(–) and Gram(+). Gram(–) bacteria, like *Escherichia coli* (*E. coli*), have a cell envelope with two lipid bilayers and a thin layer of peptidoglycan (PGN) between them (figure 3.1(a)). The outer leaflet of the outer bilayer of Gram(–) cells has a large component of lipopolysaccharide (LPS). LPS molecules are made up of a Lipid A group that serves as a membrane anchor, linked to an anionic phosphorylate oligosaccharide

doi:10.1088/978-0-7503-2532-5ch3

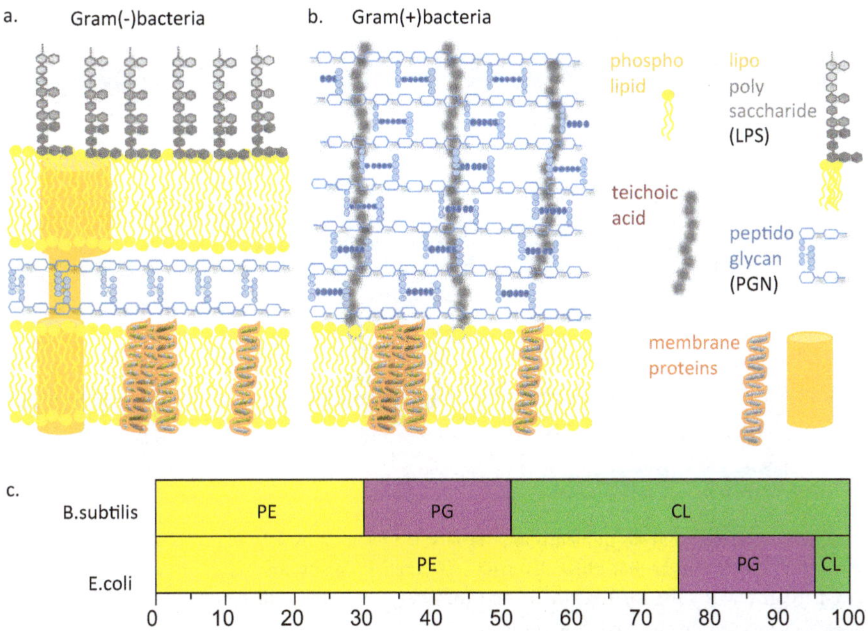


**Figure 3.1.** Cell envelope structure and composition. (a) Gram(–) bacterial cell envelope; (b) Gram(+) bacterial cell envelope; and (c) typical lipid compositions for *E. coli* and *B. subtilis* by weight (Sohlenkamp and Geiger 2016, Rowlett *et al* 2017, Lohner and Prenner 1999, Dennison *et al* 2008).

that forms the core layer, with a highly variable polysaccharide region, termed the O antigen, on the outside (Ebbensgaard *et al* 2018). PGN is composed of pairs of carbohydrate units cross-linked by a short peptide and forms a mesh-like sac around the cell (Malanovic and Lohner 2016a). Unlike with Gram(–) bacteria, which possess only about 10% by dry weight of PGN, Gram(+) bacteria are typically made up of 90% PGN by dry weight (Malanovic and Lohner 2016a) (figure 3.1(b)).

Gram(+) bacteria, such as *Bacillus subtilis* (*B. subtilis)*, have only one lipid bilayer, but have a thick PGN layer outside the single bilayer. Gram(+) cell envelopes also contain an anionic molecule called teichoic acid that can be anchored to the membrane via an acyl chain or linked covalently to the PGN itself (Malanovic and Lohner 2016b). Also present as well in the cell envelopes of both types of bacteria are integral and peripheral membrane proteins. The lipid composition of bacteria varies according to the specific species in question and can also vary with environmental conditions (Sohlenkamp and Geiger 2016). Typical values for *E. coli* and *B. subtilis* (Sohlenkamp and Geiger 2016, Rowlett *et al* 2017, Lohner and Prenner 1999, Dennison *et al* 2008) are shown in figure 3.1(c). The main component of *E. coli* cytoplasmic membranes is the zwitterionic phospholipid phosphatidylethanolamine (PE), followed by the anionic phospholipid phosphatidylglycerol (PG), as well as small amounts of the 4-acyl chain anionic lipid cardiolipin (CL). *B. subtilis,* on the other hand, has much greater amounts of CL, along with PG and PE.

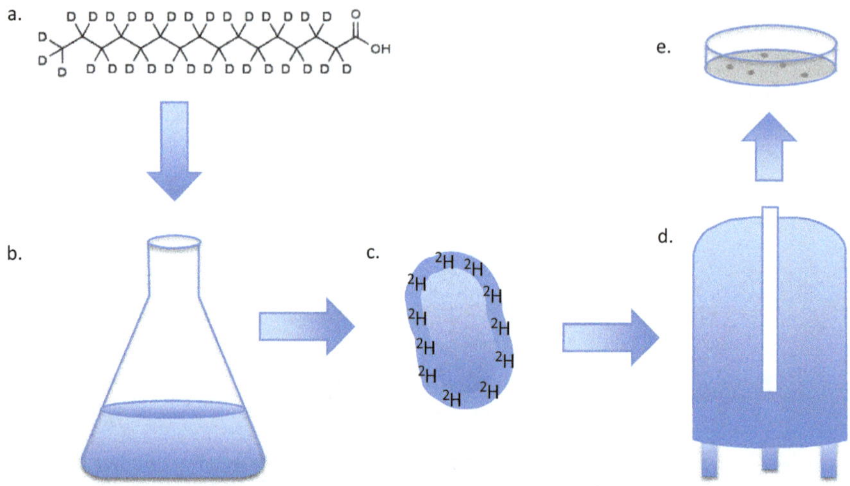


**Figure 3.2.** Workflow for analysis of membrane $^2$H-labelled bacteria by NMR: (a) $^2$H-labelled palmitic acid is incorporated into growth media; (b) cells with $^2$H-labelled membranes are grown; (c) harvested, and transferred to an NMR tube for (d) NMR; and (e) after NMR the cells can be characterized by plating, MTT assay, mass spectrometry and flow cytometry.

## 3.2 Sample preparation

To use NMR to look at $^2$H-labelled lipids in intact cells, it is of course necessary to introduce $^2$H-labels into the bacterial lipids, ideally whilst minimizing $^2$H nuclei in non-lipid components of the cells (figure 3.2). This has been done in two main ways, as detailed elsewhere (Santisteban *et al* 2017, Booth *et al* 2017). One method employs mutated bacteria with impairments in fatty acid synthesis and/or metabolism (Pius *et al* 2012, Davis 1983), while the other method employs wild type bacteria (Tardy-Laporte *et al* 2013, Booth *et al* 2017). In both methods, a deuterated saturated fatty acid, palmitic acid, is added to the bacteria growth media. In work with wild type bacteria, the deuterated fatty acids have been first complexed with the detergent dodecylphosphocholine (DPC) to maximize uptake by the bacteria. In order for the bacteria to be healthy and to maintain a normal acyl chain composition, it can also be necessary to provide non-deuterated fatty acids at the appropriate proportion for the bacterial type (Santisteban *et al* 2020, Santisteban *et al* 2017). Both mutated and non-mutated bacteria have been seen to selectively incorporate high levels of deuterons into the acyl chains of their lipids.

Once the $^2$H-enriched bacteria have been grown (using biosafety protocols appropriate for the strain) to an optical density indicating mid-log phase, they are harvested by centrifugation. Next the cells may be re-suspended in a treatment solution before NMR is performed. For example, cells can be incubated with AMPs, which can be quantified as % AMP weight to dry weight of cells (% dry weight of cells can be determined as in Santisteban *et al* (2017)). With or without treatment, the cells should be suspended in an isotonic solution, such as growth media, to prevent lysis. For NMR, the centrifuge-harvested cells, which have the consistency of a paste, can be transferred into an NMR tube (or rotor) using a spatula, or with help from a large pipette tip and a benchtop centrifuge.

$^{2}$H-NMR spectra are acquired at the desired temperature, often 37 °C, using a standard quadrupole echo pulse sequence (Santisteban *et al* 2017). The viability of the cells after the NMR experiment should be characterized, which can be done with a colony forming assay, which tests the cells' ability to divide, or and/or a tetrazolium dye, 3-(4,5-dimethylthiazol-2-yl)–2,5-diphenyltetrazolium bromide, MTT assay which indicates if the cells are metabolically active (Pius *et al* 2012, Tardy-Laporte *et al* 2013). Cell viability depends on the conditions used to grow the bacteria as well as on the temperature and length of time the samples spends in the NMR spectrometer. As examples, Pius *et al* (2012) found that ~15% of the bacteria died after 6 h in the NMR spectrometer at 37 °C and Overall *et al* (Overall *et al* 2019) measured a colony forming unit decrease of 70% after cells were in the spectrometer at 30 °C for 12 h. Additionally, flow cytometry is useful in characterizing the cells in the NMR spectrometer because it provides information about the permeabilization of the bacterial membranes (Akhoundsadegh *et al* 2019). It is also of interest to check how the lipid profile of the bacteria has been affected by the labelling strategy, which has been done by mass spectrometry (Laadhari *et al* 2016).

## 3.3 Static $^{2}$H NMR of Gram(−) and Gram(+) bacteria

Before examining the $^{2}$H-NMR spectra of bacteria, we'll first look at an example of a $^{2}$H-NMR spectrum for a model membrane comprising a single lipid species in the liquid-crystalline phase with its characteristic, symmetrical 'bat-wing' shape (figure 3.3(a)). Each deuterated methylene and methyl group in the lipid acyl chain contributes a pair of lines at a characteristic splitting between them. The detailed theory behind the spectrum can be found elsewhere in this book (in chapter 2 by Morrow and Thewalt). For our purposes the main note is that the splitting between each pair of lines is sensitive to the amount of motion at its part of the acyl chain. Thus, the methyl groups found at the end of the chain, which experience the most motion, give rise to a prominent pair of peaks with a small splitting of ~ ±2.5 kHz (labelled in figure 3.3(a)). By contrast, the prominent edges at ~ ±12.5 kHz derive from deuterons in the part of the chain close to the headgroup, which are more constrained. Decreases in the orientational ordering of the lipids with respect to the bilayer normal direction can for example, be induced by adding a membrane-perturbing AMP. Such decreases in lipid order can be expected to decrease the splittings seen in the spectra. Furthermore, the regions of the acyl chain most affected by the treatment can be deduced, e.g. changes deep in the bilayer are reflected nearer to the centre of the spectrum and changes nearer the head groups are reflected by changes in the prominent edge.

As can be seen in figures 3.3(b) and (c), $^{2}$H-NMR spectra of $^{2}$H-labelled bacteria are not nearly as esthetically pleasing as those from single lipid samples. However, they still contain much of the important information. In particular, the methyl peaks are still visible, either as resolved peaks (figure 3.3(b)), or as a shoulder (figure 3.3(c)). And, the prominent edge that derives from the acyl chain near the

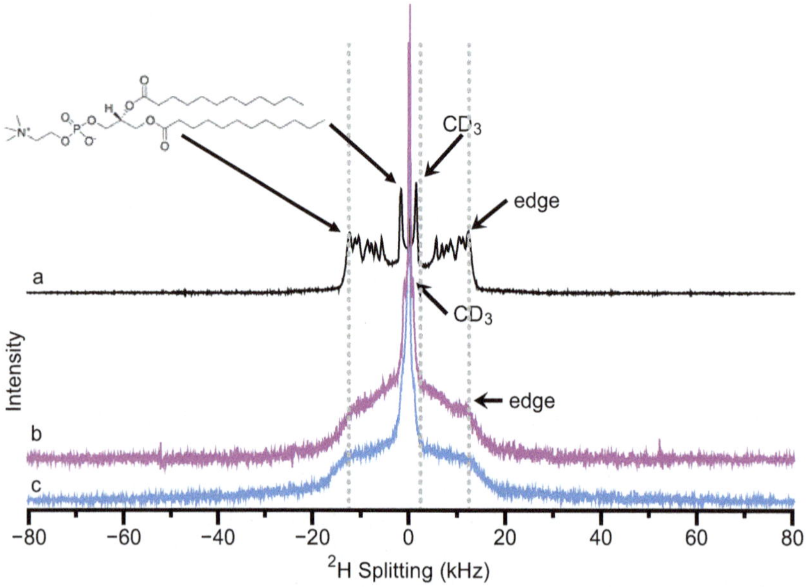


**Figure 3.3.** $^2$H NMR spectra of $^2$H-labelled bacteria share essential features with $^2$H-labelled lipids: (a) $^2$H NMR spectrum of dilauroyl phosphatidylcoholine-d$_{46}$ (DLPC-d$_{46}$) at 25 °C (black); (b) $^2$H NMR spectrum of deuterated *E. coli (LA8)* at 37 °C (purple); and (c) $^2$H NMR spectrum of deuterated *B. subtilis* at 37 °C (blue). The lipid methyl groups and prominent edge from the chain region near the lipid head groups are labelled. Dashed vertical lines at ±12.5 kHz are included to facilitate comparison of the spectra. DLPC spectrum provided by Michael Morrow and Tim Porter. *E. coli (LA8)* and *B. subtilis* spectra as in (Santisteban *et al* 2020, Santisteban 2019) provided by Nury Paula Sanisteban.

head groups is still present and has the same splitting as for the single lipid spectrum in the liquid crystalline phase. It is also possible to discern characteristic shapes between the edge and the $CD_3$ groups, which can be different for different types of bacteria. In figures 3.3(b) and (c) it can be seen that although the prominent edges in the spectra of *E. coli* and *B. subtilis* have the same splitting, there are differences in the overall shape, which can be seen by comparing the spectra at splittings between ~ 4–8 kHz.

Consideration of these subtle changes in one type of bacterium to the next leads to some natural questions: (i) how reproducible are the spectra from sample to sample? and (ii) how do the spectra and samples evolve over time in the NMR spectrometer? As can be seen in figure 3.4, with practice, it is possible to prepare fresh samples of a particular type of bacterium and get the same NMR spectrum each time. The spectra vary in signal-to-noise, likely due to variations in how many bacteria are harvested and transferred to the NMR tube, but the essential features of the spectra, i.e. the prominent edges, the methyl region and the overall shape of the spectrum in between, remain the same. The NMR acquisition time for the spectra shown in figure 3.4 was 8 h. To get a sense as to how much the spectrum changes over time, multiple back-to-back $^2$H NMR spectra have been acquired with a 2 h acquisition time, and demonstrate that the spectra are generally stable for the first 8–10 h, but by

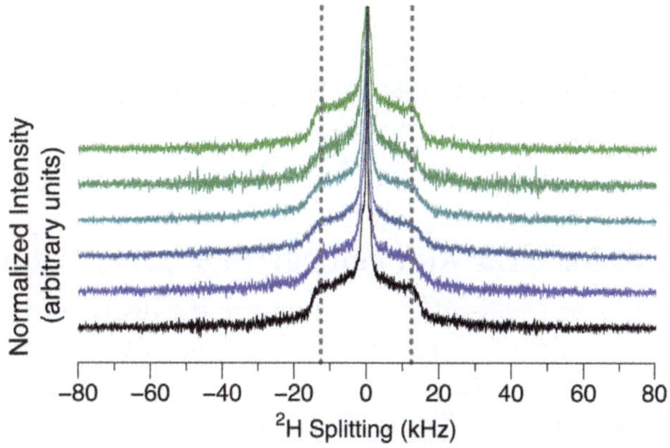


**Figure 3.4.** $^2$H spectra of bacteria are reproducible. NMR spectra at 37 °C of six different preparations of $^2$H-labelled *B. subtilis* normalized by area. Dashed vertical lines at ±12.5 kHz are included to facilitate comparison of the spectra. Data, as in Santisteban *et al* (2020), provided by Nury Paula Santisteban.

18 h significant changes in the spectra take place indicating a reduction in average chain order (Pius *et al* 2012). The same studies found that 8 h in the NMR spectrometer at 37 °C reduces the viability of the cells by 15%–20%. If the spectra are acquired with magic angle spinning (MAS) instead of statically, the acquisition times can be shortened considerably (Xavier *et al* 2016), although, at least in this study, even with the shorter acquisition time of 1 h, 30% of the cells still became unviable.

To aid in comparing different static $^2$H-NMR spectra to each other and to compare the results from static spectra to MAS spectra, it is useful to quantify the shape of the spectra. The three parameters in common usage that can be calculated from either type of spectra are the first and second moments, $M_1$ and $M_2$, and a parameter derived from $M_1$ and $M_2$, called $\Delta_2$:

$$M_1 = \frac{\int_0^\infty \omega f(\omega) d\omega}{\int_0^\infty f(\omega) d\omega} \tag{3.1}$$

$$M_2 = \frac{\int_0^\infty \omega^2 f(\omega) d\omega}{\int_0^\infty f(\omega) d\omega} \tag{3.2}$$

$$\Delta_2 = \frac{M_2}{1.35 M_1^2} - 1 \tag{3.3}$$

where $\omega$ is the angular frequency and $f(\omega)$ is the intensity. $M_1$ is proportional to the average order parameter of the $^2$H-labelled lipids in the sample, while $\Delta_2$ is sensitive to the overall shape of the spectrum (Davis *et al* 1979). $\Delta_2$ has been the parameter of most utility in comparing results from different bacterial spectra, likely in part because it is less sensitive than $M_1$ or $M_2$ to slight baseline imperfections. The values of $M_1$, $M_2$ and $\Delta_2$ for bacteria in the absence and presence of AMP are listed in table 3.1. We can see that the difference in shape between the *E. coli* and *B. subtilis* spectra is reflected by the significantly different $\Delta_2$ values. As discussed in the next section, perturbing membrane order by adding AMPs leads to increases in $\Delta_2$ values.

## 3.4 $^2$H NMR of bacteria with cell envelope perturbation

$^2$H-NMR has been used frequently to characterize the disruption AMPs induce in model membranes, i.e., lipid bilayers composed of one or two lipid types (e.g. (Sani and Separovic 2016, Lorin *et al* 2012, Salnikov *et al* 2011, Zhu *et al* 2018, Bourbigot *et al* 2009, Ramamoorthy *et al* 2006)). When $^2$H-labelled bacteria are treated with AMPs, major changes can be seen in the $^2$H-NMR spectra (Pius *et al* 2012, Booth *et al* 2017, Santisteban *et al* 2020, Tardy-Laporte *et al* 2013, Laadhari *et al* 2016). For example, *B. subtilis* treated with the AMP MSI-78 (red in figure 3.5(a)) show a dramatic decrease in intensity at the prominent edges, with a concomitant increase in intensity nearer the center of the spectrum. These changes indicate that the AMP dramatically increases the disorder of the lipid bilayers. Similar results are seen for

**Table 3.1.** $M_1$, $M_2$ and $\Delta_2$ values for different $^2$H-labelled bacteria samples.

| Sample | $M_1 \times 10^4$ (s$^{-1}$) | $M_2 \times 10^9$ (s$^{-1}$) | $\Delta_2$ |
|---|---|---|---|
| *E. coli (LA8)*[a] | 3.7 | 3.7 | 0.27 |
| *B. subtilis*[b] | 6.2 ± 0.4 | 8.4 ± 1 | 0.59 ± 0.03 |
| *B. subtilis* + 20% MSI-78[c] | 3.4 | 6.5 | 1.45 |
| *B. subtilis* + 20% BP100[c] | 3.5 | 3.7 | 1.20 |

[a] Average of 2 samples from (Pius *et al* 2012)
[b] Average and standard error from the 6 samples shown in figure 3.4
[c] Values from (Santisteban *et al* 2020)

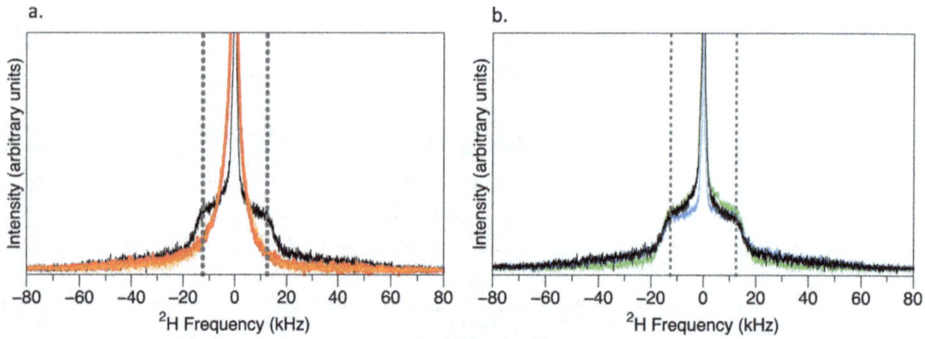


**Figure 3.5.** Treatment with AMPs alters the $^2$H NMR spectra of bacteria much more than mechanical lysis or lipid extraction: (a) $^2$H NMR spectra of untreated *B. subtilis* (black) plus 20 wt% (by dry weight of cells) of MSI-78 (red) or BP100 (orange); and (b) $^2$H spectra of untreated *B. subtilis* (black) compared to *B. subtilis* subjected to mechanical lysis (dark green), or to lipid extraction via the Bligh and Dyer method (light green). Dashed vertical lines at ±12.5 kHz are included to facilitate comparison of the spectra.

the AMP BP100 (orange in figure 3.5(a)), and for the tree frog AMPs caerin 1.1 and aurein 1.2 (Laadhari *et al* 2016). The changes in the shape of the spectra are also reflected in an increase in $\Delta_2$ values with AMP treatment (table 3.1).

One way to appreciate how dramatically AMPs affect lipid bilayers in bacteria is by comparing their effects to other ways of disrupting the cells. For example, a sample produced via mechanical lysis of membrane-deuterated bacteria (figure 3.5(b) dark green) gives a very similar spectrum to intact cells. Similarly, extraction of the lipids with organic solvent using a Bligh and Dyer protocol, followed by rehydration, also gives essentially the same NMR spectrum (figure 3.5(b), light green). Further work (not yet published) shows that treatments that leave the cells whole, but that disrupt either the LPS or the PGN layer, lead to relatively minor changes in the NMR spectra of the membrane-deuterated bacteria.

## 3.5 What have we learned from $^2$H NMR of bacteria so far?

It is satisfying to see the AMP-induced membrane disruption that has been observed so many times in NMR experiments with model lipids can also be observed in the context of whole, intact cells. Also of interest is to compare the effects of a particular AMP on different cell types. It has been found for MSI-78, caerin 1.1. and aurein 1.2, the degree of membrane disruption is quite different against *E. coli* and B. *subtilis*, most likely due to differences in cell envelope composition of Gram(−) and Gram(+) cell envelopes (Santisteban *et al* 2020, Laadhari *et al* 2016). This observation underlines the importance of studying AMP-induced membrane disruption in the context of whole cells.

With the caution that thus far, only a small number of AMPs have been examined in this way, it is noteworthy that different AMPs have similar effects on membrane order in whole bacteria. For instance, 20% (by dry weight of cells) MSI-78 has an almost identical effect on *B. subtilis* to that of 20% BP100 (figure 3.5). This may be because although MSI-78, with 22 residues, is twice as long as BP100, the two peptides still have similar charge to mass ratios (MSI-78 has a charge of +10 and BP100 has a charge of +6). When compared to the effects of very harsh treatments of the cells, like mechanical lysis, or lipid extraction, it is striking how large of an effect the AMPs have on lipid chain order. It points to a very special interaction between AMPs and lipids, that the AMPs prevent lipid bilayers from self-repairing in the same way as happens after mechanical lysis or lipid extraction.

Importantly, work on solid-state NMR of intact bacteria has the potential to link two very disparate types of results whose apparent incongruence has been a puzzle in AMP research. To see actual biological activity, for example in minimal inhibitory concentration (MIC) assays, requires very high AMP-to-lipid (L) ratios, around 100 peptides for every lipid (Wimley 2010). On the other hand, in NMR experiments with AMPs in model lipid systems, the AMP:L ratio needed to see lipid bilayer disruption is many orders of magnitude lower, about 1 AMP for every 100 lipids (e.g. (Ramamoorthy *et al* 2006)). The AMP:L ratio needed to see changes in the $^2$H spectra of whole bacteria turns out to be intermediate between these two extremes (Pius *et al* 2012). This suggests some intriguing possibilities. One is that, since the whole cell NMR experiments are done at AMP:L ratios that are expected to be sub-lethal, AMPs still induce major disruptions to the lipid bilayer that might be expected to allow AMPs access to potential intracellular targets inside the cell. On the other hand, since it is necessary to add so much more peptide per lipid in the whole cell NMR experiments compared to the model lipid NMR experiments, there may be some non-lipid components of the cells—either inside or outside—that interact with the AMPs and prevent them from binding the lipids. Alternatively, there may be other components in the membrane that help preserve the integrity of the bilayer in cells as compared to model membranes.

## References

Akhoundsadegh N, Belanger C R and Hancock R E W 2019 Outer membrane interaction kinetics of new polymyxin B analogs in gram-negative bacilli *Antimicrob. Agents Chemother.* **63** e00935–19

Booth V, Warschawski D E, Santisteban N P, Laadhari M and Marcotte I 2017 Recent progress on the application of $^2$H solid-state NMR to probe the interaction of antimicrobial peptides with intact bacteria *Biochim. Biophys. Acta* **1865** 1500–11

Bourbigot S, Fardy L, Waring A J, Yeaman M R and Booth V 2009 Structure of chemokine-derived antimicrobial peptide interleukin-8alpha and interaction with detergent micelles and oriented lipid bilayers *Biochemistry* **48** 10509–21

Davis J H 1983 The description of membrane lipid conformation, order and dynamics by $^2$H-NMR *Biochim. Biophys. Acta* **737** 117

Davis J H, Nichol C P, Weeks G and Bloom M 1979 Study of the cytoplasmic and outer membranes of *Escherichia coli* by deuterium magnetic resonance *Biochemistry* **18** 2103–12

Dennison S R, Morton L H, Harris F and Phoenix D A 2008 The impact of membrane lipid composition on antimicrobial function of an alpha-helical peptide *Chem. Phys. Lipids* **151** 92–102

Ebbensgaard A, Mordhorst H, Aarestrup F M and Hansen E B 2018 The role of outer membrane proteins and lipopolysaccharides for the sensitivity of *Escherichia coli* to antimicrobial peptides *Front. Microbiol.* **9** 2153

Laadhari M, Arnold A A, Gravel A E, Separovic F and Marcotte I 2016 Interaction of the antimicrobial peptides caerin 1.1 and aurein 1.2 with intact bacteria by $^2$H solid-state NMR *Biochim. Biophys. Acta* **1858** 2959–64

Lohner K and Prenner E J 1999 Differential scanning calorimetry and x-ray diffraction studies of the specificity of the interaction of antimicrobial peptides with membrane-mimetic systems *Biochim. Biophys. Acta* **1462** 141–56

Lorin A, Noel M, Provencher M E, Turcotte V, Cardinal S, Lague P, Voyer N and Auger M 2012 Determining the mode of action involved in the antimicrobial activity of synthetic peptides: a solid-state NMR and FTIR study *Biophys. J.* **103** 1470–9

Malanovic N and Lohner K 2016a Gram-positive bacterial cell envelopes: the impact on the activity of antimicrobial peptides *Biochim. Biophys. Acta (BBA)* **1858** 936–46

Malanovic N and Lohner K 2016b Antimicrobial peptides targeting gram-positive bacteria *Pharmaceuticals (Basel)* **9** 59

Overall S A, Zhu S, Hanssen E, Separovic F and Sani M A 2019 *In situ* monitoring of bacteria under antimicrobial stress using $^{31}$P solid-state NMR *Int. J. Mol. Sci.* **20** 181

Pius J, Morrow M R and Booth V 2012 (2)H solid-state nuclear magnetic resonance investigation of whole *Escherichia coli* interacting with antimicrobial peptide MSI-78 *Biochemistry* **51** 118–25

Ramamoorthy A, Thennarasu S, Lee D K, Tan A and Maloy L 2006 Solid-state NMR investigation of the membrane-disrupting mechanism of antimicrobial peptides MSI-78 and MSI-594 derived from magainin 2 and melittin *Biophys. J.* **91** 206–16

Rowlett V W, Mallampalli V K P S, Karlstaedt A, Dowhan W, Taegtmeyer H, Margolin W and Vitrac H 2017 Impact of membrane phospholipid alterations in *Escherichia coli* on cellular function and bacterial stress adaptation *J. Bacteriol.* **199** e00849–16

Salnikov E S, Aisenbrey C, Balandin S V, Maxim N Z, Ovchinnikova T V and Bechinger B 2011 Structure and alignment of the membrane-associated antimicrobial peptide arenicin by oriented solid-state NMR spectroscopy *Biochemistry* **50** 3784–95

Sani M A and Separovic F 2016 How membrane-active peptides get into lipid membranes *Acc. Chem. Res.* **49** 1130–8

Santisteban N P 2019 Interaction of antimicrobial peptides with bacterial cell envelopes *PhD Thesis* Memorial University of Newfoundland

Santisteban N P, Morrow M R and Booth V 2017 Protocols for studying the interaction of MSI-78 with the membranes of whole gram-positive and gram-negative bacteria by NMR *Methods Mol. Biol.* **1548** 217–30

Santisteban N P, Morrow M R and Booth V 2020 Effect of AMPs MSI-78 and BP100 on the lipid acyl chains of $^2$H-labeled intact gram positive bacteria *Biochim. Biophys. Acta Biomembr.* **1862** 183199

Sohlenkamp C and Geiger O 2016 Bacterial membrane lipids: diversity in structures and pathways *FEMS Microbiol. Rev.* **40** 133–59

Tardy-Laporte C, Arnold A A, Genard B, Gastineau R, Morancais M, Mouget J L, Tremblay R and Marcotte I 2013 A (2)H solid-state NMR study of the effect of antimicrobial agents on intact *Escherichia coli* without mutating *Biochim. Biophys. Acta* **1828** 614–22

Wimley W C 2010 Describing the mechanism of antimicrobial peptide action with the interfacial activity model *ACS Chem. Biol.* **5** 905–17

Xavier L W, Marwa L, Alexandre A A, Isabelle M and Dror E W 2016 A 2H magic-angle spinning solid-state NMR characterisation of lipid membranes in intact bacteria *Biochim. Biophys. Acta (BBA)-Biomembr.* **1858** 146–52

Zhu S, Sani M A and Separovic F 2018 Interaction of cationic antimicrobial peptides from Australian frogs with lipid membranes *Pept. Sci.* **110** e24061

**IOP** Publishing

Solid-State NMR
Applications in biomembrane structure
**Frances Separovic and Marc-Antoine Sani**

# Chapter 4

## Solid-state NMR study of microalgal membranes and cell walls

**Alexandre Poulhazan, Alexandre A Arnold, Dror E Warschawski and Isabelle Marcotte**


### Abstract

Microalgae are ubiquitous primary organisms that play key roles in the aquatic environment where they are the basis of the food web. These microorganisms also produce a variety of molecules with high added value, which find applications in human and animal nutrition, as therapeutic agents and as biofuels. Lipid membranes abound in microalgae where they protect the cell and are essential building blocks of organelles such as the chloroplast. To better understand the architecture and dynamics of microalgal cell membranes and their constituents, it is important that they are investigated in their natural environment. Solid-state nuclear magnetic resonance (SS-NMR) is emerging as a powerful tool to tackle these complex assemblies *in situ* on cell extracts and also in whole cells. Several nuclei can be studied to probe algal membranes and, moreover, microalgae can be fully $^{13}$C- or $^{15}$N-labeled in a cost-effective manner. In the case of full organism labelling, the challenge of SS-NMR is to develop and identify experiments to filter specific constituent signals. In this chapter, we show how SS-NMR can provide a wealth of information on the structure, dynamics and composition of microalgal membranes. We also summarize the main SS-NMR approaches so far applied on lipid extracts or whole cells under near to native conditions and suggest potential avenues to improve our understanding of these systems.


### 4.1 Overview

Microalgae are microscopic algal cells also known as phytoplankton. The cell wall and plasma membrane protect the organelles in the cytoplasm, such as the nucleus, the chloroplast, the reticulum, as well as the sugar and lipid reserves (figure 4.1). About 30 000 species of microalgae are found in fresh- and saltwater, where they are

doi:10.1088/978-0-7503-2532-5ch4  

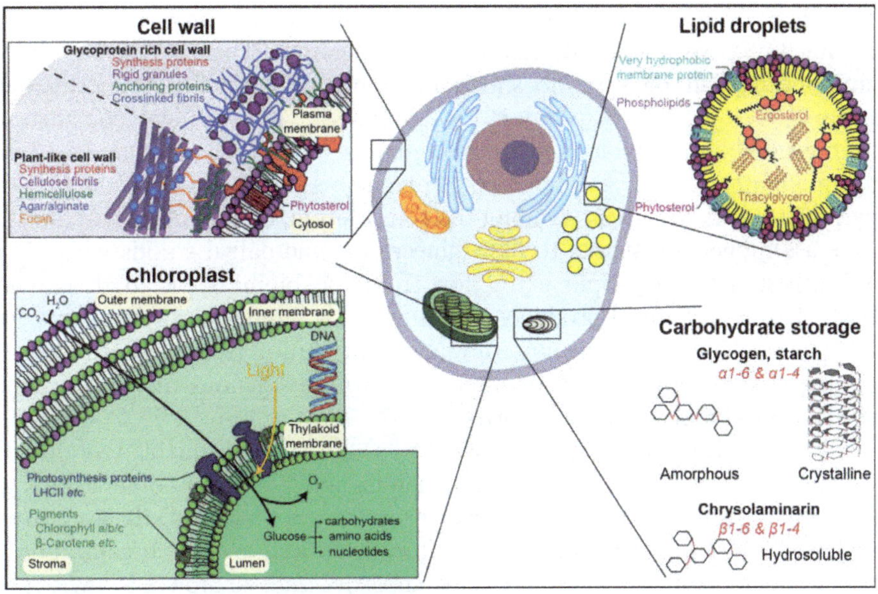


**Figure 4.1. Microalgal cell with organelles and membrane structures.** The cell wall is generally composed of glycoproteins, cellulosic material, carbonate or silica and the plasma membrane is rich in phospholipids. Chloroplast membranes have different galactolipid-to-phospholipid ratios, going from 19:81 in the chloroplast outer layer to 2:98 in thylakoid membranes (Harwood 1998). Lipid droplets incorporate TAGs and ergosterol, as a dense energy storage, surrounded by a single phospholipid monolayer. Energy can also be stored as complex carbohydrates, namely starch or chrysolaminarin.

at the basis of the aquatic food chain (Richmond and Hu 2013). As such, global warming or pollution effects on this resource can have an impact on all higher organisms.

Microalgae bear an important socio-economical potential. They are already used in human and animal nutrition, and have been exploited for their therapeutic activity related to antimicrobial (Falaise *et al* 2016), antiviral (de Jesus Raposo *et al* 2015), antioxidant, anti-inflammatory (Rasool *et al* 2007) or anti-cancer (Gutiérrez *et al* 2010) properties. Another interesting characteristic of microalgae is their ability to sequestrate and bioaccumulate heavy metals (Posadas *et al* 2017, Piccini *et al* 2019). They thus stand as an elegant avenue for contaminated water treatment. Microalgae are also used in the development of third-generation biofuels, which partially or completely come from solar instead of fossil energy (Abo *et al* 2019).

On a more fundamental level, microalgae are great tools for the study of photosynthesis. For example, carbon metabolism was deciphered in *Chlorella* by Melvin Calvin, who was awarded the Nobel Prize in 1961. The dynamics of photosynthesis has been investigated in detail in *Chlamydomonas* (Eberhard *et al* 2008). New transgene strategies are available, mainly studies of *Chlamydomonas reinhardtii* (Doron *et al* 2016, Baier *et al* 2018), which has been called the 'photosynthetic *Escherichia coli*' because of its optional heterotrophic metabolism, allowing the use of non-photosynthetic mutants.

The microalgal cell wall, membranes and lipids are key players for both biochemical processes and biotechnological applications. Their selective permeability and mechanical or osmotic shock resistance make the cell interfaces and membranes a source of inspiration for biomaterial development. Furthermore, lipids in general are valuable nutritional ingredients with long-chain and ω−3 polyunsaturated fatty acids (FAs), but also as precursors for biodiesel production from triacylglycerol (TAG). Understanding the composition, dynamics and architectures of these assemblies is essential to bioengineer new microalgal strains with efficient photosynthesis and easy lipid extraction from the biomass, or to design novel materials based on natural products with interesting biological/physical properties. Finally, microalgal membrane structure and composition (see figure 4.1) can vary depending on temperature, pH or interaction with exogenous molecules, making them excellent global warming reporters.

Solid-state nuclear magnetic resonance (SS-NMR) is an invaluable tool providing nanoscopic information on the structure, dynamics and interactions of large biological molecules with virtually no limit in molecular size. This is why SS-NMR has proven its capabilities to investigate whole cell constituents, as well as their organization and membrane interactions. However, the cell complexity leads to crowded NMR spectra in which many signals potentially overlap. Therefore, SS-NMR approaches are being developed to tackle such challenging systems (Warnet *et al* 2015, Werby and Cegelski 2019, Narasimhan *et al* 2020). Considering the environmental and economical importance of microalgae, the objective of this chapter is to present how SS-NMR can provide information on the molecular composition and organization of microalgal membranes and the cell wall. Before presenting different NMR methodologies to address these cell compartments, the composition of microalgal membranes will first be summarized.

## 4.2 Microalgal cell membrane composition and architecture

Microalgal membranes are composed of lipids, proteins, carbohydrates, phytosterols as well as pigments. Although lipids ensure membrane integrity and efficient cell impermeability, the detailed lipidome of microalgae has been described for only a handful of species such as *C. reinhardtii* and *Nannochloropsis oceanica* (Zhao *et al* 2013, Nguyen *et al* 2013). Membranes are also rich in proteins that maintain the cell structure and metabolism, and in carbohydrates bound to both lipids and proteins, allowing key processes such as cell recognition or protein activity. Pigments are found in chloroplast membranes (figure 4.1) where they are involved in light energy capture and dissipation. Indeed, a lipid membrane also protects organelles in the cytoplasm, such as the reticulum and the nucleus. Moreover, lipid droplets are also present inside the cells.

### 4.2.1 Microalgal cell wall and plasma membrane

The cell wall primarily acts as a shell and thus makes biomolecules such as TAGs harder to harvest and exploit. It also protects the cell from mechanical constraints and external stress, and is involved in cell–cell and cell–substrate adhesion, protection from bacteria as well as sexual reproduction (Domozych *et al* 2012). Although the plasma membrane is not the largest membrane in the cell, it must be studied since it faces a wide range of stresses such as temperature changes and osmotic shocks in its environment. Moreover, multiple specific transporters are found within this membrane, which is the last barrier crossed by molecules before accessing the cytosol. It is also where cell signalling is initiated. One of the first studies of the lipid composition of microalgal plasma membranes focused on the green algae *Hydrodictyon africanum*. Phospholipids, such as phosphatidylcholine (PC), phosphatidylserine (PS) and cardiolipin (CL), were mainly identified (Bailey and Northcote 1976). The microalgal cell membrane is mostly composed of polar lipids (Harwood 1998) and its composition can adapt to external stress or cell state (Mimouni *et al* 2018).

The relationship between the plasma membrane and the microalgal cell wall is so intimate that considering the membrane without the cell wall would make no sense in terms of dynamics and shape. Indeed, the cell wall synthesis and its remodelling proteins are associated to the membrane (Popper *et al* 2014) and some fibrillar assemblies anchored in the cell wall traverse the plasma membrane (Domozych 2019), making membrane and cell wall boundaries hard to define precisely (figure 4.1). For microalgae such as Chlorophyceae, the cell wall is composed of very organized carbohydrates and glycoproteins (Rashidi and Trindade 2018), so that glycan patterns and structures are important to determine. While electron microscopy (EM) has provided micrometer resolution images important for taxonomic determination, SS-NMR can potentially provide additional structural and molecular-level dynamic information of the cell membrane on different timescales.

### 4.2.2 Photosynthetic organelles (thylakoids and chloroplast)

There are several organelles in the algal cell and a membrane borders them all. The chloroplast has, by far, the largest membrane surface, with three layers of densely packed lipid membranes (figure 4.1). This explains why the lipid profile of other organelles is difficult to determine without getting any chloroplast membrane or pigment contamination during purification. The lipid and FA chain profiles of microalgae are usually given for the whole organism (figure 4.2). The membrane surface of each organelle has not been measured or estimated yet, even if a 3D reconstitution made by *in situ* cryo-electron tomography of *C. reinhardtii* has been reported (Engel *et al* 2015).

The lipid composition of chloroplasts membranes is very different from that of plasma membranes as they mostly contain galactolipids and few phospholipids. It varies according to the species but monogalactosyldiacylglycerol (MGDG, 40 to 55%), digalactosyldiacylglycerol (DGDG, 15%–35%), sulfoquinovosyldiacylglycerol (SQDG, 10%–20%) and phosphatidylglycerol (PG, 10%–20%) are the main components (Harwood 1998). In *Chlamydomonas*, the chloroplast envelopes and thylakoid membranes have slightly different lipid compositions (Mendiola-Morgenthaler *et al* 1985). On the other hand, MGDG/DGDG and SQDG/PG

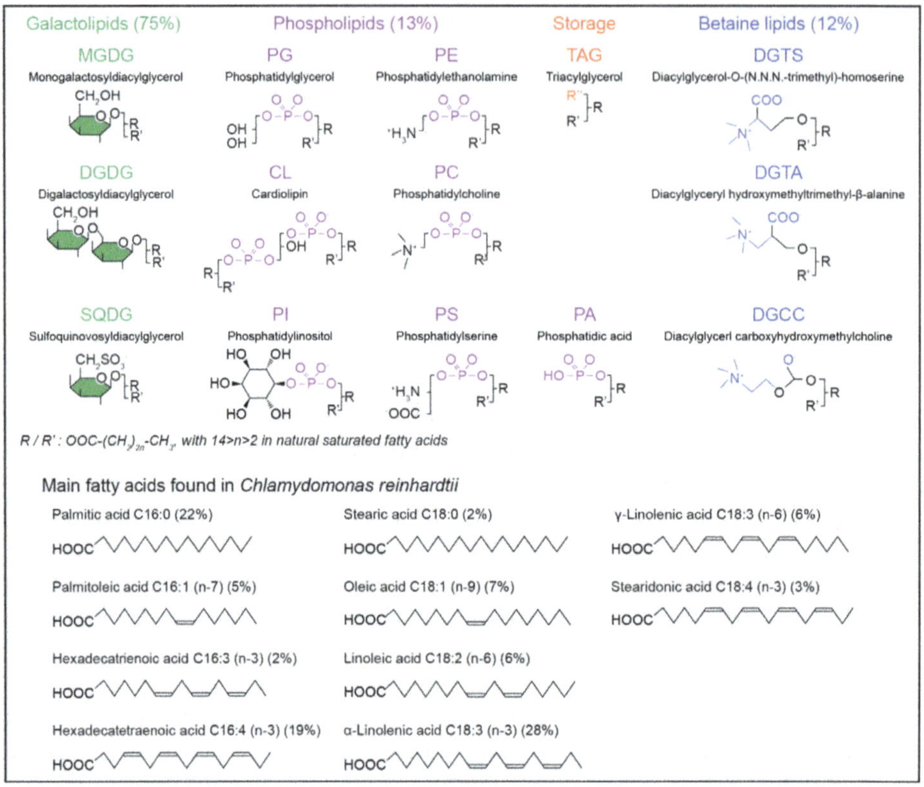


**Figure 4.2. Main lipids and fatty acids found in microalgae**. Galactolipids and betaine lipids are the most abundant in microalgae, and mainly located in photosynthetic membranes. Phospholipids are found in other membranes. Triacylglycerols (TAGs) are stored in lipid droplets. The lipid profile and fatty acid chain composition are species- and medium-dependent. As a reference, values for *C. reinhardtii* are indicated (Vieler *et al* 2007, Nguyen *et al* 2013).

ratios seem to be regulated in chloroplast membranes of various species and could be critical for proper thylakoid development (Boudière *et al* 2014). Betaine lipids have been detected in a few brown algae, such as diacylglycreyl-O-(N,N,N-trimethyl) homoserine (DGTS) and diacylglyceryl-O-(hydroxymethyl)-(N,N,N-trimethyl)β-alanine (DGTA) (Murakami *et al* 2018). Finally, recent work mentions glycolipids as microalgal membrane stabilizers because, during water or cold stress, the chloroplast membrane can be enriched in oligogalactolipids, produced by MGDG conversion (Hölzl and Dörmann 2019).

### 4.2.3 lipid droplets and free fatty acids

The lipid droplet (LD) is an important organelle for biofuel production due to the TAGs that they contain. Little is known about microalgae LDs but interest is growing due to their high potential as an alternative to fossil fuels. TAGs produce twice as much energy than starch and protein per unit weight. Moreover, microalgae are of great interest for nutrition purposes because of the essential long-chain and unsaturated FAs that are contained in the LDs (Spolaore *et al* 2006).

The number, size and composition of LDs vary greatly among microalgal cell types but also according to the life cycle and metabolic stress. For example, nitrogen, iron, sulfur and phosphorus starvation are commonly used strategies to increase lipid storage in microalgae such as *C. reinhardtii* (Hu *et al* 2008, Wang *et al* 2009, Goold *et al* 2014). The size of LDs can range from nm to μm and they are very dynamic in terms of molecular motion, anabolism and catabolism. LDs are easily observed in the cytosol using lipophilic dyes such as Nile red or Bodipy, thus facilitating their quantification by microscopy or flow cytometry, as well as transmission electron microscopy which has been used to further describe their structure (Goodson *et al* 2011). Lipid droplets share the same overall structure with a hydrophobic oil core of storage lipids that mainly comprise TAGs (80%–90% by weight) and sterol esters, surrounded by a phospholipid monolayer that also contains proteins (only 5% by weight) (figure 4.1). Biochemical and microscopy studies have shown that the endoplasmic reticulum and plastid membranes are involved in LD production (Fan *et al* 2011, Goodson *et al* 2011). LDs are also associated with lipid homeostasis and several cellular signalling processes (Farese and Walther 2009).

The FA composition of TAGs is similar to that of whole cells (Goold *et al* 2014). However, the profile of the LD enveloping monolayer is very specific. For example, in *C. reinhardtii* (Wang *et al* 2009) and *Haematococcus pluvialis* (Peled *et al* 2011), DGTS, sulfoquinovosyldiacylglycerol (SQDG) and traces of DGDG are its major component. The lipids around the LD core are thought to be intermediates for acyl chain production in TAGs. Specific proteins are associated with LDs (Yoneda *et al* 2016, Wang *et al* 2017, Wang *et al* 2019). In *C. reinhardtii*, the predominant protein is the very hydrophobic major lipid droplet protein (MLDP) (Goold *et al* 2014), but other LD-associated proteins have been identified (Vieler *et al* 2012), and complete proteomics describe more than 200 proteins linked to LDs (Nguyen *et al* 2011).

## 4.3 Technical considerations

All the microalgal cell compartments presented in the previous section have a large diversity in molecular composition and structure. However, sophisticated techniques and appropriate stable isotope labelling allow the determination of molecular structures, dynamics and interactions in complex systems such as microalgae by SS-NMR.

### 4.3.1 Observable nuclei for microalgal membranes studies

Since protons ($^{1}H$) are the most sensitive nuclei for biological applications in NMR, $^{1}H$-based techniques have been used to study microalgae, as will be illustrated below. The large abundance of protons in biological samples is an advantage in terms of sensitivity, but a drawback in terms of resolution. Therefore, other nuclei with higher specificity are frequently exploited. Phosphorous-31 ($^{31}P$), for example, has 100% natural abundance and high gyromagnetic ratio. Phosphorus is mainly found in ATP, DNA, phosphorylated proteins but also in phospholipids and, therefore, $^{31}$P NMR is an excellent probe to determine the membrane composition and dynamics (Overall *et al* 2019). Carbon is present in all organic molecules of the cell but only the $^{13}$C isotope, which has a low (1%) natural abundance, generates an NMR signal. Since its gyromagnetic ratio is one quarter that of $^{1}$H, $^{13}$C has a lower sensitivity and isotopic enrichment is required for experiments to be time efficient. Finally, $^{15}$N NMR can also be used to study microalgae, but is even less sensitive than $^{13}$C and with very low natural abundance (0.4%), so $^{15}$N NMR can only be performed with isotopic labelling.

Uniform $^{13}$C- and/or $^{15}$N-labelling of microalgae is easily carried out with affordable carbon and nitrogen sources, i.e., sodium bicarbonate ($NaH^{13}CO_3$) and ammonium chloride ($^{15}NH_4Cl$), respectively. The cost of growing uniformly labelled *C. reinhardtii*, for example, can be estimated to less than 100$/L of medium with an optimal cell concentration of 3 to $4 \times 10^6$ cells/mL. Depending on metabolism or the targeted molecule, the medium might also require more expensive $^{13}$C-labelled glucose, sodium acetate (Venters *et al* 1991) or $CO_2$ (Akhter *et al* 2016).

Targeted labelling of specific molecules of interest, such as lipids, amino acids, carbohydrates or pigments, within the whole organism is an interesting alternative approach. This strategy can be applied to $^{13}$C or $^{15}$N but also to other reporter nuclei, such as $^{2}$H (Pius *et al* 2012, Tardy-Laporte *et al* 2013, Warnet *et al* 2016, Molugu *et al* 2017, Salnikov *et al* 2019) (see chapter 4) or $^{19}$F (Matsumori *et al* 2008, Shi *et al* 2012, Afonin *et al* 2014, Kozorog *et al* 2018), often chosen to replace protons in key locations. These specific labelling approaches have rarely been applied to microalgae as they require a good knowledge of algal metabolism, but they hold great promise in particular for *in vivo* SS-NMR studies, as performed with intact bacteria (Booth *et al* 2017).

### 4.3.2 Solution and solid-state NMR

The NMR line width of a given nucleus depends on its interactions with neighbouring nuclei or electrons, including dipolar and quadrupolar interactions, chemical shift anisotropy but also anisotropic magnetic susceptibility effects. Most of these interactions are orientation-dependent but averaged out under fast and isotropic tumbling, such as for small molecules in solution, resulting in very narrow NMR lines.

In solution NMR, molecules can be discriminated according to their diffusion correlation time. Such spectral filtering hence simplifies the spectra and has been applied to study whole microalgal cells (Courtier-Murias *et al* 2012, Akhter *et al* 2016). Among the available 'diffusion filters', *diffusion editing* (DE) removes the signal of mobile molecules, while *inverse diffusion editing* (IDE) only shows the spectrum of fast diffusing molecules. Single pulse experiments will obviously show both types of signals.

Macromolecules and complex assemblies such as cell extracts, or even whole cells such as microalgae, are considered as 'solid' on the NMR time scale. The orientation dependence of the interactions presented above leads to broad lines that often overlap with neighbouring signals, resulting in the loss of spectral resolution. Such systems are thus studied using SS-NMR techniques. High resolution can be restored by using strategies that replace fast and isotropic tumbling by coherent averaging. One such strategy is magic-angle spinning (MAS), where the sample is spun at high frequency at an angle of 54.7° with respect to the magnetic field direction. Depending on the available spinning frequency, interactions can be completely averaged out, resulting in narrower lines. The required spinning frequency depends on the observed nuclei and the interactions that they experience. One may wonder if high spinning frequencies would be deleterious to soft eukaryotic cells. Microalgae are very resistant organisms, owing to their protective and rigid cell wall. It has been shown that *C. reinhardtii* can spend 24 h in an SS-NMR rotor spinning at 10 kHz, at room temperature and without any nutriments, with around 80% survival and cell morphology preserved (Poulhazan *et al* 2018).

In the case of protons in rigid solids, homonuclear dipolar interactions are very strong and often cannot be efficiently averaged out using standard MAS. Therefore, most $^{1}$H MAS SS-NMR studies are restricted to very dynamic small molecules where the remaining dipolar couplings and anisotropic magnetic susceptibility can be cancelled by MAS at moderate frequencies (typically below 10 kHz). This is the field of high-resolution magic-angle spinning NMR (HRMAS). In $^{1}$H HRMAS, as in solution NMR, the lipid droplet signals dominate those of membrane lipids, whereas higher spinning frequencies would be required to observe intense membrane phospholipid, carbohydrate or protein signals. HRMAS allows for direct measurements of non-liquid soft tissues and cells, generally by $^{1}$H NMR, and is typically used to characterize metabolites, even in intact cells, as is summarized in the in-cell section.

High-resolution SS-NMR developments over the last 30 years have provided robust methodologies to characterize the structure and dynamics of biological membrane components (lipids and proteins) by observing $^{13}$C signals under MAS, sometimes in combination with $^{15}$N since these nuclei are found in all peptide bonds. Since the internuclear interactions involved are smaller than for $^{1}$H, standard MAS is often enough to generate high-resolution one- (1D) and multidimensional spectra. In some cases, line narrowing by MAS results in membrane components spectra that are similar to those obtained by solution NMR.

While solution NMR and HRMAS are restricted to molecules with 'fast' dynamic properties, SS-NMR can reveal and exploit the dynamic heterogeneity of biological membranes. To do so, adequate 'dynamic filters' that select only rigid or dynamic components within the sample are utilized. This results not only in spectral simplification but also provides information regarding the molecular environment. The most common filters take advantage of proton-to-carbon polarization transfer schemes. For example, INEPT (nuclei enhancement by polarization transfer) will only be efficient for very dynamic molecules (with tumbling times from $10^{-15}$ to $10^{-9}$ s) and only detect carbons directly bound to protons. On the other hand, cross polarization (CP) will only show carbons from more rigid molecules (with correlation times from $10^{-9}$ to 1 s). Single pulse $^{13}$C experiments will show both types of signals. Other types of filters, that will not be detailed here, have also been developed (Reckel *et al* 2012, Renault *et al* 2012, Warnet *et al* 2015, Matlahov and van der Wel 2018).

Since cells are not pure solids or pure liquids, the research group of Simpson has developed tools to study such heterogeneous samples with an approach called *comprehensive multiphase* (CMP) NMR (Simpson *et al* 2013, Liaghati Mobarhan *et al* 2019). Indeed, the dynamical heterogeneity of intact cells requires the use of both solution and SS-NMR approaches to obtain a complete picture of their composition and function. CMP-NMR probes thus combine lock, gradients and improved shimming capabilities, typical of solution NMR probes, with access to moderate magic-angle spinning frequencies (*vide infra*) of a few kHz and high-power pulses typical of SS-NMR probes. With this type of equipment, monitoring different types of molecules is possible: from dissolved metabolites up to very rigid storage molecules or proteins.

## 4.4 NMR of lipid extracts and model membranes

Considering the complexity of microalgal cells, simple and straightforward approaches are used to garner information on membrane lipids by NMR. First, lipids can be extracted, dissolved in organic solvents, and studied by simple $^{1}$H, $^{13}$C, or $^{31}$P solution NMR. This approach has been applied, for example, on whole cell extracts of *Pavlova spp.* (Kobayashi *et al* 2007) or *Neochloris oleoabundans* (Beal *et al* 2010) for lipid assignment and quantification in combination with other techniques (Challagulla *et al* 2017). The use of lipid extracts obviously neglects important information such as the lipid environment, collective properties and membrane dynamics.

Another strategy to study microalgal membrane properties is to employ model systems with lipid composition similar to the mimicked membranes. Although part of the living organism complexity is sacrificed, model membranes provide valuable information on specific biophysical properties and interactions with the lipids. Mixtures of galactolipids such as MGDG and DGDG have been studied by $^{2}$H and $^{31}$P static SS-NMR as early as 1985 to investigate phase transitions in microalgae and chloroplast membranes (Brentel *et al* 1985, Lindblom *et al* 1986, Borovyagin *et al* 1988). Galactolipid structures and dynamics were later probed by $^{13}$C SS-NMR using MAS (Adebodun *et al* 1992, Gross *et al* 1995, Castro *et al* 2007) or in fast-tumbling bicelles by solution NMR (Howard and Prestegard 1995, Ye *et al* 2013). Figure 4.3 shows a $^{1}$H–$^{13}$C proton-detected local field (PDLF) study that allowed measurement of the order parameters at various carbon positions in model unlabelled DMPC/MGDG membranes. This revealed that MGDG reduced the order parameters along DMPC molecules (Castro *et al* 2007). Simplified thylakoid model membranes made of palmitoyl-oleoyl phosphocholine (POPC) and MGDG were used to determine the interactions of Hcf106—a peptide found in the thylakoid of microalgae—with its host membrane lipids (Zhang *et al* 2013, Zhang *et al* 2014). Finally, the light harvesting complex II (LHCII) of *C. reinhardtii* was also studied by $^{13}$C,$^{15}$N-NMR reconstituted in a thylakoid-like membrane environment made up of MGDG, DGDG, SQDG and PG (Azadi-Chegeni *et al* 2018b).

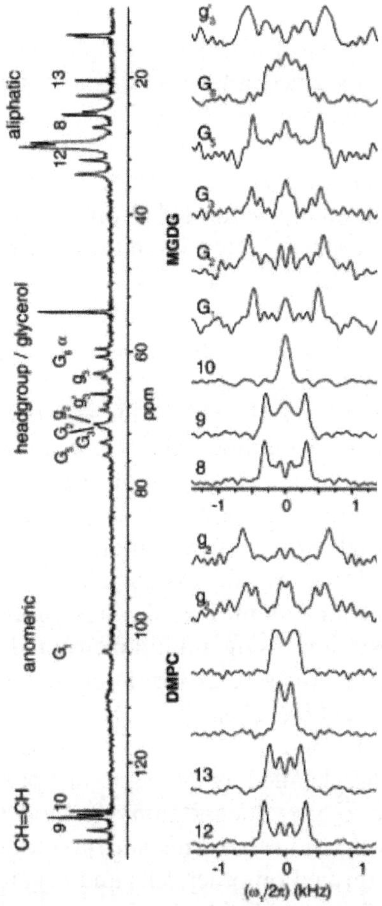

**Figure 4.3.** $^{1}H$–$^{13}C$ MAS (5 kHz) NMR PDLF spectra of a model membrane made of unlabelled DMPC and MGDG, recorded at 42 °C. The 1D $^{13}C$ NMR spectrum is displayed on the left. Reprinted from Castro *et al* (2007) with permission from Elsevier.

## 4.5 *In vivo*, in-cell and *in situ* NMR

In-cell SS-NMR tackles complex samples that are very hard to characterize using other atomic-resolution techniques such as EM, x-ray crystallography or solution NMR. While *in vivo* NMR addresses the challenge of studying living cells, *in situ* NMR examines cell extracts—the cell is no longer intact but molecules are still in their native local environment. *In vivo*, in-cell and *in situ* SS-NMR have only recently been applied to microalgae but a wide range of applications are possible, as will be presented in the next sections.

### 4.5.1 $^{1}$H-NMR based approaches

As previously explained, HRMAS provides insights into cellular metabolism by producing spectra with resolution sometimes comparable to that of solution NMR;

and also allows spectral editing according to the dynamics of the constituents. HRMAS is applied to a variety of biological samples including human or mice cell lines, or tissues for clinical studies, allowing diagnosis and follow up of organism responses to treatments (Kaebisch *et al* 2017). It is a common tool in metabolomics, especially for detecting metabolites associated with lipids such as inositol, phosphoryl choline, FAs and cholesterol. $^{1}$H HRMAS is also used in lipid profiling and characterization of the saturated/unsaturated FA ratio (Griffin *et al* 2003, Kabli *et al* 2009, Berry *et al* 2016). Terpenes found in chloroplast membranes are interesting molecules also detected by HRMAS (Gaysinski *et al* 2015). Finally, because lipids are an important part of cell metabolism, several studies have focused on alterations of lipid metabolism using HRMAS to detect lipid droplet perturbation, for example in cancer cell lines (Pan *et al* 2012).

Considering the capabilities of HRMAS, this approach has been used to study microalgae, including their membranes and the *in vivo* intracellular C/N ratio (Bondu *et al* 2008, Hashim *et al* 2013). Native membranes are difficult to study because they are rigid structures on the NMR time scale, thus leading to broad and overlapping lines, with notable exceptions (Simon *et al* 2015). Merkley and Syvitsk (Merkley and Syvitski 2011) successfully assigned lipids and carbohydrates in whole *Nannochloropsis granulata* and monitored the unsaturated/saturated FA ratio as a function of cell harvest time using multidimensional $^{1}$H–$^{13}$C HRMAS. Also, Chauton *et al* published several studies on different microalgae species, such as *C. mülleri* (Webb *et al* 2006), *Dunaliella sp.* (Chauton and Størseth 2008), *Chaetoceros muelleri* (Chauton *et al* 2004) and *Thalassiosira pseudonana* (Chauton *et al* 2003). In these works, detailed assignments of lipids were reported, allowing quantification in response to growth conditions. HRMAS was also used to classify microalgae using statistical data treatments such as principal component analysis (PCA). NMR studies performed on other cells and organisms will certainly inspire future investigations of microalgal lipids. For example, HRMAS has been used to characterize erythrocyte membranes and perform permeability studies in *Daphnia magna* (Bruno *et al* 2006, Kovacevic *et al* 2018). Recent work on living earthworms (*Eisenia fetida*) and freshwater shrimps (*Hyalella azteca*) have improved the lipid signal treatment based on relaxation and diffusion spectral filtering (Mobarhan *et al* 2016, Hassan *et al* 2019).

An alternative to HRMAS studies for $^{1}$H detection is the emerging use of ultrahigh frequency MAS. Although membrane lipids and proteins can be studied at intermediate frequencies (15 kHz (Gopinath *et al* 2017)), ultra-fast MAS (>50 kHz) allows detection of protons in rigid molecules with increased sensitivity and a resolution comparable to that of solution NMR. Combined with $^{13}$C and $^{15}$N labelling of samples, sequential assignment of membrane proteins or other complex molecules is possible, even in their native environment, e.g. proteins involved in photosynthesis (Beal *et al* 2010, Medeiros-Silva *et al* 2016). New information on biomembrane structure is expected to emerge from this approach, in particular for the study of membrane proteins in microalgae.

### 4.5.2 1D $^{13}$C solid-state NMR methods

As stated above, SS-NMR can be exploited to obtain valuable information on cell samples. Depending on the spectral editing used, both rigid and dynamic molecules can be identified and their dynamics *in situ* revealed—information that is lost using solution NMR of cell extracts, for example. Whole uniformly $^{13}$C-labelled *C. reinhardtii* were studied by our group in 2015 by 1D $^{13}$C SS-NMR with MAS, along with *Pavlova lutheri* and *Nannochloropsis oculata* (Arnold *et al* 2015). This work allowed the assignment of different lipids, carbohydrates including glycans, as well as rigid cell-wall proteins using different dynamic filters. As shown in figure 4.4, the more mobile molecules of *C. reinhardtii* are revealed by $^{13}$C SS-NMR using the INEPT filter, the most intense peaks being ascribed to methyl groups from the lipid acyl chains around 30 ppm. This spectrum is comparable to a $^{13}$C solution NMR spectrum using the IDE filter, with a slightly lower resolution.

On the exact same sample, a CP filter revealed the more rigid molecules such as starch, with C1 carbon at 101 ppm. Starch is invisible in solution NMR, even with a DE filter. These experiments highlight that many lipids are very mobile, while proteins in the cell wall are much more rigid assemblies. In 2018, Azadi-Chegeni *et al* published similar work on *C. reinhardtii* whole cells (Azadi-Chegeni *et al* 2018a, Azadi-Chegeni *et al* 2018b). For example, they assigned several component resonances in whole cell samples and quantified the dynamics in terms of correlation times and order parameters of these molecules at different temperatures (Azadi-Chegeni *et al* 2018a). They also investigated LHCII in native membranes and showed that they favored protein conformational stability. These works represent a step toward the understanding of chloroplast complexity in terms of molecular composition and dynamics.

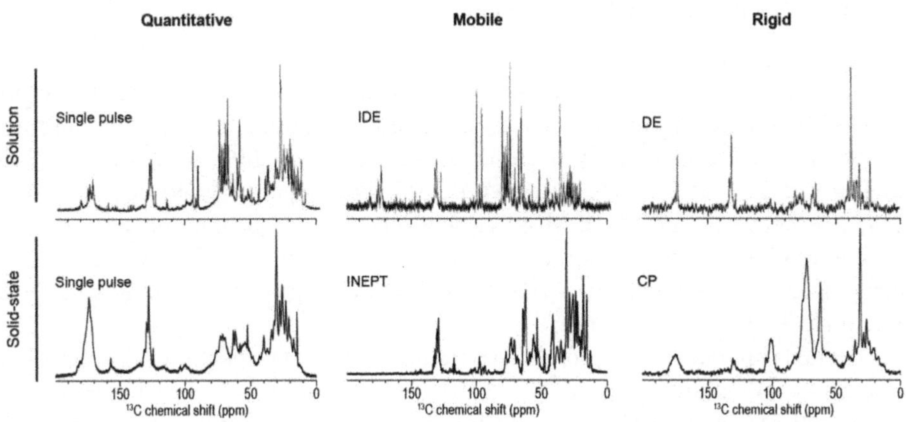


**Figure 4.4.** Comparison of $^{13}$C labelled *C. reinhardtii* whole cell spectra by solution and solid-state NMR using either single pulse or filtered experiments. Solid-state experiments were performed at 600 MHz at 25 °C with a 10 kHz MAS frequency (adapted from Arnold *et al* (2018). Solution NMR experiments were performed at 500 MHz at room temperature (adapted from Akhter *et al* (2016). Reproduced with permission from Springer.

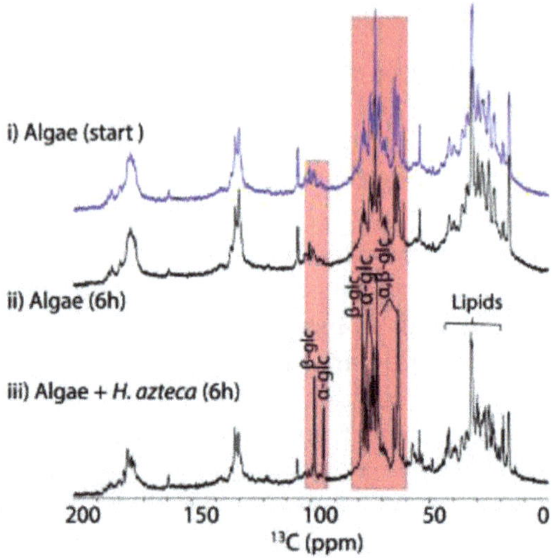


**Figure 4.5.** $^{13}C$ CMP NMR (15 °C and 1.1 kHz spinning frequency) spectra of uniformly $^{13}C$-labelled *Chlorella vulgaris* microalgae with and without *H. azteca* daphnia. The pink region highlights the glucose peaks that increase when *H. azteca* is in contact with the algae. Reprinted from Liaghati Mobarhan *et al* (2019) with permission, copyright (2019) John Wiley & Sons.

Another interesting study reports the consumption of *Chlorella vulgaris* lipids by the daphnia *Hyalella azteca* using comprehensive multiphase NMR and $^{13}C$-labelled microalgae. As shown in figure 4.5, good resolution was obtained and lipid consumption could be monitored with this new probe. Multidimensional solid-state experiments were also recorded with $^{1}H$–$^{13}C$ correlations from $^{13}C$-labelled algae and $^{2}H$–$^{13}C$ correlations from $^{2}H$–$^{13}C$ uniformly labelled algae.

### 4.5.3 2D-$^{13}C$ and $^{15}N$-NMR experiments

With isotopically-labelled samples sensitivity is greatly increased, enabling the use of multidimensional SS-NMR which also significantly enhances spectral resolution. 2D SS-NMR NMR has already proven successful in studying biological membranes *in situ* or in whole cells, mainly bacteria (Andronesi *et al* 2005, Schanda *et al* 2014, Romaniuk and Cegelski 2015, Booth *et al* 2017, Bouhlel *et al* 2019) but also bovine retinas (Verhoeven *et al* 2001). More closely related to algal membranes, the work on bacterial chlorosomal antenna by Van Rossum *et al* is the first to mention and assign galactolipid signals in a native thylakoid membrane protein study (van Rossum *et al* 2001).

As proteins are part of the membrane architecture, it is important to present some of the SS-NMR studies carried out on these proteins in microalgal membranes. Protein conformations, dynamics and interactions with different chloroplast partners in photosynthesis have been described in several model organisms in a membrane environment using multidimensional SS-NMR, for example by the De Groot group, for green and purple bacteria light harvesting complexes (Huang and McDermott 2008, Alia *et al* 2009) but also bacteriorhodopsin (Varga *et al* 2008).

They have also shown the interaction between photosynthesis-associated proteins and their chromophore, using a single amino acid labelling approach (Alia *et al* 2001). To better understand the photosynthesis machinery in an environment close to native, the LHCII protein complex was purified from *C. reinhardtii* with some residual lipids and examined by multidimensional SS-NMR (Pandit *et al* 2011, Pandit *et al* 2013, Sunku *et al* 2013). LHCII switching its conformation from light harvesting to photoprotective mode through non-photochemical quenching was studied under MAS (<15 kHz) NMR and at temperatures below −30 °C.

The only multidimensional SS-NMR study of a light harvesting system in native microalga membranes, without any purification or reconstitution steps, was performed on isolated *C. reinhardtii* thylakoid membranes (Pandit *et al* 2013). The resolution was sufficient to report LHCII conformational dynamics—a phenomenon impossible to observe using detergents or protein crystals. The results are thus specific to the dynamics in membranes of a photosynthetic organism, which prevents photo-damage. Figure 4.6 compares $^{13}C$–$^{13}C$ spectra of LHCII in its native thylakoid environment and reconstituted in detergent. This can be compared also to $^{13}C$–$^{13}C$ spectra of the same protein extracted from its membrane or reconstituted in a model thylakoid-like membrane, as done by Azadi-Chegeni *et al* (2018b).

By using *in vivo* 2D solid-state NMR and dynamic spectral edition on $^{13}C$-labelled microalgae, we improved our previous 1D NMR results for *C. reinhardtii*, showing that

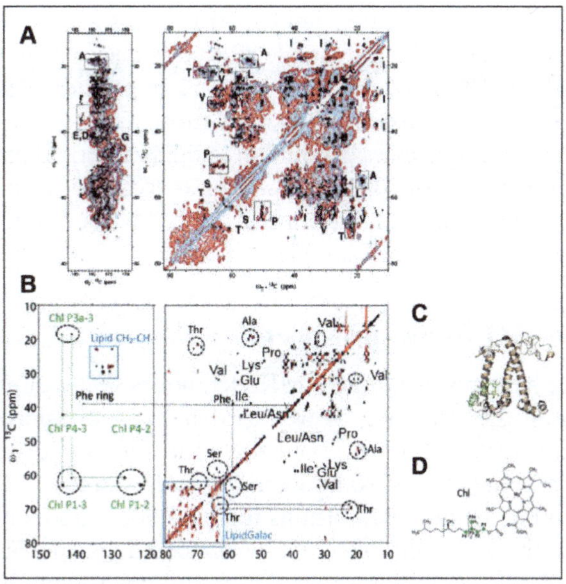


**Figure 4.6.** $^{13}C$–$^{13}C$ solid-state NMR spectra of LHCII from *C. reinhardtii* in: (**A**) its native thylakoid environment (red contours), reconstituted in β-n-dodecyl-D-maltoside detergent buffer (light blue contours); and (**B**) pure extracted protein (black contours) and in a reconstituted thylakoid membrane (red contours). (**C**) The LHCII crystal structure, and (**D**) chlorophyll molecular structure are also given. **A** is adapted from Pandit *et al* (2013) with permission from Elsevier, and **B-D** are adapted from Azadi-Chegeni *et al* (2018b), with permission.

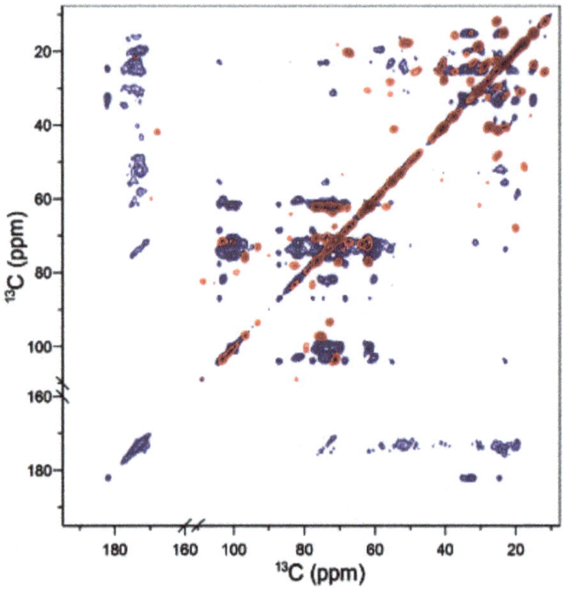


**Figure 4.7.** $^{13}$C–$^{13}$C *in vivo* solid-state NMR spectra of uniformly $^{13}$C-labelled *C. reinhardtii* recorded at a MAS frequency of 26.7 kHz using either INEPT- (in red) or CP-based (in blue) filters. From Arnold *et al* (2018) with permission from Springer.

precise assignments of carbohydrates, lipids and proteins are possible (Arnold *et al* 2018). Figure 4.7 compares rigid and mobile molecules detected by $^{13}$C–$^{13}$C SS-NMR using either INEPT- or CP-based filters. A large range of microalgal membrane lipid classes could thus be identified *in situ*, with those from thylakoid membranes dominating due to their abundance. In addition to lipids and rigid proteins, assignment and characterization of carbohydrates *in situ* is a necessary step towards a better understanding of carbohydrates architecture in the cell walls of microalgae, and their ability to face osmotic shocks, for example.

## 4.6 Conclusion and future prospects

In this chapter, we have reviewed the NMR techniques that can provide molecular-level information on the structure, as well as dynamics, of microalgal membranes constituents. So far, little is known about these membranes reflecting their overall complexity but they are a key component to understanding the photosynthetic machinery, and to bioengineer new strains for biofuel or food supplements.

We have shown the versatility of NMR, which can detect nuclei, such as $^{1}$H, $^{31}$P, $^{13}$C, $^{15}$N, $^{2}$H and $^{19}$F, in either single or multidimensional experiments. It is applicable to small or large, dynamic or rigid molecules, to metabolites, model membranes, cell extracts, whole cells and even living cells. For NMR, isotopic labelling is generally required. While uniform labelling is routinely performed, as demonstrated here with examples of fully $^{13}$C-labelled microalgae, new approaches can certainly be developed to specifically label molecules of interest within whole microalgae, providing that the biosynthetic pathways are known. This strategy has notably been applied to bacteria and to $^{2}H$ *in vivo* SS-NMR study of their membranes (Booth *et al* 2017).

Depending on the sample and cellular constituent studied, solution or SS-NMR approaches are available. In the solid state, they include static or MAS samples, spinning at moderate or high frequencies. New techniques continue to emerge, including the use of dynamic filters and comprehensive multiphase NMR, which are useful to discriminate membrane lipids from lipid droplets, for example.

Beside broad molecular identification, NMR focuses on high-resolution structure determination of biomolecules. Microalgal membrane proteins and photosynthetic systems are important targets since they are involved in metabolism and transport. Their structures will need to be determined to understand these mechanisms (Ladizhansky 2017). Revealing the structure of microalgae cell walls (including their connexion to the membrane) is also an area where SS-NMR can contribute, either within whole cells or extracts, as has been done with higher plants to obtain atomic resolution of cellulose nanostructures (Dick-Pérez *et al* 2011, Wang and Hong 2015). Such information is essential for a fundamental understanding of microalgal species and also for commercial exploitation of this aquatic resource.

Finally, dynamic nuclear polarization (DNP) coupled with SS-NMR has been shown to be very powerful for studies of biological assemblies, such as bacterial cell walls (Wang *et al* 2016), or membrane proteins (Becker-Baldus *et al* 2015), with unprecedented sensitivity. Using this approach on microalgae should improve our knowledge of the complex entanglement of proteins, lipids and carbohydrates in cellular external matrices. It is exciting to see that, while in-cell SS-NMR is still the source of many new developments, it has now reached the maturity to be applied to whole microalgae to give better understanding of living organisms at an atomic level.

## Acknowledgements

This work was supported by the Natural Sciences and Engineering Research Council (NSERC) of Canada (grant RGPIN-2018–06200 to IM) and the Centre National de la Recherche Scientifique (UMR 7203 to DEW). AP would like to thank the PROTEO network, the Faculty of Science of the Université du Québec à Montréal (UQAM) and the Fonds Québécois de la Recherche sur la Nature et les Technologies for the award of scholarships. IM is a member of Ressources Aquatiques Québec (RAQ) and PROTEO research networks.

## References

Abo B O, Odey E A, Bakayoko M and Kalakodio L 2019 Microalgae to biofuels production: a review on cultivation, application and renewable energy *Rev. Environ. Health* **34** 91

Adebodun F, Chung J, Montez B, Oldfield E and Shan X 1992 Spectroscopic studies of lipids and biological membranes: carbon-13 and proton magic-angle sample-spinning nuclear magnetic resonance study of glycolipid-water systems *Biochem.* **31** 4502–09

Afonin S, Glaser R W, Sachse C, Salgado J, Wadhwani P and Ulrich A S 2014 $^{19}$F NMR screening of unrelated antimicrobial peptides shows that membrane interactions are largely governed by lipids *Biochim. Biophys. Acta Biomembr.* **1838** 2260–8

Akhter M, Dutta Majumdar R, Fortier-Mcgill B, Soong R, Liaghati-Mobarhan Y, Simpson M, Arhonditsis G, Schmidt S, Heumann H and Simpson A J 2016 Identification of aquatically available carbon from algae through solution-state NMR of whole $^{13}$C-labelled cells *Anal. Bioanal. Chem.* **408** 4357–70

Alia A, Ganapathy S and De Groot H J 2009 Magic Angle Spinning (MAS) NMR: a new tool to study the spatial and electronic structure of photosynthetic complexes *Photosynth. Res.* **102** 415–25

Alia A, Matysik J, Soede-Huijbregts C, Baldus M, Raap J, Lugtenburg J, Gast P, Van Gorkom H J, Hoff A J and De Groot H J M 2001 Ultrahigh field MAS NMR dipolar correlation spectroscopy of the histidine residues in light-harvesting complex II from photosynthetic bacteria reveals partial internal charge transfer in the B850/His complex *J. Am. Chem. Soc.* **123** 4803–9

Andronesi O C, Becker S, Seidel K, Heise H, Young H S and Baldus M 2005 Determination of membrane protein structure and dynamics by magic-angle-spinning solid-state NMR spectroscopy *J. Am. Chem. Soc.* **127** 12965–74

Arnold A A, Bourgouin J P, Genard B, Warschawski D E, Tremblay R and Marcotte I 2018 Whole cell solid-state NMR study of *Chlamydomonas reinhardtii* microalgae *J. Biomol. NMR* **70** 123–31

Arnold A A, Genard B, Zito F, Tremblay R, Warschawski D E and Marcotte I 2015 Identification of lipid and saccharide constituents of whole microalgal cells by $^{13}$C solid-state NMR *Biochim. Biophys. Acta Biomembr.* **1848** 369–77

Azadi-Chegeni F, Schiphorst C and Pandit A 2018a *In vivo* NMR as a tool for probing molecular structure and dynamics in intact *Chlamydomonas reinhardtii* cells *Photosynth. Res.* **135** 227–37

Azadi-Chegeni F, Ward M E, Perin G, Simionato D, Morosinotto T, Baldus M and Pandit A 2018b Conformational dynamics of photosynthetic light-harvesting complex II in a native membrane environment *bioRxiv* 288860 (Article retrieved June 9, 2020)

Baier T, Wichmann J, Kruse O and Lauersen K J 2018 Intron-containing algal transgenes mediate efficient recombinant gene expression in the green microalga *Chlamydomonas reinhardtii* *Nucleic Acids Res.* **46** 6909–19

Bailey D S and Northcote D H 1976 Phospholipid composition of the plasma membrane of the green alga, *Hydrodictyon africanum* *Biochem.* J **156** 295–300

Beal C M, Webber M E, Ruoff R S and Hebner R E 2010 Lipid analysis of *Neochloris oleoabundans* by liquid state NMR *Biotechnol. Bioeng.* **106** 573–83

Becker-Baldus J *et al* 2015 Enlightening the photoactive site of channelrhodopsin-2 by DNP-enhanced solid-state NMR spectroscopy *Proc. Natl Acad. Sci. USA* **112** 9896–01

Berry J P, Roy U, Jaja-Chimedza A, Sanchez K, Matysik J and Alia A 2016 High-resolution magic angle spinning nuclear magnetic resonance of intact zebrafish embryos detects metabolic changes following exposure to teratogenic polymethoxyalkenes from algae *Zebrafish* **13** 456–65

Bondu S, Kervarec N, Deslandes E and Pichon R 2008 The use of HRMAS NMR spectroscopy to study the *in vivo* intra–cellular carbon/nitrogen ratio of *Solieria chordalis* (Rhodophyta) *J. Appl. Phycol.* **20** 673–9

Booth V, Warschawski D E, Santisteban N P, Laadhari M and Marcotte I 2017 Recent progress on the application of $^{2}$H solid-state NMR to probe the interaction of antimicrobial peptides with intact bacteria *Biochim. Biophys. Acta Proteins Proteom.* **1865** 1500–11

Borovyagin V L, Tarakhovsky Y S and Vasilenko I A 1988 Polymorphism in galactolipid/phosphatidylglycerol model membranes initiated by chlorophylls: $^{31}$P-NMR and electron-microscopy studies *Biochim. Biophys. Acta* **939** 111–23

Boudière L *et al* 2014 Glycerolipids in photosynthesis: composition, synthesis and trafficking *Biochim. Biophys. Acta Bioenerg.* **1837** 470–80

Bouhlel Z, Arnold A A, Warschawski D E, Lemarchand K, Tremblay R and Marcotte I 2019 Labelling strategy and membrane characterization of marine bacteria *Vibrio splendidus* by *in vivo* $^{2}$H NMR *Biochim. Biophys. Acta Biomembr.* **1861** 871–8

Brentel I, Selstam E and Lindblom G 1985 Phase equilibria of mixtures of plant galactolipids. The formation of a bicontinuous cubic phase *Biochim. Biophys. Acta Biomembr.* **812** 816–26

Bruno E, Digilio G, Cabella C, DE Reggi A, Baroni S, Mainero V and Aime S 2006 Water exchange across the erythrocyte plasma membrane studied by HR-MAS NMR spectroscopy *Magn. Reson. Med.* **56** 978–85

Castro V, Dvinskikh S V, Widmalm G, Sandström D and Maliniak A 2007 NMR studies of membranes composed of glycolipids and phospholipids *Biochim. Biophys. Acta Biomembr.* **1768** 2432–7

Challagulla V, Nayar S, Walsh K and Fabbro L 2017 Advances in techniques for assessment of microalgal lipids *Crit. Rev. Biotechnol.* **37** 566–78

Chauton M, Størseth T and Johnsen G 2003 High-resolution magic angle spinning $^{1}$H NMR analysis of whole cells of *Thalassiosira pseudonana* (*Bacillariophyceae*): broad range analysis of metabolic composition and nutritional value *J. Appl. Phycol.* **15** 533–42

Chauton M, Størseth T and Krane J 2004 High-resolution magic angle spinning NMR analysis of whole cells of *Chaetoceros muelleri* (*Bacillariophyceae*) and comparison with $^{13}$C-NMR and distortionless enhancement by polarization transfer $^{13}$C-NMR analysis of lipophilic extracts *J. Phycol* **40** 611–8

Chauton M and Størseth T F 2008 HR MAS NMR spectroscopy of marine microalgae, part 1: classification and metabolite composition from HR MAS $^{1}$H NMR spectra and multivariate analysis *Modern Magnetic Resonance* ed G A Webb (Dordrecht: Springer)

Courtier-Murias D *et al* 2012 Comprehensive multiphase NMR spectroscopy: Basic experimental approaches to differentiate phases in heterogeneous samples *J. Magn. Reson.* **217** 61–76

De Jesus Raposo M F, De Morais A M and De Morais R M 2015 Marine polysaccharides from algae with potential biomedical applications *Mar. Drugs* **13** 2967–3028

Dick-Pérez M, Zhang Y, Hayes J, Salazar A, Zabotina O A and Hong M 2011 Structure and interactions of plant cell-wall polysaccharides by two- and three-dimensional magic-angle-spinning solid-state NMR *Biochemistry* **50** 989–1000

Domozych D, Ciancia M, Fangel J, Mikkelsen M, Ulvskov P and Willats W 2012 The cell walls of green algae: a journey through evolution and diversity *Front. Plant Sci.* **3** 82

Domozych D S 2019 Algal cell walls *eLS* (Chichester: Wiley)

Doron L, Segal N A and Shapira M 2016 Transgene expression in microalgae—from tools to applications *Front. Plant Sci.* **7** 505

Eberhard S, Finazzi G and Wollman F A 2008 The dynamics of photosynthesis *Annu. Rev. Genet.* **42** 463–515

Engel B D, Schaffer M, Kuhn Cuellar L, Villa E, Plitzko J M and Baumeister W 2015 Native architecture of the *Chlamydomonas chloroplast* revealed by *in situ* cryo-electron tomography *eLife* **4** e04889

Falaise C *et al* 2016 Antimicrobial compounds from eukaryotic microalgae against human pathogens and diseases in aquaculture *Mar. Drugs* **14** 159

Fan J, Andre C and Xu C 2011 A chloroplast pathway for the de novo biosynthesis of triacylglycerol in *Chlamydomonas reinhardtii FEBS Lett.* **585** 1985–91

Farese R V Jr and Walther T C 2009 Lipid droplets finally get a little R-E-S-P-E-C-T *Cell* **139** 855–60

Gaysinski M, Annick O-M, Thomas O and Culioli G 2015 Extraction, purification, and NMR analysis of terpenes from brown algae *Methods Mol. Biol* **1308** 207–23

Goodson C, Roth R, Wang Z T and Goodenough U 2011 Structural correlates of cytoplasmic and chloroplast lipid body synthesis in *Chlamydomonas reinhardtii* and stimulation of lipid body production with acetate boost *Eukaryot. Cell* **10** 1592–606

Goold H, Beisson F, Peltier G and Li-Beisson Y 2014 Microalgal lipid droplets: composition, diversity, biogenesis and functions *Plant Cell Rep.* **34** 545–55

Gopinath T, Nelson S E D, Soller K J and Veglia G 2017 Probing the conformationally excited states of membrane proteins via $^1$H-detected MAS solid-State NMR spectroscopy *J. Phys. Chem.* B **121** 4456–65

Griffin J L, Lehtimaki K K, Valonen P K, Grohn O H, Kettunen M I, Yla-Herttuala S, Pitkanen A, Nicholson J K and Kauppinen R A 2003 Assignment of $^1$H nuclear magnetic resonance visible polyunsaturated fatty acids in BT4C gliomas undergoing ganciclovir-thymidine kinase gene therapy-induced programmed cell death *Cancer Res.* **63** 3195–201

Gross J D, Costa P R, Dubacq J P, Warschawski D E, Lirsac P N, Devaux P F and Griffin R G 1995 Multidimensional NMR in lipid systems. Coherence transfer through J couplings under MAS *J. Magn. Reson.* B **106** 187–90

Gutiérrez M, Tidgewell K, Capson T L, Engene N, Almanza A, Schemies J, Jung M and Gerwick W H 2010 Malyngolide dimer, a bioactive symmetric cyclodepside from the panamanian marine cyanobacterium *Lyngbya majuscula J. Nat. Prod.* **73** 709–11

Harwood J L 1998 Membrane lipids in algae ed S Paul-André and M Norio *Lipids in Photosynthesis: Structure, Function and Genetics* (Dordrecht: Springer)

Hashim F, Denis C-M, Ronald S, Wolfgang B, William M K and André J S 2013 HR-MAS NMR spectroscopy: a practical guide for natural samples *Curr. Org. Chem.* **17** 3013–31

Hassan Q, Dutta Majumdar R, Wu B, Lane D, Tabatabaei-Anraki M, Soong R, Simpson M J and Simpson A J 2019 Improvements in lipid suppression for $^1$H NMR-based metabolomics: applications to solution-state and HR-MAS NMR in natural and *in vivo* samples *Magn. Reson. Chem.* **57** 69–81

Hölzl G and Dörmann P 2019 Chloroplast lipids and their biosynthesis *Annu. Rev. Plant Biol.* **70** 51–81

Howard K P and Prestegard J H 1995 Membrane and solution conformations of monogalactosyldiacylglycerol using NMR/molecular modeling methods *J. Am. Chem. Soc.* **117** 5031–40

Hu Q, Sommerfeld M, Jarvis E, Ghirardi M, Posewitz M, Seibert M and Darzins A 2008 Microalgal triacylglycerols as feedstocks for biofuel production: perspectives and advances *Plant J.* **54** 621–39

Huang L and Mcdermott A E 2008 Partial site-specific assignment of a uniformly $^{13}C$, $^{15}N$ enriched membrane protein, light-harvesting complex 1 (LH1), by solid state NMR *Biochim. Biophys. Acta* **1777** 1098–108

Kabli S, Spaink H P, De Groot H J and Alia A 2009 *In vivo* metabolite profile of adult zebrafish brain obtained by high-resolution localized magnetic resonance spectroscopy *J. Magn. Reson. Imaging* **29** 275–81

Kaebisch E, Fuss T L, Vandergrift L A, Toews K, Habbel P and Cheng L L 2017 Applications of high-resolution magic angle spinning MRS in biomedical studies I-cell line and animal models *NMR Biomed.* 30

Kobayashi Y, Torii A, Kato M and Adachi K 2007 Accumulation of cyclitols functioning as compatible solutes in the haptophyte alga *Pavlova* sp *Phycol. Res.* **55** 81–90

Kovacevic V, Simpson A J and Simpson M J 2018 Evaluation of *Daphnia magna* metabolic responses to organic contaminant exposure with and without dissolved organic matter using $^{1}H$ nuclear magnetic resonance (NMR)-based metabolomics *Ecotoxicol. Environ. Saf.* **164** 189–200

Kozorog M, Sani M-A, Lenarčič Živković M, Ilc G, Hodnik V, Separovic F, Plavec J and Anderluh G 2018 $^{19}F$ NMR studies provide insights into lipid membrane interactions of listeriolysin O, a pore forming toxin from *Listeria monocytogenes Sci. Rep.* **8** 6894

Ladizhansky V 2017 Applications of solid-state NMR to membrane proteins *Biochim. Biophys. Acta Proteins Proteom.* **1865** 1577–86

Liaghati Mobarhan Y, Soong R, Lane D and Simpson A J 2019 *In vivo* comprehensive multiphase NMR *Magn. Reson. Chem.* **58** 427–44

Lindblom G, Brentel I, Sjolund M, Wikander G and Wieslander A 1986 Phase equilibria of membrane lipids from *Acholeplasma laidlawii*: importance of a single lipid forming non-lamellar phases *Biochemistry* **25** 7502–10

Matlahov I and van Der Wel P C A 2018 Hidden motions and motion-induced invisibility: dynamics-based spectral editing in solid-state NMR *Methods* **148** 123–35

Matsumori N, Kasai Y, Oishi T, Murata M and Nomura K 2008 Orientation of fluorinated cholesterol in lipid bilayers analyzed by $^{19}F$ tensor calculation and solid-state NMR *J. Am. Chem. Soc.* **130** 4757–66

Medeiros-Silva J, Mance D, Daniëls M, Jekhmane S, Houben K, Baldus M and Weingarth M 2016 $^{1}H$-detected solid-state NMR studies of water-inaccessible proteins *in vitro* and *in situ Angew. Chem. Int. Ed.* **55** 13606–10

Mendiola-Morgenthaler L, Eichenberger W and Boschetti A 1985 Isolation of chloroplast envelopes from Chlamydomonas. Lipid and polypeptide composition *Plant Sci.* **41** 97–104

Merkley N and Syvitski R 2011 Profiling whole microalgal cells by high-resolution magic angle spinning (HR-MAS) magnetic resonance spectroscopy *J. Appl. Phycol.* **24** 535–40

Mimouni V, Couzinet-Mossion A, Ulmann L and Wielgosz-Collin G 2018 Lipids from microalgae ed I A Levine and J Fleurence *Microalgae in Health and Disease Prevention* (New York: Academic)

Mobarhan Y L *et al* 2016 Comprehensive multiphase NMR applied to a living organism *Chem. Sci.* **7** 4856–66

Molugu T R, Lee S and Brown M F 2017 Concepts and methods of solid-state NMR spectroscopy applied to biomembranes *Chem. Rev.* **117** 12087–132

Murakami H, Nobusawa T, Hori K, Shimojima M and Ohta H 2018 Betaine lipid is crucial for adapting to low temperature and phosphate deficiency in Nannochloropsis *Plant Physiol.* **177** 181–93

Narasimhan S, Folkers G and Baldus M 2020 When small becomes too big: expanding the use of in-cell solid-state NMR spectroscopy *Chem. Plus. Chem.* **85** 760–8

Nguyen H M *et al* 2011 Proteomic profiling of oil bodies isolated from the unicellular green microalga *Chlamydomonas reinhardtii*: with focus on proteins involved in lipid metabolism *Proteomics* **11** 4266–73

Nguyen H M, Cuine S, Beyly-Adriano A, Legeret B, Billon E, Auroy P, Beisson F, Peltier G and Li-Beisson Y 2013 The green microalga *Chlamydomonas reinhardtii* has a single omega-3 fatty acid desaturase that localizes to the chloroplast and impacts both plastidic and extraplastidic membrane lipids *Plant Physiol.* **163** 914–28

Overall S A, Zhu S, Hanssen E, Separovic F and Sani M A 2019 *In situ* monitoring of bacteria under antimicrobial stress using $^{31}$P solid-state NMR *Int. J. Mol. Sci.* **20** 181

Pan X, Wilson M, Mcconville C, Arvanitis T N, Kauppinen R A and Peet A C 2012 The size of cytoplasmic lipid droplets varies between tumour cell lines of the nervous system: a $^{1}$H NMR spectroscopy study *MAGMA* **25** 479–85

Pandit A, Morosinotto T, Reus M, Holzwarth A R, Bassi R and De Groot H J M 2011 First solid-state NMR analysis of uniformly $^{13}$C-enriched major light-harvesting complexes from *Chlamydomonas reinhardtii* and identification of protein and cofactor spin clusters *Biochim. Biophys. Acta Bioenerg.* **1807** 437–43

Pandit A, Reus M, Morosinotto T, Bassi R, Holzwarth A R and De Groot H J M 2013 An NMR comparison of the light-harvesting complex II (LHCII) in active and photoprotective states reveals subtle changes in the chlorophyll a ground-state electronic structure *Biochim. Biophys. Acta Bioenerg.* **1827** 738–44

Peled E, Leu S, Zarka A, Weiss M, Pick U, Khozin-Goldberg I and Boussiba S 2011 Isolation of a novel oil globule protein from the green alga *Haematococcus pluvialis* (Chlorophyceae) *Lipids* **46** 851–61

Piccini M, Raikova S, Allen M J and Chuck C J 2019 A synergistic use of microalgae and macroalgae for heavy metal bioremediation and bioenergy production through hydrothermal liquefaction *Sustain. Energy Fuels* **3** 292–301

Pius J, Morrow M R and Booth V 2012 2H solid-state nuclear magnetic resonance investigation of whole *Escherichia coli* interacting with antimicrobial peptide MSI-78 *Biochemistry* **51** 118–25

Popper Z A, Ralet M-C and Domozych D S 2014 Plant and algal cell walls: diversity and functionality *Ann. Bot* **114** 1043–8

Posadas E *et al* 2017 Microalgae cultivation in wastewater *Microalgae-Based Biofuels and Bioproducts* ed C Gonzalez-Fernandez and R Muñoz (Cambridge: Woodhead Publishing)

Poulhazan A, Arnold A A, Warschawski D E and Marcotte I 2018 Unambiguous ex situ and in cell 2D $^{13}$C solid-state NMR characterization of starch and its constituents *Int. J. Mol. Sci.* **19** 3817

Rashidi B and Trindade L M 2018 Detailed biochemical and morphologic characteristics of the green microalga *Neochloris oleoabundans* cell wall *Algal Res.* **35** 152–9

Rasool M, Sabina E and Lavanya B 2007 Anti-inflammatory effect of *Spirulina fusiformis* on adjuvant-induced arthritis in mice *Biol. Pharm. Bull.* **29** 2483–7

Reckel S, Lopez J J, Lohr F, Glaubitz C and Dotsch V 2012 In-cell solid-state NMR as a tool to study proteins in large complexes *Chem. Bio. Chem.* **13** 534–7

Renault M, Tommassen-Van Boxtel R, Bos M P, Post J A, Tommassen J and Baldus M 2012 Cellular solid-state nuclear magnetic resonance spectroscopy *Proc. Natl Acad. Sci. USA* **109** 4863–8

Richmond A and Hu Q 2013 *Handbook of Microalgal Culture: Applied Phycology and Biotechnology* (Hoboken, NJ: Wiley)

Romaniuk J A H and Cegelski L 2015 Bacterial cell wall composition and the influence of antibiotics by cell-wall and whole-cell NMR *Philos. Trans. R. Soc. Lond. B Biol. Sci.* **370** 20150024

Salnikov E S, Aisenbrey C, Pokrandt B, Brügger B and Bechinger B 2019 Structure, topology, and dynamics of membrane-inserted polypeptides and lipids by solid-state NMR spectroscopy: investigations of the transmembrane domains of the DQ beta-1 subunit of the MHC II receptor and of the COP I protein p24 *Front. Mol. Biosci.* **6** 83

Schanda P, Triboulet S, Laguri C, Bougault C M, Ayala I, Callon M, Arthur M and Simorre J-P 2014 Atomic model of a cell-wall cross-linking enzyme in complex with an intact bacterial peptidoglycan *J. Am. Chem. Soc.* **136** 17852–60

Shi P, Li D, Chen H, Xiong Y, Wang Y and Tian C 2012 *In situ* $^{19}$F NMR studies of an *E. coli* membrane protein *Protein Sci.* **21** 596–600

Simon G, Kervarec N and Cerantola S 2015 HRMAS NMR analysis of algae and identification of molecules of interest via conventional 1D and 2D NMR: sample preparation and optimization of experimental conditions *Methods Mol. Biol.* **1308** 191–205

Simpson A J *et al* 2013 Environmental comprehensive multi-phase NMR *Emagres* **2** 399–414

Spolaore P, Joannis-Cassan C, Duran E and Isambert A 2006 Commercial applications of microalgae *J. Biosci. Bioeng.* **101** 87–96

Sunku K, De Groot H J and Pandit A 2013 Insights into the photoprotective switch of the major light-harvesting complex II (LHCII): a preserved core of arginine-glutamate interlocked helices complemented by adjustable loops *J. Biol. Chem.* **288** 19796–804

Tardy-Laporte C, Arnold A A, Genard B, Gastineau R, Morançais M, Mouget J-L, Tremblay R and Marcotte I 2013 A $^{2}$H solid-state NMR study of the effect of antimicrobial agents on intact *Escherichia coli* without mutating *Biochim. Biophys. Acta Biomembr.* **1828** 614–22

van Rossum B J, Steensgaard D B, Mulder F M, Boender G J, Schaffner K, Holzwarth A R and Degroot H J 2001 A refined model of the chlorosomal antennae of the green bacterium *Chlorobium tepidum* from proton chemical shift constraints obtained with high-field 2-D and 3-D MAS NMR dipolar correlation spectroscopy *Biochemistry* **40** 1587–95

Varga K, Aslimovska L and Watts A 2008 Advances towards resonance assignments for uniformly—$^{13}$C, $^{15}$N enriched bacteriorhodopsin at 18.8 T in purple membranes *J. Biomol. NMR* **41** 1–4

Venters R A, Calderone T L, Spicer L D and Fierke C A 1991 Uniform $^{13}$C isotope labeling of proteins with sodium acetate for NMR studies: application to human carbonic anhydrase II *Biochemistry* **30** 4491–4

Verhoeven M A, Creemers A F L, Bovee-Geurts P H M, De Grip W J, Lugtenburg J and De Groot H J M 2001 Ultra-high-field MAS NMR assay of a multispin labeled ligand bound to its G-protein receptor target in the natural membrane environment: electronic structure of the retinylidene chromophore in rhodopsin *Biochemistry* **40** 3282–8

Vieler A, Brubaker S B, Vick B and Benning C 2012 A lipid droplet protein of Nannochloropsis with functions partially analogous to plant oleosins *Plant Physiol.* **158** 1562–9

Vieler A, Wilhelm C, Goss R, Süss R and Schiller J 2007 The lipid composition of the unicellular green alga *Chlamydomonas reinhardtii* and the diatom *Cyclotella meneghiniana* investigated by MALDI-TOF MS and TLC *Chem. Phys. Lipids* **150** 143–55

Wang T and Hong M 2015 Solid-state NMR investigations of cellulose structure and interactions with matrix polysaccharides in plant primary cell walls *J. Exp. Bot.* **67** 503–14

Wang T, Yang H, Kubicki J D and Hong M 2016 Cellulose structural polymorphism in plant primary cell walls investigated by high-field 2D solid-state NMR spectroscopy and density functional theory calculations *Biomacromolecules* **17** 2210–22

Wang X, Hao T-B, Balamurugan S, Yang W-D, Liu J-S, Dong H-P and Li H-Y 2017 A lipid droplet-associated protein involved in lipid droplet biogenesis and triacylglycerol accumulation in the oleaginous microalga *Phaeodactylum tricornutum Algal Res* **26** 215–24

Wang X, Wei H, Mao X and Liu J 2019 Proteomics analysis of lipid droplets from the oleaginous alga *Chromochloris zofingiensis* reveals novel proteins for lipid metabolism *Genomics Proteomics Bioinformatics* **17** 260–72

Wang Z T, Ullrich N, Joo S, Waffenschmidt S and Goodenough U 2009 Algal lipid bodies: stress induction, purification, and biochemical characterization in wild-type and starchless *Chlamydomonas reinhardtii Eukaryot. Cell* **8** 1856–68

Warnet X L, Arnold A A, Marcotte I and Warschawski D E 2015 In-cell solid-state NMR: an emerging technique for the study of biological membranes *Biophys. J.* **109** 2461–6

Warnet X L, Laadhari M, Arnold A A, Marcotte I and Warschawski D E 2016 A $^{2}$H magic-angle spinning solid-state NMR characterisation of lipid membranes in intact bacteria *Biochim. Biophys. Acta Biomembr.* **1858** 146–52

Webb G, Størseth T, Chauton M and Krane J 2006 HR MAS NMR spectroscopy of marine microalgae, part 2: $^{13}$C and $^{13}$C HR MAS NMR analysis used to study fatty acid composition and polysaccharide structure *Modern Magnetic Resonance* ed G A Webb (Dordrecht: Springer)

Werby S H and Cegelski L 2019 Spectral comparisons of mammalian cells and intact organelles by solid-state NMR *J. Struct. Biol.* **206** 49–54

Ye W, Liebau J and Maler L 2013 New membrane mimetics with galactolipids: lipid properties in fast-tumbling bicelles *J. Phys. Chem.* B **117** 1044–50

Yoneda K, Yoshida M, Suzuki I and Watanabe M M 2016 Identification of a major lipid droplet protein in a marine diatom *Phaeodactylum tricornutum Plant Cell Physiol.* **57** 397–406

Zhang L, Liu L, Maltsev S, Lorigan G A and Dabney-Smith C 2013 Solid-state NMR investigations of peptide-lipid interactions of the transmembrane domain of a plant-derived protein, Hcf106 *Chem. Phys. Lipids* **175-176** 123–30

Zhang L, Liu L, Maltsev S, Lorigan G A and Dabney-Smith C 2014 Investigating the interaction between peptides of the amphipathic helix of Hcf106 and the phospholipid bilayer by solid-state NMR spectroscopy *Biochim. Biophys. Acta* **1838** 413–8

Zhao L X, Liu A C, Yu S W, Wang Z X, Lin X Q, Zhai G X and Zhang Q Z 2013 The permeability of puerarin loaded poly(butylcyanoacrylate) nanoparticles coated with polysorbate 80 on the blood-brain barrier and its protective effect against cerebral ischemia/reperfusion injury *Biol. Pharm. Bull.* **36** 1263–70

**IOP** Publishing

Solid-State NMR
Applications in biomembrane structure
**Frances Separovic and Marc-Antoine Sani**

# Chapter 5

# Determining the mechanism of action of host defense peptides: solid-state NMR approaches

**Suzana K Straus, Allen Takayesu and Prashant Kumar**

## Abstract

Host defense peptides (HDPs) are short, cationic, amphiphilic peptides that show promise not only as antimicrobials, but also as antibiofilm agents, immune modulators, anti-cancer agents and antivirals. To harness their potential, it is important to understand how HDPs function. Solid-state NMR is a powerful technique that can be used to characterize HDP mechanisms of action. Various examples of the application of this technique on model membranes and whole cells are presented.

## 5.1 Introduction

The high levels of bacterial antibiotic resistance currently being observed has led researchers to look for alternatives to current antibiotics (Davies 2006, Stanton 2013, Chellat *et al* 2016). Host defense peptides (HDPs) are an integral part of the innate immune defense system of many life forms (Hancock 2005, Jenssen *et al* 2006). Sometimes referred to as antimicrobial peptides or AMPs, HDPs are polypeptide sequences comprised of cationic and hydrophobic amino acids (of *ca.* 12–50 residues in length). These natural peptides or their synthetic analogues are considered to be interesting substitutes to antibiotics because they can function in a variety of ways (Hancock and Sahl 2006, Kumar *et al* 2018, Haney *et al* 2019). They display anti-cancer activity (Chernysh *et al* 2002, Hoskin and Ramamoorthy 2008, Felício *et al* 2017), as well as act against other pathogenic species such as viruses, fungi and parasites (Hancock and Falla 1996, Hancock and Sahl 2006, Travkova *et al* 2017). HDPs can also modulate the immune response (Bowdish *et al* 2005, Bowdish *et al* 2005, Haney and Hancock 2013). However, HDPs are best known for their ability to directly kill bacteria, either by selectively targeting or by targeting a broad spectrum of Gram-positive and/or Gram-negative bacteria. Most studies have

doi:10.1088/978-0-7503-2532-5ch5  

focused on free-swimming (i.e., planktonic) bacteria, though more recent work has examined how HDPs can inhibit or eradicate biofilms (de la Fuente-Núñez *et al* 2016, Yasir *et al* 2018, Raheem and Straus 2019). Microbial biofilms are surface-associated bacterial communities that grow in a protective polymeric matrix (de la Fuente-Núñez *et al* 2014). Biofilms represent the natural growth state of bacteria (Jefferson 2004) and the bacteria within them are even harder to eliminate than their planktonic counterparts.

The manner in which HDPs either inhibit bacterial growth (i.e., are bacteriostatic) or kill bacteria (i.e., are bactericidal) is varied. This diversity in modes of action is likely to contribute to the fact that HDPs give rise to a small amount of bacterial resistance (Hancock and Sahl 2006). In the different mechanisms, illustrated in figure 5.1, HDPs: (1) target and disrupt the cytoplasmic membrane (Matsuzaki *et al* 1998, Vogt and Bechinger 1999, Fernandez *et al* 2012, Wenzel *et al* 2015) or other cell wall components (Leeuw *et al* 2010, Sinha *et al* 2017); (2) interfere with DNA or RNA synthesis (Park *et al* 2009, Haney *et al* 2013); (3) interact with

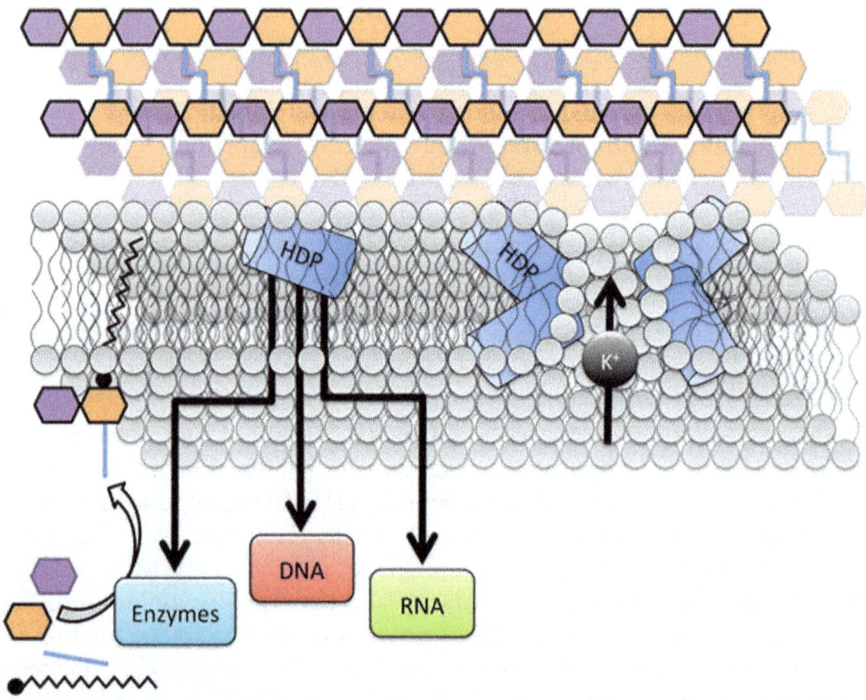


**Figure 5.1.** Illustration of the various mechanisms of HDP action against bacteria. On the left, HDPs (blue cylinder) transit through the cytoplasmic membrane of bacteria without damage to interact with enzymes (cyan) required for bacterial cell wall biosynthesis (C55P, undecaprenyl-phosphate—black; GlcNAc, N-acetyl-glucosamine—purple; MurNAc, N-acetyl-muramic acid—orange; and pentapeptide—blue stick; these elements come together to form the peptidoglycan layer seen above the membrane bilayer), DNA (red) or RNA (green). On the right, the HDPs cause damage to the membrane, leading to perturbation of the arrangement of the lipids. This perturbation is often followed by ion leakage (e.g. potassium). Additional mechanisms presented in the text are not illustrated.

enzymes required for bacterial cell wall biosynthesis (Schmitt *et al* 2010, Ulm *et al* 2012, Wenzel *et al* 2012); and (4) other mechanisms, such as autolysin activation (Yasir *et al* 2019). There are many reviews which discuss various aspects of the mechanisms of action shown in figure 5.1 (Hale and Hancock 2007, Nguyen *et al* 2011, Ulm *et al* 2012, Nuri *et al* 2015, Kumar and Sanil 2017, Kumar *et al* 2018, Lee *et al* 2019), including a recent one by ourselves (Raheem and Straus 2019), that details all the biological assays and biophysical tools which can be used to investigate these mechanisms in detail for both planktonic bacteria and biofilms. In this chapter, we will present some applications of the solid-state NMR methods reviewed in other chapters of this book as well as in the literature (Straus 2004, Toke *et al* 2006, Porcelli *et al* 2013, Opella 2014) for the elucidation of HDP mechanisms of action. Specifically, we will examine how solid-state NMR can be used on model membrane systems, as well as whole bacterial cells, to investigate specific aspects of the mechanisms shown in figure 5.1.

## 5.2 Model membranes

A particularly useful way to probe whether an HDP functions by perturbing the membrane (figure 5.1, right) is to use mechanically aligned samples (Straus 2004, Opella 2014). These typically consist of a mixture of HDPs and synthetic lipids, deposited on a solid support such as a glass or polymer slide. The deposited peptide/lipid mixture is first allowed to dry, the slides are stacked, and then the stack is rehydrated in a high humidity chamber for a given period of time (e.g. 2–4 days). Solid-state NMR experiments are conducted using a flat coil probe, with the lipid bilayer normal usually aligned parallel to the magnetic field. Samples can be prepared as a function of HDP concentration (or peptide-to-lipid ratio) or lipid composition. Over the years, a number of lipids have been used for such studies including 1,2-dimyristoyl-sn-glycero-3-phosphocholine (DMPC) (e.g. aurein 1.2 (Marcotte *et al* 2003)), 1-palmitoyl-2-oleoyl-sn-glycero-3-phosphocholine (POPC) (e.g. MSI-78 and MSI-594 (Porcelli *et al* 2006)), 1,2-diphytanoyl-sn-glycero-3-phosphatidylcholine (DPhPC) (e.g. alamethicin (Dave *et al* 2005)), and other diacylphosphatidylcholine membranes (e.g. K2(LA)xK2 (Harzer and Bechinger 2000)). Lipid mixtures have also been used: DMPC/1,2-dimyristoyl-snglycero-3-phospho-(1'-rac-glycerol) (DMPG) (e.g. PGLa (Tremouilhac *et al* 2006)), POPC/1-palmitoyl-2-oleoyl-sn-glycero-3-phospho-(1'-rac-glycerol) (POPG) (e.g. aurein 2.2 (Cheng *et al* 2010)), 1-palmitoyl-2-oleoyl-sn-glycero-3-phosphoethanolamine (POPE)/POPG (Mani *et al* 2006, Wi and Kim 2008), cardiolipin (CL)/POPG (Cheng *et al* 2011), or others (Strandberg *et al* 2020). The bilayers containing PC alone are typically used to mimic mammalian cells, whereas lipid compositions that contain components with negatively charged headgroups are used to mimic bacteria.

The membrane perturbing effect of HDPs can be tracked from the perspective of the bilayer, i.e., using $^{31}P$ or $^{2}H$ NMR. Alternatively, the orientation of the HDPs can be monitored by introducing one or more $^{15}N$ labels in the peptide backbone, or other NMR active nuclei such as $^{2}H$ in a methyl sidechain (e.g. alanine) (Zamora-Carreras *et al* 2016) or $^{19}F$ in a 4-trifluoromethyl-phenylglycine (Glaser *et al* 2004).

There are many examples in the literature that describe the types of experiments and information that can be obtained from them: in addition to the references already listed in this section, the reader is invited to consult the selection of references given here (Fernandez *et al* 2012, 2013, Grage *et al* 2014, Lee *et al* 2015, Jeong *et al* 2016, Harmouche *et al* 2017, Harmouche and Bechinger 2018, Aisenbrey *et al* 2019, Paredes *et al* 2020). As an example, solid-state $^{31}$P NMR was used to determine the mechanism of action of the HDP aurein 2.2, which functions by forming pores and causing leakage of select ions (Wenzel *et al* 2015). Cheng *et al* (2011) showed that adding aurein 2.2 to aligned bilayers composed of 1:1 CL/POPG (mol/mol) resulted in a higher percentage of signal arising from unaligned lipids, as seen from the $^{31}$P powder pattern, than when the same experiment was conducted using 1:1 POPE/POPG. These results suggested that aurein 2.2 was better at perturbing membranes that mimic bacteria such as *Staphylococcus aureus* than membranes that mimic other bacteria (e.g. *Bacillus cereus*). Indeed, the solid-state NMR data was further supported by oriented circular dichroism data, revealing that aurein 2.2 inserted much more readily into aligned bilayers composed of CL/POPG. Finally, the minimum inhibitory concentration (MIC), which is a measure of the concentration of peptide required to inhibit the growth of bacteria, was determined for aurein 2.2 against *S. aureus* and *B. cereus*: aurein 2.2 has an MIC of 16 μg/ml for the former (indicating that little is required to inhibit bacterial growth, i.e., is more active), as compared to 128 μg/ml for *B. cereus*. This study shows the importance of using the correct model membranes. It also demonstrates the strength of aligned solid-state NMR methods. It should be noted that other solid-state NMR methods that rely on magic angle sample spinning (MAS) have also been reported (Lanne *et al* 2019, Manzo *et al* 2019, Rai *et al* 2019) and clearly show how this methodology can also be applied in antimicrobial research.

## 5.3 Whole cells

An alternative to using model membranes is of course to use whole bacterial cells. Using such systems is often more complex but, of course, also more biologically relevant. In this section, examples will be given on how solid-state NMR methods can be applied to reveal different aspects of the HDP mechanism of action (figure 5.1).

The first approach involves growing whole bacterial cells with an $^{15}$N-labeled or $^{15}$N- and $^{13}$C-labeled algal amino-acid mixture. This has been demonstrated on *S. aureus* (Nygaard *et al* 2015), as well as for *Escherichia coli* and *Vibrio cholerae* (Cegelski 2015). MAS NMR techniques (i.e., CPMAS or REDOR experiments) have been used to characterize the chemical composition and architecture of cell walls in bacteria. The quantification of key components is achieved by comparing spectra obtained from e.g. isolated bacterial cell wall or ribosomes (Nygaard *et al* 2017) versus spectra obtained for intact cells. In this way, by exposing the bacterial cells to antibiotics, Cegelski and co-workers have been able to propose that fosfomycin interferes with cell wall biosynthesis by showing a decrease in polysaccharide carbon signals in the $^{13}$C NMR spectra (Nygaard *et al* 2015). Likewise, they were able to show that chloramphenicol leads to accumulation of cell wall content and hence indicate that this antibiotic functions by interfering with enzymes (figure 5.1—left). Such an approach could easily be extended to HDPs.

An alternative approach is to incorporate $^{2}$H into bacteria. This was first demonstrated in a modified strain of *E. coli* (Pius *et al* 2012), which was grown in minimal media called Medium 63, to which deuterated palmitic acid and unlabelled oleic acid were added. A minimum of two samples were made: one consisting of a bacterial pellet alone, and another where the bacteria were first exposed to the membrane perturbing HDP MSI-78 (figure 5.1—right) prior to being pelleted. Solid-state $^{2}$H NMR spectra of membrane-deuterated LA8 bacteria were recorded with increasing concentrations of MSI-78 (i.e., 4 samples + replicates). Large effects on the lineshapes were observed for HDP concentrations between 20% and 60% MSI-78 (wt/wt). Given that the MIC of MSI-78 against *E. coli* is 4 μg ml$^{-1}$ (Ramamoorthy *et al* 2006), and given the typical parameters used in an MIC assay (Wiegand *et al* 2008), the percentage range tested by Pius *et al* is well below the concentration required to kill the bacteria. Interestingly, however, in order to determine how MSI-78 affects lipid chain order, much higher peptide concentrations would be required to see a measurable effect. More recently, this approach was extended to the Gram positive bacterium *B. subtilis* (Santisteban *et al* 2020). Intriguingly, the effect of MSI-78 appears to be more pronounced on the $^{2}$H spectra of Gram positive *B. subtilis* than Gram negative *E. coli*, even though the activities are comparable (Ramamoorthy *et al* 2006). The spectra described above were acquired without MAS, but of course, this can easily be implemented as well and has been shown to yield important information on the mechanism of action of HDPs such as polymyxin B (Tardy-Laporte *et al* 2013), as well as caerin 1.1 and aurein 1.2 (Laadhari *et al* 2016).

Of course, it is also possible to exploit the natural abundance $^{31}$P NMR signal arising from the lipids and the bacterial DNA or RNA in intact cells. Recently, Overall *et al* (Overall *et al* 2019) demonstrated that the HDP maculatin 1.1 might function by affecting nucleic acid packing (figure 5.1—middle) in *E. coli*. This was shown by examining the signal arising from DNA and RNA in static $^{31}$P cross-polarization (CP) NMR spectra, as well as those obtained by direct detection. Detailed analysis of the effect of the HDP on phospholipid resonances was difficult due to overlap in both the CP and direct excitation spectra.

The examples given above examine the effect of the HDP from the perspective of the bacterial cell. To characterize the interaction from the angle of the HDP, it is necessary to introduce isotopic labels. As discussed in section 5.3, a variety of individual labels can be introduced, and this is usually easily done during peptide synthesis. Single labelling of HDPs can then be used, for example, to determine the location of an HDP in bacteria, using MAS experiments (Sani and Separovic 2018). Using fully $^{15}$N-labelled (or $^{15}$N and $^{13}$C-labelled) HDPs would also be beneficial. Due to cost considerations, such extensive isotopic labelling is usually achieved by recombinant expression. For HDPs, this is a challenge, since HDPs function by killing bacteria (Sampaio de Oliveira *et al* 2020). However, recent work has shown that specific recombinant expression systems can be utilized to produce isotopically labeled HDPs (Vidovic *et al* 2009, Ishida *et al* 2016, Meng *et al* 2019, Azari *et al* 2020). Vidovic *et al* used a construct in which HDP LAH4 was fused to the histone fold domain of the human transcription factor TAF12. Interestingly, this fusion construct was expressed within insoluble inclusion bodies and did not require any purification: it was cleaved chemically by formic acid due to the unique Asp-Pro bond in the vector. The approach used by Ishida *et al* involved the use of a calmodulin fusion construct, with a Tobacco Etch Virus (TEV) cleavage site. The authors showed that many HDPs such as indolicidin and magainin could be expressed using this fusion construct, with calmodulin effectively binding to the HDP and shielding it from the bacteria. The resulting HDPs were produced in milligram quantities and had a carboxy C-terminus, as is expected from expressed peptides and proteins.

Recently, the calmodulin construct (Ishida *et al* 2016) was used to express fully $^{15}$N-labelled aurein 2.2 (figure 5.2(a)). Unsurprisingly, the purified product, i.e., aurein 2.2 with a carboxy C-terminus (Pan *et al* 2007), was inactive against *S. aureus* because unlike many other HDPs (with high positive charge of +4 or more), aurein 2.2 only has a net +2 charge. In other words, introduction of a carboxy C-terminus (as opposed to the naturally occurring amidated C-terminus) has a big impact on activity, as it reduces the charge to +1. To recover the activity of the expressed aurein 2.2-COOH (with an MIC > 250 μg/ml), selective esterification was carried out, using an established protocol (Li and Sha 2008).

The resulting diesterification of aurein 2.2 partially recovered the antimicrobial activity of the expressed peptide with an MIC of 64 μg mL$^{-1}$ compared to the amidated aurein 2.2 (32 μg mL$^{-1}$) obtained by peptide synthesis. MS/MS analysis confirmed the addition of a $CH_3$ moiety at the C-terminus, but also showed addition of a $CH_3$ group at the aspartic acid sidechain at position 4. It was recently suggested that the aspartic acid may be involved in binding selective ions, thereby allowing aurein 2.2. to form ion selective pores in *B. subtilis* (Wenzel *et al* 2015). Consequently, a modification of this residue by esterification is likely to result in a less active HDP. This was further tested in a proof-of-concept experiment, where expressed aurein 2.2 and diesterified aurein 2.2 were mixed with ca. $10^6$ colony forming units (CFU) per mL of *S. aureus* cells and liquid state NMR HSQC spectra were recorded as a function of time (figure 5.2(b)). Expressed aurein 2.2 (aurein 2.2-COOH), which is inactive against *S. aureus*, does not show any differences after 1, 2 and 3 h of incubation, whereas in the spectra of diesterified aurein 2.2 the peak intensity decreases with respect to time. This could be explained by the fact that diesterified aurein 2.2 inserts into the membrane of *S. aureus* which leads to slower tumbling of the peptide in solution and faster relaxation of transverse magnetization (short $T_2$) resulting in line broadening (Foster *et al* 2007). The peaks seen at the end of 3 h when all the bacteria were killed (figure 5.2(b)—SEM inset) is likely from the unbound peptide that is freely tumbling in solution. In contrast, the aurein 2.2-COOH spectra does not show any significant decrease in peak intensity indicating the peptides do not bind and insert into the membrane (leaving the *S. aureus* intact, figure 5.2(b)-inset). This approach could easily be extended to other solid-state NMR experiments in the future, thereby allowing various aspects of HDP function to be probed.

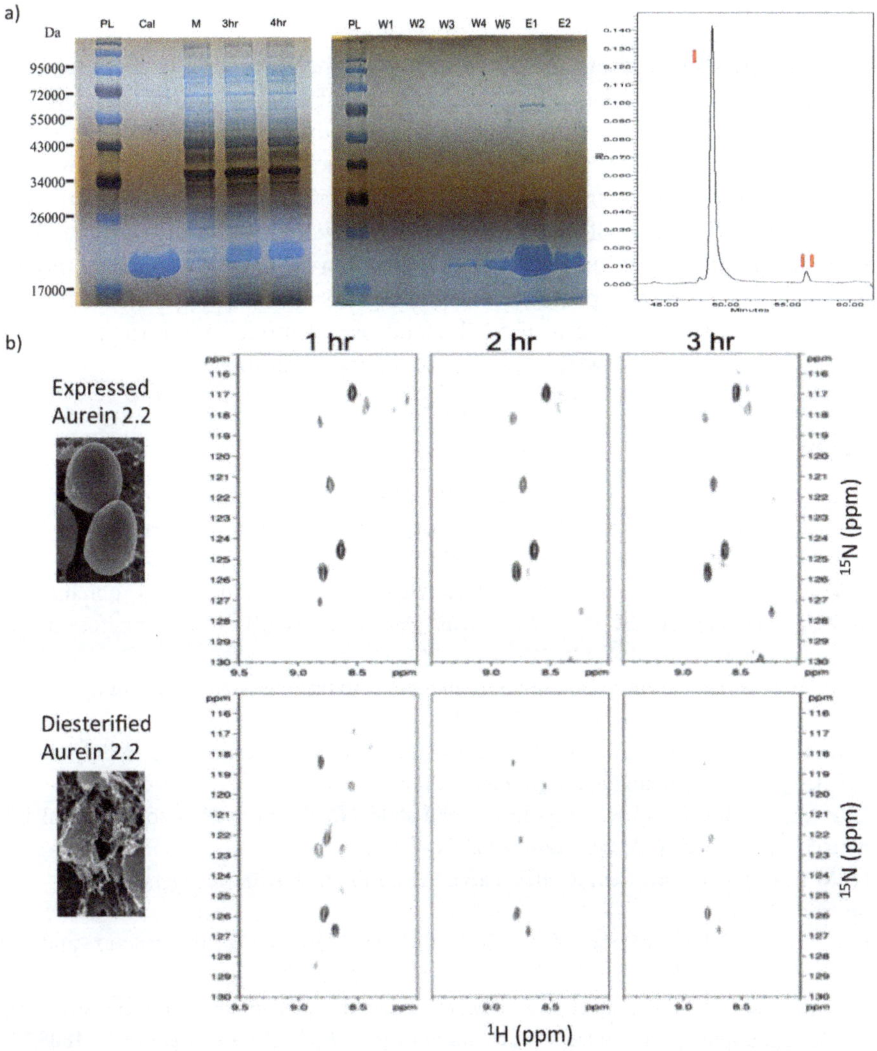


**Figure 5.2.** (a) Expression and purification of the calmodulin-aurein 2.2. SDS-PAGE (left) from *E. coli* expression taken at different times before and after IPTG induction (PL: protein ladder, Cal: commercial calmodulin, M: before IPTG induction, 3 h and 4 h: after induction). SDS-PAGE (middle) of Ni-NTA column purification. After the sample was loaded, washes (W1 to W5) of the Ni-NTA resin were collected until the absorbance measured at 280 nm was similar to the buffer only (~0.03). Elution samples (E1 and E2) were collected using 400 mM elution buffer. HPLC trace (right) of calmodulin-aurein 2.2 after cleavage with TEV. The small peak marked II corresponds to aurein 2.2, while the larger peak (I) corresponds to calmodulin. (b) $^{1}H,^{15}N$-HSQC spectra of whole *S. aureus* cells with the $^{15}N$-labelled aurein peptides. Aurein 2.2-COOH (top) and diesterified aurein 2.2 (bottom) were mixed with *S. aureus* cells. The spectra were acquired after 1, 2 and 3 h of incubation using standard parameters. The effect of these expressed peptides on *S. aureus* was further confirmed by scanning electron microscopy (SEM) (insets).

## 5.4 Conclusions

Host defense peptides are unique molecules that are promising candidates to treat multidrug resistant organisms. Interestingly, HDPs have many mechanisms of action and can also display antiviral and anti-cancer activities (Kumar *et al* 2018). In order to develop HDPs as potential therapeutics to combat the emergence of antibiotic resistance, it is imperative to characterize their mechanism of action. Solid-state NMR is one of the most powerful techniques that can be used to elucidate the detailed structural interactions between HDPs with potential molecular targets such as membranes, proteins and nucleotides. Various approaches have been presented here illustrating the well-established use of model membranes, as well as the promising whole cell NMR experiments being developed.

## References

Aisenbrey C, Marquette A and Bechinger B 2019 The mechanisms of action of cationic antimicrobial peptides refined by novel concepts from biophysical investigations *Advances in Experimental Medicine and Biology* (New York: Springer) *LLC* 33–64

Azari M, Asad S and Mehrnia M R 2020 Heterologous production of porcine derived antimicrobial peptide PR-39 in *Escherichia coli* using SUMO and intein fusion systems *Protein Express. Purif.* **169** 105568

Bowdish D M E *et al* 2005 Immunomodulatory activities of small host defense peptides *Antimicrob. Agents Chemother.* **49** 1727–32

Bowdish D M E, Davidson D J and Hancock R E W 2005 A re-evaluation of the role of host defence peptides in mammalian immunity *Curr. Protein Pept. Sci.* **6** 35–51

Cegelski L 2015 Bottom-up and top-down solid-state NMR approaches for bacterial biofilm matrix composition *J. Magn. Reson.* **253** 91–7

Chellat M F, Raguž L and Riedl R 2016 Targeting antibiotic resistance *Angew. Chem.—Int. Ed.* **55** 6600–26

Cheng J T J *et al* 2010 Importance of residue 13 and the C-terminus for the structure and activity of the antimicrobial peptide aurein 2.2 *Biophys. J.* **99** P2926–35

Cheng J T J *et al* 2011 The importance of bacterial membrane composition in the structure and function of aurein 2.2 and selected variants *Biochim. Biophys. Acta—Biomembr.* **1808** 622–33

Chernysh S *et al* 2002 Antiviral and antitumor peptides from insects. *Proc. Natl Acad. Sci. USA.* **99** 12628–32

Dave P C *et al* 2005 Interaction of alamethicin with ether-linked phospholipid bilayers: oriented circular dichroism, $^{31}$P solid-state NMR, and differential scanning calorimetry studies *Biophys. J.* **89** P2434–42

Davies J 2006 Where have all the antibiotics gone? *Can. J. Infect. Dis. Med. Microbiol.* **17** 287–90

de la Fuente-Núñez C *et al* 2014 Broad-spectrum anti-biofilm peptide that targets a cellular stress response *PLoS Pathog.* **10** e1004152

de la Fuente-Núñez C *et al* 2016 Synthetic antibiofilm peptides *Biochim. Biophy. Acta—Biomembr.* **1858** 1061–9

de Leeuw E *et al* 2010 Functional interaction of human neutrophil peptide-1 with the cell wall precursor lipid II *FEBS Lett.* **584** 1543–8

Felício M R *et al* 2017 Peptides with dual antimicrobial and anticancer activities. *Front. Chem.* **5** 5

Fernandez D I *et al* 2012 The antimicrobial peptide aurein 1.2 disrupts model membranes via the carpet mechanism *Phys. Chem. Chem. Phys.* **14** 15739–51

Fernandez D I *et al* 2013 Membrane defects enhance the interaction of antimicrobial peptides, aurein 1.2 versus caerin 1.1 *Biochim. Biophys. Acta—Biomembr.* **1828** 1863–72

Foster M P, McElroy C A and Amero C D 2007 Solution NMR of large molecules and assemblies *Biochemistry* **46** 331–40

Glaser R W *et al* 2004 Orientation of the antimicrobial peptide PGLa in lipid membranes determined from $^{19}$F-NMR dipolar couplings of 4-$CF_3$-phenylglycine labels *J. Magn. Reson.* **168** 153–63

Grage S L *et al* 2014 $^{19}$F-labeling of peptides revealing long-range NMR distances in fluid membranes *J. Phys. Chem. Lett.* **5** 4256–9

Hale J D and Hancock R E 2007 Alternative mechanisms of action of cationic antimicrobial peptides on bacteria *Expert. Rev. Anti. Infect. Ther.* **5** 951–9

Hancock R E W 2005 Mechanisms of action of newer antibiotics for Gram-positive pathogens *Lancet Inf. Dis.* **5** 209–18

Hancock R E W and Falla T J 1996 Antimicrobial peptides: broad-spectrum antibiotics from nature *Clin. Microbiol. Infect.* **1** 226–9

Hancock R E W and Sahl H-G 2006 Antimicrobial and host-defense peptides as new anti-infective therapeutic strategies *Nat. Biotechnol.* **24** 1551–7

Haney E F *et al* 2013 Mechanism of action of puroindoline derived tryptophan-rich antimicrobial peptides *Biochim. Biophys. Acta—Biomembr.* **1828** 1802–13

Haney E F and Hancock R E W 2013 Peptide design for antimicrobial and immunomodulatory applications. *Biopolymers* **100** 572–83

Haney E F, Straus S K and Hancock R E W 2019 Reassessing the host defense peptide landscape *Front. Chem.* **7** 43

Harmouche N *et al* 2017 Solution and solid-state nuclear magnetic resonance structural investigations of the antimicrobial designer peptide GL13K in membranes *Biochemistry* **56** 4269–78

Harmouche N and Bechinger B 2018 Lipid-mediated interactions between the antimicrobial peptides Magainin 2 and PGLa in bilayers *Biophys. J.* **115** 1033–44

Harzer U and Bechinger B 2000 Alignment of lysine-anchored membrane peptides under conditions of hydrophobic mismatch: a CD, $^{15}$N and $^{31}$P solid-state NMR spectroscopy investigation *Biochemistry* **39** 13106–14

Hoskin D W and Ramamoorthy A 2008 Studies on anticancer activities of antimicrobial peptides *Biochim. Biophys. Acta* **1778** 357–75

Ishida H *et al* 2016 Overexpression of antimicrobial, anticancer, and transmembrane peptides in *Escherichia coli* through a calmodulin-peptide fusion system *JACS* **138** 11318–26

Jefferson K K 2004 What drives bacteria to produce a biofilm? *FEMS Microbiol. Lett.* **236** 163–73

Jenssen H, Hamill P and Hancock R E W 2006 Peptide antimicrobial agents *Clin. Microbiol. Rev.* **19** 491–511

Jeong J H *et al* 2016 NMR structural studies of antimicrobial peptides: LPcin analogs *Biophys. J.* **110** 423–30

Kumar P, Kizhakkedathu J N and Straus S K 2018 Antimicrobial peptides: diversity, mechanism of action and strategies to improve the activity and biocompatibility *in vivo Biomolecules* **8** 4

Kumar T V V and Sanil G 2017 A review of the mechanism of action of amphibian antimicrobial peptides focusing on peptide-membrane interaction and membrane curvature *Curr. Protein Pept. Sci.* **18** 1263–73

Laadhari M *et al* 2016 Interaction of the antimicrobial peptides caerin 1.1 and aurein 1.2 with intact bacteria by $^2$H solid-state NMR *Biochim. Biophys. Acta—Biomembr.* **1858** 2959–64

Lanne A B M *et al* 2019 Molecular recognition of lipopolysaccharide by the lantibiotic nisin *Biochim. Biophys. Acta—Biomembr.* **1861** 83–92

Lee D K *et al* 2015 Detergent-type membrane fragmentation by MSI-78, MSI-367, MSI-594, and MSI-843 antimicrobial peptides and inhibition by cholesterol: a solid-state nuclear magnetic resonance study *Biochemistry* **54** 1897–907

Lee T H *et al* 2019 The role of bacterial lipid diversity and membrane properties in modulating antimicrobial peptide activity and drug resistance *Curr. Opin. Chem. Biol* **52** 85–92

Li J and Sha Y 2008 A convenient synthesis of amino acid methyl esters *Molecules* **13** 1111–9

Mani R *et al* 2006 Membrane-dependent oligomeric structure and pore formation of a β-hairpin antimicrobial peptide in lipid bilayers from solid-state NMR *Proc. Natl Acad. Sci. USA* **103** 16242–7

Manzo G *et al* 2019 Minor sequence modifications in temporin B cause drastic changes in antibacterial potency and selectivity by fundamentally altering membrane activity *Sci. Rep.* **9** 1385

Marcotte I *et al* 2003 Interaction of antimicrobial peptides from Australian amphibians with lipid membranes *Chem. Phys. Lipids* **122** 107–20

Matsuzaki K *et al* 1998 Relationship of membrane curvature to the formation of pores by Magainin 2 *Biochemistry* **37** 11856–63

Meng D M *et al* 2019 Expression, purification and characterization of a recombinant antimicrobial peptide Hispidalin in *Pichia pastoris Protein Expres. Purif.* **160** 19–27

Nguyen L T, Haney E F and Vogel H J 2011 The expanding scope of antimicrobial peptide structures and their modes of action *Trends Biotechnol.* **29** 464–72

Nuri R, Shprung T and Shai Y 2015 Defensive remodeling: how bacterial surface properties and biofilm formation promote resistance to antimicrobial peptides *Biochim. Biophys. Acta—Biomembr.* **1848** 3089–100

Nygaard R *et al* 2015 Spectral snapshots of bacterial cell-wall composition and the influence of antibiotics by whole-cell NMR *Biophys. J.* **108** 1380–9

Nygaard R *et al* 2017 Whole ribosome NMR: dipolar couplings and contributions to whole cells *J. Phys. Chem.* B **121** 9331–5

Opella S J 2014 The development of solid-state NMR of membrane proteins *Biomed. Spectrosc. Imaging* **3** 81–105

Overall S A *et al* 2019 In situ monitoring of bacteria under antimicrobial stress using $^{31}$P solid-state NMR *Int. J. Mol. Sci.* **20** 181

Pan Y-L *et al* 2007 Characterization of the structure and membrane interaction of the antimicrobial peptides aurein 2.2 and 2.3 from Australian southern bell frogs *Biophys. J.* **92** P2854–64

Paredes S D *et al* 2020 Enhancing the membrane activity of Piscidin 1 through peptide metallation and the presence of oxidized lipid species: implications for the unification of host defense mechanisms at lipid membranes *Biochim. Biophys. Acta—Biomembr.* **1862** 183236

Park K H *et al* 2009 Cell specificity, anti-inflammatory activity, and plausible bactericidal mechanism of designed Trp-rich model antimicrobial peptides *Biochim. Biophys. Acta—Biomembr.* **1788** 1193–203

Pius J, Morrow M R and Booth V 2012 $^2$H solid-state nuclear magnetic resonance investigation of whole escherichia coli interacting with antimicrobial peptide MSI-78 *Biochemistry* **51** 118–25

Porcelli F *et al* 2006 Structures of the dimeric and monomeric variants of magainin antimicrobial peptides (MSI-78 and MSI-594) in micelles and bilayers, determined by NMR spectroscopy *Biochemistry* **45** 5793–9

Porcelli F *et al* 2013 On the role of NMR spectroscopy for characterization of antimicrobial peptides *Methods Mol. Biol.* **1063** 159–80

Raheem N and Straus S K 2019 Mechanisms of action for antimicrobial peptides with antibacterial and antibiofilm functions *Front. Microbiol.* **10** 1–14

Rai R K *et al* 2019 Metal-ion binding to host defense peptide piscidin 3 observed in phospholipid bilayers by magic angle spinning solid-state NMR *ChemPhysChem* **20** 295–301

Ramamoorthy A *et al* 2006 Solid-state NMR investigation of the membrane-disrupting mechanism of antimicrobial peptides MSI-78 and MSI-594 derived from magainin 2 and melittin *Biophys. J.* **91** 206–16

Sampaio de Oliveira K B *et al* 2020 Strategies for recombinant production of antimicrobial peptides with pharmacological potential *Exp. Rev. Clin. Pharmacol.* **13** 367–90

Sani M A and Separovic F 2018 Antimicrobial peptide structures: from model membranes to live cells *Chem.—A Eur. J.* **24** 286–91

Santisteban N P, Morrow M R and Booth V 2020 Effect of AMPs MSI-78 and BP100 on the lipid acyl chains of $^2$H-labeled intact Gram positive bacteria *Biochim. Biophys. Acta—Biomembr.* **1862** 183199

Schmitt P *et al* 2010 Insight into invertebrate defensin mechanism of action: oyster defensins inhibit peptidoglycan biosynthesis by binding to lipid II *J. Biol. Chem.* **285** 29208–16

Sinha S *et al* 2017 Structure and interactions of a host defense antimicrobial peptide thanatin in lipopolysaccharide micelles reveal mechanism of bacterial cell agglutination *Sci. Rep.* **7** 1–13

Stanton T B 2013 A call for antibiotic alternatives research *Trends Microbiol.* **21** 111–3

Strandberg E *et al* 2020 Chiral supramolecular architecture of stable transmembrane pores formed by an α-helical antibiotic peptide in the presence of lyso-lipids *Sci. Rep.* **10** 4710

Straus S K 2004 Recent developments in solid-state magic-angle spinning, nuclear magnetic resonance of fully and significantly isotopically labelled peptides and proteins *Philos. Trans. R Soc. Lond. B Biol. Sci.* **359** 1446

Tardy-Laporte C *et al* 2013 A 2 H solid-state NMR study of the effect of antimicrobial agents on intact *Escherichia coli* without mutating *BBA—Biomembr.* **1828** 614–22

Toke O, Cegelski L and Schaefer J 2006 Peptide antibiotics in action: investigation of polypeptide chains in insoluble environments by rotational-echo double resonance *Biochim. Biophys. Acta—Biomembr* **1758** 1314–29 .

Travkova O G, Moehwald H and Brezesinski G 2017 The interaction of antimicrobial peptides with membranes *Adv. Colloid Interface Sci.* **247** 521–32

Tremouilhac P *et al* 2006 Synergistic transmembrane alignment of the antimicrobial heterodimer PGLa/magainin *J. Biol. Chem.* **281** 32089–94

Ulm H *et al* 2012 Antimicrobial host defensins—specific antibiotic activities and innate defense modulation. *Front. Immunol.* **3** 249

Vidovic V *et al* 2009 Production and isotope labeling of antimicrobial peptides in *Escherichia coli* by means of a novel fusion partner that enables high-yield insoluble expression and fast purification *J. Pept. Sci.* **15** 278–84

Vogt T C B and Bechinger B 1999 The interactions of histidine-containing amphipathic helical peptide antibiotics with lipid bilayers. The effects of charges and pH *J. Biol. Chem.* **274** 29115–21

Wenzel M *et al* 2012 Proteomic response of *Bacillus subtilis* to lantibiotics reflects differences in interaction with the cytoplasmic membrane *Antimicrob. Agents Chemother.* **56** 5749–57

Wenzel M *et al* 2015 Antimicrobial peptides from the aurein family form ion-selective pores in *Bacillus subtilis ChemBioChem* **16** 1101–8

Wi S and Kim C 2008 Pore structure, thinning effect, and lateral diffusive dynamics of oriented lipid membranes interacting with antimicrobial peptide protegrin-1: $^{31}P$ and $^{2}H$ solid-state NMR study *J. Phys. Chem.* B **112** 11402–14

Wiegand I, Hilpert K and Hancock R E W 2008 Agar and broth dilution methods to determine the minimal inhibitory concentration (MIC) of antimicrobial substances *Nat. Protoc.* **3** 163–75

Yasir M, Dutta D and Willcox M D P 2019 Mode of action of the antimicrobial peptide $Me_{14}$ is independent of *Staphylococcus aureus* cell membrane permeability *PLoS One* **14** e0215703

Yasir M, Willcox M D P and Dutta D 2018 Action of antimicrobial peptides against bacterial biofilms *Materials* **11** 2468

Zamora-Carreras H *et al* 2016 Alanine scan and $^{2}H$ NMR analysis of the membrane-active peptide BP100 point to a distinct carpet mechanism of action *Biochim. Biophys. Acta—Biomembr.* **1858** 1328–38

**IOP** Publishing

Solid-State NMR
Applications in biomembrane structure
**Frances Separovic and Marc-Antoine Sani**

# Chapter 6

# $^{19}$F NMR of biomembranes

**Stephan L Grage, Sergii Afonin and Anne S Ulrich**

## Abstract

In this chapter we give an overview of the use of $^{19}$F NMR for the study of biomembranes. Due to its unique sensitivity and small size, fluorine may be considered to bridge the gap between NMR on the one hand, and ESR and fluorescence on the other hand. Conventional NMR isotope labels do not alter the studied molecules but possess only low sensitivity, while ESR and fluorescence labels provide higher sensitivity but also represent a larger modification of the system of interest. The high sensitivity to its environment, large anisotropies and strong dipolar couplings render $^{19}$F a useful NMR label to probe structure and dynamics in biomembranes, and to study intermolecular interactions. Fluorine has been introduced into numerous constituents of biomembranes, such as lipids, sterols as well as membrane associated peptides and integral membrane proteins. In particular, a growing set of fluorine-labeled amino acids has allowed one to address selected structural information in membrane proteins and peptides. The large anisotropy of $^{19}$F NMR interactions in the solid-state has been used to measure molecular orientations with respect to the membrane normal, allowing characterization, for example, of the function of membrane-active peptides. Distance measurements exploring the large distance range of $^{19}$F have been beneficial in the study of intermolecular interactions, to reveal oligomeric architectures of membrane proteins and/or transmembrane peptides. The sensitivity of the $^{19}$F chemical shift to the environment makes it possible to probe the molecular environment and detect even solvent isotope effects, this way revealing conformational changes and solvent/lipid accessibilities. Given the constraints in placing fluorine into biomembranes and embedded molecules, $^{19}$F NMR may not be a universal technique, but it certainly provides a powerful tool to answer specific questions regarding the structure and dynamics in biomembranes that would be difficult to solve otherwise.

doi:10.1088/978-0-7503-2532-5ch6  

## 6.1 Introduction

Fluorine plays a special role in NMR studies of biomembranes. Its favorable NMR properties have been used early on. Already in the 1970s, at the time when the first $^{1}$H NMR structural studies of proteins were on their way, fluorine had been introduced into lipids and membrane proteins to gain specific structural and biophysical insight. In their pioneering research, Gent and Ho, and later Post *et al*, plus the groups of Sykes and McElhaney, used fluorine-labeled phospholipids to explore the order parameter in model membranes and even in bacteria (Engelsberg *et al* 1982, Gent *et al* 1978, McDonald *et al* 1983, Post *et al* 1981). Membrane proteins were also addressed already in the 1970s and 1980s by $^{19}$F NMR. Fluorine labeled tyrosine or tryptophan has been introduced via biosynthesis into proteins such as M13 coat protein and LDH (Hagen *et al* 1978, Ho *et al* 1986, Wilson and Dahlquist 1985). Arseniev used $^{19}$F NMR to explore bacteriorhodopsin in 1987 (Arseniev *et al* 1987). Also, a first hint for the head-to-head dimer structure of gramicidin A came from $^{19}$F NMR in 1979 (Weinstein *et al* 1979).

Today, $^{19}$F NMR has developed into a versatile tool in membrane research, which, due to its use of a few specific labels, targets local structural or dynamic information rather than overall protein structures. Several approaches have been applied: (1) using the large anisotropies inherent in the NMR interactions of fluorine, $^{19}$F NMR has been particularly powerful in determining the orientation of molecules in the lipid bilayer. For example, using uniformly stacked membrane samples, the alignment of antimicrobial peptides and cyclotides in the lipid bilayer has been determined from orientational constraints obtained by solid-state $^{19}$F NMR (Afonin *et al* 2008a, Glaser *et al* 2004, Glaser *et al* 2005, Grage *et al* 2017, Salgado *et al* 2001, Strandberg *et al* 2006, Strandberg and Ulrich 2018, Ulrich 2007). (2) Due to the strong dipolar couplings of $^{19}$F with other nuclei, $^{19}$F has often been the nucleus of choice in distance measurements by $^{19}$F NMR. This way, similar to solution-state NMR NOEs, solid-state NMR MAS techniques such as REDOR, CODEX, and rotational resonance have provided distance constraints in peptides and membrane proteins. Important structural clues have been obtained this way, for example $^{19}$F CODEX was instrumental in exploring the oligomeric structure of the influenza A proton ion channel M2 in membranes (Luo and Hong 2006). Most of these solid-state NMR distance measurements are applied to frozen samples and rely on recoupling under MAS. In a different manner, the CPMG experiment on oriented samples at ambient temperatures is particularly suited to address inter-nuclear distances within the mobile environment of fluid membranes. This way, even distances beyond 10 Å have been detected in antimicrobial peptides bound to fully hydrated liquid-crystalline membranes (Grage *et al* 2014). (3) The $^{19}$F chemical shift demonstrates a particularly high dispersion and sensitivity to the environment. This property has been utilized already in early applications of $^{19}$F NMR on membrane proteins (Arseniev *et al* 1987, Ho *et al* 1986), and refined in remarkable applications recently. For example, by exploring the small difference of $^{19}$F chemical shift of a label exposed to $H_2O$ or $D_2O$, it was possible to study the accessibility of membrane protein surfaces to water (Didenko *et al* 2013, Hou *et al* 2017).

In this book chapter, we present an overview of the use of $^{19}$F NMR to the study of biomembranes. The advantages and limitations of $^{19}$F NMR will be discussed in the context of biomembranes, i.e., how fluorine can be introduced into these biological systems, which techniques are useful in a membrane or membrane-mimetic environment, and what kind of questions are typically answered this way.

## 6.2 $^{19}$F NMR

### 6.2.1 Advantages of $^{19}$F as NMR probe

The gyromagnetic ratio of $^{19}$F is the second highest of all stable isotopes, and only 6% lower than that of $^{1}$H. As a result, $^{19}$F NMR possesses—for NMR circumstances—a particularly high sensitivity, plus large dipolar couplings. In practice, the sensitivity of $^{19}$F can be reduced by relatively short $T_2$ relaxation times and the associated line broadening. Nonetheless, it is usually significantly higher than that of other nuclei that are typically employed such as $^{13}$C or $^{15}$N. The signal/noise ratio scales as

$$S/N \sim n\,\gamma_e I(I+1)(\gamma_d B_0/T)^{3/2}(t\;T_2/T_1)^{1/2},$$

with the number of spins $n$, the gyromagnetic ratios $\gamma_e$ and $\gamma_d$ of the excited and detected spin, the spin number of the excited spin $I$, the magnetic field $B_0$, the temperature $T$, the longitudinal relaxation time of the excited spin $T_1$, and the transversal relaxation time of the detected spin $T_2$, as well as the measurement time $t$ (Cavanagh *et al* 1996). Not considering differences in relaxation times and assuming 100% isotope labeling, $^{19}$F thus possesses a sensitivity which is 86% of $^{1}$H, 27 times higher than that of $^{13}$C, and 263 times higher than for $^{15}$N. Considering the typically used cross-polarization from $^{1}$H to enhance the sensitivity of $^{13}$C and $^{15}$N, there is still a 6.8- and 26.6-fold advantage in signal/noise for $^{19}$F compared to $^{13}$C and $^{15}$N, respectively. Translated into measurement time needed to acquire a spectrum of the same number of spins, a $^{13}$C or $^{15}$N experiment would last ~50 or ~700 times longer. Even assuming a ~10 fold faster $T_2$ relaxation for $^{19}$F, its sensitivity would still be higher than that of $^{13}$C or $^{15}$N by a factor of ~2 or ~8, respectively, leading to 4 or 64 fold shorter measurement times. Hence, by using $^{19}$F NMR, much lower sample amounts in the ~10–100 nmol range become accessible, thus extending the sensitivity limit of NMR considerably.

The distance range accessible by $^{19}$F can reach up to ~15–18 Å for $^{19}$F-$^{19}$F or $^{19}$F-$^{1}$H couplings (Wi *et al* 2004, Hong and Schmidt-Rohr 2013, Grage *et al* 2014), hence it is not surprising to find distance measurements among the major applications of $^{19}$F NMR.

Also, the large chemical shift dispersion and high sensitivity of $^{19}$F chemical shift to the molecular environment of the label adds to the advantages of $^{19}$F. In particular, the dominating contribution of the chemical shift from the lone electron pairs renders $^{19}$F rather susceptible to van der Waals interactions with its surroundings (Danielson and Falke 1996). In organic compounds, $^{19}$F chemical shifts span a range from about 25 ppm (some -CO-F) to −275 ppm (MeF), and an even larger range considering more exotic compounds. In particular the strong dependence on the environment has been useful, for example, in distinguishing multiple labels of the same amino acid type in membrane proteins. For example, all eight 5F-tryptophan residues introduced into bacteriorhodopsin showed distinct chemical shifts over a range of ~1 ppm (Arseniev *et al* 1987), and similarly, five or seven 5F-tryptophan labels were resolved in the pore-forming toxins equinatoxin II and listeriolysin O (Anderluh *et al* 2005, Kozorog *et al* 2018). Even exposure to $H_2O$ and $D_2O$ leads to appreciable chemical shift differences of ~0.15 ppm (Hou *et al* 2017). This wide dispersion of chemical shifts renders $^{19}$F one of the most sensitive nuclei for probing the chemical environment.

Furthermore, $^{19}$F possesses particularly large chemical shift anisotropies (CSA), which present a further source of structural information such as molecular or segmental orientation. The large CSA of fluorine in aromatic compounds of 50–60 ppm (Dürr *et al* 2008), for example, has been used to determine the orientation of tryptophan sidechains in gramicidin A (Grage *et al* 2002, Sternberg *et al* 2009). A further practical advantage of $^{19}$F in biomembrane applications is the lack of fluorine in native biological systems. Hence, other than, e.g., as in $^{13}$C NMR, no background signals have to be considered. Also, due to its natural abundance of 100%, labeling with $^{19}$F does not require any enrichment of the desired isotope, as is the case in $^{13}$C, $^{2}$H, $^{15}$N or $^{17}$O NMR.

### 6.2.2 Limitations of $^{19}$F NMR

A major concern of $^{19}$F NMR of biological systems like membranes is the incorporation of this rather non-biological element into biomolecules. As mentioned above, the absence of fluorine in most of the living world allows background-free measurements of only the labeled sites, but it also requires modifications of the studied molecules in order to introduce fluorine. Due to its high electronegativity, fluorine-labeling imposes a significant change of some physical properties, such as the polarity, around the labeled position. The electron-withdrawing power of fluorine, for example, can influence the electronic structure of the remaining molecule. This impact may be more complex than simply described by an increase in hydrophobicity, and it may even lead to a third type of polarity besides hydrophobic and hydrophilic, namely fluorophilic (Cametti *et al* 2012, Marsh 2000, Yoder and Kumar 2002).

Furthermore, the electronegativity imposes restrictions on the chemistry, limiting the sites which can be labeled or substituted. For example, $^{19}$F-labeling in the polypeptide backbone has been achieved via an amide bond isostere, but the folding of the model peptides was compromised. Therefore, in most cases fluorine is introduced via bulky hydrophobic sidechains. 5F-tryptophan and 4F-phenylalanine can even be incorporated recombinantly into proteins, as they are recognized by the corresponding tRNA synthethase. Nonetheless, some creative chemistry has led to recent tailor-made labeling approaches, and specific artificial amino acids have been designed for conservative substitutions of polar or hydrophobic residues, either with rigid or flexible linkage of the fluorine to the peptide backbone (Komarov *et al* 2019).

Even the physical properties of the labeled molecule can be altered by fluorine substitutions. For example, a single fluorine-labeled methylene segment in a phospholipid reduces the acyl chain order parameter significantly by 30% (Oldfield *et al* 1980). When introduced into the terminal methyl group, fluorine labeled lipids can even interdigitate instead of forming a tail-to-tail bilayer (Hirsh *et al* 1998, Sanii *et al* 2010). Hence, each labeling strategy with even single fluorine substitutions needs careful consideration of the accompanying changes in chemistry, as well as in the physical and biological properties of the obtained analogue.

The relaxation behavior of $^{19}$F presents a further limitation of $^{19}$F NMR. In particular, the fast $T_2$ relaxation leads to considerable line broadening and partially reduces the high sensitivity advantage of $^{19}$F. Typical $T_2$-relaxation times of $^{19}$F-labeled peptides in biomembranes, for example, are on the 1–10 ms timescale, leading to rather broad lines of often several 100 Hz, even under line-narrowing conditions such as magic angle spinning (MAS). Also, applications relying on the persistence of transversal magnetization are compromised by the fast $T_2$ relaxation. For example, one of the most frequently applied methods for the determination of internuclear distances in solid-state NMR, the REDOR experiment, is particularly susceptible to such short $T_2$ relaxation. In this experiment, which is used to probe the coupling between nuclei of different types (such as $^{13}$C and $^{19}$F), the signal of one nucleus is observed after a time period, during which this nucleus is exposed to the dipolar coupling (introduced by RF pulses at the Larmor frequency of the second nucleus), but also suffers from $T_2$ relaxation. Therefore, in the vast majority, if not all, applications of REDOR to measure distances between $^{19}$F and another nucleus, $^{19}$F has not been chosen as the observed nucleus, despite possessing the higher sensitivity (e.g. in a pair of $^{13}$C and $^{19}$F).

Another technical concern, in particular in solid-state $^{19}$F NMR, is presented by the proximity of the Larmor frequencies of $^{1}$H and $^{19}$F. Other than in membrane mimetics such as detergent micelles, the molecular mobility in lipid membranes is not high enough to average dipolar couplings to zero. Hence in most applications on biomembranes, proton decoupling is necessary to reduce the line broadening from $^{1}$H–$^{19}$F dipolar couplings. Due to the close Larmor frequencies, the necessary double-resonant probes need special electronic circuits and high-quality RF filters are required to perform $^{1}$H-decoupled $^{19}$F NMR experiments. In practice, arcing is often observed during decoupling, limiting the accessible decoupling strengths, and making it difficult to achieve similar decoupling performance as for $^{13}$C or $^{15}$N.

### 6.2.3 $^{19}$F NMR bridges the gap

Overall, a $^{19}$F substituent or a $CF_3$ group can be seen as a compromise between the very bulky labels used for fluorescence or EPR spectroscopy on one hand, and the entirely non-perturbing isotope labels that are conventionally used in NMR, such as $^{13}$C or $^{15}$N, on the other hand. Fluorescence dyes and electron spin labels are frequently used in biomembranes because of their high sensitivity. They allow one to examine the exposure of molecular segments to lipidic or aqueous environments, to measure molecular mobilities, and to obtain structural parameters such as intermolecular distances or orientations. However, due to their considerable size (figure 6.1) they constitute a significant change of the studied molecules, and are certainly able to modify their behavior. For example, fluorescein has a molecular weight of 332 Da, which is about half the weight of a lipid molecule, and ~15% of the molecular weight of a typical antimicrobial peptide. This fluorescence label adds considerably to the mass of many molecules of interest, and will change the physical properties, like polarity, of the labeled molecule. Likewise, electron spin labels have to be incorporated into a suitable molecular scaffold like TEMPO, and this substituent cannot be arbitrarily placed. On the other hand, EPR can detect nanomolar amounts, and large distances of up to 4 or 10 nm can be reached using EPR and fluorescence spectroscopy, respectively (King *et al* 2014, Yang *et al* 2014).

In contrast, conventional isotope labels used in biomolecular NMR, such as $^{13}C$, $^{2}H$ or $^{15}N$, allow one to address structural information without changing the molecules of interest. In structural studies of biomembranes, NMR using these

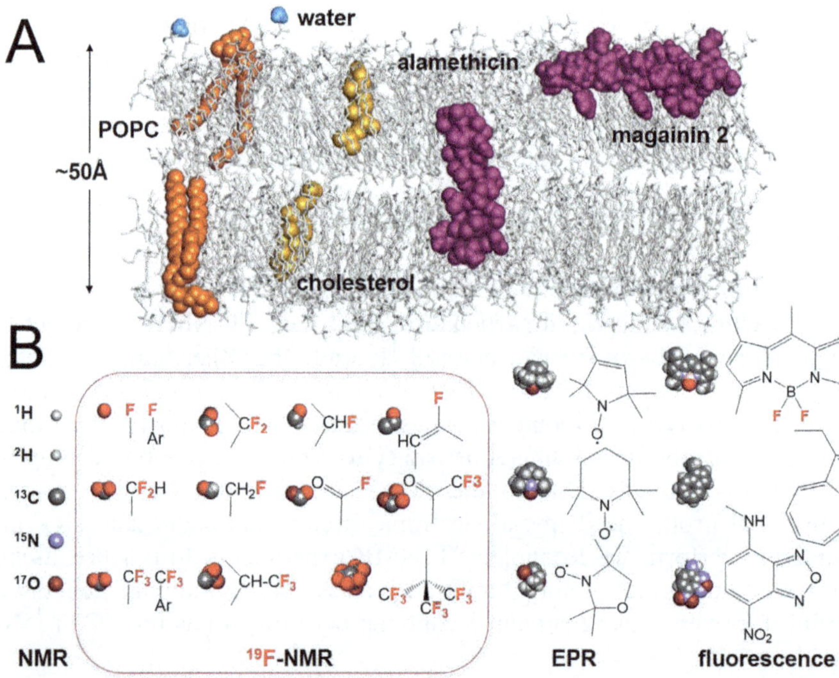


**Figure 6.1.** Comparison of fluorine NMR labels with ESR and fluorescence labels, all of which are employed in structural investigations in biomembranes. (A) $^{19}F$-labels have been introduced into numerous membrane constituents, such as lipids, sterols or membrane-active peptides. (B) Fluorine labels bridge the gap between non-perturbing but comparatively insensitive isotope labels used in conventional NMR on the one hand, and very bulky labels used in EPR and fluorescence spectroscopy on the other hand. All molecules and segments in (A) and (B) are drawn to scale, demonstrating that the larger labeling groups can make up a considerable fraction of the labeled molecule.

nuclei has, therefore, become an established and versatile tool. Nonetheless, due to their low sensitivity usually comparatively large amounts of sample material are typically needed. Also, the weak dipole–dipole interaction of $^{13}C$ or $^{15}N$ limits distance measurements to about 5–7 Å.

$^{19}F$ NMR may thus be considered to bridge the gap between the two labeling approaches of using either non-invasive but low sensitivity labels or highly sensitive but potentially interfering labels in several aspects. The fluorine atom has a van der Waals radius of 144–146 pm, which is only moderately larger than that of a hydrogen atom of 110–120 pm (Rowland and Taylor 1996). Hence, a hydrogen substitution by fluorine does not require significantly more space and increases the contact distance to neighboring molecules only by 20%–30%. Of course, labeling with the highly electronegative fluorine can induce further changes in the molecular structure than just the hydrogen substitution. Nonetheless, compared to the bulky labels used in EPR or fluorescence spectroscopy, fluorine labeling can be designed to keep the modification of the molecule of interest acceptable. In particular, there are far fewer restrictions imposed on the chemical nature of the $^{19}F$-label as compared to an ESR spin label that needs to be stabilized, or a fluorescence label in which optical properties have to be taken into account. To date, numerous artificial amino acids have been designed to carry a $^{19}F$-label for NMR analysis, which occupy very similar volumes and have similar polarity as the designated natural amino acid to be replaced (Kokhan *et al* 2016, Komarov *et al* 2019).

Furthermore, solid-state NMR is a very versatile technique in general, allowing characterization of molecular properties, structure and dynamics in great detail, with far more experimental possibilities than EPR or fluorescence. Hence $^{19}F$ NMR offers a way to expand the repertoire of conventional isotope labels, offering a high sensitivity along with this versatility. Regarding distance measurements, $^{19}F$ NMR covers a distance range and also accuracy between NMR with conventional selective labels (such as $^{2}H$, $^{13}C$ or $^{15}N$) and methods like EPR or fluorescence. Whilst, for example solid-state $^{13}C$ NMR or $^{15}N$ NMR are limited to rather short distances of 5–7 Å, they nonetheless allow very accurate distance measurements with ~0.1 Å precision. On the other hand, the large distance ranges of EPR or fluorescence spectroscopy-based methods of up to 100 Å are often compromised by a lower definition of this structural information. $^{19}F$ NMR falls in between, providing distances over 15–18 Å with good accuracy of a few tenths of Ångstroms.

### 6.2.4 Hardware requirements

Standard NMR spectrometers are laid out for the detection of $^{1}H$, and of further nuclei with lower Larmor frequencies such as $^{13}C$ or $^{15}N$. As the $^{19}F$ Larmor frequency is close to the $^{1}H$ frequency, the hardware components used for $^{1}H$ can be used also for $^{19}F$ NMR experiments, provided the bandwidth of the involved electronic components is large enough to cover both frequencies. While conventional spectrometer components usually impose no problem, dedicated $^{19}F$ NMR probes are often needed. That is because most probes cover only a narrow bandwidth for $^{1}H$

detection, and they are not built from fluorine-free materials that would be essential to avoid background signals.

Dedicated hardware, in particular a suitable double-tuned probe, is definitely needed for experiments which, besides $^{19}$F detection, also require manipulation of the $^{1}$H spins. Especially in solid-state NMR, $^{1}$H-decoupling is needed to reduce the line broadening arising from $^{19}$F-$^{1}$H dipolar couplings. Due to the close proximity of the $^{1}$H and $^{19}$F resonance frequencies, the usual circuit design for double-tuned probes cannot be applied, and dedicated double-tuned designs are usually employed (Gor'kov *et al* 2007, Matson 1977). The aqueous buffer of many biological samples, including biomembranes, furthermore imposes another difficulty, in particular for high-gamma nuclei such as $^{1}$H and also $^{19}$F. The salt content leads to sample heating due to the electric field component of the RF field irradiation during an NMR experiment. Specialized low-E probes have been designed to cope with this issue, and also been adapted to $^{19}$F/$^{1}$H double resonance NMR experiments (Gor'kov *et al* 2007).

Though pulse sequences are available to reduce background signals, in particular in static applications of $^{19}$F NMR without magic angle spinning (MAS), a $^{19}$F background signal originating from the probe can be a problem. Therefore, a careful choice of fluorine-free materials to be used in the probe parts is necessary. Fluoropolymers such as Teflon and KelF have to be avoided, but even materials made of non-fluorinated polymers can contain fluorinated additives leading to a fluorine background signal.

In some special cases, nonetheless, $^{19}$F NMR can be done using conventional probe hardware. For example, for the measurement of $^{2}$H–$^{19}$F dipolar couplings a conventional double-tuned HX probe can be used to detect the $^{2}$H signal, and the $^{1}$H channel can be tuned to the $^{19}$F frequency, provided the bandwidth of the high-frequency channel is sufficient (Grage *et al* 2004).

### 6.2.5 Typical $^{19}$F NMR chemical shifts

A property rendering $^{19}$F NMR particularly advantageous is its large range of chemical shifts and the high sensitivity of the chemical shift to the environment. The chemical shift range of $^{19}$F is about 10 times that of $^{1}$H, spanning more than 200 ppm. Due to the strong dependence of the chemical shift on the lone electron pairs, $^{19}$F is very sensitive to the environment, in particular sensing changes in its van der Waals interactions. Typical $^{19}$F chemical shifts found in biological samples are around −75 ppm for $CF_3$ groups, around −120 ppm for $^{19}$F-labeled aromatic groups, and around −220 ppm for mono-fluorine substitutions in aliphatic segments. The $^{19}$F chemical shift anisotropy is around 40–50 ppm in $CF_3$ groups, 10–30 ppm for single $^{19}$F in aliphatic positions, and 50–60 ppm in aromatic rings (Dürr *et al* 2008, Grage *et al* 2008).

## 6.3 Introducing $^{19}$F into biomembranes

### 6.3.1 $^{19}$F-labels

Fluorine labels can consist of a single fluorine, but also of groups carrying multiple fluorines, such as $CF_2$ or $CF_3$ groups, all of which have been favorably employed.

Single fluorine substitutions keep the chemical changes to a minimum and have often been employed on aromatic carbons. The lack of $^{19}F$-$^{19}F$ couplings leads to easily interpretable singlet signals, without any additional broadening from homonuclear couplings. Using several fluorines tends to increase the sensitivity, even though, depending on the employed NMR method, this gain in signal may be lost due to the distribution of the intensity over a multiplet or additional line broadening. On the other hand, the $^{19}F-^{19}F$ dipolar couplings within a label can provide very valuable additional structural information.

$CF_2$ groups have been used, for example, in the alkyl chains of $^{19}F$-labeled lipids, or in aliphatic positions of amino acids such as methanoprolines (Gagnon *et al* 2018b, Kubyshkin *et al* 2012b, Kubyshkin *et al* 2013, Post *et al* 1981, Oldfield *et al* 1980, Sturtevant *et al* 1979).

In many situations, labeling with $CF_3$ groups has turned out to be particularly fruitful (Kubyshkin *et al* 2012a). In particular for the analysis of peptide orientation benefits from labeling by $CF_3$ groups. The $^{19}F$–$^{19}F$ dipolar coupling within the $CF_3$ group can be used here as a straightforward reporter for segmental orientation, whereas $^{19}F$ chemical shift requires accurate chemical shift referencing and knowledge of the parameters of the chemical shift anisotropy tensor.

On the other hand, monofluoro-substituted amino acids tend to be most favorable for some applications such as $^{19}F$–$^{19}F$ distance measurements, which are based on the observation of dipolar spectra. In this case, multiple spin systems like that of two $CF_3$ groups would yield complicated multiplet spectra, which can be difficult to resolve, and where the intensity is spread over many signals. Using singly fluorine substituted amino acids has thus been the preferred choice for distance measurements (Grage *et al* 2014, Kokhan *et al* 2016, Tkachenko *et al* 2014). We note at this point, however, that MAS methods for distance measurements are not limited by the resolution of multiplets and may, therefore, benefit from the higher sensitivity of the $CF_3$ group.

In general, we may distinguish two ways of probing biomembranes using $^{19}F$ NMR. Firstly, $^{19}F$-labeled molecules can serve as a probe by themselves, or secondly, $^{19}F$ carrying reporter groups can be covalently linked to the molecule of interest, such as a membrane protein. Early applications of $^{19}F$ NMR on biomembranes have used $^{19}F$-labeled lipid molecules or fatty acids. One fluorine or a $CF_2$ group were introduced into DMPC to examine the molecular mobility in the lipid bilayer (Engelsberg *et al* 1982, McDonald *et al* 1983, McDonough *et al* 1983, Oldfield *et al* 1980, Post *et al* 1981). Even native biological membranes in bacteria have been studied using $^{19}F$-labeled lipids or lipophilic molecules (Gent *et al* 1981, McDonald *et al* 1983, Macdonald *et al* 1985, McDonough *et al* 1983). Also, $^{19}F$-labeled steroids and cholesterol derivatives, as well as detergents like TFOM have been used to characterize the behavior of guest molecules in a lipid bilayer and to determine their immersion into the membrane (Devoisselle *et al* 1992, Kauffman *et al* 2000, Prosser *et al* 2001). Drug molecules, in particular, present an interesting target for $^{19}F$ NMR. That is because a considerable number of small molecule drugs carry fluorine to influence the physical properties of the drug, such as its hydrophobicity, in order to improve their uptake into the cell across the membrane. Here,

$^{19}F$ NMR presents a unique way to study the interaction of such drugs with biomembranes, which is an important question in pharmacokinetics (Chekmenev *et al* 2008). For example, the anti-inflammatory agents flufenamic acid and diflucortolonvalerate were characterized in model lipid bilayers using $^{19}F$ NMR (Grage and Ulrich 1999, Grage *et al* 2000b). Even membrane proteins can be characterized without the need of introducing a covalently linked $^{19}F$-label. As many drug compounds carry fluorine, and a large fraction of drug targets are membrane proteins, a $^{19}F$ probe can be introduced in the form of a ligand that binds to the membrane protein of interest. This way, for example, the binding pocket of the gastric $H^+/K^+$ ATPase has been elucidated using drug compounds (Watts *et al* 2001).

In most studies, the aim is to resolve the structure, dynamics and function of a membrane protein or peptide as the major player in biomembranes. Except for the special case where a suitable fluorinated ligand is available as $^{19}F$ probe, direct $^{19}F$-labeling of the polypeptide is then usually needed for $^{19}F$ NMR. Particularly attractive are $^{19}F$-labels in the peptide backbone, as they allow the most immediate access to protein or peptide structure. Such labeling schemes have been achieved using isosteres of the peptide bond, such as monofluoro-alkenes. By placing fluorine in lieu of oxygen, this isostere preserves the polarity characteristics of the peptide backbone and remains chemically stable (Drouin *et al* 2020). However, the monofluoro-alkene backbone isostere is not fully compatible with all secondary structures. Therefore, in most cases of protein labeling, $^{19}F$ has to be inserted in an amino acid side chain to reduce the impact of the fluorine on the native protein structure.

Fortunately, fluorine has been introduced in quite a number of amino acids, and many tailor-made $^{19}F$ reporters based on artificial amino acids have been developed to satisfy the boundary conditions (figure 6.2). Fluorine could be introduced into several natural amino acids, keeping the chemical changes to a minimum. Frequently used are Phe, Trp or Tyr analogues with single fluorine substitutions, such as 4F-Phe, 5F-Trp or 3F-Tyr (Anderluh *et al* 2005, Arias *et al* 2020, Arseniev *et al* 1987, Grage *et al* 2002, Hagen *et al* 1978, Ho *et al* 1986, Hou *et al* 2017, Kozorog *et al* 2018, Luo and Hong 2006, Wilson and Dahlquist 1985), or occasionally with a $CF_3$ substitution in $CF_3$-Phe (Shi *et al* 2011, Shi *et al* 2012). Also, aliphatic amino acids like alanine or leucine have been $^{19}F$-labeled with a single fluorine (Toke *et al* 2004a, Toke *et al* 2004b, Gerling *et al* 2014), or aliphatic analogues with $CF_3$ such as trifluoro-ethylglycine (Auger *et al* 2019). Even the special case of proline has been subjected to versatile fluorine labeling. Introducing a $CF_3$ group into proline or methanoproline (Mykhailiuk *et al* 2008, Kubyshkin *et al* 2015, Kubyshkin *et al* 2018, Kubyshkin *et al* 2019), or two fluorines into methanoproline (Kubyshkin *et al* 2012b, Kubyshkin *et al* 2013), specific peptide secondary structures involving proline could be addressed. For example, $CF_3$-proline or $F_2$-methanoproline were inserted into the β-turn of the cyclic peptide gramicidin S (Kubyshkin *et al* 2013, Kubyshkin *et al* 2015). Using $CF_3$-methanoproline, it has even been possible to label helical polyproline-II structures (Afonin *et al* 2017, Kubyshkin *et al* 2018, Kubyshkin *et al* 2019).

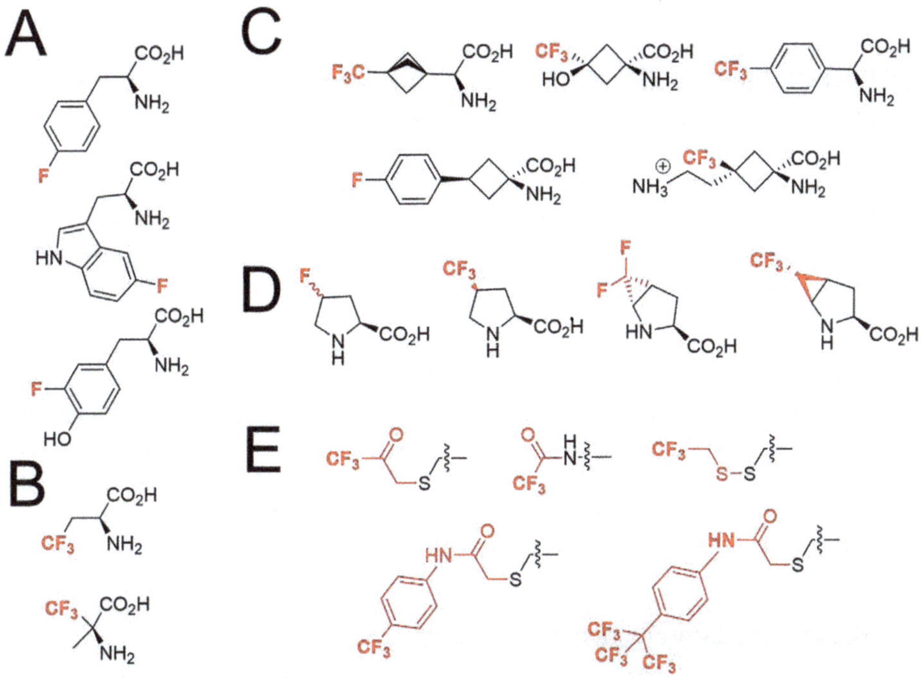


**Figure 6.2.** Selected examples of $^{19}$F-labeled amino acids used for $^{19}$F NMR on membrane associated peptides or proteins. (A) Aromatic amino acids with fluorine substitutions, (B) aliphatic amino acids with fluorine substitutions, (C) artificial amino acids in which the $^{19}$F-label is rigidly connected to the peptide backbone, designed to possess either a hydrophobic or a polar character, (D) $^{19}$F-labeled proline analogues, (E) $^{19}$F-labels attached chemically via the functional groups of cysteine or lysine.

All these $^{19}$F-labels, derived from natural amino acids mostly by a single hydrogen-to-fluorine substitution in a larger side chain, represent conservative ways of introducing $^{19}$F without much change. Nonetheless, the versatility and applicability of $^{19}$F-labeling by these amino acid analogues is limited: labeling is mostly restricted to hydrophobic residues, and the label is connected to the backbone via a series of rotatable bonds. Hence, polar regions of proteins, for example in amphiphilic peptides, are difficult to probe with $^{19}$F. Furthermore, if precise structural information is desired, a rigid connection between the $^{19}$F-label to the protein backbone is required. That is, the ambiguity of rotamers has to be strictly avoided, if we want to be able to translate the structural information obtained on the side chain into the context of the entire protein. To provide such rigid connection to the backbone, and to extend the set of $^{19}$F-labeled amino acids to polar residues, several designated artificial amino acids have thus been developed. The first amino acids with a rigid connection between the $^{19}$F-label and the peptide backbone were based on phenylglycine (fluoro-phenylglycine and trifluoromethyl-phenylglycine), and later based on bicyclopentylglycine (fluoro-bicyclopentylglycine and trifluoromethyl-bicyclopentylglycine). These hydrophobic $^{19}$F-labels have been instrumental in the structural analysis of amphiphilic peptides, such as antimicrobial agents and the cyclotide kalata B1. The three-dimensional orientation of these peptides with respect to the lipid bilayer could be determined by translating the anisotropy of the $^{19}$F NMR signal via the rigid $^{19}$F connection to the folded polypeptide backbone (Glaser *et al* 2004, Glaser *et al* 2005, Mykhailiuk *et al* 2010, Komarov *et al* 2019, Mikhailiuk *et al* 2006, Grage *et al* 2017). More recently, this approach has been expanded to polar amino acids, including analogues of lysine, arginine, serine and threonine (Michurin *et al* 2016, Michurin *et al* 2018, Tkachenko *et al* 2013). In particular in amphiphilic membrane-active peptides, also the hydrophilic face can now be labeled this way, which allows for the measurement of orientational constraints from both hydrophobic and hydrophilic positions in a peptide.

A further elegant way of introducing fluorine even into functional, polar groups in amino acid sidechains, is based on fluoroallyl derivatives (Laue *et al* 1999). The fluoroallyl group constitutes an isostere of a glutamine or asparagine side chain, where the oxygen is replaced by fluorine and the amine is substituted by a vinyl group. The hydrogen bonding characteristics of this isostere have been shown to be preserved (Lübke *et al* 2013), rendering this $^{19}$F-label particularly interesting to probe intermolecular contacts via H-bonded sidechains.

### 6.3.2 Biosynthetic incorporation of $^{19}$F

Even though fluorine is alien to biology, $^{19}$F has been incorporated into biomembranes and membrane proteins using the biosynthetic machinery of bacteria. Already in early $^{19}$F NMR studies of bacterial membranes, using fluorine-substituted fatty acids, $^{19}$F could be incorporated into lipids employing natural biosynthetic routes (Gent *et al* 1978, Gent *et al* 1981, McDonald *et al* 1983, Macdonald *et al* 1985, McDonough *et al* 1983). It turned out that bacterial strains that are auxotroph in unsaturated lipids are able to take up fluorine-containing fatty acids (Gent *et al* 1978). This way, first insights into the fluidity and order parameters of native bacterial membranes was possible.

Biosynthetic labeling of membrane proteins was also achieved early on. Certain fluorine substituted amino acids are non-toxic and compatible with the natural tRNA of the respective non-fluorinated amino acid. This way, biosynthetic labelling with tryptophan analogues (4F-, 5F- or 6F-Trp), tyrosine analogues (2F-Tyr, 3F-Tyr) or 4F-Phe has been possible by adding the respective $^{19}$F-labeled amino acid to the growth medium of *E. coli* (Ho *et al* 1986, Arseniev *et al* 1987, Hagen *et al* 1978, Wilson and Dahlquist 1985, Arias *et al* 2020). Glyphosate has been used to suppress the biosynthesis of aromatic amino acids, and to enforce the incorporation of the extraneous $^{19}$F-labeled amino acids. Membrane proteins like bacteriorhodopsin, LDH or M13 coat protein have been labelled this way with $^{19}$F in Trp or Tyr positions (Ho *et al* 1986, Arseniev *et al* 1987, Hagen *et al* 1978, Wilson and Dahlquist 1985). 4F-Phe has been introduced into over-expressed proteins like the GPCR ligand glucagon (Hou *et al* 2017). Advantageous can be the use of mutant strains which are auxotroph in the amino acid to be substituted by the fluorine-labeled analogue (Anderluh *et al* 2005, Kozorog *et al* 2018).

In some cases where the $^{19}$F-labeled amino acid is not compatible with the native tRNA recognition machinery of the expressing bacteria, the amber stop codon or nonsense codon has been successfully used to encode the $^{19}$F-labeled amino acid (Jackson *et al* 2007, Shi *et al* 2011, Shi *et al* 2012). This could be achieved by co-transforming DNA sequences encoding for the extraneous tRNA as well as for the respective aminoacyl tRNA synthetase. This way, $CF_3$-Phe has been introduced into membrane proteins such as the DAGK (Shi *et al* 2011, Shi *et al* 2012).

Recent advances in cell-free expression lend hope to the possibility of labeling even with amino acids that are not compatible with native tRNA, or to label single selective residue positions. This way, for example, a retinol binding protein has been labeled by 6F-Trp in selected positions (Neerathilingam *et al* 2005).

### 6.3.3 Chemical introduction of $^{19}$F

When biosynthetic $^{19}$F-labeling is not possible, there are several chemical ways of incorporating fluorine into proteins. A side chain or terminus of a protein can be $^{19}$F-labeled by chemical linkage of a reactive fluorinated compound after synthesis or expression of the protein. Most frequently, cysteines are modified this way by attaching a substance like trifluoroethylthiol (TET) or bromo-trifluoroacetone (BTFA) to the reactive -SH group. This way, GPCRs and mammalian rhodopsin have been labeled with $^{19}$F (Didenko *et al* 2013, Klein-Seetharaman *et al* 1999). Also, the sidechain amino group of lysine and its homologues is suitable for chemical labeling of peptides or proteins. For example, trifluoro-acetic acid and monofluoro-acetic acid have been used to introduce a $CF_3$-group or a single $^{19}$F-label, respectively, into membrane-active peptides (Salnikov *et al* 2016). Chemical modification of a peptide or protein to introduce a $^{19}$F-label, however, is not limited to the sidechains. Also, the N- or C-terminus can be labeled chemically. In an early example, p-fluorobenzoyl or 4F-Phe were attached to the C- or N-terminus of gramicidin A, respectively, for $^{19}$F NMR analysis (Weinstein *et al* 1985).

Using chemical peptide synthesis opens many further possibilities to incorporate fluorine in the form of artificial or fluorine-substituted amino acids. Nonetheless, several issues need to be considered in the synthesis of fluorine-containing peptides: due to the high electron withdrawing effect of fluorine, a rearrangement of the ligands at the alpha-carbon is possible, e.g., in 4-fluoro-phenylglycine, leading to racemization and a loss of stereoselectivity (Glaser *et al* 2004). Also, when fluorine is introduced near the backbone, as e.g., in monofluoro-alanine, H-F elimination can render the product instable. To combat these challenges, methods to separate the stereoisomers, or to avoid H-F elimination and racemization by the choice of suitable $^{19}$F-labeled amino acids, have been established (Glaser *et al* 2004, Komarov *et al* 2019).

## 6.4 Line narrowing techniques for $^{19}$F NMR of biomembranes

Lipid membranes are self-assembled mesoscopic structures, and this property has significant implications on the application of NMR techniques to biomembranes. As the observed NMR frequency in general depends on the local molecular orientation with respect to the external magnetic field, NMR spectra usually contain contributions from molecules in all orientations and hence cover a large frequency range. Such powder spectra are particularly broad in the case of $^{19}$F NMR, because of the large chemical shift anisotropy and dipolar couplings of $^{19}$F. In standard applications of NMR in solution, these anisotropies are averaged to their isotropic values as a consequence of fast tumbling of the molecules in solution. In lipid bilayers, on the other hand, the molecular mobility tends to be much slower than the necessary timescale of ~0.1–10 ms that leads to averaging of the NMR frequencies. The lateral diffusion of lipids, for example, allows averaging only in vesicles with very small diameters of below ~100 nm. Even less mobile are membrane embedded proteins and peptides.

In early $^{19}$F NMR experiments on biomembranes, which focussed on bacterial membranes of *A. ladlawii* and *E. coli*, solid-state NMR lineshapes were obtained without the need for any special techniques. At those relatively low magnetic fields, and thanks to the comparatively high degree of mobility of the lipids, for example the order parameter could be determined form the linewidth and lineshape of the $^{19}$F NMR signal of the labeled lipids (McDonald *et al* 1983, Macdonald *et al* 1985, McDonough *et al* 1983, Gent *et al* 1978, Gent *et al* 1981).

Nonetheless, in general, an improvement of linewidth and hence of resolution and signal/noise is desirable, and the molecular mobility has to be reintroduced or the orientation distribution has to be reduced by suitable choice of the lipid system and sample geometry, or by dedicated NMR techniques. Most of the experimental procedures for 'soft-state' NMR analysis of liquid crystalline membranes are adopted from classical solid-state NMR. The time scales of molecular averaging in biomembranes are slower than the NMR timescale of most anisotropic parameters, hence from an NMR perspective, biomembranes appear to be 'solid'-like in many aspects. Several approaches have thus been applied in $^{19}$F NMR studies of biomembranes, as discussed in the following (figure 6.3).

### 6.4.1 Multipulse solid-state $^{19}$F NMR

Not only the anisotropies, but also the convolution of different NMR interactions, lead to the broad and complex lineshapes observed in solid-state NMR. This way, structural information inherent in the chemical shift or dipolar coupling is often obscured, and spectra are difficult to interpret. $^{19}$F NMR spectra are particularly affected by such complexity, because at typical magnetic fields of around 500 MHz, the $^{19}$F chemical shift and $^{19}$F–$^{19}$F dipolar couplings of close fluorines are of similar magnitude. To simplify such spectra, RF pulse sequences can be applied, which allow one to selectively average one or several of the NMR interactions to zero. This approach is routinely applied in recoupling techniques used in MAS experiments. But, also in static NMR, for example, the Carr Purcell Meiboom Gill (CPMG) experiment has been applied to reduce the complexity of NMR spectra and keep only the contributions arising from dipolar couplings. This multipulse sequence consists of a series of 180° pulses, which cause a series of Hahn echoes. The 180° pulses refocus the chemical shift and heteronuclear couplings to surrounding

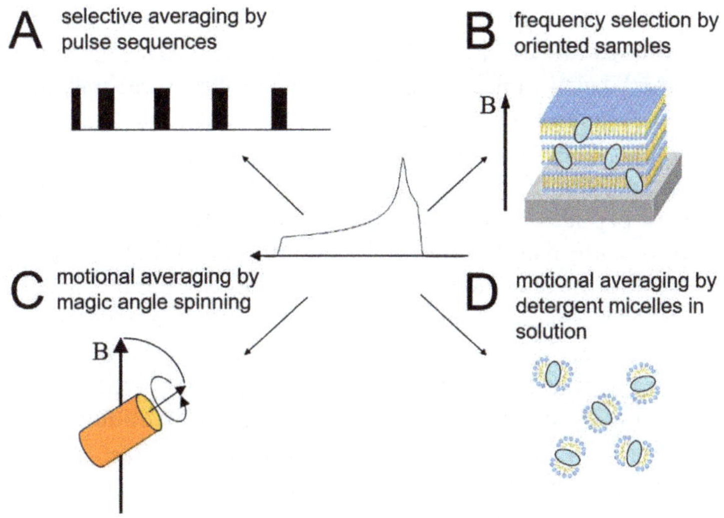


**Figure 6.3.** Methods for line narrowing employed in $^{19}$F NMR on biomembranes. (A) Pulse sequences can selectively average undesired NMR interactions, allowing one to reduce their contributions to the NMR spectrum. (B) Using oriented membranes, a single frequency is selected from the anisotropy range by preparing a macroscopically aligned sample with a uniaxial orientation. (C) Magic angle spinning averages the anisotropies, resulting into narrow solution-NMR-like signals. (D) Reconstituting membrane proteins or peptides into detergent micelles as membrane mimetics, fast molecular tumbling can be achieved which averages the NMR anisotropies.

protons, but homonuclear dipolar couplings are left to evolve. Hence, the resulting series of echoes varies according to the homonuclear dipolar interaction and can be converted into a spectrum of homonuclear dipolar couplings in which other sources of broadening are removed. As a result, for two spins, for example, a doublet reflecting the homonuclear dipolar coupling is obtained without broadening by chemical shift or couplings to $^1$H. The CPMG experiment has been frequently applied in $^{19}$F NMR studies of biomembranes. Early on, Post *et al* and other groups were able to reveal the doublet of lipids carrying a $CF_2$ segment, where without such techniques the $^{19}$F NMR spectrum would consist of a featureless line (Engelsberg and Yannoni 1990, Grage and Ulrich 1999, Post *et al* 1981, Post *et al* 1982, Post *et al* 1984). The experiment was taken up later on for the study of small molecule drugs in membranes, as well as for distance measurements in membranes and membrane-bound peptides (Grage and Ulrich 1999, 2000a, Grage *et al* 2000b, Grage *et al* 2004, Grage *et al* 2014, Salgado *et al* 2001). Notably, the CPMG method has been particularly useful in combination with oriented samples to obtain readily interpretable dipolar splittings.

### 6.4.2 Oriented samples for solid-state $^{19}$F NMR

Oriented samples provide an elegant way of achieving line narrowing and at the same time obtaining structural information. Here, the molecules of interest are uniformly oriented with respect to the magnetic field, so their orientation dependent

NMR signals all coalesce into a single signal, leading to increased resolution and signal/noise. Moreover, the position of this signal in the NMR spectrum encodes the orientation of the labeled molecular segment, hence structural information can be directly addressed in oriented samples. This approach has been very fruitful, especially in using $^{19}$F NMR with its large anisotropies. It provides a powerful technique to obtain the conformation, the molecular orientation, and the mobility of membrane-bound peptides and integral membrane proteins.

Oriented samples are readily produced exploiting the ability of lipids to spontaneously self-assemble into lamellar bilayers. Providing a smooth surface like a glass cover slide is sufficient to cause lipids to form multiple bilayers which are aligned parallel to the template. After spreading lipids (typically ~1 mg per slide) in organic solvent onto such a surface and subsequent removal of the solvent, stacks of several thousands of uniformly aligned bilayers can form upon hydration. This procedure works especially well for lipid molecules with a low spontaneous curvature, such as zwitterionic phosphatidylcholines or anionic phosphatidylglycerols that have a more or less cylindrical molecular shape. Such a sample for oriented solid-state NMR analysis thus consists of a sandwich-like structure of about a dozen glass slides, carrying several 1000s of hydrated bilayers between them, sealed by parafilm or cling foil. Typically, ~10 mg of lipids with ~0.1–1 mg peptide can be oriented this way. The stack is usually placed into the static magnetic field of the NMR spectrometer such that the bilayer normal is aligned parallel to $B_0$.

Besides mechanically aligned bilayers on a glass support, membranes can also be oriented magnetically. The magnetic susceptibility anisotropy of a lipid leads to a preferable orientation such that the bilayer normal becomes aligned perpendicular to the magnetic field. This intrinsic alignment propensity can be exploited in bicelles, which consist of a mixture of long-chain and short-chain lipids like DMPC and DHPC. When mixed in an appropriate ratio, they form small discs (or swiss-cheese like layers), where the short-chain lipids shield the edges of membrane patches formed by the long-chain lipids. Due to the preferred perpendicular alignment, the anisotropy range is reduced (to ½ of that of mechanically aligned samples) and mobility issues as well as challenges in reconstituting amphiphilic components may arise.

Oriented samples can be easily combined with static multipulse experiments such as the CPMG method introduced above. While most conventional solid-state NMR techniques rely on MAS, where uniform orientation cannot be maintained, CPMG on oriented samples allows one to select the $^{19}$F–$^{19}$F dipolar coupling and evaluate it in terms of orientation (Grage and Ulrich 1999). Also, the sensitivity and resolution benefits from this combination of two simultaneous line-narrowing methods. It was possible this way to resolve complex $^{19}$F–$^{19}$F dipolar coupling spectra of multispin systems with a resolution of ~10 Hz in lipid dissolved small molecule drugs and of membrane bound peptides such as gramicidin S in lipid bilayers (Dürr *et al* 2012, Grage *et al* 2006). In spin pairs, doublets of small drug molecules in membranes and of membrane-active peptides of ~100 Hz could be obtained (Dürr *et al* 2012, Grage and Ulrich 1999, Grage and Ulrich 2000a, Grage *et al* 2000b, Grage *et al* 2014).

Many structural questions, such as order parameters, distance constraints, distinguishing stereoisomers, could be addressed this way.

### 6.4.3 Magic angle spinning $^{19}$F NMR

Insufficient molecular mobility within biomembranes and of vesicles or other lipid structures as a whole is the reason for line broadening in solid-state NMR. Besides the intrinsic $T_2$ relaxation, anisotropies and dipolar couplings dominate the frequency dispersion in the NMR spectra. These effects are usually averaged out in solution NMR by molecular tumbling. Such rapid molecular motion can be mimicked by spinning a solid sample around an axis that is inclined at the so-called magic angle of 54.7° with respect to the static magnetic field $B_0$. Biomembrane samples of vesicle suspensions and reconstituted membrane proteins can be easily subjected to magic angle spinning (MAS). Even though biomembranes do not actually have a solid consistency, but form rather viscous liquids, they are easily spun in practice.

For MAS to be effective, the spinning speed needs to exceed the size of the anisotropic linewidths of the underlying static spectrum. Hence, for $^{19}$F NMR with comparatively large anisotropies and dipolar couplings of up to ~30 kHz, relatively high spinning speeds would be needed. However, the molecular mobility in fluid membranes results in partial motional averaging and a reduction of the anisotropies. Therefore, lower spinning speeds are usually sufficient, as long the samples do not have to be frozen for the applied NMR technique. Also, as $^{19}$F is usually introduced in the form of selective labels, there is no $^{19}$F dipolar network to be expected, as for example in fluoropolymers. For biomembrane samples, additional multipulse techniques for line narrowing, such as CRAMPS, are therefore not required.

### 6.4.4 Solution NMR of membrane proteins in membrane mimetics

Solution $^{19}$F NMR techniques are attractive in terms of resolution and signal/noise. Membrane proteins can be studied by solution NMR when reconstituted into suitable membrane mimetics, which allow sufficient tumbling and at the same time keep the protein in a proper lipid-like environment. Detergent micelles are the most frequently applied membrane mimetic, and were employed already in the early $^{19}$F NMR studies of membrane proteins. Due to the large chemical shift dispersion of $^{19}$F, the requirements for narrow linewidths are actually less severe than for $^{1}$H NMR. Therefore, even sizeable protein/detergent complexes can be addressed by solution $^{19}$F NMR in detergent micelles, and up to 10 distinct $^{19}$F signals of the same type of amino acid have been resolved (Arseniev *et al* 1987, Wilson and Dahlquist 1985). However, it has to be noted that great care has to be taken when using detergent micelles for the study of smaller membrane proteins or peptides. While the structure of large membrane proteins is defined mostly by interactions within the protein, smaller systems depend much more on interactions with the lipid environment. These critical features and the dynamics thereof, cannot usually be represented by detergent micelles, because they lack important aspects such as planarity or lateral pressure profile. An alternative for detergent micelles can be isotropic bicelles, which have been used for example for $^{19}$F NMR studies of $O_2$ accessibilities of membrane-associated peptides (Prosser *et al* 2001). However, even in this case there is a risk that an amphiphilic peptide is not embedded in the center of the bicelle but rather intermingles with the short-chain lipids at the edge of the planar disk. Peptide re-alignment and oligomerization are therefore difficult to interpret in a bicelle environment.

## 6.5 Solid-state $^{19}$F NMR distance measurements in biomembranes

Due to its high gyromagnetic ratio, $^{19}$F possesses strong dipolar couplings with neighboring nuclei. Hence, accurate distance measurements or the qualitative evaluation of interatomic contacts is one of the most potent applications of $^{19}$F NMR. The dipolar coupling strength of two nuclei can be quantified by the splitting of the signal

$$2\pi\,\Delta = \mu_0\gamma_1\gamma_2\hbar/(4\pi\,r^3)\ \tfrac{1}{2} < 3\cos^2\theta - 1 > f,$$

with the gyromagnetic ratios $\gamma_1$ and $\gamma_2$, the internuclear distance $r$, the angle between the internuclear vector and the magnetic field $\theta$, <> denotes the time average, and $f = 2$ or 3 for hetero- or homonuclear couplings.

This can be put into numbers by considering two fluorine nuclei 1 Å apart:

$$\Delta = 108\text{ kHz}\,(\gamma_1/\gamma_F)(\gamma_2/\gamma_F)(r/1\text{Å})^{-3}\ \tfrac{1}{2} < 3\cos^2\theta - 1 > f.$$

$^{19}$F couplings can be well in the 10 kHz range, as for example the intramolecular coupling strength of a $CF_3$ group is ~15 kHz (when oriented along the magnetic field direction). The large dipolar couplings of $^{19}$F lead to a large accessible distance range of $^{19}$F NMR, compared to other isotope labels. Assuming that, e.g., a dipolar splitting of 30 Hz can be resolved, distances between $^{19}$F and $^{1}$H as large as 20 Å are experimentally accessible. For couplings to $^{31}$P, $^{13}$C, and $^{15}$N, the corresponding distance ranges are still 14, 12, and 9 Å, respectively. $^{19}$F–$^{19}$F homonuclear couplings have a range of 22 Å in these considerations. In practice, of course some other factors such as $T_2$ relaxation and MAS have to be considered, so the actual distance ranges achievable are shorter. Nonetheless, $^{19}$F NMR has been used to detect distances well beyond 10 Å.

A rich repertoire of techniques that was originally developed for solid-state NMR applications to other systems, has been transferred to $^{19}$F and was successfully used to reveal dipolar couplings, in order to determine internuclear distances and contacts, and to count the number of units in an oligomer.

### 6.5.1 Distance measurements in oriented samples

With the help of oriented samples, resolution can be increased by concentrating the signals from all orientations to that corresponding to a single molecular orientation. In the absence of line broadening due to anisotropies, any doublet or multiplet signature of dipolar couplings can become well resolved. Moreover, in fluid membranes, the dipolar splittings adopt the highest value out of the powder spectrum range when bilayers are oriented with their normal parallel to the magnetic field.

Multipulse sequences like the CPMG experiment can further improve the resolution of dipolar splittings. As described above, this experiment removes contributions from chemical shift and couplings to $^{1}$H, leaving a $^{19}$F NMR lineshape that reveals purely the dipolar couplings. Using this approach, long-range distances of ~11 Å, spanning two helix turns within an antimicrobial peptide, were measured in fluid membranes (Grage *et al* 2014). For the distance measurement, the peptide PGLa was labeled with fluoro-phenylglycine in two positions in the sequence, 7 residues apart (figure 6.4A). With the help of the CPMG experiment (figure 6.4B), a dipolar coupling of ~80 Hz (figure 6.4C) could be resolved, which was not visible in the regular spectrum. It should be noted that fluid membranes are beneficial in this approach as the partial averaging of chemical shifts due to the residual mobility in the membranes increase the performance of the static multipulse experiment. This way, it is possible to measure long-range distances in fluid membranes, whereas in contrast most MAS methods require frozen samples to reduce molecular mobility.

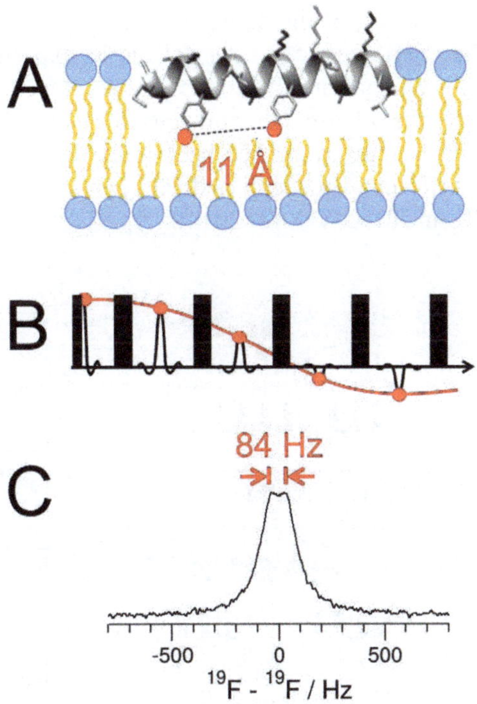


**Figure 6.4.** Long-range $^{19}$F–$^{19}$F distance measurement in fluid bilayers using CPMG multipulse narrowing on oriented membranes. (A) PGLa was selectively labeled with $^{19}$F in two positions, two helix turns apart, and reconstituted into oriented bilayers. (B) The CPMG pulse sequence refocuses the chemical shift and heteronuclear couplings in a series of echoes, whereas the homonuclear dipolar coupling can evolve. (C) In the resulting dipolar spectrum (Grage *et al* 2014), the $^{19}$F–$^{19}$F dipolar splitting can be directly resolved. Reprinted with permission from Grage *et al* (2014). Copyright (2014) American Chemical Society.

The use of fluid membranes often is highly relevant in order to obtain biologically meaningful data, because freezing tends to expel many amphiphilic solutes from the lipid bilayer and can affect the alignment and oligomeric state of peptides in other fundamental ways.

### 6.5.2 Distance measurements under MAS

Dipolar couplings are averaged to zero when applying magic angle spinning to reduce linewidth. However, this averaging process can be disrupted by the application of suitable RF pulse sequences, and this way, the dipolar coupling can be specifically recovered for distance measurements under MAS. One of the most robust and frequently used methods for such recoupling is the rotational echo double resonance (REDOR) experiment (figure 6.5A). It measures the dipolar coupling between two nuclei of different types, by observing the spectrum of one of these nuclei after a delay period of variable length. During this delay, the dipolar coupling is recoupled by a series of RF pulses that is applied at the frequency of the other nucleus. If the two spins are close enough and coupled, the observed NMR signal is reduced. This reduction can be quantified by analyzing a 'REDOR curve' of the loss of signal intensity as a function of the dephasing delay length.

Especially in conjunction with $^{19}$F, the REDOR experiment has been very successful in structural investigations of biomembranes. Due to the short transverse relaxation of $^{19}$F, usually the other partner nucleus, for example $^{31}$P, $^{13}$C, $^{15}$N or $^{2}$H, is observed. A particularly long range has been achieved for $^{19}$F–$^{1}$H distances by reading out the $^{1}$H NMR signal through a third spin, such as the amide nitrogen (Hong and Schmidt-Rohr 2013, Wi *et al* 2014).

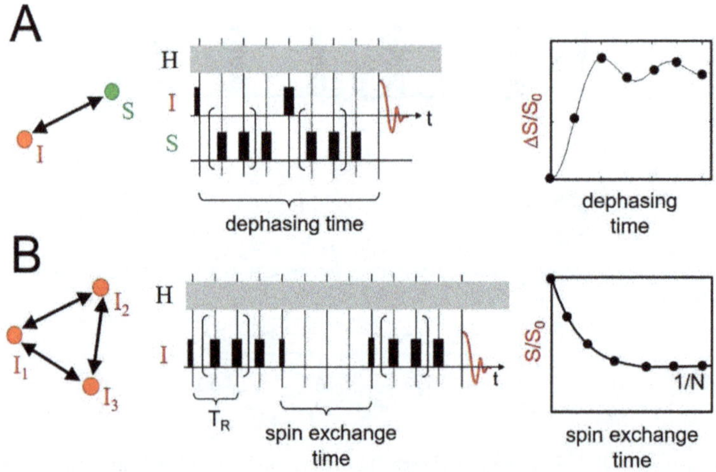


**Figure 6.5.** Distance measurements using MAS $^{19}$F NMR. (A) The REDOR experiment re-introduces the dipolar coupling between unlike spins. The dipolar coupling can be deduced from the reduction of the signal intensity after a dephasing time. (B) The CODEX experiment uses spin exchange between alike spins. The number of spins and dipolar network geometry can be assessed from the signal reduction as a function of the exchange time.

To explore $^{19}F$–$^{19}F$ homonuclear distances in biomembranes, and to count numbers of spins, the center-band only detection of exchange (CODEX) experiment (figure 6.5B) has been particularly fruitful. This method was developed by Klaus Schmidt-Rohr initially to investigate slow molecular dynamics. Later, it was adapted by Mei Hong to evaluate internuclear distances and to count the number of molecular units in oligomeric peptide bundles in membranes (Buffy *et al* 2005, Luo and Hong 2006). CODEX relies on polarization exchange between, e.g., $^{19}F$-labels in an oligomeric complex, which possess different orientations and hence different time-dependent chemical shifts. As this exchange process, also referred to as spin diffusion, is caused by the dipolar coupling between the spins, it encodes the internuclear distances and spin geometry. It is measured by a pulse sequence resembling a stimulated echo sequence. An echo is formed, provided that no exchange happens during a mixing time delay. If spin diffusion leads to a transfer of polarization to another site, the echo refocusing is incomplete and the signal will be reduced. Interestingly, the experiment readily allows one to determine the number of spins involved, by observing the reduction of the signal after long mixing times. For sufficiently long mixing times, each spin retains only its own fraction of the total polarization, and hence the signal intensity decays to $1/N$, with $N$ the number of spins in the molecular assembly. Therefore, in oligomers of singly fluorine-labeled peptides, CODEX can be used to measure the oligomer number directly.

### 6.5.3 Multidimensional solid-state $^{19}F$ NMR experiments

The use of multidimensional NMR experiments is well established for $^{13}C$ and $^{15}N$ NMR, yielding a large number of constraints in a single experiment as a basis for elucidating entire protein structures. Recently, also 2D $^{19}F$ NMR experiments have been designed to determine multiple distance constraints. By combining $^{19}F$–$^{1}H$ REDOR with $^{15}N$–$^{1}H$ HSQC, it has been possible to obtain about 30 $^{19}F$–$^{1}H$ distances related to individual amide nitrogens in the 3F-Tyr labeled model protein GB1 (Shcherbakov and Hong 2018). When $^{19}F$-$^{13}C$ REDOR has been combined with $^{13}C$–$^{13}C$ correlation experiments, the same authors could detect clusters of the influenza B M2 transmembrane domain (Shcherbakov *et al* 2020).

The approach based on spin exchange, as used in CODEX, has also been explored further by Mei Hong *et al* (Roos *et al* 2018). Spin diffusion, as routinely used in 2D correlation experiments for $^{13}C$–$^{13}C$ distance measurements, has been applied to $^{19}F$–$^{19}F$ and successfully revealed distances of up to 15–20 Å.

All these methodological developments make $^{19}F$ NMR a promising tool for structural investigations of sizeable membrane proteins and membrane bound peptides.

### 6.5.4 Ligands bound to membrane proteins

Solid-state $^{19}F$ NMR distance measurements played an important role in elucidating the conformation of fluorine-substituted drugs when bound to their binding site in membrane proteins. As many drugs carry fluorine, this inherent 'label' provides a straightforward way to bring fluorine into membrane proteins, and also to direct the label right to their center of activity. Using REDOR, for example, the structures of inhibitors of the renal $Na^{+}/K^{+}$ ATPase and the gastric $H^{+}/K^{+}$ ATPase was elucidated in their bound state using $^{19}F$–$^{13}C$ and $^{19}F$–$^{2}H$ REDOR distance constraints (Watts *et al* 2001, Middleton *et al* 2000, Kim 2006).

### 6.5.5 Intermolecular interactions of oligomeric membrane proteins and peptides

Long-range distances, as they can be obtained using $^{19}F$ NMR, are essential to characterize intermolecular interactions between peptides or proteins in lipid membranes. Knowledge of oligomeric contacts are particularly informative to get insight into the structure of protein or peptide assemblies in the membrane, such as transmembrane pores or enzyme complexes. The proton channel M2 of the influenza virus is an example where $^{19}F$ NMR distance measurements contributed to a better understanding of its oligomeric structure and function. Using $^{19}F$ CODEX, the tetrameric state of the M2 channel could be validated (figure 6.6), and models of the oligomeric architecture could be derived on the basis of $^{19}F$–$^{19}F$ distance constraints (Luo and Hong 2006). Also, large receptor complexes have been addressed with $^{19}F$ NMR distance measurements. For example, the bacterial chemoreceptor complex regulating chemotaxis was studied in *E. coli* membranes using $^{13}C$–$^{19}F$ REDOR (Fowler *et al* 2010). Two models for the assembly of the receptor monomers were tested by introducing a $^{13}C$-labeled cysteine and a 4F-phenylalanine into the monomer in putative interface positions. Based on the $^{13}C$–$^{19}F$ distances, both the hedge-row and the trimer-of-dimer arrangement could be ruled out. This example demonstrates the potential of solid-state $^{19}F$ NMR distance measurements to obtain structural information in transmembrane receptor complexes, which would not be accessible otherwise.

The question of molecular assembly in the membrane is also crucial for the function of diverse antimicrobial peptides. Various mechanisms have been described in the literature involving discrete pores, such as a barrel-stave or a toroidal wormhole, while other models suggest that the lipid bilayer is simply perturbed by a peptidic carpet. Distance measurements using $^{19}F$ NMR can help to understand the structural basis for the action of such membrane-active peptides. For example, the barrel-stave pore of alamethicin, an antimicrobial peptide of fungal origin, could be confirmed using $^{19}F$ NMR distance measurements (Salnikov *et al* 2016). An interesting labelling scheme, where fluorine was introduced into alamethicin by

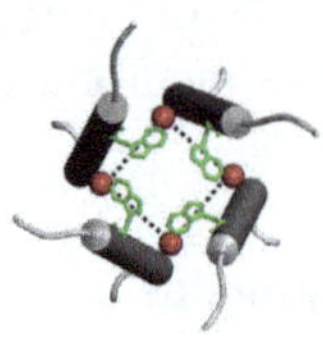


**Figure 6.6.** The oligomeric structure of the proton channel M2 from influenza virus has been determined using $^{19}F$ NMR, employing the CODEX experiment to measure the oligomer number and intermolecular distances (Luo and Hong 2006).

coupling trifluoroacetic acid to diaminobutyric acid sidechains, was employed here. Applying CODEX to the dipolar coupling network between the $CF_3$-labels, the pentameric structure of the alamethicin pore was then confirmed for high peptide: lipid ratios. Below a peptide:lipid ratio of 1:30, however, a dimeric structure was found.

In another example, intermolecular interactions of the antimicrobial peptide K3 have been studied in lipid bilayers using solid-state $^{19}$F NMR (Toke *et al* 2004a, Toke *et al* 2004b). This model peptide, which had been derived from the native antibiotic PGLa, was labeled with $^{13}$C in several positions, or with $^{19}$F by substituting an alanine in the peptide center by 3F-Ala. On the basis of $^{13}$C–$^{19}$F distances obtained in REDOR experiments, K3 was found to form parallel dimers in the membrane with an intermolecular crossing angle of ~20°. Together with additional $^{13}$C–$^{19}$F distances between the peptide and lipids (labeled with $^{19}$F in their tails), a model for the interaction of K3 in the form of pores could be proposed. In this model, monomeric and dimeric K3 peptides, as well as lipids, line a transmembrane pore.

While the typical examples for antimicrobial peptides are α-helices, also β-stranded peptides have been reported to assemble in the membrane. For example, Protegrin-1, an antimicrobial peptide found in porcine leucocytes, was shown to form dimers in the membrane, based on $^{19}$F–$^{19}$F and $^{13}$C–$^{19}$F distances obtained from CODEX and REDOR (Mani *et al* 2006, Buffy *et al* 2005). The β-stranded hairpin structure can immerse into the bilayer by forming dimers, though it does not form pores.

Membrane pores or channels are also discussed for non-peptidic antibiotics. Solid-state $^{19}$F NMR has been used to elucidate, e.g., the transmembrane ion channel formed by amphotericin B, a polyene macrolide from *S. nodosus* acting against fungi. Exploring $^{13}$C–$^{19}$F distances via REDOR, interactions between amphotericin B molecules but also between amphotericin B and ergosterol were found. This way, $^{19}$F NMR could confirm models of a hetero-oligomeric channel consisting of amphotericin B and ergosterol or cholesterol, thereby contributing information also to the general architecture of lipidic channels (Umegawa *et al* 2012, Murata *et al* 2009, Tsuchikawa *et al* 2018).

Furthermore, distance measurements employing $^{19}$F NMR have been a powerful tool to monitor contacts between proteins or peptides with the surrounding lipid environment. For example, interactions with cholesterol play an important role for the function of many membrane proteins. For example, the influenza M2 channel seems to recognize target membranes via specific interaction with cholesterol. Indeed, $^{13}$C–$^{19}$F REDOR distances measurements between $^{13}$C-labelled M2 and a $^{19}$F-labeled cholesterol analogue, revealed four different binding sites for cholesterol (Elkins *et al* 2017).

### 6.5.6 Distance measurements in solution

Also in solution, $^{19}$F NMR can be a useful tool to obtain distance constraints in large membrane proteins reconstituted in detergent micelles, which are difficult to address with routine solution-state NMR techniques. As demonstrated on rhodopsin in detergent micelles, $^{19}$F nuclear Overhauser effects (NOEs) can provide selective distance constraints, which can help to define the tertiary structure (Loewen *et al* 2000). In another example, heteronuclear NOEs were used to reveal interactions of detergent-solubilized gramicidin A, an ion-channel forming antimicrobial peptide, with $^{19}$F-labeled anesthetics (Tang *et al* 1999).

## 6.6 Solid-state $^{19}$F NMR orientation measurements in biomembranes

### 6.6.1 Orientation dependence in solid-state $^{19}$F NMR

The strength of most NMR spin interactions, such as dipolar coupling or chemical shift, depends on the orientation of the labeled segment with respect to the external static magnetic field. This anisotropy of NMR provides a rich source of structural information, especially in $^{19}$F NMR where these anisotropies are comparatively large.

Furthermore, in biomembranes the structural information inherent in NMR anisotropies can be accessed very easily, as they provide a convenient liquid crystalline alignment medium for embedded peptides or proteins. In oriented bilayer samples all molecules are uniformly aligned, and hence their orientation is straightforward encoded in the NMR signal.

Mathematically the orientation dependence can be described by a tensor, such as the CSA tensor $\boldsymbol{\delta}$ of the $^{19}$F-labeled segment, which is defined in the molecular framework by an associated principal axis system (PAS). With the convention of the external field being aligned along the *z*-axis of the laboratory frame, the $\delta_{zz}{}^{LAB}$ component of the CSA tensor in laboratory frame coordinates then represents the observed chemical shift. Hence, to predict the chemical shift, all that is needed is to transform the CSA tensor from the principal axis coordinates to laboratory frame coordinates. This way, the observed chemical shift is:

$$\delta_{zz}{}^{LAB} = \delta_{xx}{}^{PAS}(\cos\alpha \sin\beta)^2 + \delta_{yy}{}^{PAS}(\sin\alpha \sin\beta)^2 + \delta_{zz}{}^{PAS}(\cos\beta)^2,$$

with $\alpha$ and $\beta$ being the Euler angles relating the principal axis frame to the coordinate system. If there is axial symmetry, the orientation dependence simplifies further to

$$\delta_{zz}{}^{LAB} = \delta_{iso} + (\delta_{zz}{}^{PAS} - \delta_{iso})(3\cos^2\beta - 1)/2,$$

with the isotropic chemical shift $\delta_{iso} = 1/3\ (\delta_{xx} + \delta_{yy} + \delta_{zz})$.

The fluorine chemical shift anisotropy of $^{19}$F-labeled amino acids, such as 4F-phenylglycine or 5F-tryptophan, has been used to characterize the orientation of membrane embedded peptides (Afonin *et al* 2003, Grage *et al* 2002). A prerequisite for this analysis is the knowledge of the CSA tensor and how its principal axes are aligned with respect to the amino acid scaffold.

Likewise, the $^{19}$F–$^{19}$F dipolar coupling has been used for orientation determination. The intra-segmental dipolar coupling of a $CF_3$-label is particularly useful for determining molecular orientations. Using $CF_3$-labeled amino acids, such as the ones depicted in figure 6.2, numerous peptides have been characterized in terms of their alignment by solid-state $^{19}$F NMR in oriented membranes (Glaser *et al* 2005, Komarov *et al* 2019, Strandberg and Ulrich 2018, 2019). By measuring the local orientations of at least four separate molecular segments that are fixed to the peptide backbone, the conformation and dynamical behavior of virtually any folded peptide backbone can be determined (Grage *et al* 2017, Komarov *et al* 2019, Strandberg and Ulrich 2019).

In oriented bilayer samples, the $CF_3$ dipolar coupling results in a triplet with two equal splittings of

$$\Delta = \Delta_0(3\cos^2\beta - 1)/2,$$

leading to a maximum splitting between an outer and the center line of $\Delta_0 = 15$ kHz. From the value splitting $\Delta$, the angle $\beta$ between the methyl axis and the external field can be directly determined.

### 6.6.2 Mobility and orientation

Membranes present a highly mobile environment, and molecular motion has to be considered in the analysis of molecular orientation in the bilayer. In fluid membranes, even large membrane proteins undergo rotational diffusion around the membrane normal on time scales that lead to motional averaging. In addition, molecules embedded in the lipid bilayer, and even more so the lipids themselves, do not adopt a single orientation, but assume a distribution of alignments which is motionally averaged. The degree of motional averaging can be characterized by a set of order parameters, which describe how the different tensor components of an interaction tensor contribute to the observed NMR frequency (Nehring and Saupe 1970). For rotational diffusion and axially symmetric wobble, the effect of mobility can be described by a single order parameter $S_{zz}$, which scales the anisotropic part of the chemical shift or the dipolar interaction:

$$\delta_{zz}{}^{LAB} = \delta_{iso} + S_{zz}\left(\delta_{xx}{}^{PAS}(\cos\alpha\sin\beta)^2 + \delta_{yy}{}^{PAS}(\sin\alpha\sin\beta)^2 + \delta_{zz}{}^{PAS}(\cos\beta)^2 - \delta_{iso}\right)$$

$$\Delta = S_{zz}\,\Delta_0(3\cos^2\beta - 1)/2$$

$S_{zz}$ can assume values between $-1/2$ and 1. $S_{zz} = 0$ corresponds to isotropic motion, whereas in the absence of motion reducing the uniaxial order, $S_{zz} = 1$.

### 6.6.3 Molecular orientation in membranes from $^{19}$F NMR on oriented samples

Using oriented samples provides direct access to the orientation of molecules in membranes. When aligning the bilayer normal parallel to the magnetic field, all embedded proteins or peptides assume the same alignment with respect to $B_0$. The observed chemical shift of a specific $^{19}$F-label, or the dipolar coupling within an F–F pair or a $CF_3$ group, is directly related to the orientation of the corresponding tensor with respect to the membrane normal. This orientation-dependent chemical shift or dipolar coupling, however, does not encode the orientation of the molecule of interest, but only that of the label represented by the tensor. For example, from the triplet splitting observed for a $CF_3$ group, the orientation of this particular trifluoromethyl group can be deduced, but not yet the orientation of the labeled molecule.

If, by the design of the amino acid, the $^{19}$F-label is rigidly attached to the peptide backbone, the observable chemical shift or dipolar splitting then provides information on the local orientation of the labeled polypeptide segment with respect to the membrane normal. If the peptide assumes a well-defined fold, such as an $\alpha$-helix, the structure and mobility of the entire molecule can be defined from only few orientational constraints.

In order to translate the orientation information obtained for the labels to the orientation of the entire molecule of interest, the following approach has been used.

Firstly, $^{19}$F-labels are attached to the backbone through rigid linkers, to avoid ambiguity in orientation due to multiple conformations. For example, artificial amino acids such as F-phenylglycine, $CF_3$-phenylglycine and $CF_3$-bicyclopentylglycine have been used, which connect a fluorine or $CF_3$-label directly to the peptide backbone without additional rotatable bonds (Afonin *et al* 2007, Glaser *et al* 2004, Glaser *et al* 2005, Mühlhäuser *et al* 2017, Komarov *et al* 2019, Strandberg and Ulrich 2018, 2019).

Secondly, the NMR data from a set of several such labels is combined, in order to pin down the molecular orientation. The orientation found for a single label, namely, is compatible with a range of molecule orientations. Therefore, additional labels are required to narrow down the range of molecular orientations which agree with the data. Typically, at least four labels are necessary to determine the orientation of a well-folded secondary structure element. In the case of peptides, the selective $^{19}$F-labels are introduced one at a time, in order to avoid spectral overlap and assignment ambiguities, resulting eventually in the combined analysis of four or more individual samples.

In practice, the data from the individual labels is combined in an rmsd (root of mean square deviation) calculation, which reveals the molecular orientation. For this, the orientation of the molecule of interest is parameterized by two suitable angles, e.g., Euler angles $\alpha$ and $\beta$, or the tilt angle $\tau$ and an azimuthal angle $\rho$ in the case of a helical peptide, which define the orientation of a molecule fixed coordinate system with respect to the membrane normal. Together with the order parameter $S_{zz}$, which describes the extent of motional averaging due to molecular mobility, three parameters ($\alpha$, $\beta$, $S_{zz}$) need to be determined from the rmsd analysis. The best fitting peptide orientation and order parameter can be found by calculating the $^{19}$F NMR data for the molecule in a given orientation described by $\alpha$, $\beta$ and $S_{zz}$. This essentially involves coordinate transformations of the PAS systems characterizing the orientation of the individual labels to the laboratory frame of the oriented membrane. A well-defined backbone-fold has to be assumed at this stage, which is often known from circular dichroism. Alternatively, several different secondary structures can be assumed and tested, as to which conformation is best represented by the set of NMR data. By comparing the predicted $^{19}$F NMR parameters with the experimental data in the form of an rmsd, and by systematically repeating the calculation for all possible combinations of $\alpha$, $\beta$ and $S_{zz}$, the best fit can be found from the minimum in the rmsd plot.

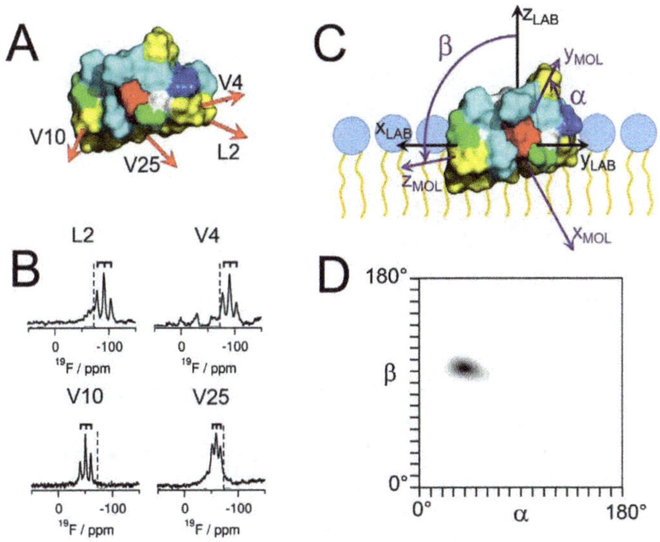


**Figure 6.7.** Orientation of the knotted cyclotide kalata B1 as determined by solid-state $^{19}$F NMR in oriented phospholipid (DMPC/DMPE) bilayers (Grage *et al* 2017, reprinted with permission from Elsevier). (A) Four separate $CF_3$-bicyclopentylglycine labels were incorporated into kalata B1, to sample different directions with respect to the membrane normal. (B) All solid-state $^{19}$F NMR spectra of the four $CF_3$-groups gave well-resolved dipolar triplets. (C) The orientation of kalata B1 can be described by two angles $\alpha$ and $\beta$, relating a molecule fixed frame to the membrane normal. (D) The best fit molecular orientation is determined as the minimum in the rmsd plot, which compares the agreement between the experimental NMR splittings and the calculated splittings for any combination of $\alpha$ and $\beta$ (as well as $S_{zz}$, which is not illustrated here).

The data evaluation procedure is illustrated in figure 6.7 on the example of kalata B1, which presents a distinctly folded cyclic backbone that is held together by three disulfide bridges. Kalata B1 is a cyclotide of plant origin with multiple activities, of which many are speculated to involve interaction with membranes. To get insight into its interactions with biomembranes, the orientation of kalata B1 was determined using solid-state $^{19}$F NMR in oriented membranes (Grage *et al* 2017). Four $CF_3$-bicyclopentylglycine labels were placed into the rigid scaffold of this knotted protein at different positions one by one. Based on the available solution NMR structure, these positions were chosen such that the resulting $CF_3$-methyl directions would not yield redundant structural information (figure 6.7A). $^{19}$F NMR spectra were acquired of the four labeled kalata B1 analogues reconstituted in oriented bilayers (figure 6.7B). The best fit solution of the rmsd analysis (figure 6.7C), showed an orientation by which several residues involved in lipid interactions are indeed favorably oriented towards the bilayer for contacts with the lipid headgroups.

The alignment approach has been particularly useful for the analysis of $\alpha$-helical membrane-active peptides. The helix alignment, given by a tilt angle $\tau$ and azimutal rotation angle $\rho$ (figure 6.8A), is of fundamental importance to understand the function of such peptides, e.g., to reveal pore formation or dimerization. A helical structure generally leads to a periodic repetition of the orientation chemical shift or

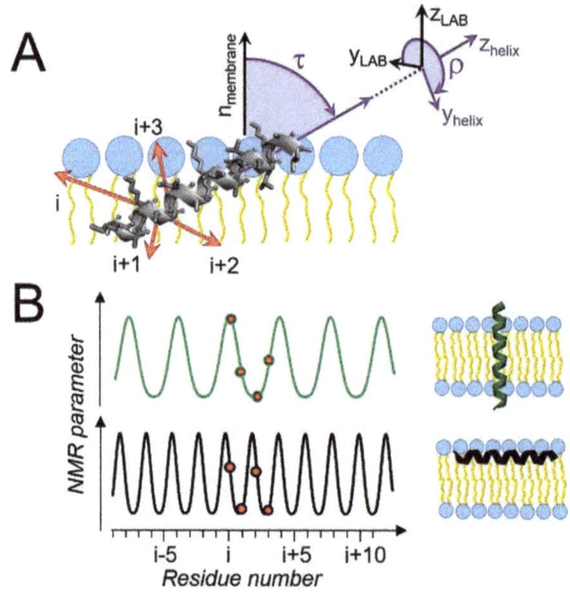


**Figure 6.8.** The orientation of an $\alpha$-helix is from solid-state $^{19}$F NMR in oriented bilayers, as illustrated for the antimicrobial peptide PGLa. (A) At least four $^{19}$F-labels are introduced into the peptide, sampling different local orientations. The peptide orientation can be described by a tilt angle $\tau$ (turn around $y_{\text{helix}}$) and an azimuthal rotation angle $\rho$ (turn around the helix axis $z_{\text{helix}}$). (B) The $^{19}$F NMR signals, e.g. the orientation-dependent splittings of four $CF_3$-labels, follow a wave-like pattern as a function of the labeled residue position. From the shape and phase of this dipolar wave, the tilt and azimuthal rotation angles can be extracted.

dipolar coupling when moving the label position along the helix. As a result, the measured NMR frequencies follow a wave-like behavior as a function of the position along the helix, from which $\tau$, $\rho$ and an order parameter $S_{zz}$ can be deduced (figure 6.8B).

In regular peptide structures like helices, however, it can be difficult to apply the $^{19}$F NMR approach, because there may not be enough independent directions that can be sampled by a single type of hydrophobic label such as $CF_3$-bicyclopentyl-glycine. This issue arises particularly in amphiphilic helices, where it has been difficult to label the polar face. A further limitation is presented by the fact that most artificial amino acids used for $^{19}$F NMR will sample the same direction, namely the vector along the C$\alpha$–C$\beta$ bond. A solution can be orthogonal labeling, whereby geometrically different labels are combined to probe different local directions. Such an approach has been used on the peptaibols harzianin and alamethicin (Grage *et al* 2018, Maisch *et al* 2009). To analyze their orientations in membranes, both L- and D-stereoisomers of a $^{19}$F-labeled aminoisobutyric acid derivative were used to double the number of sampled directions. Similarly, $CF_3$-methanoproline and a $CF_3$-labeled cyclobutane-derived lysine analogue have been combined with $CF_3$-bicyclopentylglycine to obtain a set of orthogonal labels (Afonin *et al* 2017, Michurin *et al* 2016)

The question may arise as to how reliable the determination of orientations is when using $^{19}$F NMR, given the chemical modifications needed to introduce fluorine into the molecule of interest. In comparative studies, the solid-state $^{19}$F NMR approach has been shown to give the same results as the respective non-fluorinated compound. For example, 6F-cholesterol was found to adopt the same orientation in the bilayer as normal cholesterol (Matsumori *et al* 2008). Also, for the peptide PGLa, the orientation obtained from largely non-perturbing $^{2}$H NMR labels and from $^{19}$F NMR was found to be identical within experimental error (Strandberg *et al* 2006, Wang 2014). Tailor-made other artificial amino acids that were designed to introduce a $^{19}$F-label rigidly linked to the backbone, have been found to give comparable results on $\alpha$-helical membrane-bound peptides, in full agreement with $^{2}$H NMR using conservative labels (Afonin *et al* 2007, Michurin *et al* 2016).

### 6.6.4 Peptide re-alignment transitions in membranes revealed by solid-state $^{19}$F NMR

When determining the orientation of membrane-active peptides using solid-state $^{19}$F NMR, some interesting dynamic behavior has been observed in several instances. Of particular interest has been the alignment of amphiphilic peptides, often $\alpha$-helices that show antimicrobial, fusogenic or cell-penetrating activity. Whether the alignment of the such membrane-active peptides is parallel to the surface, obliquely tilted or inserted transmembrane, can give important clues on their functional mechanism, and reveal for example their propensity for pore formation. In many cases a transition between different alignment states of such peptides has been observed. Many amphiphilic $\alpha$-helical helices such as PGLa have been found by $^{19}$F NMR to adopt a flat orientation parallel to the bilayer surface at low concentration, which changes to a tilted or even fully inserted transmembrane state above a concentration threshold (Glaser *et al* 2005). Various physicochemical factors, like temperature, pH, or any concomitant changes of the lipid phase state, can affect the re-alignment (Afonin *et al* 2008b, Misiewicz *et al* 2015b).).

PGLa, for example, was found to be aligned parallel to the membrane surface at high temperature, to assume a tilted orientation near the lipid phase transition, and to become inserted in a transmembrane alignment in the lipid gel phase (figure 6.9). Also, many other peptides have been observed to respond to the membrane phase state, according to $^{19}$F NMR. For example, the cyclic peptide gramicidin S can insert into the membrane at temperatures near the lipid phase transition, possibly forming a pore of hydrogen-bridged $\beta$-strands (Afonin *et al* 2008a). The lipid composition also influences the orientation of amphiphilic peptides. Using solid-state $^{19}$F NMR amongst other techniques, the spontaneous curvature of lipids has been found to have a significant influence on peptide insertion. For example, unsaturated lipids with a negative spontaneous curvature prevent the insertion of PGLa or gramicidin S (Afonin *et al* 2014). Lyso-lipids with a positive spontaneous curvature, on the other hand, tend to promote insertion of peptides in a tilted or transmembrane alignment (Strandberg and Ulrich 2018).

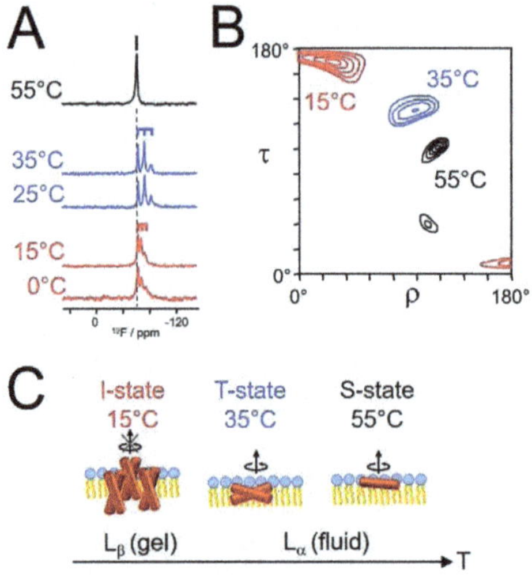


**Figure 6.9.** Temperature dependent re-alignment of amphiphilic peptides. (A) The solid-state [19]F NMR spectra of PGLa (labeled with $CF_3$-bis-cyclopentylglycine and reconstituted in oriented DMPC/DMPG bilayers) were found to change as a function of temperature. (B) Three different orientations could be identified at different temperatures. (C) The peptide assumes an inserted state (I-state) at temperatures below the lipid phase transition, which changes to an obliquely tilted state (T-state) above the phase transition temperature, and finally assumes an alignment parallel to the membrane surface (S-state) at higher temperatures.

### 6.6.5 Peptide mobility in membranes

The anisotropy inherent in $^{19}$F dipolar couplings or the chemical shift carries both information on peptide orientation and peptide mobility. This way, $^{19}$F NMR has also been used to get insight into the molecular mobility of peptides associated with membranes. For example, monitoring the dynamics of peptides by $^{19}$F NMR in oriented bilayers has revealed an interesting aggregation behavior of $\beta$-strand forming peptides on the membrane surface. The model peptides MAP or $(KIGAKI)_3$, which can form amphiphilic $\beta$-strands, for example, have been found to aggregate to form immobilized $\beta$-sheets on the bilayer above a threshold concentration, but to be a flexible chain at low concentrations (Wadhwani *et al* 2012, Wadhwani *et al* 2013). The two forms have been distinguishable using a $CF_3$-bicyclopentylglycine label, which gives rise to a triplet splitting of different size depending on the molecular mobility. Interestingly, the aggregation could be prevented by introducing D-amino acids, where also the $CF_3$-label itself could be used as $\beta$-sheet blocker (Wadhwani *et al* 2013).

### 6.6.6 Orientation and mobility of lipids and small lipid-soluble molecules

Solid-state $^{19}$F NMR has been particularly useful to determine the orientation and mobility of small molecules in membranes. The question of membrane behavior is of interest, for example, for many drug molecules, as they may act in the membrane or need to cross cellular membranes to reach their target. Often small molecule drugs carry fluorine atoms to endow them with sufficient lipophilicity, which renders them ideal candidates for $^{19}$F NMR. But, also the orientation of lipids and sterols such as cholesterol is important to understand, e.g. the packing of lipids in various lipid phases.

In order to apply the strategy to determine the orientation in the bilayer described above, a rigid molecular scaffold is necessary, which for many molecules like sterols is the case. Furthermore, a set of orientation constraints is needed to pin down the molecular orientation and order parameter. This has been achieved, for example, in an orientation study of 6F-cholesterol by combining $^{19}$F chemical shift anisotropy, $^{2}$H quadrupolar couplings, and $^{13}$C–$^{19}$F dipolar couplings (Matsumori *et al* 2008). The orientation of 6F-cholesterol has been found to be quite similar to that of unmodified cholesterol, with an order parameter of 0.85.

The $^{19}$F dipolar coupling multiplets of multiply $^{19}$F-labeled guest molecules has provided another way to analyze molecular orientation and mobility. This way drugs like diflucortolone valerate or flufenamic acid have been characterized in oriented bilayers by $^{19}$F NMR (Dürr *et al* 2012, Grage and Ulrich 1999, Grage *et al* 2000b). A combination of $^{19}$F chemical shift anisotropy and $^{19}$F–$^{1}$H dipolar couplings, obtained in $^{19}$F–$^{1}$H PISEMA experiments, has been used in another example to determine the orientation of a fluorinated aniline analogue (Hughes *et al* 2018).

$^{19}$F-labeled lipids were of interest early on as probes for molecular mobility of model and native membranes. Using fluorine-substituted fatty acids, they have been incorporated into bacterial membranes of *E. coli* and *A. ladlawii* (Gent *et al* 1978, Gent *et al* 1981, McDonald *et al* 1983, Macdonald *et al* 1985, McDonough *et al* 1983). Here, lipids with $CF_2$ groups have turned out to possess order parameters which are significantly lower than non-fluorinated ones, whereas lipids with only a single fluorine present suitable probes for lipid mobility (Oldfield *et al* 1980, Gagnon *et al* 2018a). A number of different fluorine-substituted lipid analogues, mostly of DMPC or DPPC, has been synthesized and characterized in terms of their changes of physical membrane properties (Gent *et al* 1978, Post *et al* 1984, Gagnon *et al* 2018b). It has been found that the impact of the fluorine substitution on the mobility and phase behavior of the lipid depends crucially on the number of fluorine atoms and their position along the chain (Gagnon *et al* 2018a, Gagnon *et al* 2018b, Hirsh *et al* 1998, Sanii *et al* 2010).

## 6.7 Probing the environment

The high sensitivity of the $^{19}$F chemical shift to the chemical surroundings of the fluorine label provides a unique and simple tool to probe the molecular environment. Due to the lone pair electrons, the $^{19}$F chemical shift responds to small changes in the surrounding. This way, both solution and solid-state $^{19}$F NMR has been able to detect conformational changes in membrane proteins and the proximity to lipids or the aqueous environment.

### 6.7.1 Conformational changes

Conformational changes have been observed on the basis of $^{19}$F chemical shifts in various membrane proteins, such as receptors, channels, or even GPCRs. For example, 5F-Trp labels exhibited chemical shift changes of up to ~2 ppm in the leucine receptor protein upon ligand binding (Salopek-Sondi and Luck 2002). Binding of amantadine and associated conformational changes have been monitored through $^{19}$F chemical shift differences of 5F-Trp in the influenza A proton channel M2 of similar size (Kolocouris *et al* 2007). In the chloride channel CLC-ec1, substrate-dependent changes of the $^{19}$F chemical shift yielded insight into conformational changes (Elvington *et al* 2009).

Thanks to its high sensitivity, $^{19}$F NMR provides a possibility to tackle even large and challenging membrane proteins such as GPCRs. Using site directed $^{19}$F-labeling of cysteines with TET, conformational differences between the active and inactive state have been determined for the $\beta$-adrenergic receptor and for rhodopsin (Liu *et al* 2012, Klein-Seetharaman *et al* 1999). Interestingly, the observed shift changes in the region of a few tenths of a ppm up to 2 ppm could be attributed to ring current shifts of aromatic residues close to the fluorine label (Liu and Wüthrich 2016), which opens up new strategies for structural studies of large membrane proteins. Several conformers could also be revealed for the adenosin receptor $A_{2AA}R$ on the basis of different $^{19}$F chemical shifts of site-directed TET labels (Sušac *et al* 2018). Four different populations were found, distinguishable by chemical shift differences of a couple of ppm, and an exchange between the conformers on the timescale of ~1 s was detected using 2-dimensional $^{19}$F NMR exchange spectroscopy. Much larger changes in $^{19}$F chemical shift of up to 35 ppm have been observed for $^{19}$F-labeled retinal in rhodopsin (Brinkmann *et al* 2018). This extraordinary shift was attributed to the isomerization of retinal during the transition between the light-adapted and dark state of rhodopsin.

### 6.7.2 Probing solvent environment

Many structural problems in biomembranes involve the question as to whether a molecule or part of a molecule is exposed to the lipid or to the aqueous phase. Here, the $^{19}$F chemical shift has proven to be a valuable reporter for the solvent environment, helping to identify lipid or water exposed molecular surfaces. The sensitivity of $^{19}$F to distinguish different environments, for example, became apparent in an interesting observation, when a fluorine-carrying inhibitor of the ATP citrate lyase was reconstituted into small unilamellar vesicles. Two $^{19}$F NMR signals with a chemical shift difference of ~0.5 ppm were observed and attributed to the inhibitor being accommodated either in the inner or outer leaflet of the SUV. Due to the high curvature of SUVs, the lipid packing seems to be sufficiently different to create a distinct environment or to lead to a different immersion depth of the inhibitor (Reid *et al* 1995).

In favorable cases, different $^{19}$F chemical shifts in lipid and aqueous environment can be directly compared if the exposure to either environment can be triggered, e.g. by a conformational change. Such a change in $^{19}$F chemical shift has been observed for the neuronal calcium sensing protein recoverin. This soluble protein carries a lipid anchor, which is extruded and able to bind to membranes in the absence of calcium. Using $^{19}$F-labels in the lipid anchor, different binding modes could be distinguished on the basis of environment dependent $^{19}$F chemical shifts (Potvin-Fournier *et al* 2018). The hydrophobic environment of lipid bilayers was also found to be responsible for $^{19}$F chemical shift changes observed in a model study of peptide–lipid interactions (Suzuki *et al* 2011). The antimicrobial peptide MSI-78, representing a typical amphiphilic cationic helix, was labeled with trifluoroethanol-glycine in eight positions in order to examine chemical shift changes upon binding to bicelles. A reduction of the chemical shift of up to ~ −1 ppm upon binding was observed for most label positions and attributed to an exposure to the hydrophobic environment. Similarly, residues involved in interactions with lipid membranes were identified on the basis of $^{19}$F chemical shift changes in the cytolysins listeriolysin O and equinatoxin II (Anderluh *et al* 2005, Kozorog *et al* 2018). Several 5F-tryptophan $^{19}$F-NMR signals showed differences in chemical shift of up to ~1 ppm when the toxin was either in aqueous solution, or reconstituted into cholesterol-rich vesicles or sphingomyelin-containing detergent micelles, which gave important clues on the protein regions involved in membrane binding and the recognition of specific lipids such as cholesterol or sphingomyelin.

Another way to probe for exposure to a lipid environment has been the use of oxygen as a paramagnetic shift reagent. Due to its high lipid solubility, $O_2$ accumulates in the membrane, leading to a paramagnetic gradient in the bilayer. The interaction between $O_2$ and $^{19}$F-labels thus leads to changes in $^{19}$F chemical shift and $T_1$ relaxation, depending on the immersion depth of the $^{19}$F-label. Such $^{19}$F chemical shift changes and $T_1$ relaxation enhancement have been attributed to contact interaction and dipolar interaction, respectively (Prosser and Luchette 2004). Using oxygen induced chemical shift and relaxation changes in high pressure $^{19}$F NMR, the immersion depth of several $^{19}$F-labeled amphiphiles such as cholesterol were determined (Prosser *et al* 2001). This approach has also been applied to elucidate the structure and topology of membrane proteins. In diacylglycerolkinase (DAGK) labeled with bromo-trifluoro-propanone on cysteine residues in transmembrane helix 1 (TM1), chemical shift changes of ~1 ppm were found to be induced by 100 atm $O_2$. From the resulting $^{19}$F chemical shift variations along TM1, the helix structure and immersion could be determined.

To evaluate contact to water, even some subtle changes in $^{19}$F chemical shift due to isotope effects have been explored. Replacing $H_2O$ with $D_2O$ has been found to have a measurable effect on $^{19}$F chemical shifts on the order of some 0.1 ppm, which provides a versatile tool to probe water accessibility of, e.g., membrane-active peptides. For example, residues exposed to water or to the lipid environment have been characterized in the antimicrobial peptide tritrpticin by means of different $^{19}$F chemical shifts in $H_2O$ or $D_2O$ of 4F-Phe and 3F-Tyr labels (Arias *et al* 2020). Water exposed residues have been also revealed for the model antimicrobial peptide MSI-78 labeled with trifluoroethanol-glycine bound to bicelles (Suzuki *et al* 2011). Here, solvent-isotope induced differences in chemical shifts of ~0.1 ppm could be observed, which were found to be decreased in the case of lipid exposed residues upon binding to bicelles.

## 6.8 $^{19}$F NMR in native membranes

Given the high sensitivity of the $^{19}$F nucleus and the absence of a $^{19}$F background, $^{19}$F NMR studies of membrane proteins or peptides in native membranes are in reach. Questions like how membrane-associated peptides or membrane proteins bind to and are structured in native membranes, or even in the context of whole cells, have been addressed. $^{19}$F NMR thus provides a sound basis for future developments towards *in situ* and *in vivo* NMR.

In most NMR studies, biological membranes are usually substituted by model lipid bilayers or even by highly curved membrane mimetics. Hence, the question arises as to whether the interaction with such models indeed represents the genuine behavior in native cell membranes. In particular, in the case of smaller molecules such as membrane-active peptides, the conformation, the membrane alignment, and the state of oligomerization is known to depend strongly on the actual membrane environment. Therefore, using solid-state $^{19}$F NMR in oriented bilayers, the structural behavior of antimicrobial peptides has been addressed in native bacterial as well as eukaryotic membranes. Here, it could be shown that many representative examples, such as PGLa, gramicidin S or BP100, adopt the same orientation in bacterial membranes and erythrocyte ghosts as they do in fluid model membranes (Ieronimo *et al* 2010, Koch *et al* 2012, Misiewicz *et al* 2015a). The re-alignment observed for many amphiphilic peptides in model membranes may, however, be less clearly visible in native membranes. For example, the antibiotics PGLa and gramicidin S have been described to tilt obliquely or insert in model membranes near the lipid phase transition, but such behavior was not observed in bacterial membranes or erythrocyte ghosts (Koch *et al* 2012).

$^{19}$F NMR has also provided insights into the interaction of larger proteins with cellular plasma membranes. For example, the binding process of alpha-synuclein coupled to a cell penetrating peptide has been monitored in eukaryotic cells using $^{19}$F NMR. By introducing several 3F-Tyr labels into the protein, the regions of the protein interacting with membranes could be identified, as their intensity has been reduced upon binding (Zigoneanu and Pielak 2012). In another study, DAGK was studied in *E. coli* membranes using *in situ* solid-state MAS $^{19}$F NMR. By comparing the $^{19}$F chemical shift obtained for $CF_3$-phenylalanine labels in detergent micelles and in *E. coli* membranes, the influence of the lipid environment could be characterized (Shi *et al* 2012).

## 6.9 Conclusions

The unique properties of the $^{19}$F nucleus, such as its high sensitivity, long-range dipolar couplings, sensitive chemical shifts, and the lack of a natural abundance background in biological systems, render $^{19}$F NMR a powerful tool for structural studies of biomembranes. As the introduction of a $^{19}$F-label can lead to slight modifications in the system of interest, and is limited by chemical constraints, the use of $^{19}$F NMR may not be as universal as, for example, $^{1}$H, $^{13}$C or $^{15}$N NMR. Also, depending on the applied technique, some spectroscopic properties such as large $^{19}$F anisotropies or short relaxation times may have a limiting effect. However, $^{19}$F NMR has proven to be a valuable option for many questions whose answer can be addressed by targeted selective labeling and by exploiting the particular strengths of fluorine. Especially in structural biology and biophysics of biomembranes, where it is often difficult to find applicable techniques, such specific pinpointed insights from $^{19}$F NMR can be very valuable.

## Acknowledgements

We acknowledge the Helmholtz Association (Program BIF-TM) and a DFG instrument grant INST 121384/58–1 for financial support.

## References

Afonin S, Dürr U H N, Wadhwani P, Salgado J B and Ulrich A S 2008a Solid state NMR structure analysis of the antimicrobial peptide Gramicidin S in lipid membranes: concentration-dependent re-alignment and self-assembly as a β-barrel *Topics Curr. Chem.* **273** 139–54

Afonin S, Glaser R W, Berditchevskaja M, Wadhwani P, Gührs K H, Möllmann U and Ulrich A S 2003 4-Fluoro-phenylglycine as a label for $^{19}$F-NMR structure analysis of membrane-associated peptides *ChemBioChem* **4** 1151–63

Afonin S, Mikhailiuk P, Komarov I V and Ulrich A S 2007 Evaluating the amino acid $CF_3$-bicyclopentylglycine as a new label for solid-state $^{19}$F-NMR structure analysis of membrane-bound peptides *J. Pept. Sci.* **13** 614–23

Afonin S, Grage S L, Ieronimo M, Wadhwani P and Ulrich A S 2008b Temperature-dependent transmembrane insertion of the amphiphilic peptide PGLa in lipid bilayers observerd by solid state $^{19}$F-NMR *J. Am. Chem. Soc.* **130** 16512–4

Afonin S, Glaser R W, Sachse C, Salgado J, Wadhwani P and Ulrich A S 2014 $^{19}$F NMR screening of unrelated antimicrobial peptides shows that membrane interactions are largely governed by lipids *Biochim. Biophys. Acta—Biomembr* **1838** 2260–8

Afonin S, Kubyshkin V, Mykhailiuk P K, Komarov I V and Ulrich A S 2017 Conformational plasticity of the cell-penetrating peptide SAP as revealed by solid-state $^{19}$F-NMR and circular dichroism spectroscopies *J. Phys. Chem.* B **121** 6479–91

Anderluh G, Razpotnik A, Podlesek Z, Macek P, Separovic F and Norton R S 2005 Interaction of the eukaryotic pore-forming cytolysin equinatoxin II with model membranes: $^{19}$F NMR studies *J. Mol. Biol.* **347** 27–39

Arias M, Aramini J M, Riopel N D and Vogel H J 2020 Fluorine-19 NMR spectroscopy of fluorinated analogs of tritrpticin highlight a distinct role for Tyr residues in antimicrobial peptides *Biochim. Biophys. Acta* **1862** 183260

Arseniev A S, Kuryatov A B, Tsetlin V I, Bystrov V F, Ivanov V T and Ovchinnikov Y A 1987 $^{19}$F NMR study of 5-fluoro-tryptophan labeled bacteriorhodopsin *FEBS Lett.* **213** 283–8

Auger M, Lefèvre T, Otis F, Voyer N and Auger M 2019 Lipid membrane interactions of a fluorinated peptide with potential ion channel-forming ability *Pept. Sci.* **111** e24051

Brinkmann A, Sternberg U, Bovee-Geurts P H M, Fernández I F, Lugtenburg J, Kentgens A P M and DeGrip W J 2018 Insight into the chromophore of rhodopsin and its Meta-II

photointermediate by $^{19}$F solid-state NMR and chemical shift tensor calculations *Phys. Chem. Chem. Phys.* **20** 30174–88
Buffy J J, Waring A J and Hong M 2005 Determination of peptide oligomerization in lipid bilayers using $^{19}$F spin diffusion NMR *J. Am. Chem. Soc.* **127** 4477–83
Cavanagh J, Fairbrother W J, Palmer A G III and Skelton N J 1996 *Protein NMR Spectroscopy* (New York: Academic)
Cametti M, Crousse B, Metrangolo P, Milani R and Resnati G 2012 The fluorous effect in biomolecular applications *Chem. Soc. Rev.* **41** 31–42
Chekmenev E Y, Chow S K, Tofan D, Weitekamp D P, Ross B D and Bhattacharya P 2008 Fluorine-19 NMR chemical shift probes molecular binding to lipid membranes *J. Phys. Chem.* B **112** 6285–7
Danielson M A and Falke J J 1996 Use of $^{19}$F NMR to probe protein structure and conformational changes *Annu. Rev. Biophys. Biomol. Struct.* **25** 163–95
Devoisselle J M, Vion-Dury J, Confort-Gouny S, Coustaut D and Cozzone P J 1992 Liposomes containing fluorinated steroids: an analysis based on photon correlation and fluorine-19 nuclear magnetic resonance spectroscopy *J. Pharm. Sci.* **81** 249–54
Didenko T, Liu J J, Horst R, Stevens R C and Wüthrich K 2013 Fluorine-19 NMR of integral membrane proteins illustrated with studies of GPCRs *Curr. Opin. Struct. Biol.* **23** 740–47
Drouin M *et al* 2020 Monofluoroalkene-isostere as a $^{19}$F NMR label for the peptide backbone: Synthesis and evaluation in membrane-bound PGLa and $(KIGAKI)_3$ *Chem. Eur. J.* **26** 1511–7
Dürr U H N, Grage S L, Witter R and Ulrich A S 2008 Solid-state $^{19}$F-NMR parameters of fluorine-labelled amino acids (Part I): aromatic substituents *J. Magn. Res* **191** 7–15
Dürr U H N, Afonin S, Hoff B, de Luca G, Emsley J W and Ulrich A S 2012 Alignment of druglike compounds in lipid bilayers analyzed by solid-state $^{19}$F-NMR and molecular dynamics, based on dipolar couplings of adjacent $CF_3$ groups *J. Phys. Chem.* B **116** 4769–82
Elkins M R, Williams J K, Gelenter M D, Dai P, Kwon B, Sergeyev I V, Pentelute B L and Hong M 2017 Cholesterol-binding site of the influenza M2 protein in lipid bilayers from solid-state NMR *Proc. Natl Acad. Sci.* **114** 12946–51
Elvington S M, Liu C W and Maduke M C 2009 Substrate-driven conformational changes in ClC-ec1 observed by fluorine NMR *EMBO J.* **28** 3090–102
Engelsberg M, Dowd S R, Simplaceanu V, Cook B W and Ho C 1982 Nuclear magnetic resonance line-shape analysis of fluorine-19-labeled phospholipids *Biochemistry* **21** 6985–9
Engelsberg M and Yannoni C S 1990 The determination of bond lengths in solids using the Carr–Purcell Sequence *J. Magn. Reson.* **88** 393–400
Fowler D J, Weis R M and Thompson L K 2010 Kinase-active signaling complexes of bacterial chemoreceptors do not contain proposed receptor-receptor contacts observed in crystal structures *Biochemistry* **49** 1425–34
Gagnon M C, Strandberg E, Ulrich A S, Paquin J F and Auger M 2018a New insights into the influence of monofluorination on dimyristoylphosphatidylcholine membrane properties: a solid-state NMR study *Biochim. Biophys. Acta—Biomembr.* **1860** 654–63
Gagnon M C, Auger M and Paquin J F 2018b Progress in the synthesis of fluorinated phosphatidylcholines for biological applications *Org. Biomol. Chem.* **27** 4925–41
Gent M P N, Cottam P F and Ho C 1978 Fluorine-19 nuclear magnetic resonance studies of *Escherichia coli* membranes *Proc. Natl Acad. Sci.* **75** 630–4

Gent M P, Cottam P F and Ho C 1981 A biophysical study of protein-lipid interactions in membranes of *Escherichia coli.* Fluoromyristic acid as a probe. *Biophys. J* **33** 211–24

Gerling U I M *et al* 2014 Fluorinated amino acids in amyloid formation: a symphony of size, hydrophobicity and α-helix propensity *Chem. Sci.* **5** 819–30

Glaser R W, Sachse C, Dürr U H N, Wadhwani P and Ulrich A S 2004 Orientation of the antimicrobial peptide PGLa in lipid membranes determined from $^{19}$F-NMR dipolar couplings of 4-$CF_3$-phenylglycine labels *J. Mag. Res* **168** 153–63

Glaser R W, Sachse C, Dürr U H N, Wadhwani P, Afonin S, Strandberg E and Ulrich A S 2005 Concentration-dependent re-alignment of the peptide PGLa in lipid membranes observed by solid state $^{19}$F-NMR *Biophys. J.* **88** 3392–7

Gor'kov P L, Witter R, Chekmenev E Y, Nozirov F, Fu R and Brey W W 2007 Low-E probe for $^{19}$F-$^{1}$H NMR of dilute biological samples *J. Magn. Res* **190** 17–24

Grage S L and Ulrich A S 1999 Structural parameters from $^{19}$F homonuclear dipolar couplings, obtained by multipulse solid state NMR in static oriented systems *J. Mag. Res* **138** 98–106

Grage S L and Ulrich A S 2000a Orientation-dependent $^{19}$F dipolar couplings within a trifluoromethyl-group are revealed by multipulse solid state NMR *J. Mag. Res* **146** 81–8

Grage S L, Gauger D, Selle C, Pohle W, Richter W and Ulrich A S 2000b The amphiphilic drug flufenamic acid can induce hexagonal lipid phases in DMPC - a $^{31}$P and $^{19}$F-NMR study *Phys. Chem. Chem. Phys.* **2** 4574–9

Grage S L, Wang J, Cross T A and Ulrich A S 2002 Structure analysis of fluorine-labelled tryptophan side-chains in gramicidin A by solid state $^{19}$F-NMR *Biophys. J.* **83** 3336–50

Grage S L, Watts J A and Watts A 2004 $^{2}$H{$^{19}$F} REDOR for distance measurements in biological solids using a double resonance spectrometer *J. Magn. Reson.* **166** 1–10

Grage S L, Suleymanova A V, Afonin S, Wadwhani P and Ulrich A S 2006 Solid state NMR analysis of the dipolar couplings within and between distant $CF_3$-groups in a membrane-bound peptide *J. Magn. Reson.* **183** 77–86

Grage S L, Dürr U H N, Afonin S, Mikhailiuk P, Komarov I V and Ulrich A S 2008 Solid-state $^{19}$F-NMR parameters of fluorine-labelled amino acids (Part II): aliphatic substituents *J. Magn. Res* **191** 16–23

Grage S L, Xu X, Schmitt M, Wadhwani P and Ulrich A S 2014 $^{19}$F-labeling of peptides revealing long-range NMR distances in fluid membranes *J. Phys. Chem. Lett.* **5** 4256–9

Grage S L *et al* 2017 Orientation and location of the cyclotide kalata B1 in phospholipid bilayers revealed by solid-state NMR *Biophys. J.* **112** 630–42

Grage S L *et al* 2018 Orthogonal $^{19}$F-labeling for solid-state NMR spectroscopy reveals the conformation and orientation of short peptaibols in membranes *Chem. Eur. J.* **24** 4328–35

Hagen D S, Weiner J H and Sykes B D 1978 Fluorotyrosine M13 coat protein: fluorine-19 nuclear magnetic resonance study of the motional properties of an integral membrane protein in phospholipid vesicles *Biochemistry* **17** 3860–6

Hirsh D J, Lazaro N, Wright L R, Boggs J M, McIntosh T J, Schaefer J and Blazyk J 1998 A new monofluorinated phosphatidylcholine forms interdigitated bilayers *Biophys. J.* **75** 1858–68

Ho C, Rule G S and Pratt E A 1986 Nuclear magnetic resonance, biochemical, and molecular genetic studies of the membrane-bound d-lactate dehydrogenase of *Escherichia coli Biophys. J.* **49** 113–5

Hong M and Schmidt-Rohr K 2013 Magic-angle-spinning NMR techniques for measuring long-range distances in biological macromolecules *Acc. Chem. Res.* **46** 2154–63

Hou Y, Hu W, Li X, Skinner J J, Dongsheng Liu D and Wüthrich K 2017 Solvent-accessibility of discrete residue positions in the polypeptide hormone glucagon by $^{19}$F-NMR observation of 4-fluorophenylalanine *J. Biomol. NMR* **68** 1–6

Hughes E, Griffin J M, Coogan M P and Middleton D A 2018 Average orientation of a fluoroaromatic molecule in lipid bilayers from DFT-informed NMR measurements of $^{1}$H–$^{19}$F dipolar couplings *Phys. Chem. Chem. Phys.* **20** 18207–15

Ieronimo M, Afonin S, Koch K, Berditsch M, Wadhwani P and Ulrich A S 2010 F-NMR analysis of the antimicrobial peptide PGLa bound to native cell membranes from bacterial protoplasts and human erythrocytes *J. Am. Chem. Soc.* **132** 8822–4

Jackson J C, Hammill J T and Mehl R A 2007 site-specific incorporation of a $^{19}$F-amino acid into proteins as an NMR probe for characterizing protein structure and reactivity *J. Am. Chem. Soc.* **129** 1160–6

Kauffman J M, Westerman P W and Carey M C 2000 Fluorocholesterols, in contrast to hydroxycholesterols, exhibit interfacial properties similar to cholesterol *J. Lipid Res.* **41** 991–1003

Kim C G 2006 *PhD Thesis* University of Oxford

King C, Sarabipour S, Byrne P, Leahy D J and Hristova K 2014 The FRET signatures of noninteracting proteins in membranes: simulations and experiments *Biophys. J.* **106** 1309–17

Klein-Seetharaman J, Getmanova E V, Loewen M C, Reeves P J and Khorana H G 1999 NMR spectroscopy in studies of light-induced structural changes in mammalian rhodopsin: applicability of solution $^{19}$F NMR *Proc. Natl Acad. Sci.* **96** 13744–9

Koch K, Afonin S, Ieronimo M, Berditsch M and Ulrich A S 2012 Solid -state $^{19}$F-NMR of peptides in native membranes *Top. Curr. Chem.* **306** 89–119

Kolocouris A, Zikos C and Broadhurst R W 2007 $^{19}$F NMR detection of the complex between amantadine and the receptor portion of the influenza A M2 ion channel in DPC micelles *Bioorg. Med. Chem. Lett.* **17** 3947–52

Kokhan S O *et al* 2016 Design, synthesis, and application of an optimized monofluorinated aliphatic label for peptide studies by solid-state $^{19}$F NMR spectroscopy *Angew. Chem. Int. Ed.* **55** 14788–92

Komarov I V, Afonin S and Ulrich A S 2019 $^{19}$F-labelled amino acids for NMR structure analysis of membrane-bound peptides *Fluorine in Life Sciences: Pharmaceuticals, Medicinal Diagnostics, and Agrochemicals* (Progress in Fluorine Science series vol 4) ed G Haufe and F Léroux (London: Academic Press) pp 349–95

Kozorog M, Sani M A, Živković M L, Ilc G, Hodnik V, Separovic F, Plavec J and Anderluh G 2018 $^{19}$F NMR studies provide insights into lipid membrane interactions of listeriolysin O, a pore forming toxin from *Listeria monocytogenes Sci. Rep.* **8** 6894

Kubyshkin V S, Komarov I V, Afonin S, Mykhailiuk P K, Grage S L and Ulrich A S 2012a Trifluoromethyl-substituted α-amino acids as solid state $^{19}$F-NMR labels for structural studies of membrane-bound peptides *Fluorine in Pharmaceutical and Medicinal Chemistry: From Biophysical Aspects to Clinical Applications* ed V Gouverneur and K Müller (London: Imperial College Press) pp 91–138

Kubyshkin V S, Mykhailiuk P K, Afonin S, Ulrich A S and Komarov I V 2012b Incorporation of *cis*- and *trans*-4,5-difluoromethanoprolines into polypeptides *Org. Lett.* **14** 5254–7

Kubyshkin V S, Mykhailiuk P K, Afonin S, Grage S L, Komarov I V and Ulrich A S 2013 Incorporation of labile trans-4,5-difluoromethanoproline into a peptide as a stable label for $^{19}$F-NMR structure analysis *J. Fluor. Chem.* **152** 136–43

Kubyshkin V, Afonin S, Kara S, Budisa N, Mykhailiuk P K and Ulrich A S 2015 gamma-(S)-Trifluoromethyl proline: evaluation as a structural substitute of proline for solid state $^{19}$F-NMR peptide studies *Org. Biomol. Chem.* **13** 3171–81

Kubyshkin V, Grage S L, Bürck J, Ulrich A S and Budisa N 2018 Transmembrane polyproline helix *J. Phys. Chem. Lett.* **9** 2170–4

Kubyshkin V, Grage S L, Ulrich A S and Budisa N 2019 Bilayer thickness determines the alignment of model polyproline helices in lipid membranes *Phys. Chem. Chem. Phys.* **21** 22396–408

Laue K W, Mück-Lichtenfeld C and Haufe G 1999 Enantioselective syntheses of 2-amino-4-fluoropent-4-enoic acids. Isosteres of asparagine *Tetrahedon* **55** 10413–24

Liu J J, Horst R, Katritch V, Stevens R C and Wüthrich K 2012 Biased signaling pathways in $\beta_2$-adrenergic receptor characterized by $^{19}$F-NMR *Science* **335** 1106–10

Liu D S and Wüthrich K 2016 Ring current shifts in F-19-NMR of membrane proteins 2016 *J. Biomol. NMR* **65** 1–5

Loewen M C, Klein-Seetharaman J, Getmanova E V, Reeves P J, Schwalbe H and Khorana H G 2000 Solution $^{19}$F nuclear overhauser effects in structural studies of the cytoplasmic domain of mammalian rhodopsin *Proc. Natl Acad. Sci.* **98** 4888–92

Lübke M, Jung M and Haufe G 2013 New histone deacetylase inhibitors based on 4-fluoro-2-amino acid esters: synthesis and activity *J. Fluor. Chem.* **152** 144–56

Luo W and Hong M 2006 Determination of the oligomeric number and intermolecular distances of membrane protein assemblies by anisotropic $^1$H-driven spin diffusion NMR spectroscopy *J. Am. Chem. Soc.* **128** 7242–51

Maisch D, Wadhwani P, Afonin S, Böttcher C, Koksch B and Ulrich A S 2009 Chemical labeling strategy with (R)- and (S)-trifluoromethylalanine for solid state $^{19}$F-NMR analysis of peptaibols in membranes *J. Am. Chem. Soc.* **131** 15596–7

Mani R, Tang M, Wu X, Buffy J J, Waring A J, Sherman M A and Hong M 2006 Membrane-bound dimer structure of a β-hairpin antimicrobial peptide from rotational-echo double-resonance solid-state NMR *Biochemistry* **45** 8341–9

Marsh E N G 2000 Towards the nonstick egg: designing fluorous proteins *Chem. Biol.* **7** R153–7

Matson G B 1977 NMR probe double-tuned matching boxes for double-resonance FT experiments on hydrogen and fluorine *J. Magn. Reson.* **25** 481–3

Matsumori N, Kasai Y, Oishi T, Murata M and Nomura K 2008 Orientation of fluorinated cholesterol in lipid bilayers analyzed by $^{19}$F tensor calculation and solid-state NMR *J. Am. Chem. Soc.* **130** 4757–66

McDonald P M, McDonough B, Sykes B D and McElhaney R N 1983 Fluorine-19 nuclear magnetic resonance studies of lipid fatty acyl chain order and dynamics in *Acholeplasma laidlawii B* membranes. Effects of methyl-branch substitution and of trans unsaturation upon membrane acyl-chain orientational order *Biochemistry* **22** 5103–11

Macdonald P M, Sykes B D and McElhaney R N 1985 Fluorine-19 nuclear magnetic resonance studies of lipid fatty acyl chain order and dynamics in *Acholeplasma laidlawii B* membranes. A direct comparison of the effects of *cis* and *trans* cyclopropane ring and double-bond substituents on orientational order *Biochemistry* **24** 4651–9

McDonough B, Macdonald P M, Sykes B D and McElhaney R N 1983 Fluorine-19 nuclear magnetic resonance studies of lipid fatty acyl chain order and dynamics in *Acholeplasma laidlawii B* membranes. A physical, biochemical, and biological evaluation of monofluoropalmitic acids as membrane probes *Biochemistry* **22** 5097–03

Michurin O M, Afonin S, Berditsch M, Daniliuc C G, Ulrich A S, Komarov I V and Radchenko D S 2016 Delivering structural information on the polar face of membrane-active peptides: $^{19}$F-NMR labels with a cationic side chain *Angew. Chem. Int. Ed.* **55** 14595–9

Michurin O M, Tolmachova K, Afonin S, Babii O, Grage S L, Ulrich A S, Komarov I V and Radchenko D S 2018 Conformationally constrained mono-fluorinated arginine as a cationic label for solid-state $^{19}$F NMR analysis of membrane-bound peptides *Eur. J. Org. Chem.* **27-28** 3826–33

Middleton D A, Rankin S, Esmann M and Watts A 2000 Structural insights into the binding of cardiac glycosides to the digitalis receptor revealed by solid-state NMR *Proc. Natl Acad. Sci.* **97** 13602–7

Misiewicz J, Afonin S, Grage S L, van den Berg J, Strandberg E, Wadhwani P and Ulrich A S 2015a Action of the multifunctional peptide BP100 on native biomembranes examined by solid-state NMR *J. Biomol. NMR* **61** 287–98

Misiewicz J, Afonin S and Ulrich A S 2015b Control and role of pH in peptide-lipid interactions in oriented membrane samples *Biochim. Biophys. Acta, Biomembr.* **1848** 833–41

Mikhailiuk P K, Afonin S, Chernega A N, Rusanov E B, Platonov M O, Dubinina G G, Berditsch M, Ulrich A S and Komarov I V 2006 Conformationally rigid trifluoromethyl-substituted α-amino acid designed for peptide structure analysis by solid state $^{19}$F-NMR *Angew. Chem. Int. Ed.* **45** 5659–61

Mühlhäuser P, Wadhwani P, Strandberg E, Bürck J and Ulrich A S 2017 Structure analysis of the membrane-bound dermcidin-derived peptide SSL-25 from human sweat *Biochim. Biophys. Acta* **1859** 2308–18

Murata M, Kasai Y, Umegawa Y, Matsushita N, Tsuchikawa H, Matsumori N and Oishi T 2009 Ion channel complex of antibiotics as viewed by NMR *Pure Appl. Chem.* **81** 1123–9

Mykhailiuk P K, Afonin S, Palamarchuk G V, Shishkin O V, Ulrich A S and Komarov I V 2008 Synthesis of trifluoromethyl-substituted proline analogues as $^{19}$F NMR labels for peptides in the polyproline II conformation *Angew. Chem. Int. Ed* **47** 5765–7

Mykhailiuk P K, Voievoda N M, Afonin S, Ulrich A S and Komarov I V 2010 An optimized protocol for the multigram synthesis of 3-(trifluoromethyl)bicyclopent-[1.1.1]-1-ylglycine ($CF_3$-Bpg) *J. Fluor. Chem.* **131** 217–20

Neerathilingam M, Greene L H, Colebrooke S A, Campbell I D and Staunton D 2005 Quantitation of protein expression in a cell-free system: efficient detection of yields and $^{19}$F NMR to identify folded protein *J. Biomol. NMR* **31** 11–9

Nehring J and Saupe A 1970 Anisotropies of the $^{19}$F chemical shifts in fluorobenzene compounds from NMR in liquid crystals *J. Chem. Phys.* **52** 1307–10

Oldfield E, Lee R W, Meadows M, Dowd S R and Ho C 1980 Deuterium NMR of specifically deuterated fluorine spin probes *J. Biol. Chem.* **255** 11652–5

Post J F M, De Ruiter E E J and Berendsen H J C 1981 A fluorine NMR study of model membranes containing $^{19}$F-labeled phospholipids and an intrinsic membrane protein *FEBS Lett.* **132** 257–60

Post J F M, James E and Berendsen H J C 1982 Multipulse fluorine nmr experiments on lyotropic liquid crystals. I. Determination of order parameter tensor 1982 *J. Magn. Reson.* **47** 251–63 1982

Post J F M, Cook B W, Dowd S R, Lowe I J and Ho C 1984 Fluorine-19 nuclear magnetic resonance investigation of fluorine-19-labeled phospholipids. 1. A multiple-pulse study *Biochemistry* **1984** 6138–41

Potvin-Fournier K, Paillard G V, Gagnon M-C, Lefèvre T, Audet P, Cantin L, Paquin J F, Salesse C and Auger M 2018 Novel approaches to probe the binding of recoverin to membranes *Eur. Biophys. J.* **47** 679–91

Prosser R S, Luchette P A, Westerman P W, Rozek A and Hancock R E W 2001 Determination of membrane immersion depth with $O_2$: a high-pressure $^{19}F$ NMR study *Biophys. J.* **80** 1406–16

Prosser R S and Luchette P A 2004 An NMR study of the origin of dioxygen-induced spin-lattice relaxation enhancement and chemical shift perturbation *J. Magn. Reson.* **171** 225–32

Reid D R, John Jackson J, Tew D G and Gribble A D 1995 A fluorine-19 NMR study of interactions between the ATP-citrate lyase inhibitor SB-200517 and model membranes *Chem. Phys. Lipids* **75** 93–6

Roos M, Wang T, Shcherbakov A A and Hong M 2018 Fast magic-angle-spinning $^{19}F$ spin exchange NMR for determining nanometer $^{19}F$–$^{19}F$ distances in proteins and pharmaceutical compounds *J. Phys. Chem.* B **122** 2900–11

Rowland R S and Taylor R 1996 Intermolecular nonbonded contact distances in organic crystal structures: comparison with distances expected from van der Waals radii *J. Phys. Chem.* **100** 7384–91

Salgado J, Grage S L, Kondejewski L H, McElhany R N, Hodges R S and Ulrich A S 2001 Alignment of the antimicrobial β-sheet peptide gramicidin S in membranes: a solid state $^{19}F$-NMR study in oriented lipid bilayers *J. Biomol. NMR* **21** 191–208

Salnikov E S, Raya J, De Zotti M, Zaitseva E, Peggion C, Ballano G, Toniolo C, Raap J and Bechinger B 2016 Alamethicin supramolecular organization in lipid membranes from $^{19}F$ solid-state NMR *Biophys. J.* **111** 2450–9

Salopek-Sondi B and Luck L A 2002 $^{19}F$ NMR study of the leucine-specific binding protein of *Escherichia coli*: mutagenesis and assignment of the 5-fluorotryptophan-labeled residues *Protein Eng.* **15** 855–9

Sanii B, Szmodis A W, Bricarello D A, Oliver A E and Parikh A N 2010 Frustrated phase transformations in supported, interdigitating lipid bilayers *J. Phys. Chem.* B 2010 **114** 215–9

Shcherbakov A A and Hong M 2018 Rapid measurement of long-range distances in proteins by multidimensional $^{13}C$–$^{19}F$ REDOR NMR under fast magic-angle spinning *J. Biomol. NMR* **71** 31–43

Shcherbakov A A, Roos M, Kwon B and Hong M 2020 Two-dimensional $^{19}F$–$^{13}C$ correlation NMR for $^{19}F$ resonance assignment of fluorinated proteins *J. Biomol. NMR* **74** 193–204

Shi P, Wang H, Xi Z, Shi C, Xiong Y and Tian C 2011 Site-specific $^{19}F$ NMR chemical shift and side chain relaxation analysis of a membrane protein labeled with an unnatural amino acid *Prot. Sci.* **20** 224–8

Shi P, Li D, Chen H, Xiong Y, Wang Y and Tian C 2012 In situ $^{19}F$ NMR studies of an E. coli membrane protein *Protein Sci.* 2012 **21** 596–600

Sternberg U, Klipfel M, Grage S L, Witter R and Ulrich A S 2009 Calculation of fluorine chemical shift tensors for the interpretation of oriented $^{19}F$-NMR spectra of gramicidin A in membranes *Phys. Chem. Chem. Phys.* **11** 7048–60

Strandberg E, Wadhwani P, Tremouilhac P, Dürr U H N and Ulrich A S 2006 Solid state NMR analysis of the PGLa peptide orientation in DMPC bilayers: structural fidelity of $^2H$- versus high sensitvity $^{19}F$-NMR *Biophys. J.* **90** 1676–86

Strandberg E and Ulrich A S 2018 Solid-state $^{19}F$-NMR analysis of peptides in oriented biomembranes *Modern Magnetic Resonance* ed G A Webb (Cham: Springer) pp 651–67

Strandberg E and Ulrich A S 2019 Flow charts for the systematic solid-state 19F/2H-NMR structure analysis of membrane-bound peptides *Ann. Rep. NMR Spect* **99** 79–118

Sturtevant J M, Ho C and Reimann A 1979 Thermotropic behavior of some fluorodimyristoyl-phosphatidylcholines *Proc. Natl Acad. Sci.* **76** 2239–43

Sušac L, Eddy M T, Didenko T, Stevens R C and Wüthrich K 2018 $A_{2A}$-adenosine receptor functional states characterized by $^{19}$F-NMR *Proc. Natl Acad. Sci.* **115** 12733–8

Suzuki Y, Buer B C, Al-Hashimi H M and Marsh E N G 2011 Using fluorine nuclear magnetic resonance to probe changes in the structure and dynamics of membrane-active peptides interacting with lipid bilayers *Biochemistry* **50** 5979–87

Tang P, Hu J, Liachenko S and Xu Y 1999 Distinctly different interactions of anesthetic and nonimmobilizer withtransmembrane channel peptides *Biophys. J.* **77** 739–46

Tkachenko A N, Mykhailiuk P K, Afonin S, Radchenko D S, Kubyshkin V S, Ulrich A S and Komarov I V 2013 A $^{19}$F NMR label to substitute polar amino acids in peptides: a $CF_3$-substituted analogue of serine and threonine *Angew. Chem. Int. Ed* **52** 1486–9

Tkachenko A N, Mykhailiuk P K, Radchenko D S, Babii O, Afonin S, Ulrich A S and Komarov I V 2014 Design and synthesis of a monofluoro-substituted aromatic amino acid as a conformationally restricted $^{19}$F NMR label for membrane-bound peptides *Eur. J. Org. Chem.* **17** 3584–91

Toke O, O'Connor R D, Weldeghiorghis T K, Maloy W L, Glaser R W, Ulrich A S and Schaefer J 2004a Structure of $(KIAGKIA)_3$ aggregates in phospholipid bilayers by solid-state NMR *Biophys. J.* **87** 675–8

Toke O, Maloy W L, Kim S J, Blazyk J and Schaefer J 2004b Secondary structure and lipid contact of a peptide antibiotic in phospholipid bilayers by REDOR *Biophys. J.* **87** 662–74

Tsuchikawa H, Umegawa Y, Murata M and Oishi T 2018 A synthetic approach to the channel complex structure of antibiotic in a membrane: backbone $^{19}$F labeled amphotericin B for solid state NMR analysis *J. Syn. Org. Chem. Jpn.* **76** 1197–05

Ulrich A S 2007 Solid state $^{19}$F-NMR analysis of oriented biomembranes *Modern Magnetic Resonance* ed G A Webb (Berlin: Springer) pp 257–63

Umegawa Y, Adachi T, Matsumori N and Murata M 2012 Possible conformation of amphotericin B dimer in membrane-bound assembly as deduced from solid-state NMR *Bioorg. Med. Chem.* **20** 5699–704

Wadhwani P, Strandberg E, Heidenreich N, Bürck J, Fanghänel S and Ulrich A S 2012 Self-assembly of flexible β-strands into immobile amyloid-like β-sheets in membranes as revealed by solid-state $^{19}$F NMR *J. Am. Chem. Soc.* **134** 6512–5

Wadhwani P *et al* 2013 Stereochemical effects on the aggregation and biological properties of the fibril-forming peptide $[KIGAKI]_3$ in membranes *Phys. Chem. Chem. Phys.* **15** 8962–71

Wang Y, Zhao T, Dongqing Wei D, Erik Strandberg E, Ulrich A S and Ulmschneider J P 2014 How reliable are molecular dynamics simulations of membrane active antimicrobial peptides *Biochim. Biophys. Acta* **1838** 2280–8

Watts J A, Watts A and Middleton D A 2001 A model of reversible inhibitors in the gastric H+/K+-ATPase binding site determined by rotational echo double resonance NMR *J. Biol. Chem.* **276** 43197–204

Weinstein S, Durkin J T, Veatch W R and Blout E R 1985 Conformation of the gramicidin A channel in phospholipid vesicles: A fluorine-19 nuclear magnetic resonance study *Biochemistry* **24** 4374–82

Weinstein S, Wallace B A, Blout E R, Morrow J S and Veatch W 1979 Conformation of gramicidin A channel in phospholipid vesicles: a $^{13}C$ and $^{19}F$ nuclear magnetic resonance study *Proc. Natl Acad. Sci.* **76** 4230–4

Wi S, Sinha N and Hong M 2014 Long-range $^{1}H$–$^{19}F$ distance measurement in peptides by solid-state NMR *J. Am. Chem. Soc.* **126** 12754–5

Wilson M L and Dahlquist F W 1985 Membrane protein conformational change dependent on the hydrophobic environment *Biochem* **24** 1920–8

Yang Z, Jiménez-Osés G, López C J, Bridges M D, Houk K N and Hubbell W L 2014 Long-range distance measurements in proteins at physiological temperatures using saturation recovery EPR spectroscopy *J. Am. Chem. Soc.* **136** 15356–65

Yoder N C and Kumar K 2002 Fluorinated amino acids in protein design and engineering *Chem. Soc. Rev.* **31** 355–41

Zigoneanu I G and Pielak G J 2012 Interaction of α-synuclein and a cell penetrating fusion peptide with higher eukaryotic cell membranes assessed by $^{19}F$ NMR *Mol. Pharmaceutics* **9** 1024–9

# Chapter 7

# Structure, topology and dynamics of membrane-associated peptides by solid-state NMR

**Burkhard Bechinger and Evgeniy Salnikov**

## Summary

Oriented solid-state NMR has proven to provide valuable information about the conformation, topology and dynamics of peptides when inserted into extended phospholipid bilayers. Membranes have been aligned either mechanically as stacks of supported lipid bilayers or magnetically when prepared as anisotropic bicelles. Whereas liquid crystalline membranes of many different lipid compositions have been oriented along glass or polymer surfaces, the magnetic alignment of bicelles imposes more restrictions on the lipid composition and conditions. However, the internal motions of and inside the bicelles more easily result in well-resolved solid-state NMR spectra.

Whereas the angular constraints obtained by measuring anisotropic chemical shifts, dipolar and quadrupolar couplings have been used for a full-fledged structure determination of membrane-associated polypeptides, these are highly complementary to and have been combined with distance information from multidimensional solution or MAS solid-state NMR spectroscopy. Finally, recent developments using dynamic nuclear polarization to enhance the signal intensity of spectra from polypeptides in oriented membrane samples are presented.

## 7.1 Introduction

Membrane-associated peptides have many important functions in biological systems and in medical applications. They act as hormones, have regulatory function, exhibit antimicrobial activities, and have been developed as cell penetrating peptides for research and biomedicine. In order to understand their mechanisms of action detailed structural studies that reveal the interactions with and passage through membranes are necessary (Aisenbrey *et al* 2019b). On a supramolecular level a wide

doi:10.1088/978-0-7503-2532-5ch7  

variety of complex structural arrangements involving peptides and lipids are formed. On a molecular level both lipids and peptides can adopt a range of conformations which in most cases are best described by a dynamic distribution of conformational states rather than static models. Therefore, concepts that take into consideration flexibility and exchange are best suited for their analysis. For example, decades ago the concept of lipid order parameters has been introduced to describe lipid fatty acyl chains in bilayer assemblies, rather than referring to a defined molecular conformation (Seelig and Seelig 1980). In a related manner, membrane-associated peptides undergo conformational and topological changes. Matters are even more complex as lipids and polypeptides mutually influence each other which makes it difficult to predict their supramolecular arrangement.

As peptide association and structure are dependent on the physical state of the lipid membrane these investigations should preferentially be performed in extended liquid crystalline bilayers. While a number of biophysical investigations, such as fluorescence and CD spectroscopies can be used for this purpose, solid-state NMR spectroscopy is probably the only technique that can provide information at atomic resolution (Marquette and Bechinger 2018).

## 7.2 Solid-state NMR spectroscopic approaches

Most NMR interactions are strongly dependent on the molecular alignment relative to the magnetic field of the NMR spectrometer ($B_o$) (figure 7.1, first row). This anisotropy is averaged in solution whereby the molecules undergo fast rotational diffusion and thus the spectra are characterized by their isotropic contributions. However, in the case of large molecular complexes, such as vesicles or extended lipid bilayers, that realign slowly in $B_o$, the chemical shifts, dipolar and quadrupolar interactions are all anisotropic. When these samples are merely inserted into the NMR coil, broad overlapping powder pattern line shapes are observed (figure 7.1, second row).

Therefore, to obtain well-resolved NMR spectra from solid samples two fundamentally different approaches are used. On the one hand fast spinning around the magic angle (MAS) mechanically averages the orientation-dependent NMR interactions. With most advanced technologies MAS frequencies of up to 170 kHz are applied while at the same time the size of the samples becomes smaller with the spinning speed (Samoson 2019). Thus, narrow line shapes of proteins in the solid state have been obtained (figure 7.1, second row) and well-resolved multidimensional high-resolution MAS spectra recorded. These are rich in angular and distance information and can be used in a related manner as in solution NMR spectroscopy for protein structure determination. Recent examples include the influenza B M2 channels (Mandala *et al* 2020), the hepatitis NS4B membrane protein (Jirasko *et al* 2020) or the lipid II complex with the nisin antimicrobial peptide (Medeiros-Silva *et al* 2018). In many cases cross peaks from within spin systems allowed assigning the MAS spectra but because of spectral crowding, inter-residual distance information remained sparse. Therefore, in a recent approach long-range distances have been evaluated using $^{19}$F labelling schemes (Shcherbakov *et al* 2019). It is then also possible to derive secondary structure information from characteristic $^{13}$C and

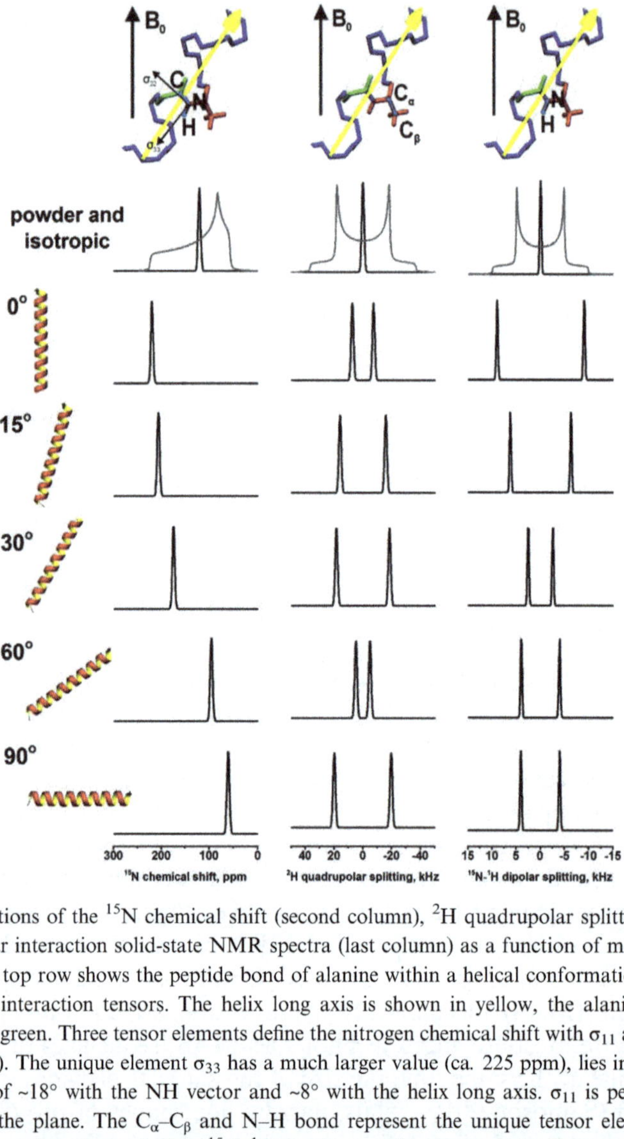


**Figure 7.1.** Simulations of the $^{15}N$ chemical shift (second column), $^{2}H$ quadrupolar splitting (third column) and $^{1}H$–$^{15}N$ dipolar interaction solid-state NMR spectra (last column) as a function of molecular alignment (first column). The top row shows the peptide bond of alanine within a helical conformation and sketches of the corresponding interaction tensors. The helix long axis is shown in yellow, the alanine in red and the previous residue in green. Three tensor elements define the nitrogen chemical shift with $\sigma_{11}$ and $\sigma_{22}$ being close in size (ca. 70 ppm). The unique element $\sigma_{33}$ has a much larger value (ca. 225 ppm), lies in the peptide plane and forms angles of ~18° with the NH vector and ~8° with the helix long axis. $\sigma_{11}$ is perpendicular to $\sigma_{22}$ and $\sigma_{33}$ and exits the plane. The $C_\alpha$–$C_\beta$ and N–H bond represent the unique tensor elements of the $CD_3$ quadrupolar splitting of alanine and of the $^{15}N$–$^{1}H$ dipolar coupling, respectively. For comparison the second row shows the isotropic peak position observed in solution or under MAS conditions as well as the powder pattern line shape that is obtained when adding up all possible alignments when they occur in a random manner. The figure shows the orientation dependence when a helix is rotated in the magnetic field of the NMR spectrometer ($B_o$), but similar considerations apply when other conformations are considered.

$^{1}H$ chemical shifts (Cornilescu *et al* 1999, Tamiola and Mulder 2012). Notably, the optimization of sample conditions is an essential step for high-resolution MAS spectra as narrow lines are only observed when the samples exhibit a high degree of local order (Pauli *et al* 2000).

Since the MAS approach uses known concepts from multidimensional solution NMR and will be discussed in other places (see chapter 9 and chapter 15), here we focus on a conceptually different alternative, which relies on structural information from samples that are uniaxially oriented relative to the $B_o$ field. The approach is particularly useful for lipid membranes which can be aligned either mechanically along glass or polymer surfaces (Aisenbrey *et al* 2010, Cornell *et al* 1988, Sizun and Bechinger 2002), or magnetically in the $B_o$ field of the NMR spectrometer, e.g. (Gentry *et al* 2019). Under these circumstances the chemical shifts, quadrupolar and dipolar interactions are all dependent on the molecular orientation with respect to $B_o$ (figure 7.1), and thereby also the normal of the oriented membrane. Thus, all of these NMR interactions allow one to derive structural information in terms of angular restraints from which the conformation and membrane topology of lipids and membrane-associated polypeptides are calculated (Bechinger and Sizun 2003, Cross 1997). The approach has been proven valuable in the structural analysis of e.g. bilayer-associated CrgA from Mycobacterium tuberculosis (Das *et al* 2015), sarcolipin (Gopinath *et al* 2015), cytochrome C (Ravula *et al* 2017), antimicrobial peptides (Aisenbrey *et al* 2019b, Comert *et al* 2019) or specific lipid interaction domains of the MHC II receptor (Salnikov *et al* 2019b). Notably the approach reveals not only the structure and dynamics but also the molecular alignment relative to the membrane surface (cf. below). The MAS and oriented sample approaches have been combined when supported lipid bilayers were spun slowly around the magic angle (Glaubitz and Watts 1998, Sizun and Bechinger 2002). Furthermore, mixing information from MAS or solution NMR with angular restraints from oriented samples has proven highly valuable (Das *et al* 2015, Michalek *et al* 2013, Traaseth *et al* 2009), similar to the use of residual dipolar couplings in multidimensional solution NMR spectroscopy (Brunner 2001).

## 7.3 The anisotropy of solid-state NMR interactions

In this section the inherent anisotropy of some chemical shift, dipolar and quadrupolar interactions is presented in more detail because their angular dependence is key to using spectra from uniaxially oriented samples for structural analysis (Bechinger and Sizun 2003). Notably, in order to make use of these interactions they first need to be characterized in a quantitative fashion. The $^{15}$N chemical shift of peptide bonds is one of the fundamental parameters used in the structural analysis of membrane-associated peptides. It exhibits an anisotropy of about 170 ppm (Salnikov *et al* 2009a). The chemical shift angular dependence is $\sigma_{zz} = \sigma_{11} \sin^2 \Theta \cos^2 \Phi + \sigma_{22} \sin^2 \Theta \sin^2 \Phi + \sigma_{33} \cos^2 \Theta$, where $\sigma_{zz}$ is the chemical shift, $\sigma_{11}$, $\sigma_{22}$ and $\sigma_{33}$ are the main tensor elements, and the Euler angles $\Theta$ and $\Phi$ position the chemical shift tensor relative to the magnetic field direction of the NMR spectrometer ($B_o$, figure 7.1, second column). In the case of the $^{15}$N amide bond the tensor has been extensively studied and the data are reviewed in Salnikov *et al* (2009a). The main tensor elements $\sigma_{11}$, $\sigma_{22}$ and $\sigma_{33}$ are about 56, 81 and 223 ppm (typically ±5 ppm). Furthermore, their alignments relative to the molecular coordinate system have been characterized in various studies, which converge to $\sigma_{33}$ covering an angle of about 18° relative to the N–H vector (reviewed in (Salnikov *et al* 2009a)) (figure 7.1).

The dipolar splitting is described by $D_{ZZ} = \frac{\mu_o \gamma_1 \gamma_2 \hbar}{4\pi r^3}\left(3\cos^2\theta - 1\right)$, where $r$ is the distance between the two interacting nuclei and $\Theta$ the angle between the distance vector and $B_o$ (figure 7.1, last column). The dipolar coupling constant $\frac{\mu_o \gamma_1 \gamma_2 \hbar}{4\pi r^3}$ is dependent on the magnetogyric ratios $\gamma_1$ and $\gamma_2$ of the interacting nuclear dipoles and $\mu_o$ is the free magnetic inductivity. Thus the $^1$H–$^{15}$N coupling within an amide bond typically varies between −20 kHz and 10 kHz (Bechinger and Sizun 2003), where in general only the absolute values are measured (figure 7.1).

Finally, the deuterium quadrupolar splitting of the alanyl methyl group shall be introduced because the $CD_3$ group is directly linked to the peptide backbone (figure 7.1) where the $C_\alpha$–$C_\beta$ bond is aligned at 59.4° relative to the helix long axis (Vostrikov *et al* 2010). Due to fast rotation around this bond at ambient temperatures a single quadrupolar splitting is observed for the three methyl deuterons (Batchelder *et al* 1983). The orientation dependence of this interaction is $\Delta\nu_Q = \frac{3}{2}\left(\frac{e^2qQ}{h}\right)\left(\frac{3\cos^2\theta - 1}{2}\right)$, with $\Theta$ being the angle between the $C_\alpha$–$C_\beta$ bond and $B_o$ (figure 7.1, third column). Depending on the molecular alignment, quadrupolar splittings of up to 80 kHz are observed (Batchelder *et al* 1983). Notably, the $^2$H measurements are enhanced by the short $T_1$ relaxation time of this nucleus and the equivalence of the three deuterons which cause overlap of their resonances.

## 7.4 Sample preparation

### 7.4.1 Peptides

As with other NMR spectroscopic approaches the preparation of quantitative amounts of isotopically labelled polypeptides is a prerequisite also for solid-state NMR investigations. Peptides up to 30 amino acids in length have been prepared routinely and in good yields in our laboratory using automated solid-phase peptide synthesis and reverse-phase HPLC purification, while longer and/or very hydrophobic sequences tend to be more difficult to prepare and purify, e.g. (Aisenbrey *et al* 2019c). By using (commercially available) amino acid precursors incorporating stable NMR isotopes, these sequences have been labelled at specific sites.

Furthermore, uniformly or selectively labelled polypeptides can be prepared by bacterial over-expression, usually by fusing the peptides to a large soluble protein and/or by expressing it as non-soluble inclusion bodies ((Vidovic *et al* 2009) and references cited therein). The best strategy for the preparation of membrane polypeptides is much dependent on the polypeptide properties but this shall not be elaborated on in the context of this chapter (reviewed in Verardi *et al* (2012)).

### 7.4.2 Sample preparation: bicelles and supported lipid bilayer

Many structural investigations have been performed using multidimensional solution NMR spectroscopy of membrane-associated polypeptides reconstituted into detergent micelles or into isotropic bicelles, e.g. (Gentry *et al* 2019). Bicelles are patches of bilayers made from long chain lipids stablilized by a rim of short chain lipids (typically DHPC; figure 7.2(A)), detergents (Marcotte and Auger 2005, Sanders and Landis 1995), polymers (Ravula *et al* 2018) or peptides (Gentry *et al* 2019, Salnikov *et al* 2019a). In a related manner, scaffolding proteins assure a more homogenous and reproducible size of the 'nanodiscs' (Hagn *et al* 2018). Whereas the highly curved micellar surface can significantly affect structural characteristics of the polypeptides, the bilayer patch within an isotropic bicelle may be too limited to assure unperturbed insertion and the cooperative properties of a true bilayer structure (Martinez *et al* 2017). Therefore, many questions can only be solved from solid-state NMR studies of extended lipid bilayers that are oriented as stacks on solid supports (figure 7.2(B)) or of large 'anisotropic' bicelles that spontaneously align in the magnetic field (*see* ultra).

The reconstitution of the peptide into (oriented) supported membranes can be achieved by first preparing a film of peptide and lipid from organic solvent as described in detail in (Aisenbrey *et al* 2010). In short, a homogenous solution of lipid and peptide is prepared. The solvent is slowly evaporated until a viscous solution is obtained, which is applied onto the surfaces of ultra-thin glass plates fitting the NMR coil (Salnikov *et al* 2015). The sample is thoroughly dried, hydrated through the gas phase, stacked and sealed (Aisenbrey *et al* 2010). Alternatively, when exposure to organic solvent is to be avoided proteins have been reconstituted into vesicles or nanodiscs (Das *et al* 2013, Gentry *et al* 2019, Nelson *et al* 2018), which are then aligned mechanically (protocol in Aisenbrey *et al* (2019d)), or pelleted by ultra-centrifugation and transferred into MAS solid-state NMR rotors (Nelson *et al* 2018).

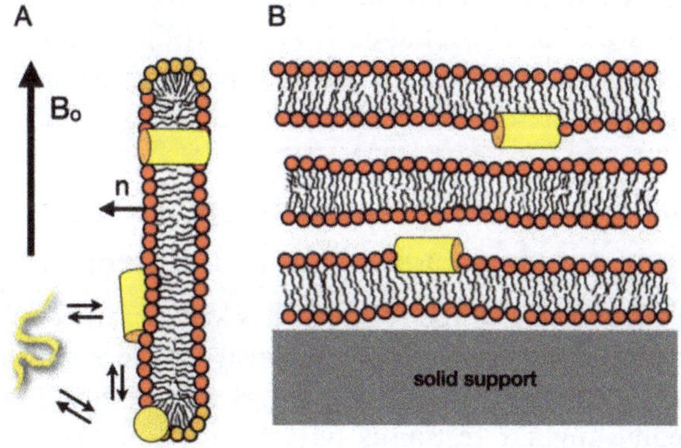


**Figure 7.2.** Shows oriented membrane preparations used for solid-state NMR spectroscopic investigations. (A). Bicelles are patches of bilayer (red head groups) with a rim of short chain lipids (yellow heads). The rim can also be formed from polymers, amphipathic peptides or, in the case of 'nanodiscs', by scaffolding proteins (not shown). While it is believed that transmembrane polypeptides stably insert into the bilayer (top), amphipathic sequences tend to exchange with the bulk solvent and the rim (bottom). The magnetic interactions of large bicelles naturally result in an orientation of the normal *n* perpendicular to $B_0$ (as shown in A) but can be 'flipped' by 90°, e.g. by addition of lanthanides, e.g. Gopinath *et al* (2013). (B). Lipid bilayers have been uniaxially oriented along the surface of solid supports (glass, mica, polymers, etc). Oriented solid-state NMR sample stacks of membranes typically contain thousands of bilayers that are supported from the bottom and the top.

Notably, in cases where fast rotational averaging occurs around the membrane normal it is also possible to deduce the alignment of peptides from the 'powder pattern' solid-state NMR spectra obtained from non-oriented samples (Bechinger and Sizun 2003, Raya *et al* 2011).

Reconstitution into bicelles can be achieved by dissolving peptide powder directly in the bicellar medium or by detergent removal approaches (Das *et al* 2013, Marcotte and Auger 2005, Nelson *et al* 2018). For solid-state NMR magnetic alignment the bicelles have to be >30 nm in diameter (Ravula *et al* 2017), i.e., a ratio of long- and short-chain lipids around $q = 3$, and be made up predominantly of lipids carrying saturated fatty acyl chains (Loudet *et al* 2007, Marcotte and Auger 2005, Sanders and Landis 1995). Bicelles have mostly been used to investigate hydrophobic peptides by solid-state NMR while the presence of bulk water may result in dynamic exchange of amphipathic sequences between the membrane and the water phase (figure 7.2(A)). Exchange probably also occurs between the flat central areas and the rim structure. This is of particular concern because amphipathic peptides have been shown to form the rim of bicelles and have even been used as alternatives to replace the detergents (Bechinger 2005, Ravula *et al* 2020, Salnikov *et al* 2018, Wolf *et al* 2017).

While on the one hand the orientation along the $B_o$ field and the well-resolved oriented solid-state NMR spectra make bicelles a highly attractive system in particular for transmembrane domains, motions of the bicelles themselves cause partial averaging of the anisotropies which have to be taken into account during structure calculations (Marcotte and Auger 2005, Salnikov *et al* 2014, Sanders and Landis 1995). On the other hand, mechanically oriented samples work well also for amphipathic sequences and allow for the study of a wide range of membrane lipid compositions (Glattard *et al* 2016, Salnikov *et al* 2019b, Traaseth *et al* 2006) or at very low temperatures (Salnikov *et al* 2016). The lipid composition of oriented bilayers is discussed in more detail below.

## 7.5 Oriented solid-state NMR spectroscopy to study the conformation, topology and dynamics of peptides in membranes

The orientational dependence of the solid-state NMR parameters provides angular information which has then been used to reconstruct the structure of membrane-associated peptides (Bechinger *et al* 1993, Cross 1997). A full structure determination requires many angular restraints but if the secondary structure is already known from multidimensional solution NMR or MAS solid-state NMR spectroscopy, a few additional solid-state NMR measurements from oriented samples allow aligning this structure relative to the bilayer normal and/or to refine the structures obtained by other methods (Mandala *et al* 2018, Michalek *et al* 2013, Traaseth *et al* 2009).

For structural analysis the most convenient alignment of the bilayer is with the normal parallel to the magnetic field direction ('0° alignment') because in this configuration rotational or lateral diffusion does not affect the angle of the peptides relative to the $B_o$ field (figure 7.2(B)). As large phospholipid bicelles naturally align at an angle of 90° (figure 7.2(A)), which results in a wide range of molecular alignments relative to $B_o$ (cf. figure 7.1 in (Aisenbrey and Bechinger 2004a)), a variety of modifications have been introduced that 'flip' them into a parallel orientation (Loudet *et al* 2007, Marcotte and Auger 2005).

To illustrate the anisotropy effect of NMR interactions the orientation dependence of the $^{15}N$ chemical shift of a labelled peptide bond is described in more detail (Bechinger and Sizun 2003). At a sample alignment with the normal parallel to $B_o$ (figure 7.2(B)), a $^{15}N$ chemical shift <100 ppm is indicative that the $^{15}N$–$^{1}H$ bond is approximately perpendicular to the membrane normal/$B_o$ field (figure 7.1). In contrast, a parallel alignment of the $^{15}N$–$^{1}H$ bond results in values around 200 ppm. Since the $^{15}N$–$^{1}H$ vector is close to parallel to the long axis of a helical secondary structure, a single $^{15}N$ chemical shift measurement at the same time reveals if a helix is transmembrane (ca. 200 ppm) or oriented parallel to the bilayer surface (<100 ppm, (Bechinger and Sizun 2003)). This will be elaborated in more detail in the context of figure 7.5.

Figure 7.3 shows $^{15}N$ and $^{2}H$ solid-state NMR spectra of the antimicrobial peptide [$^{2}H_3$-Ala10; $^{15}N$-Ala14]-PGLa. The peptide has been labelled with $^{15}N$ at residue 14 and $^{2}H_3$-alanine at position 10 (sequence: GMASK AGAIA GKIAK VALKA L-NH$_2$) and reconstituted into thin membranes made of di-C10-PC that are uniaxially oriented along the surfaces of ultra-thin microscope cover glasses. The $^{15}N$ chemical shift of 170 ± 10 ppm corresponds to previously reported values in these thin membranes (Salnikov and Bechinger 2011), and the $^{2}H$ quadrupolar splitting is 35 ± 2 kHz. The isotropic deuterium signal (0 kHz) is from the naturally abundant deuterons in membrane-associated water (ca. 0.02%). As PGLa shows enhanced activity in the presence of equimolar magainin, this second antimicrobial

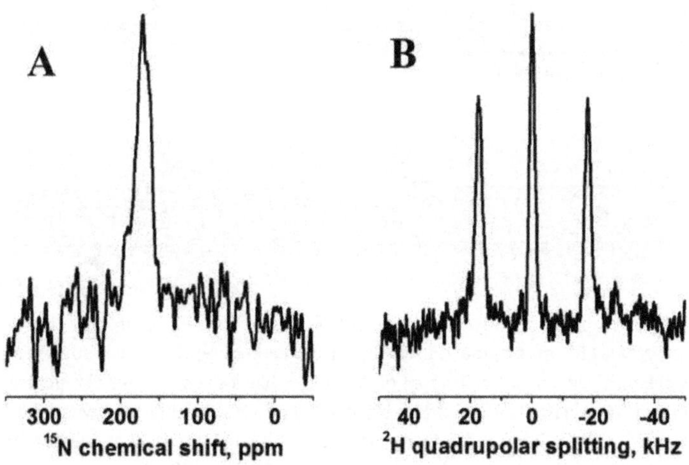


**Figure 7.3.** $^{15}N$ and $^{2}H$ solid-state NMR spectra of the antimicrobial peptide [$^{2}H_3$-Ala10, $^{15}N$-Ala14]-PGLa reconstituted into uniaxially oriented di-C10-PC at a PGLa/lipid ratio of 1 mole% in the presence of 1 mole% unlabelled magainin 2. The normal of the bilayer was inserted into the NMR coil at alignments parallel to the magnetic field of the NMR spectrometer (cf. figure 7.2(B)). The sample was equilibrated in an atmosphere of 93% RH and measured at room temperature. Experimental details are given in Salnikov and Bechinger (2011).

peptide with the sequence GIGKF LHSAK KFGKA FVGEI MNS-COOH was added in equimolar quantities but unlabelled. More details about the structural and biological findings with these peptides can be found in recent publications (Aisenbrey *et al* 2020, submitted, Aisenbrey *et al* 2019b).

The data are analysed in more detail using a restriction plot analysis (figure 7.4). For these calculations a starting configuration is defined. For an amphipathic helical peptide this can be a perfect transmembrane orientation (tilt angle 0°) where the hydrophilic/hydrophobic separation is aligned along the *x*-axis (pitch angle 0°). The molecule can be represented by PDB coordinates either representing an ideal helix or a known structure which may have been obtained from multidimensional solutions state NMR or another structural approach (Michalek *et al* 2013, Resende *et al* 2014, Salnikov *et al* 2011). In the following all molecular orientations are tested by step-wise rotations (1° intervals, 100 ×100 steps) around the helical long (*z*-axis) and the *y*-axis, thereby screening all possible tilt and rotational pitch angles. For each orientation thus obtained the chemical shifts, dipolar couplings and quadrupolar splittings of the labelled sites are calculated and compared to the experimental values and its error bars (figure 7.3). Whenever there is agreement the tilt/rotational pitch angular pair is labelled in the restriction plot where each NMR measurement is

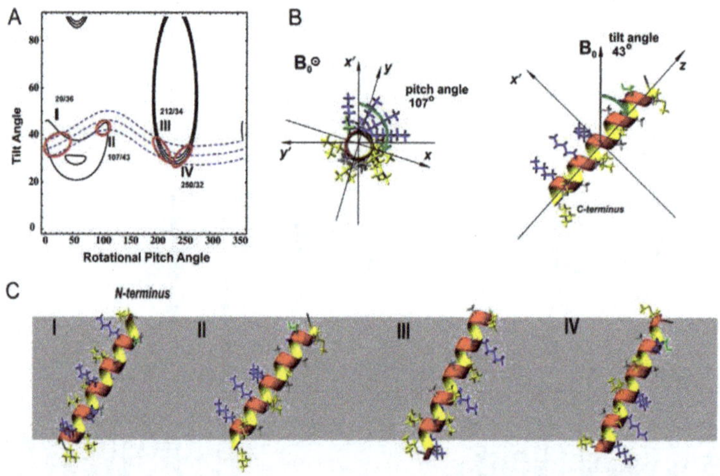


**Figure 7.4.** (A). Restriction plot analysis of 1 mole% PGLa in di-C10-PC in the presence of equimolar magainin 2 using the data shown in figure 7.3. The hatched blue lines represent the tilt/pitch angular pairs that agree with a $^{15}N$ chemical shift of 170 ± 10 ppm of Ala14, the black lines with the $^{2}H$ quadrupolar splitting is 35 ± 2 kHz obtained for Ala10. The red circles highlight the PGLa orientations where both measurements agree. For the simulations, wobbling and rocking motions of tilt and azimuthal angles described by Gaussian distributions of 8° and 20°, respectively, were taken into consideration. (B). Representations of the topology II of panel A. The magnetic field of the NMR spectrometer is shown by the $B_o$ vector. The definitions of tilt ($\angle B_o, z$) and pitch angles ($\angle x,x'$) are also indicated. (C). Structural models of four topologies resulting from panel A. The lysine side chains are shown in blue, serine in green and hydrophobic amino acids in yellow. The grey slab sketches an unperturbed lipid bilayer. Other experiments provide information about bilayer deformations, interactions between magainin 2 and PGLa or peptide oligomerization.

given its own colour (figure 7.4). Whereas the central line in the plots represents the main value, the flanking traces represent the angles covered by the error bars.

It becomes obvious from figure 7.4 (blue traces) that in the case of helical peptides the $^{15}N$ chemical shift defines a narrow range of possible tilt angles while being insensitive to the pitch angle. Therefore, for a more precise determination of the bilayer topology, additional measurements are required (Gopinath *et al* 2015, Michalek *et al* 2013). This data can be obtained e.g. from the highly complementary $^{2}H$ quadrupolar splitting of alanine deuterated at the methyl group (Aisenbrey and Bechinger 2004b). The black traces in figure 7.4 show that this parameter is strongly dependent on both the tilt and the pitch angle. This complementarity arises from the fact that the most distinct element of the $^{15}N$ chemical shift tensor, i.e., $\sigma_{33}$ (cf. previous section), is close to parallel to the helix long axis (ca. 8°) whereas the $C_\alpha$–$C_\beta$ bond of alanine defines the unique element for the quadrupolar splitting tensor and is oriented quite differently, namely at an angle of 59.4° (Vostrikov *et al* 2010).

The restriction analysis strongly relies on the accuracy of the main tensor elements (cf. previous section on the anisotropy of NMR parameters) as well as the precise determination of the NMR parameters where a 5 ppm deviation can translate into a change in topology by a few degrees (Michalek *et al* 2013). To take into account such systematic errors, the uncertainties in the chemical shift tensor (Salnikov *et al* 2009a), the chemical shift reference (Bertani *et al* 2014) and/or the effect of the motional regime have been investigated in great detail (Michalek *et al* 2013).

## 7.6 Sample heterogeneity and orientational distributions

Notably, the $^{2}H$ line shape has proven to be very sensitive to the realignment of the $C_\alpha$–$C_\beta{}^2H_3$ group. Reorienting this bond by as little as 1° can translate into changes in the deuterium quadrupole splitting of up to 2 kHz, which are easily detectable (Aisenbrey and Bechinger 2004b, Bechinger *et al* 2011). It has thus been possible to monitor minor topological rearrangements of a model peptide when the palmitoyl chain of POPC lipid is exchanged with a second oleoyl chain in DOPC (Aisenbrey and Bechinger 2004b). This same effect causes pronounced alterations of the deuterium line shape in cases where the deuterated site exhibits mosaic spread (Aisenbrey and Bechinger 2004b).

The effects of sample heterogeneity on the $^{15}N$ chemical shift spectra are illustrated in figure 7.5. A first observation made from the first two columns is that for a helical domain the $^{15}N$ chemical shift changes with tilt angle in a continuous manner from ca. 225 ppm (transmembrane, tilt angle 0°) to ca. 60 ppm (in-plane, tilt angle 90°), thereby the peptide bond $^{15}N$ chemical shift provides a direct read-out of the approximate helix topology (Bechinger and Sizun 2003). Furthermore, figure 7.5 illustrates that a perfectly transmembrane peptide is relatively insensitive (first row) to rotations around the helix long axis. In contrast such heterogenous distributions by slow rotations can have a large line broadening effect on helices being oriented parallel to the membrane surface (last row). When

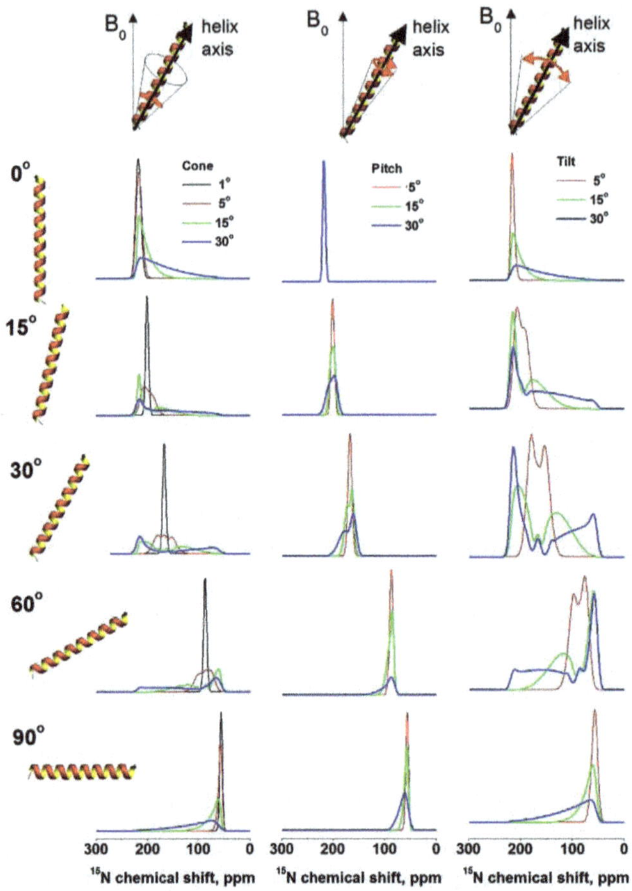


**Figure 7.5.** Spectral simulations of $^{15}$N chemical shift line shapes of a singly labelled peptide bond when orientational distributions occur. Helices with different tilt angles, shown in the first column, are considered. The second column shows the signal when the helix long axis is oriented along cones with opening angles of 1° (black), 5° (red), 15° (green) or 30° (blue). The third column shows the spectra when a Gaussian distribution of pitch angles occurs with amplitudes of 5° (red), 15° (green) or 30° (blue). The last column shows the spectra when a Gaussian distribution of tilt angles occurs with amplitudes of 5° (red), 15° (green) or 30° (blue). Spectra are shown normalized to the integral intensities in each panel.

significant Gaussian distributions of the tilt angle exist, this can result in resonances covering the full chemical shift anisotropy (last column).

The $^{15}$N and $^{2}H_3$ labels can be incorporated together in a single peptide sequence and the corresponding NMR spectra recorded from the same sample (figure 7.3). It is interesting to note that, for some doubly labelled peptides, well-oriented $^{15}$N but broad $^{2}$H solid-state NMR spectra were observed (Itkin *et al* 2017) or vice versa (Marquette *et al* 2020). Because these spectra were obtained from the same sample the peptide as a whole must be well aligned while at the same time a localized set of amino acids exhibits conformational heterogeneity.

As a full structure determination requires NMR spectra with high resolution, sample conditions are often optimized for narrow lines. However, it should be kept in mind that spectral broadening can be an indicator of conformational heterogeneity, peptide oligomerization and equilibria that may all be of importance for biological and biomedical function (Aisenbrey *et al* 2019c, Itkin *et al* 2017, Marquette *et al* 2020, Salnikov *et al* 2019b).

Without additional labelling efforts, two-dimensional separated local field spectra can reveal the $^{1}H$–$^{15}N$ dipolar coupling in addition to the $^{15}N$ chemical shift (Ramamoorthy *et al* 2004). Since the main tensor element $\sigma_{33}$ of the $^{15}N$ chemical shift is close to parallel to the $^{1}H$–$^{15}N$ dipolar vector (cf. above and figure 7.1), the angular restraints from these two measurements closely match each other (Michalek *et al* 2013, Resende *et al* 2014, Salnikov *et al* 2011). However, in samples labelled uniformly with $^{15}N$, these PISEMA spectra result in helical wheel patterns that allow an analysis of the secondary structure and of the helix topology (Kim and Cross 2004, Salnikov *et al* 2009c, Tang *et al* 2006, Traaseth *et al* 2006). Thereby, which of the measurements provides the best result depends on how the peptides were prepared. Multidimensional solid-state NMR spectra are required to spread the signals of uniformly labelled peptides obtained by bacterial overexpression (Verardi *et al* 2012), whereas introducing a combination of selective $^{2}H$ and $^{15}N$ labels is a favoured approach in the case of selective labelling schemes by chemical peptide synthesis.

## 7.7 Membrane lipid composition

In recent years solid-state NMR investigations have revealed interesting details about the lipid-dependent membrane topology of a number of peptides. Early on, magainin 2 antimicrobial peptides were reconstituted in a wide variety of phospholipid bilayers, made of POPE/POPG 3:1 (molar ratio), POPC/POPG 3:1, POPC or POPC/POPG/cholesterol 3:1:4, and always found to align parallel to the membrane surface (Bechinger *et al* 1992). This alignment was confirmed in many other phospholipid bilayers, including DMPC and DMPC/DMPG 3:1 or DMPE/DMPG 3:1 (Harmouche and Bechinger 2018, Salnikov and Bechinger 2011, Strandberg *et al* 2013).

This contrasts the membrane interactions of its relative PGLa, which when associated with fully saturated lipid membranes can adopt a wide range of tilt angles. The detailed topology of this peptide turns out to be dependent on lipid saturation, the membrane hydrophobic thickness and the degree of bilayer hydration (Salnikov and Bechinger 2011, Salnikov *et al* 2016, Strandberg *et al* 2013). Furthermore, in saturated membranes and when mixed with magainin 2, tilt angles suggestive of transmembrane helices are measured for PGLa (Harmouche and Bechinger 2018, Salnikov and Bechinger 2011, Strandberg *et al* 2013) (figures 7.3 and 7.4). However, the topology of PGLa remains always in-plane in membranes made of phospholipids carrying a significant fraction of unsaturated chains, including the bacterial membrane mimetic POPE/POPG 3:1 or *E. coli* lipid extracts (Glattard *et al* 2016, Harmouche and Bechinger 2018).

Further studies of magainin 2 suggest that this peptide also adopts a somewhat (by 30°) more tilted orientation in DPPC or in DMPC/DMPG in the presence of equimolar amounts of PGLa (Strandberg *et al* 2016, Strandberg *et al* 2013). Thus, lipid saturation has a pronounced influence on the topology of these antimicrobial sequences. It has been demonstrated by $^{2}$H- solid-state NMR that a disordering of the deuterated fatty acyl chain occurs when these amphipathic peptides partition into the membrane interface, an effect which is energetically particularly unfavourable in the presence of fully saturated fatty acyl chains (Harmouche and Bechinger 2018).

An interesting lipid dependent alignment has also been observed for alamethicin F50/7, a hydrophobic peptide encompassing 20 residues belonging to the family of peptaibols (sequences rich in Aib (α-aminoisobutyric acid)). This sequence has been investigated in great detail by biophysical approaches and is considered the prime example for channels formed by transmembrane helical bundles. Indeed, when investigated in POPC or DMPC membranes the $^{15}$N chemical shifts and $^{1}$H–$^{15}$N dipolar interactions are characteristic for helices that span the lipid bilayer (Bak *et al* 2001, North *et al* 1995, Salnikov *et al* 2009b, Salnikov *et al* 2009c). However, when studied in POPE/POPG 3:1 model membranes, that are thought to better mimic bacterial membranes, the $^{15}$N chemical shifts are $<100$ ppm indicative of helices oriented parallel to the membrane surface (Salnikov *et al* 2010a).

Notably, ampullosporin A and zervamicin II, other sequences of the peptaibol family, have also been investigated by $^{15}$N solid-state NMR spectroscopy (Bechinger *et al* 2001, Salnikov *et al* 2009c). These consist of only 15 and 16 residues, respectively, exhibit in-planar alignments in POPC but transmembrane alignments in thin bilayers made from C10- or C12-PC. Thus, the peptide sequence, hydrophobic mismatch, the size and composition of the head group, lipid saturation and interactions with other peptides all contribute to the peptide–membrane interactions.

Finally, recent $^{15}$N solid-state investigation of the transmembrane anchors of the p24, DQA1 and DQB1 membrane proteins reveal that these peptides associate in the membrane in a lipid- and concentration-dependent manner concomitant with changes in topology and/or dynamics (Aisenbrey *et al* 2019a, Aisenbrey *et al* 2019c, Salnikov *et al* 2019a, Salnikov *et al* 2019b). While the p24 protein is involved in the sorting of lipids and proteins inside cells, DQA1 and DQB1 are building blocks of the MHC II receptor which exerts key functions in immunity. These proteins have elicited interest because an amino acid motif has been described in the p24 transmembrane domain recognising sphingomyelin but only when it carries a C18 fatty acyl chain (SM-C18) (Contreras *et al* 2012). A homology search has revealed similar motifs in the MHC II receptor (Bjorkholm *et al* 2014). Therefore, we have investigated the membrane interactions of these polypeptides by oriented solid-state NMR, analysed their complex $^{15}$N line shapes in quantitative detail and suggest the existence of equilibria of mono- and dimers (Aisenbrey *et al* 2019c, Salnikov *et al* 2019b). Notably, whereas composite peaks with interesting dynamic features were detected in POPC or POPC/SM-18 membranes, a single resonance was observed in DMPC bilayers. Thus, the equilibrium could not be detected in the DMPC lipid bicelle system.

Valuable complementary information is also obtained from $^{31}$P and $^{2}$H solid-state NMR spectra of the lipids which reveal the global alignment of the lipids, conformational changes, the macroscopic phase properties of the membranes, the order parameter of the fatty acyl chains, membrane thinning and the lateral phase separation in mixed lipid membranes (reviewed in (Bechinger and Salnikov 2012)). Thus, solid-state NMR spectroscopy provides a comprehensive view of the peptide–lipid interactions, the response of the soft membranes to peptide association and ultimately the mechanisms of action of many biologically relevant peptides.

## 7.8 Taking peptide motions into consideration

Solid-state NMR investigations allow investigation of polypeptides at atomic resolution in native liquid disordered membrane environments. In this regard the technique is unique but requires dedicated instrumentation (e.g. probe) and sample preparations. At physiological temperatures the membranes are 'soft' and the peptides and lipids exhibit conformational and topological flexibility up to the extent that local and global phase transitions can occur (Bechinger 2015). During the analysis of the peptides' membrane topologies such motions should be taken into consideration (Aisenbrey *et al* 2019c, Holt *et al* 2010, Michalek *et al* 2013, Salnikov *et al* 2019b, Strandberg *et al* 2009).

During recent years wobbling and rocking motions have been considered during the analysis of oriented solid-state NMR spectra of membrane-associated helical peptides (Aisenbrey *et al* 2019c, Bertelsen *et al* 2009, Holt *et al* 2010, Michalek *et al* 2013, Salnikov *et al* 2019b, Strandberg *et al* 2009). As liquid crystalline membranes provide a soft and flexible environment for the peptides it is reasonable to assume that they exhibit considerable motions. These were taken into account in more recent topological studies by simulating Gaussian distributions of alignments including the restriction analysis shown in figure 7.4. In the case of in-plane aligned amphipathic peptides rocking and wobbling motions by $\approx 10°$ (Gaussian distribution) and 18° matched well the experimental constraints (Michalek *et al* 2013, Strandberg *et al* 2009), while different motional regimes apply to transmembrane domains where the line shape of $^{15}$N oriented solid-state NMR spectra of the DQB1 subunit of the MHC II receptor revealed two contributions with different motional averaging (Salnikov *et al* 2019b).

The $^{15}$N chemical shift and $^{1}$H–$^{15}$N dipolar coupling anisotropies under different motional regimes and the resulting PISEMA spectra have been simulated for helical and $\beta$–sheet proteins (Aisenbrey and Bechinger 2004a, Bleile *et al* 2005, Straus *et al* 2003). Whereas the evaluation of the helical tilt angle from the amide bond $^{15}$N chemical shift is relatively unaffected by motional averaging, the $^{2}H_3$-alanine deuterium quadrupolar splitting is particularly sensitive to wobbling or rotational motions around the helix long axis (Aisenbrey and Bechinger 2004a, Salnikov *et al* 2010a) and, when these are ignored the calculated tilt angles may suffer from large systematic deviations (Ozdirekcan *et al* 2007, Vostrikov *et al* 2010).

$^{2}$H quadrupolar splitting and $^{15}$N chemical shift spectra have also been valuable in analysing the domain structure and site-specific dynamics within the Alzheimer precursor protein (Itkin *et al* 2017). They have also revealed a second slowly exchanging conformer/ topology in distinctin, a heterodimeric antimicrobial peptide (SI in (Resende *et al* 2009)). In addition, sidechain dynamics can be explored using oriented or MAS solid-state NMR spectroscopy (e.g. (Aisenbrey *et al* 2019a) and references cited therein).

A conceptually different approach to test dynamics and oligomerisation is based on tilting the samples by 90° which aligns the bilayer normal perpendicular to the magnetic field direction (Aisenbrey and Bechinger 2004a). This experiment is usually performed with the same samples used for structural analysis. In the 'tilted' arrangement rotational diffusion scans many different orientations of the amide bonds or the $C_\alpha$–$C_\beta$ vector relative to $B_o$. If rotational diffusion is fast a single averaged resonance appears, when diffusion is slow resonances from a wide range of orientations contribute to the corresponding solid-state NMR spectrum (cf. figure 7.5) and result in powder pattern line shape weighted by a circular distribution (Aisenbrey and Bechinger 2004a).

## 7.9 Signal enhancement by DNP

A decade ago solid-state NMR instrumentation for dynamic nuclear polarisation (DNP) experiments has become commercially available (Can *et al* 2015, Michaelis *et al* 2018) and been used for the investigation of membrane peptides and proteins (Bechinger 2018, Mao *et al* 2019). By transferring polarisation from electrons to $^1H$, theoretical gains in sensitivity by 660-fold are possible and have almost been achieved for some reference samples under MAS conditions (Kubicki *et al* 2014). This magnetisation is then transferred to other nuclei by cross polarisation. In the case of oriented phospholipid bilayers, a number of factors have so far prevented such highest enhancements. However, systematic developments have resulted in up to 49-fold improvements (Salnikov *et al* 2010b, Salnikov *et al* 2017) which makes the acquisition of two-dimensional solid-state NMR spectra from oriented membrane samples possible in minutes rather than days (Salnikov *et al* 2016).

Although at such high sensitivity the application of three- and four-dimensional NMR experiments, including extensive structural investigations on oriented membranes, should become more feasible (Wang *et al* 2019) the technique requires the presence of biradicals in the sample and works best at very low temperatures. Thus, motions that average sample heterogeneities at ambient temperatures are frozen and lines concomitantly broadened (Salnikov *et al* 2016). Despite the cryotemperature, oriented bilayer structures (Salnikov *et al* 2010b) and the topology of selected membrane-associated peptides are maintained (Salnikov *et al* 2016).

Optimized sample compositions, protocols for reconstitution of polypeptides and biradicals into oriented bilayers (Salnikov *et al* 2010b, Salnikov *et al* 2012), new membrane anchored biradicals (Salnikov *et al* 2017, Sani *et al* 2019), a dedicated DNP-solid-state NMR probe and sample geometries for oriented samples (Salnikov *et al* 2015, Salnikov *et al* 2019c), all work together and also improve the performance for oriented membrane samples. It can thus be expected that the DNP technique will help to further push structural investigations of membrane-associated peptides by

MAS (Mao *et al* 2019, Medeiros-Silva *et al* 2018) and oriented solid-state NMR spectroscopy (Salnikov *et al* 2016).

## Acknowledgement

We are grateful for the contributions from, discussions and exchange with our co-workers and colleagues all over the world which have helped to advance the field. Our work is only possible due to the financial contributions of the Agence Nationale de la Recherche (recent projects membraneDNP 12-BSV5–0012, MemPepSyn 14-CE34–0001–01, InMembrane 15-CE11–0017–01, Biosupramol 17-CE18–0033–3, Naturalarsenal 19-AMRB-0004–02 and the LabEx Chemistry of Complex Systems 10-LABX-0026_CSC), CEFIPRA (62T10–1), the University of Strasbourg (including IDEX grants), the CNRS, the Région Grand-Est and the RTRA International Center of Frontier Research in Chemistry.

## References

Aisenbrey C, Amaro M, Pospisil P, Hof M and Bechinger B 2020 Investigation of the supramolecular complexes formed by magainin 2, PGLa and lipids (unpublished)

Aisenbrey C and Bechinger B 2004a Investigations of peptide rotational diffusion in aligned membranes by $^{2}H$ and $^{15}N$ solid-state NMR spectroscopy *J. Am. Chem. Soc.* **126** 16676–83

Aisenbrey C and Bechinger B 2004b Tilt and rotational pitch angles of membrane-inserted polypeptides from combined $^{15}N$ and $^{2}H$ solid-state NMR spectroscopy *Biochemistry* **43** 10502–12

Aisenbrey C, Bertani P and Bechinger B 2010 Solid-state NMR investigations of membrane-associated antimicrobial peptides ed A Guiliani and A C Rinaldi *Antimicrobial Peptides* (New York: Humana Press, Springer)

Aisenbrey C, Kemayo-Koumkoua P, Salnikov E S, Glattard E and Bechinger B 2019a Investigations of the structure, topology and interactions of the transmembrane domain of the lipid sorting protein p24 being highly selective for sphingomyelin-C18 *Biochemistry* **58** 2782–95

Aisenbrey C, Marquette A and Bechinger B 2019b The mechanisms of action of cationic antimicrobial peptides refined by novel concepts from biophysical investigations *Adv. Exp. Med. Biol.* **1117** 33–64

Aisenbrey C, Salnikov E S and Bechinger B 2019c Solid-state NMR investigations of the MHC II transmembrane domains: topological equilibria and lipid interactions *J. Membr. Biol.* **252** 371–84

Aisenbrey C, Salnikov E S, Raya J, Michalek M and Bechinger B 2019d Solid-state NMR approaches to study protein structure and protein-lipid interactions *Methods Mol. Biol.* **2003** 563–98

Bak M, Bywater R P, Hohwy M, Thomsen J K, Adelhorst K, Jakobsen H J, Sorensen O W and Nielsen N C 2001 Conformation of alamethicin in oriented phospholipid bilayers determined by N-15 solid-state nuclear magnetic resonance *Biophys.* J **81** 1684–98

Batchelder L S, Niu H and Torchia D A 1983 Methyl reorientation in polycrystalline amino acids and peptides: a $^{2}H$ NMR spin lattice relaxation study *J. Am. Chem. Soc.* **105** 2228–31

Bechinger B 2005 Detergent-like properties of magainin antibiotic peptides: a $^{31}P$ solid-state NMR study *Biochim. Biophys. Acta* **1712** 101–8

Bechinger B 2015 The SMART model: soft membranes adapt and respond, also transiently, to external stimuli *J. Peptide Sci.* **21** 346–55

Bechinger B 2018 DNP solid-state NMR spectroscopy of biological membranes *eMagRes* **7** emrstm1558

Bechinger B, Resende J M and Aisenbrey C 2011 The structural and topological analysis of membrane-associated polypeptides by oriented solid-state NMR spectroscopy: established concepts and novel developments *Biophys. Chem.* **153** 115–25

Bechinger B and Salnikov E S 2012 The membrane interactions of antimicrobial peptides revealed by solid-state NMR spectroscopy *Chem. Phys. Lipids* **165** 282–301

Bechinger B and Sizun C 2003 Alignment and structural analysis of membrane polypeptides by $^{15}$N and $^{31}$P solid-state NMR spectroscopy *Concepts Magn. Reson.* **18A** 130–45

Bechinger B, Skladnev D A, Ogrel A, Li X, Swischewa N V, Ovchinnikova T V, O'neil J D J and Raap J 2001 $^{15}$N and $^{31}$P solid-state NMR investigations on the orientation of zervamicin II and alamethicin in phosphatidylcholine membranes *Biochemistry* **40** 9428–37

Bechinger B, Zasloff M and Opella S J 1992 Structure and interactions of magainin antibiotic peptides in lipid bilayers: a solid-state NMR investigation *Biophys. J.* **62** 12–4

Bechinger B, Zasloff M and Opella S J 1993 Structure and orientation of the antibiotic peptide magainin in membranes by solid-state NMR spectroscopy *Protein Sci.* **2** 2077–84

Bertani P, Raya J and Bechinger B 2014 $^{15}$N chemical shift referencing in solid state NMR *Solid-State NMR Spec* **61-62** 15–8

Bertelsen K, Paaske B, Thogersen L, Tajkhorshid E, Schiott B, Skrydstrup T, Nielsen N C and Vosegaard T 2009 Residue-specific information about the dynamics of antimicrobial peptides from $^{1}$H–$^{15}$N and $^{2}$H solid-state NMR spectroscopy *J. Am. Chem. Soc.* **131** 18335–42

Bjorkholm P, Ernst A M, Hacke M, Wieland F, Brugger B and Von Heijne G 2014 Identification of novel sphingolipid-binding motifs in mammalian membrane proteins *Biochim. Biophys. Acta* **1838** 2066–70

Bleile D W, Scott W R and Straus S K 2005 Can PISEMA experiments be used to extract structural parameters for mobile beta-barrels? *J. Biomol. NMR* **32** 101–11

Brunner E 2001 Residual dipolar couplings in protein NMR *Concepts Magn. Reson.* **13** 238–59

Can T V, Ni Q Z and Griffin R G 2015 Mechanisms of dynamic nuclear polarization in insulating solids *J. Magn. Reson.* **253** 23–35

Comert F *et al* 2019 The host-defense peptide piscidin P1 reorganizes lipid domains in membranes and decreases activation energies in mechanosensitive ion channels *J. Biol. Chem.* **294** 18557–70

Contreras F X *et al* 2012 Molecular recognition of a single sphingolipid species by a protein's transmembrane domain *Nature* **481** 525–9

Cornell B A, Separovic F, Baldassi A J and Smith R 1988 Conformation and orientation of gramicidin A in oriented phospholipid bilayers measured by solid state carbon-13 NMR *Biophys. J.* **53** 67–76

Cornilescu G, Delaglio F and Bax A 1999 Protein backbone angle restraints from searching a database for chemical shift and sequence homology *J. Biomol. NMR* **13** 289–302

Cross T A 1997 Solid-state nuclear magnetic resonance characterization of gramicidin channel structure *Methods Enzymol.* **289** 672–96

Das N, Dai J, Hung I, Rajagopalan M R, Zhou H X and Cross T A 2015 Structure of CrgA, a cell division structural and regulatory protein from *Mycobacterium tuberculosis*, in lipid bilayers *Proc. Natl Acad. Sci. USA* **112** E119–26

Das N, Murray D T and Cross T A 2013 Lipid bilayer preparations of membrane proteins for oriented and magic-angle spinning solid-state NMR samples *Nat. Protoc.* **8** 2256–70

Gentry K A, Anantharamaiah G M and Ramamoorthy A 2019 Probing protein–protein and protein–substrate interactions in the dynamic membrane-associated ternary complex of cytochromes P450, b5, and reductase *Chem. Commun.* **55** 13422–5

Glattard E, Salnikov E S, Aisenbrey C and Bechinger B 2016 Investigations of the synergistic enhancement of antimicrobial activity in mixtures of magainin 2 and PGLa *Biophys. Chem.* **210** 35–44

Glaubitz C and Watts A 1998 Magic angle oriented sample spinning (MAOSS) – a new approach toward biomembrane studies *J. Magn. Reson.* **130** 305–16

Gopinath T, Mote K R and Veglia G 2013 Sensitivity and resolution enhancement of oriented solid-state NMR: application to membrane proteins *Prog. Nucl. Magn. Reson. Spectrosc.* **75** 50–68

Gopinath T, Mote K R and Veglia G 2015 Simultaneous acquisition of 2D and 3D solid-state NMR experiments for sequential assignment of oriented membrane protein samples *J. Biomol. NMR* **62** 53–61

Hagn F, Nasr M L and Wagner G 2018 Assembly of phospholipid nanodiscs of controlled size for structural studies of membrane proteins by NMR *Nat. Protoc.* **13** 79–98

Harmouche N and Bechinger B 2018 Lipid-mediated interactions between the amphipathic antimicrobial peptides magainin 2 and PGLa in phospholipid bilayers *Biophys. J.* **115** 1033–44

Holt A, Rougier L, Reat V, Jolibois F, Saurel O, Czaplicki J, Killian J A and Milon A 2010 Order parameters of a transmembrane helix in a fluid bilayer: case study of a WALP peptide *Biophys. J.* **98** 1864–72

Itkin A, Salnikov E S, Aisenbrey C, Raya J, Raussens V, Ruysschaert J M and Bechinger B 2017 Evidence for heterogeneous conformations of the gamma cleavage site within the amyloid precursor proteins transmembrane domain *ACS Omega* **2** 6525–34

Jirasko V, Lakomek N A, Penzel S, Fogeron M L, Bartenschlager R, Meier B H and Bockmann A 2020 Proton-detected solid-state NMR of the cell-free synthesized alpha-helical transmembrane protein NS4B from hepatitis C virus *Chem. Bio. Chem.* **21** 1453–60

Kim S and Cross T A 2004 2D solid state NMR spectral simulation of $3_{10}$, $\alpha$, and $\pi$-helices *J. Magn. Reson.* **168** 187–93

Kubicki D J, Rossini A J, Purea A, Zagdoun A, Ouari O, Tordo P, Engelke F, Lesage A and Emsley L 2014 Amplifying dynamic nuclear polarization of frozen solutions by incorporating dielectric particles *JACS* **136** 15711–8

Loudet C, Manet S, Gineste S, Oda R, Achard M F and Dufourc E J 2007 Biphenyl bicelle disks align perpendicular to magnetic fields on large temperature scales: a study combining synthesis, solid-state NMR, TEM, and SAXS *Biophys. J.* **92** 3949–59

Mandala V S, Loftis A R, Shcherbakov A A, Pentelute B L and Hong M 2020 Atomic structures of closed and open influenza B M2 proton channel reveal the conduction mechanism *Nat. Struct. Mol. Biol.* **27** 160–7

Mandala V S, Williams J K and Hong M 2018 Structure and dynamics of membrane proteins from solid-state NMR *Annu. Rev. Biophys.* **47** 201–22

Mao J, Aladin V, Jin X, Leeder A J, Brown L J, Brown R C D, He X, Corzilius B and Glaubitz C 2019 Exploring protein structures by DNP-enhanced methyl solid-state NMR spectroscopy *J. Am. Chem. Soc.* **141** 19888–901

Marcotte I and Auger M 2005 Bicelles as model membranes for solid- and solution-state NMR studies of membrane peptides and proteins *Concepts Magn. Reson. Part* A **24A** 17–37
Marquette A and Bechinger B 2018 Biophysical investigations elucidating the mechanisms of action of antimicrobial peptides and their synergism *Biomolecules* **8** 18
Marquette A, Leborgne C, Schartner V, Salnikov E, Bechinger B and Kichler A 2020 Peptides derived from the C-terminal domain of HIV-1 viral protein R in lipid bilayers: structure, membrane positioning and gene delivery *Biochim. Biophys. Acta, Biomembr.* **1862** 183149
Martinez D, Decossas M, Kowal J, Frey L, Stahlberg H, Dufourc E J, Riek R, Habenstein B, Bibow S and Loquet A 2017 Lipid internal dynamics probed in nanodiscs *Chem. Phys. Chem.* **18** 2651–7
Medeiros-Silva J, Jekhmane S, Paioni A L, Gawarecka K, Baldus M, Swiezewska E, Breukink E and Weingarth M 2018 High-resolution NMR studies of antibiotics in cellular membranes *Nat. Commun.* **9** 3963
Michaelis V K, Griffin R G, Corzilius B and Vega S 2018 *Handbook of High Field Dynamic Nuclear Polarization (English Edition)* 1st edn (New York: Wiley)
Michalek M, Salnikov E, Werten S and Bechinger B 2013 Structure and topology of the huntingtin 1-17 membrane anchor by a combined solution and solid-state NMR approach *Biophys. J.* **105** 699–710
Nelson S E D, Ha K N, Gopinath T, Exline M H, Mascioni A, Thomas D D and Veglia G 2018 Effects of the Arg9Cys and Arg25Cys mutations on phospholamban's conformational equilibrium in membrane bilayers *Biochim. Biophys. Acta* **1860** 1335–41
North C L, Barranger-Mathys M and Cafiso D S 1995 Membrane orientation of the N-terminal segment of alamethicin determined by solid-state $^{15}$N NMR *Biophys. J.* **69** 2392–7
Ozdirekcan S, Etchebest C, Killian J A and Fuchs P F 2007 On the orientation of a designed transmembrane peptide: toward the right tilt angle? *J. Am. Chem. Soc.* **129** 15174–81
Pauli J, Van Rossum B, Forster H, De Groot H J and Oschkinat H 2000 Sample optimization and identification of signal patterns of amino acid side chains in 2D RFDR spectra of the alpha-spectrin SH3 domain *J. Magn. Reson.* **143** 411–6
Ramamoorthy A, Wei Y and Lee D 2004 PISEMA solid-state NMR spectroscopy *Annu. Rep. NMR Spectrosc.* **52** 1–52
Ravula T, Hardin N Z, Ramadugu S, Cox S J and Ramamoorthy A 2018 Formation of pH resistant monodispersed polymer-lipid nanodiscs *Angew. Chem. Int. Ed. Engl.* **57** 1342–5
Ravula T, Kim J, Lee D-K and Ramamoorthy A 2020 Magnetic alignment of polymer nanodiscs probed by solid-state NMR spectroscopy *Langmuir* **36** 1258–65
Ravula T, Ramadugu S K, Di Mauro G and Ramamoorthy A 2017 Bioinspired, size-tunable self-assembly of polymer-lipid bilayer nanodiscs *Angew. Chem. Int. Ed. Engl.* **56** 11466–70
Raya J, Perrone B, Bechinger B and Hirschinger J 2011 Chemical shift powder spectra obtained by using ROtor-Directed Exchange of Orientations Cross-Polarization (RODEO-CP) *Chem. Phys. Lett.* **508** 155–64
Resende J M, Moraes C M, Munhoz V H D O, Aisenbrey C, Verly R M, Bertani P, Cesar A, Pilo-Veloso D and Bechinger B 2009 Membrane structure and conformational changes of the antibiotic heterodimeric peptide distinctin by solid-state NMR spectroscopy *Proc. Natl Acad. Sci. USA* **106** 16639–44
Resende J M, Verly R M, Aisenbrey C, Amary C, Bertani P, Pilo-Veloso D and Bechinger B 2014 Membrane interactions of Phylloseptin-1, -2, and -3 peptides by oriented solid-state NMR spectroscopy *Biophys. J.* **107** 901–11

Salnikov E, Aisenbrey C, Balandin S V, Zhmak M N, Ovchinnikova A Y and Bechinger B 2011 Structure and alignment of the membrane-associated antimicrobial peptide arenicin by oriented solid-state NMR spectroscopy *Biochemistry* **50** 3784–95

Salnikov E, Aisenbrey C, Raya J and Bechinger B 2014 Investigations of the structure,topology and dynamics of membrane-associated polypeptides by solid-state NMR spectroscopy *New Developments in NMR: Advances in Biological Solid-State NMR, Proteins and Membrane-Active Peptides* ed F Separovic and A Naito (Cambridge: Royal Society of Chemistry)

Salnikov E, Aisenbrey C, Vidovic V and Bechinger B 2010a Solid-state NMR approaches to measure topological equilibria and dynamics of membrane polypeptides *Biochim. Biophys. Acta* **1798** 258–65

Salnikov E and Bechinger B 2011 Lipid-controlled peptide topology and interactions in bilayers: structural insights into the synergistic enhancement of the antimicrobial activities of PGLa and magainin 2 *Biophysical J.* **100** 1473–80

Salnikov E, Bertani P, Raap J and Bechinger B 2009a Analysis of the amide $^{15}N$ chemical shift tensor of the C(alpha) tetrasubstituted constituent of membrane-active peptaibols, the alpha-aminoisobutyric acid residue, compared to those of di- and tri-substituted proteinogenic amino acid residues *J. Biomol. NMR* **45** 373–87

Salnikov E, Rosay M, Pawsey S, Ouari O, Tordo P and Bechinger B 2010b Solid-state NMR spectroscopy of oriented membrane polypeptides at 100 K with signal enhancement by dynamic nuclear polarization *J. Am. Chem. Soc.* **132** 5940–1

Salnikov E *et al* 2015 Solid-state NMR/dynamic nuclear polarization of planar supported lipid bilayers *J. Phys. Chem.* B **119** 14574–83.

Salnikov E S, Abel S, Karthikeyan G, Karoui H, Aussenac F, Tordo P, Bechinger B and Ouari O 2017 Dynamic nuclear polarization/solid-state nmr spectroscopy of membrane polypeptides: free-radical optimization for matrix-free lipid bilayer samples *Chem. Phys. Chem.* **18** 2103–13

Salnikov E S, Aisenbrey C, Anantharamaiah G M and Bechinger B 2019a Solid-state NMR structural investigations of peptide-based nanodiscs and of transmembrane helices in bicellar disc arrangements *Chem. Phys. Lipids* **219** 58–71

Salnikov E S, Aisenbrey C, Aussenac F, Ouari O, Sarrouj H, Reiter C, Tordo P, Engelke F and Bechinger B 2016 Membrane topologies of the PGLa antimicrobial peptide and a transmembrane anchor sequence by dynamic nuclear polarization/solid-state NMR spectroscopy *Sci. Rep.* **6** 20895

Salnikov E S, Aisenbrey C, Pokrandt B, Brugger B and Bechinger B 2019b Structure, topology, and dynamics of membrane-inserted polypeptides and lipids by solid-state NMR spectroscopy: investigations of the transmembrane domains of the DQ beta-1 subunit of the MHC II receptor and of the COP I protein p24 *Front. Mol. Biosci.* **6** 83

Salnikov E S, Anantharamaiah G M and Bechinger B 2018 Supramolecular organization of apolipoprotein-A-I-derived peptides within disc-like arrangements *Biophys. J.* **115** 467–77

Salnikov E S, Aussenac F, Abel S, Purea A, Tordo P, Ouari O and Bechinger B 2019c Dynamic nuclear polarization/solid-state NMR of membranes. Thermal effects and sample geometry *Solid State Nucl. Magn. Reson.* **100** 70–6

Salnikov E S, De Zotti M, Formaggio F, Li X, Toniolo C, O'Neil J D, Raap J, Dzuba S A and Bechinger B 2009b Alamethicin topology in phospholipid membranes by oriented solid-state NMR and EPR spectroscopies: a comparison *J. Phys. Chem.* B **113** 3034–42

Salnikov E S, Friedrich H, Li X, Bertani P, Reissmann S, Hertweck C, O'Neil J D, Raap J and Bechinger B 2009c Structure and alignment of the membrane-associated peptaibols ampullosporin A and alamethicin by oriented $^{15}N$ and $^{31}P$ solid-state NMR spectroscopy *Biophys. J.* **96** 86–100

Salnikov E S *et al* 2012 Developing DNP/solid-state NMR spectroscopy of oriented membranes *Appl. Magn. Reson.* **43** 91–106

Samoson A 2019 H-Mas *J. Magn. Reson.* **306** 167–72

Sanders C R 2ND and Landis G C 1995 Reconstitution of membrane proteins into lipid-rich bilayered mixed micelles for NMR studies *Biochemistry* **34** 4030–40

Sani M A, Zhu S, Hofferek V and Separovic F 2019 Nitroxide spin-labeled peptides for DNP-NMR in-cell studies *FASEB J.* **33** 11021–7

Seelig J and Seelig A 1980 Lipid conformation in model membranes and biological membranes *Q. Rev. Biophys* **13** 19–61

Shcherbakov A A, Mandala V S and Hong M 2019 High-sensitivity detection of nanometer $^{1}H$–$^{19}F$ distances for protein structure determination by $^{1}H$-detected fast MAS NMR *J. Phys. Chem.* B **123** 4387–91

Sizun C and Bechinger B 2002 Bilayer samples for fast or slow magic angle oriented sample spinning solid-state NMR spectroscopy *J. Am. Chem. Soc.* **124** 1146–7

Strandberg E, Esteban-Martin S, Salgado J and Ulrich A S 2009 Orientation and dynamics of peptides in membranes calculated from $^{2}H$-NMR data *Biophys. J.* **96** 3223–32

Strandberg E, Horn D, Reisser S, Zerweck J, Wadhwani P and Ulrich A S 2016 $^{2}H$-NMR and MD simulations reveal membrane-bound conformation of magainin 2 and its synergy with PGLa *Biophys. J.* **111** 2149–61

Strandberg E, Zerweck J, Wadhwani P and Ulrich A S 2013 Synergistic insertion of antimicrobial magainin-family peptides in membranes depends on the lipid spontaneous curvature *Biophys. J.* **104** L09–11

Straus S K, Scott W R and Watts A 2003 Assessing the effects of time and spatial averaging in $^{15}N$ chemical shift/$^{15}N$–$^{1}H$ dipolar correlation solid state NMR experiments *J. Biomol. NMR* **26** 283–95

Tamiola K and Mulder F A 2012 Using NMR chemical shifts to calculate the propensity for structural order and disorder in proteins *Biochem. Soc. Trans.* **40** 1014–20

Tang M, Waring A J, Lehrer R I and Hong M 2006 Orientation of a beta-hairpin antimicrobial peptide in lipid bilayers from two-dimensional dipolar chemical-shift correlation NMR *Biophys. J.* **90** 3616–24

Traaseth N J, Buffy J J, Zamoon J and Veglia G 2006 Structural dynamics and topology of phospholamban in oriented lipid bilayers using multidimensional solid-state NMR *Biochemistry* **45** 13827–34

Traaseth N J, Shi L, Verardi R, Muller D, Barany G and Veglia G 2009 Structure and topology of monomeric phospholamban in lipid membranes determined by a hybrid solution and solid-state NMR approach *Proc. Natl Acad. Sci. USA* **106** 10165–70

Verardi R, Traaseth N J, Masterson L R, Vostrikov V V and Veglia G 2012 Isotope labeling for solution and solid-state NMR spectroscopy of membrane proteins *Adv. Exp. Med. Biol.* **992** 35–62

Vidovic V, Prongidi-Fix L, Bechinger B and Werten S 2009 Production and isotope labeling of antimicrobial peptides in *Escherichia coli* by means of a novel fusion partner that enables high-yield insoluble expression and fast purification *J. Pept. Sci.* **15** 278–84

Vostrikov V V, Hall B A, Greathouse D V, Koeppe R E 2ND and Sansom M S 2010 Changes in transmembrane helix alignment by arginine residues revealed by solid-state NMR experiments and coarse-grained MD simulations *J. Am. Chem. Soc.* **132** 5803–11
Wang S, Gopinath T and Veglia G 2019 Improving the quality of oriented membrane protein spectra using heat-compensated separated local field experiments *J. Biomol. NMR* **73** 617–24
Wolf J, Aisenbrey C, Harmouche N, Raya J, Bertani P, Voievoda N, Süss R and Bechinger B 2017 pH-dependent membrane interactions of the histidine-rich cell penetrating peptide LAH4-L1 *Biophys. J.* **113** 1290–300

# Chapter 8

# Solid-state NMR and dynamic nuclear polarization studies of molecular interactions in membranes

**Vivien Yeh and Boyan B Bonev**

## Abstract

Membranes confine the microenvironment that defines a cell as the smallest unit of life. They play a complex and crucial part in the life of cells in maintaining vital electrochemical and solute gradients, mediating cell homeostasis and transport, cell signaling and cellular interactions with the environment. The complex composition and planar topology of membranes hinders structural studies and is limiting current understanding of membrane organization, structure and assembly. In Gram-negative bacteria, mitochondria and plastids the envelope complexity is increased further by the presence of outer membranes that have unique composition and protein structure and modulate further solute transport and interactions with xenobiotics. Solid-state NMR is a versatile and uniquely suited analytical tool for monitoring individual atomic environments and reporting molecular interactions and dynamics. We review a toolkit of solid-state NMR approaches in the context of their applications to studies of a diverse range of molecular interactions involving membranes. Some challenges in the application of solid-state NMR to membranes are discussed along with available solutions and the use of dynamic nuclear polarization (DNP) magic-angle spinning (MAS) NMR in signal enhancement from low molarity membrane systems is described.

## 8.1 Introduction

Lipid membranes form compartments in the microenvironment that define the interface of cells, the basic units of life. Cell membranes are asymmetric complex structures composed of lipids and proteins at a roughly equal measure that perform numerous functions that are vital to cells, such as transport, signaling, recognition

doi:10.1088/978-0-7503-2532-5ch8  

and electrical polarization. They also contain around 15% of all cellular proteins (Lopez-Campistrous *et al* 2005). Biological membranes are versatile and can respond to cellular and environmental stimuli by evolving and adapting their structure, composition and dynamics.

Lipid and protein composition vary between prokaryotic and eukaryotic plasma membranes but in both types a phospholipid bilayer, functionalized with helical transmembrane domain (TMD) proteins, is responsible for creating and maintaining proton motive force (PMF) and solute gradients that are essential for maintaining cells alive. Cholesterol hallmarks eukaryotic membranes and opens the possibility of lateral compartmentalization along with zwitterionic sphingomyelin and phosphatidylcholine. Prokaryotes, on the other hand, have largely negatively charged membranes that contain unique lipid chains and cell envelope markers, characteristic of the type and architecture of the species.

The membranes in Gram-positive bacteria like *Staphylococcus* and *Bacillus spp* are enclosed in a thick peptidoglycan layer forming a rigid, protective cell wall. In Gram-negative bacteria the complexity of the cell envelope is increased further by the presence of an outer membrane (OM) that encloses a thinner peptidoglycan layer and the inner membrane (IM). The outer OM has a characteristic structure, as well as lipid, lipopolysaccharide (LPS) and protein composition (Silhavy *et al* 2010). Some eukaryotic organelles, such as mitochondria and chloroplasts, also have a double membrane devoid of LPS, which reflects their bacterial evolutionary origin and adaptation to life within the cells of multicellular organisms.

The outer membranes of Gram-negative bacteria are highly asymmetric, with a phospholipid inner leaflet that is largely identical in composition to the IM and an outer leaflet of tightly packed LPS molecules. The OM outer leaflet is stabilized by divalent cations bridging anionic phosphorylates and pyrophosphates on key LPS saccharide residues. Outer membranes are also rich in membrane proteins, but in contrast to IMs the transmembrane domains of almost all OM proteins (OMPs) follow the beta-barrel folding topology. The outer membranes are rich in porins and serve as size-selective filters that permit the free flow of small solutes but restrict the passage of larger molecules. They protect bacteria from xenobiotic compounds and serve as a recognition interface, essential for the correct bacterial identification in complex co-colonization environments, such as the mammalian gut. The periplasm, enclosed between the IM and the OM, confines the peptidoglycan layer, as well as essential enzymatic machinery for post-translational modification of proteins.

The LPS that coats the exterior of Gram-negative bacteria has broadly tri-partite organization that consists of lipid A, a saturated hexa- or hepta-acyl glucosamine disaccharide, which acts as the membrane-embedded hydrophobic anchor; a polysaccharide core region of up to 15 sugars typically characterized by the presence of 3-deoxy-d-manno-oct-2-ulosonic acid (Kdo); and an O-antigen, a polysaccharide of up to 40 oligosaccharide units that is highly diverse and characteristic of the bacterial species. Smooth chemotypes LPS contain all three elements, while rough chemotypes lack the O-antigen. Lipid A and the core region can be phosphorylated, pyrophosphorylated or may contain phosphoethanolamine groups, typically in positions 1′ and 4′ on lipid A and in the core region. Phosphorylation mediates divalent cation coordination that is essential to maintaining outer membrane stability and protects bacteria from antimicrobials and xenobiotics. Chelation of divalent cations weakens the OM and renders bacteria more susceptible to antimicrobial and antibiotic attack (Clifton *et al* 2015). At the same time, co-evolution of bacteria has yielded antimicrobial proteins and peptides (AMPs) that use the pyrophosphates and monophosphates in LPS as docking platforms to destabilize and breach the OM (Bonev *et al* 2004, Johnson *et al* 2014). In some bacteria the cells are also encapsulated in exopolysaccharide or capsular polysaccharide, which are loose or covalently tethered to the OM but have similar polyphosphorylated polysaccharide architecture that confers additional protection to the cell (Dertli *et al* 2013).

The phospholipid composition of biological membranes has been studied in the past using thin-layer chromatography (Dunnick and O'Leary 1970, Skipski *et al* 1967) and recently by mass spectroscopy (Bernat *et al* 2016, Gidden *et al* 2009, Oursel *et al* 2007). Bacterial membrane composition is important for taxonomic classification, understanding the structure and stability of cell membranes, studying lipid–protein interactions and for elucidating antimicrobial mechanisms (Epand *et al* 2007).

The phospholipid composition of bacterial plasma membranes depends on the type of the bacteria (Gram-positive/negative) and varies with species but the phospholipid head groups are primarily zwitterionic phosphatidyl ethanolamine (PE), anionic phosphatidyl glycerol (PG) with some diphosphatidylglycerol (cardiolipin, CL) and minor contributions from phosphatidylinositol (PI), phosphatidic acid (PA) and phosphatidylserine (PS). In Gram-positive bacteria, such as *Bacillus subtilis*, the headgroup composition was found to comprise 20%–40% PG, 20%–30% PE, 20%–40% lysylphosphatidylglycerol (LPG), 10%–15% CL and a small amount of PA (Bernat *et al* 2016, den Kamp *et al* 1969). In Gram-negatives, such as *Escherichia coli*, the composition of IM and the inner leaflet of OM is 72%–80% PE, PG and 4%–7% CL (Kaneda, 1991, Morein *et al* 1996) with a major contribution to the overall negative charge from LPS. In eukaryotic systems sphingomyelin and cholesterol permit lateral phase separation into fluid and raft membrane environments (Brown and London 2000, Lingwood and Simons, 2010). These carry endogenous markers such as gangliosides (Sanghera *et al* 2011), or exogenous compounds during activation of signalling cascades such as LPS (Ciesielski *et al* 2013, 2012).

In addition to different phospholipid headgroups, there is a variety of phospholipid hydrophobic chains that underpin and modulate the stability, fluidity and other characteristics of membranes. The fatty acid profile of bacteria normally consists of mixed saturated and mono or polyunsaturated chains. Gram-negative organisms may also have odd number carbon chains containing cyclopropane rings (Grogan and Cronan 1997, Loffhagen *et al* 2007, Morein *et al* 1996), whereas Gram-positives have odd number carbon chains primarily containing branched, methylated side chains (Kaneda 1977, Nickels *et al* 2017, Uttlová *et al* 2016). Both headgroup and fatty acid composition change with growth phase (Gidden *et al* 2009) and as an adaptive mechanism to environmental changes (Morein *et al* 1996, Siliakus *et al* 2017, Sullivan 1979), antibiotic pressure (Uttlová *et al* 2016) or serves as a toxic response (Ramos *et al* 1997, Weber and de Bont 1996).

## 8.2 Membranes in the presence of small organic molecules

Membranes and cell viability may be disrupted by lipophilic compounds that accumulate in the membrane, as such compounds alter the physical or chemical properties of the membrane leading to change of lipid polymorphic assembly or stability. A major recent development towards sustainable 'green' manufacture of commodity chemicals has prompted studies of the toxic effects on membrane that result from high levels of organic solvents, such as cyclic hydrocarbons and alcohols (Sikkema *et al* 1995). For example, toluene has been shown to disrupt membranes by inducing inverted hexagonal phases that destabilize membranes of Gram-negative industrial organisms like *E. coli* (De Smet *et al* 1978) and *P. putida* (Heipieper *et al* 1994, Inoue and Horikoshi 1989).

Alcohol-induced membrane stress has been studied extensively (Ingram 1986). Ethanol has been known to fluidize membranes (Dombek and Ingram 1984, Ingram 1976), causing an increased passive efflux of proton and the loss of PMF (Leão and Van Uden 1984, Sikkema *et al* 1995). It has been suggested that toxicity of alcohols depends on the alcohol chain length, as butanol shows higher cytotoxicity than ethanol and, as well as surface intercalation, butanol can disrupt the membrane by partitioning into the hydrophobic core (Ly and Longo 2004).

While addition of organic solvents leads to cell death, various organisms show adaptation during growth in the presence of toxic compounds (Sardessai and Bhosle 2002). Certain strains of *P. putida* and *E. coli* have been reported to adjust membrane composition in response to changes in membrane order or fluidity caused by the presence of solvent (Ramos *et al* 1997, Sardessai and Bhosle 2002). Some responses are fast, such as changes in the ratio of unsaturated and saturated fatty acid chains (Buttke and Ingram 1978, Dombek and Ingram 1984), *cis*- to *trans*-isomerization of unsaturated fatty acid (Heipieper *et al* 1994, Heipieper and Bont 1994, Weber and de Bont 1996), increased conversion of *cis*-unsaturation to cyclopropyl chains (Ramos *et al* 1997) and either increase or decrease of fatty acid chain length (Ingram 1976, Ramos *et al* 1997). In surviving cells further alteration in membrane composition, such as elevation in level of CL (Ramos *et al* 1997), changes in ratio of the phospholipid head groups (Mishra and Prasad 1988, Weber and de Bont 1996) and modification of lipopolysaccharides have been reported (Pinkart *et al* 1996).

## 8.3 Membrane-active proteins and peptides

In addition to their major compositional characteristics, membranes of cells can carry a number of highly specific molecular markers, which serve as docking and recognition receptors for cellular regulators and for cytotoxic xenobiotics. In Gram-positive bacteria peptidoglycan intermediates are used as selective targets by glycopeptide antibiotics (Lebreton and Cattoir 2019) and by numerous antimicrobial peptides, such as nisin (Bonev *et al* 2004, Breukink *et al* 1999, Hyde *et al* 2006, Mulholland *et al* 2016), subtilin (Parisot *et al* 2008), mersacidin (Brötz *et al* 1998) and teixobactin (Shukla *et al* 2020). Gram-negative LPS is subverted by colicins for gaining entry across the bacterial OM (Johnson *et al* 2014, 2013) and serves a dual role as a portal and decoy for antimicrobial peptides (Lanne *et al* 2019).

Besides serving as lateral compartments in membranes, cholesterol-containing eukaryotic membranes have co-evolved alongside a class of exogenous cholesterol-activated cytolysins. These can include bacterial exotoxins that destroy host cells during infection to release cellular contents and provide nutritious milieu for the colonizing bacteria, e.g. pneumolysin from *Streptococcus pneumoniae* or perfringolysin from *Clostridium perfringens* (Rossjohn *et al* 1997), as well as eukaryotic defence proteins, such as equinatoxin from the sea anemone *Actinia equina* (Bonev *et al* 2003). Such cholesterol-dependent toxins rely on the availability of cholesterol in target membranes for self-assembly and membrane disruption (Bonev *et al* 2001a, Bonev *et al* 2001b).

## 8.4 Transmembrane proteins

Membrane proteins account for 30% of protein encoded in most genomes and are responsible for a wide variety of cellular functions. Transmembrane proteins have a wide hydrophobic region that allows them to position across the hydrophobic core of membranes. At the same time, this makes membrane proteins insoluble and challenging to study experimentally. A variety of membrane mimetic carrier systems has been developed to permit the removal of membrane proteins from the lipid membrane into stable suspensions whilst largely maintaining their original fold (Warschawski *et al* 2011).

Membrane proteins have been studied in detergent micelles, where the hydrophobic protein band that normally faces the membrane interior is protected by small toroidal assemblies of short chain detergent molecules that preserve protein fold in a relatively small soluble proteodetergent complex (Garavito and Ferguson-Miller 2001, le Maire *et al* 2000). However, the use of detergent micelles in membrane protein studies can affect protein stability, structure and function (Hagn *et al* 2013, Raschle *et al* 2009). Thus, membrane models better represent the physical properties of native membranes such as charge, hydrophobic thickness, and fluidity. Membrane bilayer can also provide specific lipid partners for function and interactions of membrane proteins, and may be essential for oligomerization (Amin and Hazelbauer 2012, Lee 2004). Cholesterol is often required for membrane protein function and lipid carbonyls provide anchor points that support the correct protein fold within lipid bilayers (de Planque *et al* 2003).

In one example, bacteriorhodopsin (bR), a seven helical transmembrane domain protein from the halophilic archaea *Halobacterium salinarum*, functions as a light-driven proton pump that energizes bacterial membranes by pumping proton across the bacterial membrane. When using purified bR reconstituted in nanodiscs, the activity of proton pumping is modulated by the presence of different phospholipids *in vitro* (Lee *et al* 2015) and the headgroup of the phospholipid may be directly involved in the process of proton pumping (Yeh *et al* 2016). Native lipids can also have a significant influence on the function of membrane proteins (Yeh *et al* 2019), which is supported by x-ray structure of bR with lipids incorporated into the protein exterior and interior (Grigorieff *et al* 1996), suggesting strong lipid–protein interaction.

Potassium channel KcsA governs $K^+$ ion permeability of membranes and is essential for regulation of biological processes such as neuron and cardiac action potential. It was demonstrated that refolding KcsA *in vitro* into functional tetramer requires lipids and that negative charge is required for function (Valiyaveetil *et al* 2002). Studies of KcsA reconstituted into liposomes have shown that membrane lipid compositions affects the KcsA activation states and that CL-rich membranes may stabilize the open conductive conformation of the channel (Cruijsen *et al* 2017).

## 8.5 Membrane studies by NMR

Despite the importance of biological membranes, they remain poorly understood due to their complexity, insolubility and two-dimensional topology. A variety of experimental approaches have been employed to overcome membrane complexity, sample heterogeneity and the need to preserve the intricate molecular interactions within membranes. In that respect, solid-state NMR is ideally placed as a non-perturbing tool for obtaining atomic level information on molecular structure and dynamics from model and native membranes, as well as from whole cells.

Solution NMR has been used to study isolated membrane components outside their native membrane environment, such as lipids in organic solvents and membrane proteins in detergent micellar systems. Rapid molecular tumbling of small molecular complexes and assemblies can lead to orientationally averaged high-resolution spectra that provide valuable information on molecular structure, composition and interactions. However, studies of molecular function and collective properties such as mobility and molecular interactions are limited by the membrane-free model environment.

One way of obtaining atomic level information from membrane systems is to use recombinant model systems in combination with high-resolution solid-state NMR. Solid-state NMR can manage the anisotropic broadening arising from orientation-dependent second-order interactions or use these interactions as reporters of structure. The method is sensitive to molecular dynamics and can be employed to characterize membrane assembly, structure, phase state and behavior, molecular and segmental mobility, as well as changes or perturbation in the local environment within membranes (Bonev 2013). Broadly speaking, there are two main approaches in studying membranes and membrane proteins by solid-state NMR: wideline and high resolution magic angle spinning (MAS) NMR. In addition, oriented samples can be used in NMR studies where lipid membranes can be aligned uniaxially with respect to the external magnetic field using the lipid diamagnetic susceptibility anisotropy in bicellar formulations (Sanders and Landis 1995, Sanders and Schwonek 1992) or in mechanically oriented samples using planar physical supports (Hallock *et al* 2002). Such oriented samples provide information on the orientation of molecular phosphate effective chemical shift anisotropy (CSA) tensors. Using $^{15}N$-labelling in peptides and proteins, aligned membranes allow determination of protein NH-bond angles through $^{15}N$ chemical shift and $^{15}N{-}^{1}H$ dipolar correlations, providing orientational restraints for conformational analysis of transmembrane peptides and protein domains (Marassi and Opella 2000, Opella and Marassi 2004), or inform on the orientation of membrane active peptides (Bechinger *et al* 1993). Hybrid approaches using sample alignment in combination with mechanical rotation have also been explored (Glaubitz and Watts 1998, Sizun and Bechinger 2002).

High resolution spectra in the solid state can be obtained by mechanical sample rotation about the [111] direction that makes a 'magic' angle of 54.7° with the external magnetic field. At this angle the second Legendre polynomial $P_2(\cos\theta) = \frac{1}{2}(3\cos^2\theta - 1)$, which scales second order tensor interactions, vanishes and one can observe scalar, isotropic chemical shift-dominated spectra from motion-restricted systems.

High-resolution solid-state NMR enables studies of membrane proteins and other membrane constituents in their native or near native functional environment. Recombinantly expressed and isotopically labelled membrane proteins can be reconstituted into multilamellar liposome systems for solid-state NMR studies while maintaining close to physiological conditions (Etzkorn *et al* 2007, Good *et al* 2014, Ladizhansky 2017, Park *et al* 2012, Retel *et al* 2017, Shi *et al* 2011). For example, the helical TMD ion channel protein KcsA, has been studied by solid-state NMR within a lipid membrane environment to yield structural information (Schneider *et al* 2008), lipid dependence (Cruijsen *et al* 2017) and allosteric coupling (Wylie *et al* 2014).

Solid-state NMR also provides an opportunity to study large complexes. For example, the multicomponent beta-barrel assembly machinery (BAM) from *E. coli* has been studied in lipid bilayers as a complex (Pinto *et al* 2017), while also permitting individual characterization of its large central hub, BamA (Sinnige *et al* 2015).

## 8.6 Native membranes

Solid-state NMR is a versatile technique, which provides insights into membrane structure, organization and interactions beyond the membrane models. Bacterial membrane extract studies offer insights into the interactions between antimicrobials and intact cellular LPS (Lanne *et al* 2019) and antimicrobial binding to membrane targets has been studied in native membrane vesicles (Medeiros-Silva *et al* 2018). Solid-state NMR has also been employed to study native membrane extracts (Lanne *et al* 2019, Pinto *et al* 2017) and cell lysate (Viennet *et al* 2016). Bacterial whole cells have also been studied using solid-state NMR for lipids and LPS content (Arnold *et al* 2015), cellular integrity and interaction with AMPs (Overall *et al* 2019), and outer membrane proteins (Renault *et al* 2012).

Other biological systems can be studied using solid-state NMR such as biofilms (Reichhardt *et al* 2015), peptidoglycan (Romaniuk and Cegelski 2015), amyloid fibrils formed from Alzheimer's related β amyloid peptide (Aβ) (Tycko 2011) and Parkinson's related α-synuclein (Heise *et al* 2005).

## 8.7 Wideline and oriented membranes NMR

In the solid state, molecular reorientation is restricted, which leads to spectra dominated by large orientation-dependent second-order tensor interactions. The magnitude of these interactions significantly exceeds the isotropic chemical shift dispersion that characterizes high resolution spectra. As a result, in the solid state the NMR spectra have characteristic wideline distributions that reflect the organization and dynamics of the mesoscopic architecture of the NMR samples. Spectral broadening due to anisotropic NMR effects such as CSA, dipolar coupling and quadrupole couplings exceed the isotropic shifts by orders of magnitude, and can lead to anisotropic or isotropic lineshapes that hold information about the molecular organization within membranes or within membrane constituents. Using such wideline spectroscopy to observe strategically incorporated $^{2}$H or naturally occurring $^{31}$P reporters has spawned a number of techniques for studying the organization and dynamics of membrane lipids, proteins and their interactions, as well as changes in membrane assembly.

### 8.7.1 Deuterium NMR

Deuterium NMR is a well-developed tool for studying membrane order, stability and surface interactions (Seelig 1977). Isomorphic substitution of $^{2}$H for $^{1}$H offers a non-perturbing reporter system that can be introduced selectively into various molecular systems or different parts of the same membrane component to provide molecule or segment-specific information on orientation and dynamics. In addition to targeted labelling, natural abundance of $^{2}$H is low at 0.01% and combined with its relatively low gyromagnetic ratio ($\gamma$) and quadrupole effect-dominated wideline intensity distribution necessitates isotope-enrichment.

As a spin $I = 1$ nuclear system, the ground state of deuterium in magnetic field is split into three levels that are further offset by the quadrupole effect with a resulting quadrupole-split doublet:

$$\Delta \upsilon_Q = \frac{3}{2}\frac{e^2qQ}{h}\frac{3\cos^2\theta - 1}{2}$$

where $e^2qQ/h$ is the deuterium quadrupole coupling constant and $\theta$ is the angle between the electric field gradient tensor at the nucleus and the magnetic field. In randomly oriented samples with rapid axially rotating molecules, spherically symmetric orientational distribution leads to characteristic Pake or powder spectral intensity patterns (Pake 1948). These can be de-Paked to calculate the corresponding aligned sample spectra (Lafleur *et al* 1989).

Uniform deuteration of lipid chains offers a sensitive measure of the segmental order and dynamics along the chains within lipid membranes, as the first spectral moment is directly proportional to the mean segmental CD order (Bonev and Morrow 1997, Davis 1983). Molecular interactions affecting the hydrophobic membrane interior, such as peptide insertion (Bonev *et al* 2000), as well as membrane organization and dynamics variation in response to environmental changes, such as hydrostatic pressure (Bonev and Morrow 1997) or protein–lipid interactions (Soubias *et al* 2008), are reflected in the deuterium spectra from uniformly chain-deuterated lipids (Bonev and Morrow 1997, Davis 1983, Seelig 1977).

Headgroup deuterated lipid reporters inform on changes in surface charge density that result from molecular binding at the polar/apolar membrane interface. Deuteration at choline positions α and β leads to two distinctive doublets, which change in response to lipid headgroup tilt (Akutsu and Seelig 1981). Counter-directional change in these splittings offers a quantitative measure of peptide binding to membrane surfaces, as reported for antimicrobial peptide nisin (Bonev *et al* 2000), also reflected in isotropic chemical shift changes monitored by $^{31}$P MAS NMR during amyloid pro-peptide and pentalysin interactions with membranes (Bonev *et al* 2001a).

### 8.7.2 Lipid phase analysis by $^{31}$P wideline NMR

Often, molecular interactions within or with membranes lead to changes in lipid phase state or dynamics, in which cases wideline $^{31}$P NMR can be used as a sensitive tool to monitor and quantify results of such interactions. The presence of $^{31}$P at 100% natural abundance in phospholipid headgroups offers insights into the organization and dynamics of membranes without the need to introduce reporter nuclei. In fluid membrane and cellular systems the $^{31}$P NMR spectra are normally reflective of spherically symmetric distributions and give rise to Pake patterns (Pake 1948), which are dominated by the effective CSA of the lipid phosphates, and are sensitive to headgroup orientation and local dynamics. Changes in the lyotropic phase of hydrated phospholipid mixtures give rise to characteristic intensity distribution patterns readily observed by wideline $^{31}$P NMR (Dufourc *et al* 1992, Seelig 1978), paralleled by x-ray scattering (Rappolt *et al* 2003) and differential scanning calorimetry (DSC) (Seddon *et al* 1983).

Bilayer lipid membranes in the liquid crystalline phase give rise to spherically symmetric orientational distributions and Pake or powder spectra (Pake 1948) with overall effective CSA of 40–45 ppm (Bonev 2013). The full phosphate CSA is modulated by axially symmetric molecular rotation within the membranes and the observed effective CSA is influenced by membrane phases (Malcolm *et al* 2005), presence of cholesterol (Brown and Seelig 1978) and degree of hydration (Griffin 1976).

The structure of inverted hexagonal lipid phases ($H_{II}$) is dominated by hexagonally packed aqueous cylinders lined by lipidic superstructures with negative surface curvature. Lipids with packing parameter larger than one, where head group cross section is smaller than the cross sectional of the acyl chains (Seddon 1990), such as unsaturated phosphatidyl-ethanolamines (Cullis and Kruijff 1978), have been observed to form inverted hexagonal phase assembly spontaneously in low hydration. Lipid molecules in the $H_{II}$ phase undergo additional motional averaging due to fast lateral diffusion about the small aqueous channels. As a result, wideline $^{31}$P NMR spectra from $H_{II}$ phases show characteristic inversion in the Pake distribution with half the observed effective CSA (Cullis and Kruijff 1979).

Lastly, small molecular structures, such as micelles and small unilamellar vesicle (SUV) undergo rapid reorientational motions on the NMR timescale that collapse orientation-dependent interactions. Wideline $^{31}$P NMR reports the presence of such small structures as a sharp, symmetrical peak centered at the isotropic chemical shift of the particular phosphate (Moreau *et al* 1999). Fast lateral diffusion in lipidic cubic phases also leads to rapid angular excursions of individual reporter molecules that leads to isotropic lineshapes (Yang *et al* 2015).

Wideline $^{31}$P NMR has been used extensively to study the structure and dynamics of phospholipid multilamellar vesicle (MLV) systems (Seelig 1978) and is a sensitive reporter of changes in membrane organization following the addition of peptides (de Planque *et al* 2003, Nomura *et al* 2005), protein toxins and cytochrome c (Bonev *et al* 2001b, Bonev *et al* 2003, Pinheiro and Watts 1994), figure 8.1, as well as small organic molecules (Yeh *et al* 2020). Due to the large distinction between the characteristic static $^{31}$P NMR lineshape of lipid bilayer and non-bilayer membranes, wideline $^{31}$P NMR has been useful in monitoring disruption of model membrane in

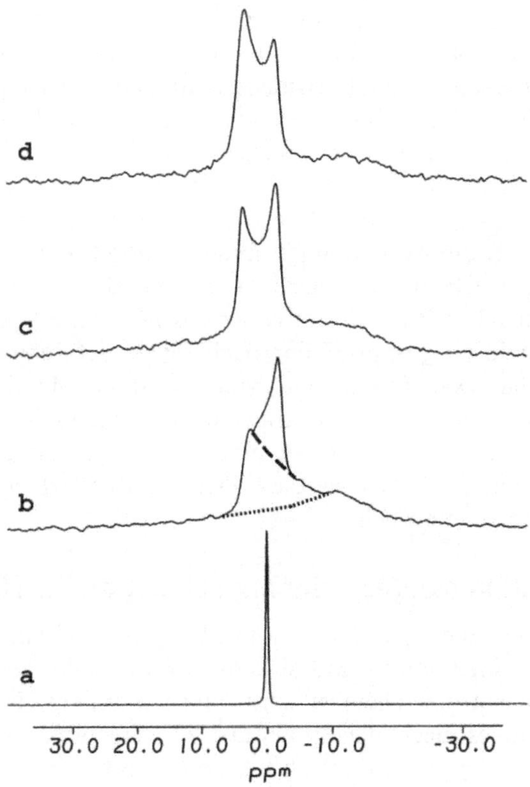


**Figure 8.1.** Wideline $^{31}$P NMR spectra of hydrated DOPC/cholesterol/dicetyl PC 100 nm large unilamellar vesicles (LUVs) (a) and following addition of (b) 0.5, (c) 1.25, and (d) 1.5 mol% pneumolysin. Vesicle tumbling averages CSA in (a) while pneumolysin-mediated aggregation develops CSA-dominated lineshapes revealing lamellar phase to $H_{II}$ conversion in (b–d). Reproduced with permission from Bonev *et al* (2001a). Copyright 2001 American Society of Biochemistry and Molecular Biology.

the presence of increasing concentrations of amphipathic and hydrophobic compounds. In one example, wideline $^{31}P$ NMR has revealed stability in membrane systems in the presence of up to 3:1 butyl methacrylate (BMA)/lipid and membrane disruption at 4:1 molar excess of methacrylate (Yeh *et al* 2020).

### 8.7.3 High resolution NMR in oriented membranes – $^{15}N$ PISEMA

Besides wideline $^{31}P$ and $^{2}H$ experiments, solid-state NMR can be used to study uniaxially oriented membrane samples to obtain orientational constraints and to follow changes in molecular orientation that report on interactions within lipid membranes. In such studies, lipid bilayers need to be oriented with respect to the external magnetic field either by mechanical alignment on glass plates or by making lipid bicelles. Bicelles are mixed lipid–detergent systems, in which a small membrane patch is collared laterally by detergent that stabilizes the interface between hydrophobic lipid core and the surrounding aqueous environment. In the presence of magnetic field, hydrated bicellar diskoid assemblies orient spontaneously with their normal perpendicular to the magnetic field due to the negative anisotropy in lipid diamagnetic susceptibility (Prosser *et al* 2006, Sanders and Landis 1995, Sanders and Schwonek 1992). Bicellar assemblies can also be prepared to align along the magnetic field in the presence of trivalent lanthanide ions (Angelis and Opella 2007) or by using lipids with a biphenyl moiety (Loudet *et al* 2007).

Nitrogen-15 labelled peptides and helical protein TMDs, reconstituted into oriented bilayers, have been studied by polarization inversion spin-exchange at the magic angle (PISEMA) (Wu *et al* 1994). Correlation between $^{15}N$ chemical shifts and associated $^{15}N$–$^{1}H$ dipolar couplings obtained from this experimental approach follow a pattern reflective of the helical orientation through polarity index slant angle (PISA) wheels. The PISA wheel is dependent on the insertion angle of the alpha-helix in the lipid bilayer, providing backbone structure and protein topology (Marassi and Opella 2000, Opella and Marassi 2004). Membrane proteins and membrane-associated peptides, such as transmembrane domain of Vpu from HIV-1 (Park *et al* 2003), M2 segment of the nicotinic acetylcholine receptor (AChR) (Opella *et al* 1999), and CXCR1 from the GPCR family (Park *et al* 2006) have been characterized using this approach.

## 8.8 High resolution sample spinning solid-state NMR

Solid-state NMR is a technique that is uniquely capable of extracting atomic level detail on molecular interactions and structure from model and biological membranes by accessing isotropic chemical shifts and J-couplings. This can be achieved by rotating the samples mechanically about [111] direction, which makes a 54.7° angle with the static magnetic field. Such rotation leads to averaging of second order tensor interactions, like CSA and dipolar couplings, leading to observation of high-resolution scalar $CS_{iso}$-dominated spectra from membranes and cells. As the space and spin parts of the nuclear interactions are separable, MAS is normally combined with multipulse sequences that manage heteronuclear or homonuclear dipolar couplings.

Proton detection is ideal for biomolecular NMR studies due to the high natural abundance and high gyromagnetic ratio of protons. However, proton detection in solid-state NMR remains a challenge as the homonuclear dipolar couplings within proton networks in membrane proteins can exceed 100 kHz and are only partially averaged at moderate MAS speed. Also, $CS_{iso}$ for protons has smaller chemical shift dispersion than $^{13}C$. Thus, to achieve higher resolution in such situations ultrafast MAS may be required for complete homonuclear decoupling along with high power RF methods. Proton detection of deuterated proteins with partial back-exchange has been reported at fast MAS frequency (~50 kHz) (Barbet-Massin *et al* 2014), while recent developments of ultrafast MAS probes (>100 kHz) have allowed high resolution proton solid-state NMR studies of fully back-exchanged proteins (Agarwal *et al* 2014), as well as from protonated protein samples (Andreas *et al* 2016).

### 8.8.1 Membrane studies by $^{13}C$ MAS NMR

The most commonly observed nuclear systems in high-resolution MAS NMR studies of membranes are $^{13}C$. Carbon is a ubiquitous element in membrane systems and the $^{13}C$ spectra in biological macromolecules are characterized by large chemical shift dispersion (ca. 200 ppm), which makes $^{13}C$ an excellent reporter for studying molecular interactions. Under moderate decoupling (ca. 60–90 kHz), $^{13}C$ MAS NMR can provide high-resolution spectra with high information content under low to moderate MAS speed (~10 kHz). Carbon-13 MAS NMR spectroscopy is also sensitive to changes in chemical shift and in segmental mobility resulting from specific molecular interactions (Sanghera *et al* 2011) and can report on conformational polymorphism (figure 8.2).

The natural abundance of $^{13}C$ in biological molecules is only 1.1%, which leads to lower signal intensity compared to $^{1}H$. The upshot of this is that in natural abundance samples the probability of spontaneous $^{13}C$–$^{13}C$ couplings and background coherence transfers is low. This leads to higher resolution in natural abundance spectra (Zorin *et al* 2010) and opens the possibility of engineering and introducing $^{13}C$ labels that deliver highly specific low background information from strategically or uniformly labelled samples. Carbon-13 labels can be introduced in membrane proteins using recombinant protein production from cellular systems grown on engineered, designer media or using uniform labelling from carbon-limited culture media.

Cross-polarization (CP) is almost universally used for excitation in solid-state MAS NMR (Pines *et al* 1972). In CP, thermodynamically warmer rare nuclei are coupled to colder proton bath via Hartmann–Hahn contact (Hartmann and Hahn 1962) that relies on strong $^{13}C$–$^{1}H$ coupling. As a result, the rare ($^{13}C$) nuclear system is equilibrated to the spin temperature of the proton reservoir, which leads to signal enhancement reflecting the higher proton polarization (up to four-fold). Within the same experimental time, the signal is further enhanced compared to direct excitation by reduction of the interpulse repeat time, which under CP is limited by shorter proton $T_1$ instead of the longer $T_1$ of weakly coupled $^{13}C$ spins.

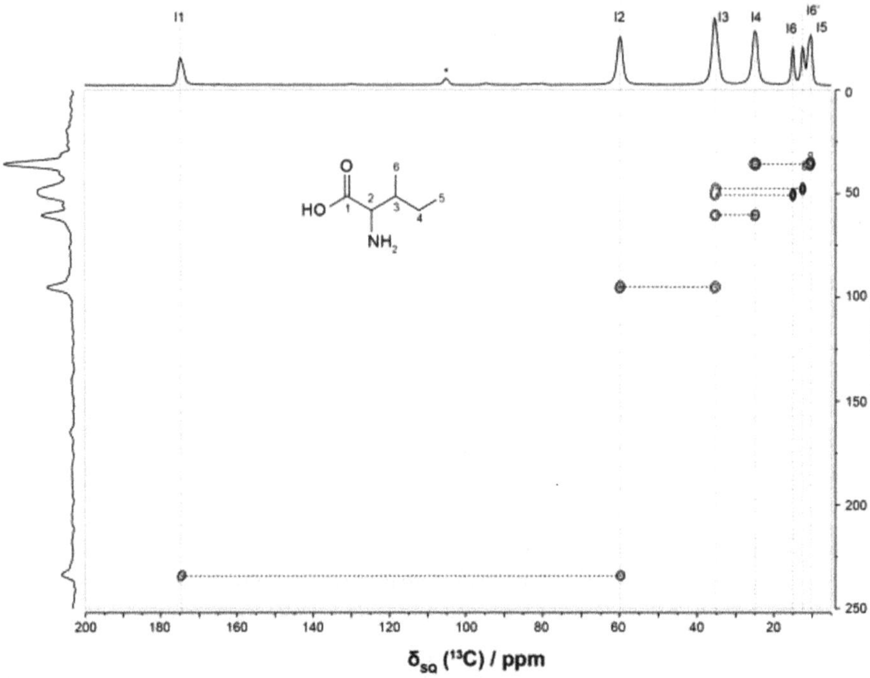


**Figure 8.2.** A $^{13}C$–$^{13}C$ double quantum-single quantum (DQSQ) solid-state MAS NMR R14 (Carravetta *et al* 2000) correlation spectrum from uniformly $^{13}C$-labelled solid isoleucine showing distinct conformational isoforms at Ile-$C_\beta$. * denotes CO spinning sideband.

In membrane studies, CP experiments can be used as a motion filter. The polarization transfer between $^1H$ and $^{13}C$ that leads to $^{13}C$ coherence buildup requires dipolar coupling between rare and abundant spins that is only observed in relatively rigid molecular systems. In more mobile molecular systems, the efficiency of dipolar coupling is attenuated as the rates of molecular motion can approach or exceed the dipolar coupling strength and the resulting spectral contributions are also attenuated. J-coupling-based transfer schemes, such as INEPT (Morris and Freeman 1979) can be used in such cases (see chapter 2).

This can be used when following the interactions of soluble membrane-active compounds with membranes, which can be observed as an increase of compound-specific signals in the presence of lipid membranes but not in solution. For example, the docking site for prion protein on membrane-embedded ganglioside GM1 is revealed as attenuation of signal from contact sites due to selective increase in relaxation, figure 8.3 (Sanghera *et al* 2011). Conversely, small amphipathic molecules such as BMA that partition selectively into lipid membranes can be observed in $^{13}C$ CP MAS NMR spectra as motional restrictions within membranes enhance CP efficiency, while free BMA in solution is spectroscopically silent under CP excitation (Yeh *et al* 2020).

The transition temperatures ($T_m$) between different bilayer phases, such as disordered liquid crystalline ($L_\alpha$) and rippled gel phase ($P_{\beta'}$), to planar gel ($L_\beta$),

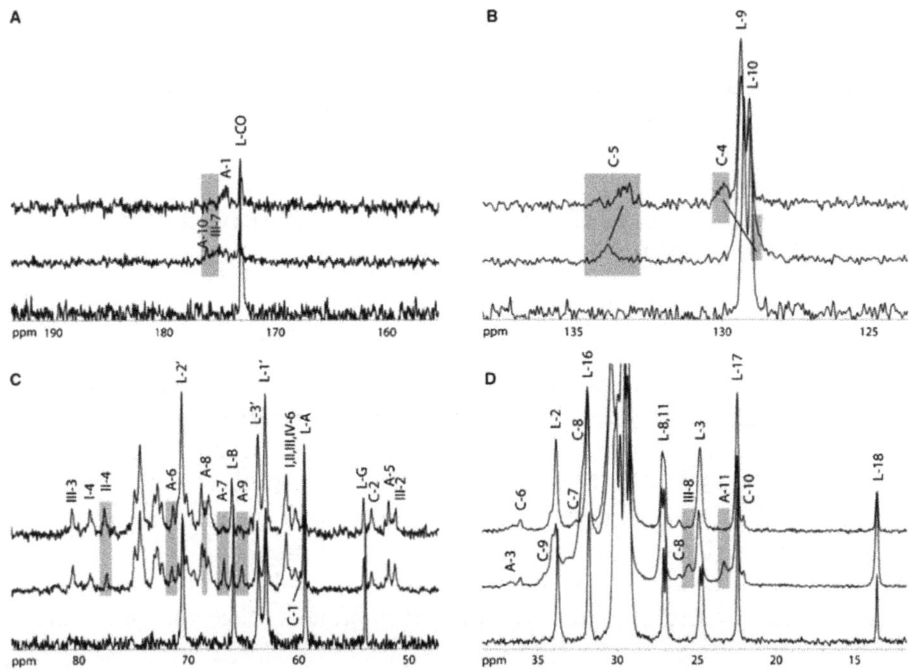


**Figure 8.3.** Carbon-13 CP MAS NMR spectra from membranes containing raft marker, ganglioside GM1; specific resonance attenuation reveals the binding epitope for prion proteins on GM1. Reproduced with permission from Sanghera *et al* (2011). Copyright 2011 Elsevier.

and to crystalline sub gel ($L_C$) is governed by freezing segmental motions in the chains, headgroups and additional lateral packing. Bilayer transition temperatures are affected by changes in the conformation and packing of lipid acyl chains and their cooperative chain–chain intermolecular interactions (Eliasz *et al* 1976, Koynova and Caffrey 1998, Weber and de Bont 1996). Often, molecular interactions in membranes can affect lipid segmental mobility and phase behavior.

One way of monitoring such changes simultaneously for all membrane components and without additional labelling is to use natural abundance $^{13}$C or $^{1}$H together with high resolution MAS NMR. While $^{1}$H MAS NMR can report directly on proximal interactions between membrane peptides and lipids (de Planque *et al* 2003), high resolution $^{13}$C NMR requires heteronuclear proton decoupling during observation delivered either via continuous wave (CW) irradiation or using multipulse sequences such as TPPM (Bennett *et al* 1995) or SPINAL (Fung *et al* 2000). Additional removal of homonuclear proton couplings under Lee–Goldburg decoupling (Bielecki *et al* 1989, Lee and Goldburg 1965, Vinogradov *et al* 1999) also delivers J-coupling modulated chemical shift-resolved spectra, which are exquisitely sensitive to changes in lipid segmental mobility (Ciesielski *et al* 2009, Zorin *et al* 2010) and show effective proximal J-couplings scaled by $1/\sqrt{3}$. Molecular mobility in fluid membranes aids decoupling and mobile molecular groups show well-resolved spectra, while restricted dynamics in lower temperature phases leads to segment-selective loss of J-resolution that can report on specific interactions and changes in cooperative behavior of lipid membranes.

The incorporation of organic solvents, such as BMA, in lipid membranes alters chain packing that affects lipid chain melting and uncouples *cis-trans* isomerization from collective chain interactions. When monitored by high resolution $^{13}C$ CP MAS NMR, this reveals a depression of DMPC chain melting from 24 °C to between 14 °C and 10 °C as the small BMA molecules intercalate between the lipid chains, figure 8.4 (Yeh *et al* 2020).

### 8.8.2 Membrane interactions monitored by $^{31}P$ MAS NMR

Phospholipids and other important membrane components contain phosphate or pyrophosphate groups that can provide information on molecular interactions on the membrane surface and changes in lipid dynamics when studied by high-resolution $^{31}P$ MAS NMR. High-resolution MAS NMR can be used to resolve individual lipid species in multicomponent mixed lipid membranes by their $^{31}P$ isotropic chemical shifts or $^{31}P$–$^{1}H$ shift correlations (Bonev *et al* 2001a, Bonev *et al* 2004, 2000, Holland *et al* 2006, Pinheiro and Watts 1994). Phosphorus-31 MAS NMR has been used to monitor changes in membrane surface charge density and

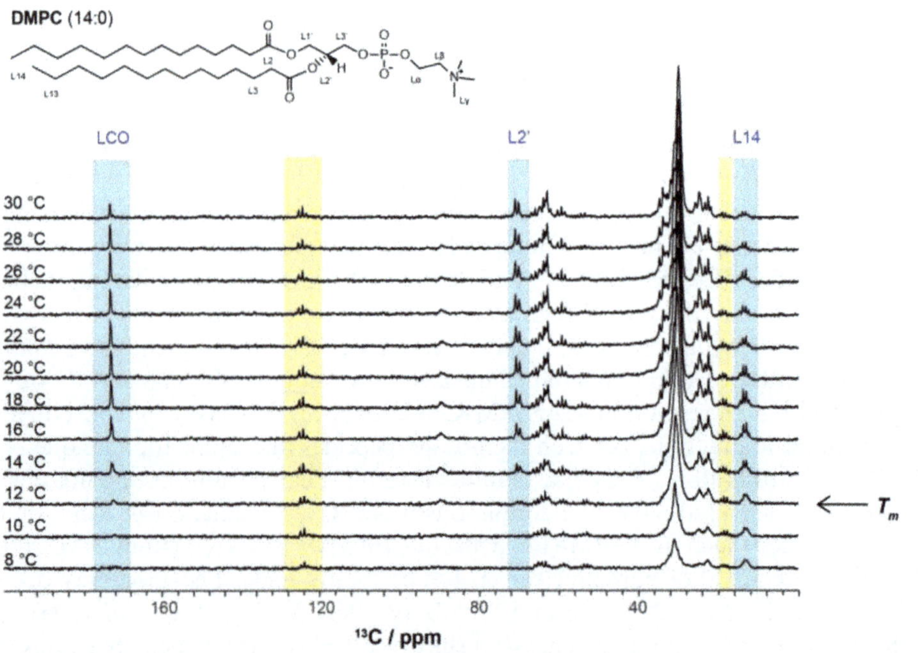


**Figure 8.4.** High resolution J resolved CP MAS spectra of DMPC MLV with 75% mol BMA, recorded under FSLG decoupling at decreasing temperature show DMPC transition temperature lowered from 24 °C to about 12 °C. Blue panels highlight key DMPC resonances used to determine the main transition temperature, whereas yellow panels highlights BMA resonances. Reproduced with permission from Yeh *et al* (2020). Copyright 2020 American Chemical Society.

lipid headgroup environment in response to peptide binding by following changes in $^{31}P$ isotropic chemical shifts (Bonev *et al* 2001b, Bonev *et al* 2003).

Phosphorus-31 MAS NMR has been used to study the effects of membrane active cytolysins on lipid membranes in a quantitative way. Membrane disruption and sphingomyelin-dependence has been observed during the action of equinatoxin, a hemolytic toxin from the sea anemone *Actinia equina* (Bonev *et al* 2003), using $^{31}P$ MAS NMR spectroscopy. Pneumolysin, a hemolytic toxin from *Streptococcus pneumoniae*, sequesters membrane lipid into proteolipid toroidal structures with 28:1 lipid/protein ratio in a cholesterol-dependent way (Bonev *et al* 2001a, Bonev *et al* 2001b).

Chemical shift resolved high-resolution $^{31}P$ MAS NMR offers unique insights into the structure, organization and interactions of biological molecules. Bacterial LPS is phosphorylated and pyrophosphorylated with $^{31}P$ pyrophosphate resonances between −8 and −12 ppm, while monophosphate gives rise to chemical shifts between 3 and −1 ppm. LPS from different bacterial species show different characteristic patterns of phosphorylation and pyrophosphorylation, for example in smooth type LPS from *Brucella melitensis*, *Klebsiella pneumoniae*, *Salmonella enterica* and *E. coli* reconstituted in model membranes and within membrane extracts (Lanne *et al* 2019). Using $^{31}P$ MAS NMR, LPS has been shown to partition preferentially into ordered cholesterol-rich lipid domains (Ciesielski *et al* 2013).

Bacterial LPS is used as a Trojan horse by antimicrobial peptides and proteins to gain access through the bacterial outer membrane defenses. Studies of LPS containing model membranes by solid-state and solution NMR have shown that colicin N relies on specific pyrophosphate mediated LPS recognition to bind and disrupt bacterial OMs (Johnson *et al* 2014). Antimicrobial peptides with receptor-mediated molecular action, such as nisin, also bind specifically to monophosphates in bacterial LPS in a chemotype and species dependent way. While rough LPS is used by nisin as a facilitator for membrane disruption, some smooth chemotypes serve as decoys preventing nisin from reaching its molecular targets on the peptidoglycan biosynthetic pathway (Lanne *et al* 2019).

Nisin is an antimicrobial peptide that targets peptidoglycan intermediates lipid II and undecaprenyl pyrophosphate, which was revealed using intermolecular heteronuclear dipolar recoupling MAS NMR, such as rotational echo double resonance (REDOR) (Gullion and Schaefer 1989) between $^{15}N$-labelled or $^{19}F$-labelled peptide and $^{31}P$ in cell wall intermediates (Bonev *et al* 2004, Kim *et al* 2002). The late stages of bacterial peptidoglycan assembly take place on the outer surface of bacterial membranes where key intermediates are targeted by a number of antimicrobials, such as nisin (Bonev *et al* 2004), subtilin (Parisot *et al* 2008), mersacidin (Brötz *et al* 1998), teixobactin (Shukla *et al* 2020), as well as glycopeptide antibiotics such as vancomycin (Kim *et al* 2002) and teicoplanin (Treviño *et al* 2014). In addition to membranes, high resolution $^{31}P$ MAS NMR has been used as a versatile approach to study other phosphorylated cell envelope components, such as bacterial peptidoglycan (Kim *et al* 2015) and bacterial exopolysaccharide (Dertli *et al* 2013). REDOR technique can also be applied to other nuclei such as $^{2}H$, $^{13}C$, $^{15}N$ and $^{19}F$

to study peptide–membrane and protein–membrane interactions (Cegelski 2013, Jia *et al* 2015).

Changes in phospholipid dynamics can also be used to detect molecular interactions in membranes by following $^{31}$P nuclear relaxation. Phosphorus relaxation times in membranes vary from milliseconds to seconds and can be studied conveniently by inversion recovery, as phosphate $^{31}$P has no directly bonded protons and nuclear coherences are preserved over seconds of undecoupled evolution. In addition, most phosphates in lipid membranes are motionally decoupled and from a NMR perspective behave as isolated systems giving rise to single exponential decays.

Inversion recovery MAS NMR is a commonly used tool for measuring longitudinal relaxation times, $T_1$, in weakly coupled nuclear systems (Tauskela and Thompson 1992), such as $^{31}$P from phosphorylated membrane systems. Molecular interactions can be detected through changes in $T_1$ and in relation to a minimum value, inversely proportional to the nuclear Larmor frequency (Slichter 1990). Phosphorus-31 within phosphate groups of different phosphorous species in membranes provides a sensitive assay of molecular interactions and informs on the rate of fast molecular motions in membranes, membrane fluidity and can report on formation of supramolecular complexes. Longitudinal relaxation of membrane lipids has also been used to resolve changes in lipid headgroup mobility in the presence of membrane-reconstituted membrane proteins (Smith *et al* 1983) or to determine partitioning of LPS into ordered, cholesterol-rich lipid rafts, figure 8.5 (Ciesielski *et al* 2013).

### 8.8.3 Correlation NMR spectroscopy in membranes

Natural abundance $^{1}$H and $^{13}$C-enrichment in membrane molecular components reveal intra and intermolecular coupling networks that can yield structural constraints and molecular interaction patterns. Akin to solution NMR, multidimensional correlation NMR experiments may be sub-divided into preparation, evolution, mixing and detection stages, not all of which are present in all experiments. The evolution period is required to build the indirect dimension(s) in all multidimensional experiments, while the mixing period often allows fine tuning of the interaction propagation beyond nearest neighbors.

Heteronuclear correlation spectroscopy can be used to combine $^{1}$H chemical shift dispersion with natural abundance $^{15}$N, $^{13}$C and $^{31}$P to provide better resolution from membrane systems. Correlation of abundant and rare spin chemical shift can be created using CP (through dipolar coupling) or by refocusing J-coupled evolution of rare nuclei using INEPT (through J coupling). A two-dimensional wideline separation (WISE) (Schmidt-Rohr *et al* 1992) type heteronuclear correlation spectroscopy can provide heteronuclear chemical shift correlations and proximal J-coupling resolved spectra from mixed lipid membranes (Zorin *et al* 2010). Undecoupled indirect evolution in $t_1$ is possible due to fast axial rotation of membrane lipids at GHz rates (Ciesielski *et al* 2013), which reduces dipolar relaxation. In more motion-restricted systems proton decoupling, such as Lee–Goldburg, may be used during $t_1$ to preserve chemical shift evolution and proton

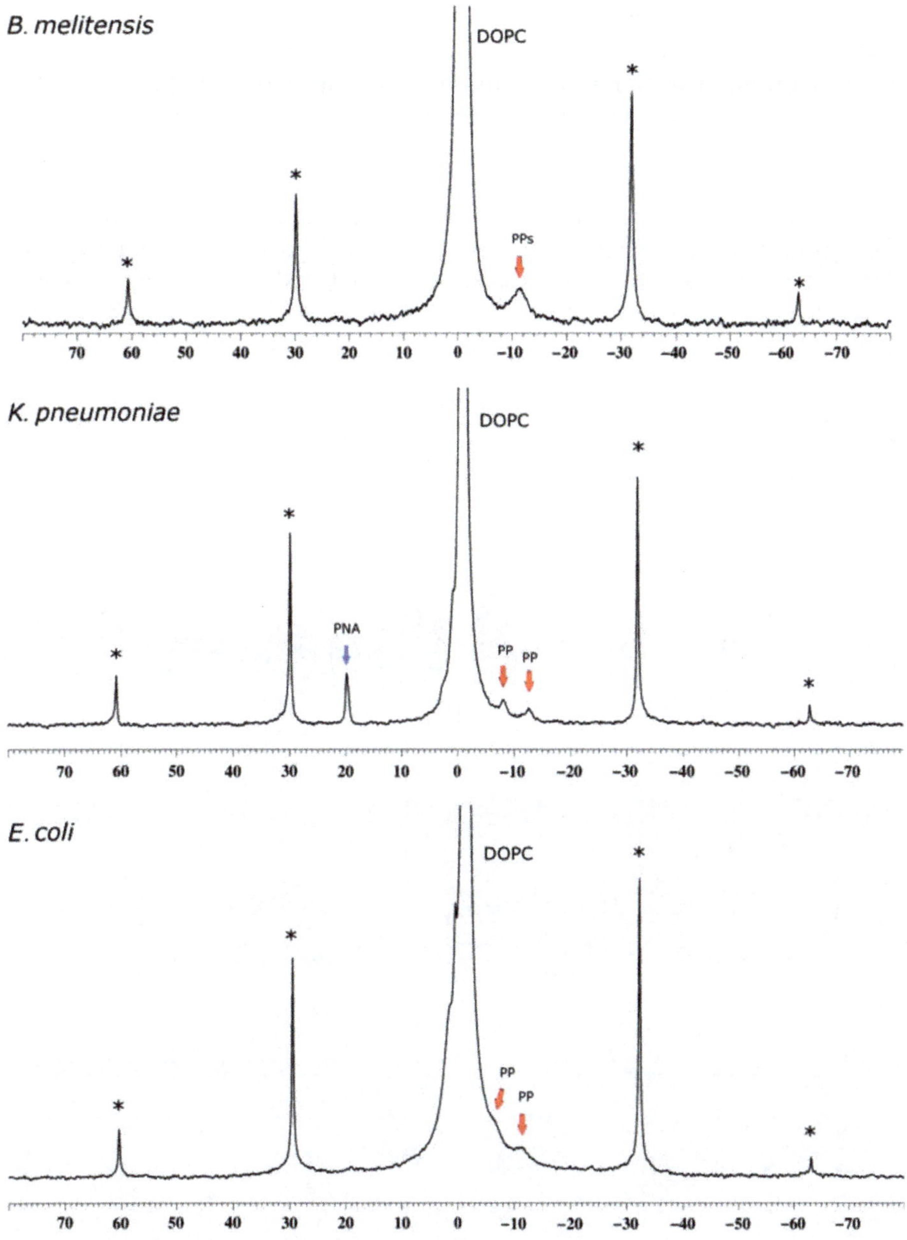


**Figure 8.5.** Phosphorous-31 MAS NMR spectra of lipid membranes containing LPS from *B. melitensis*, *K. pneumoniae* and *E. coli*. Mono and pyrophosphates are marked with arrows and * indicate rotational sidebands. Reproduced with permission from Ciesielski *et al* (2013). Copyright 2013 Elsevier.

coherence during heteronuclear correlation spectroscopy (HETCOR) (Alam and Holland 2006, Holland and Alam 2008, 2006, Zorin *et al* 2010). Heteronuclear decoupling during acquisition leads to chemical shift-resolved spectra, while the addition of homonuclear decoupling can retain scaled J-resolved multiplet patterns, figure 8.6 (Zorin *et al* 2010).

Homonuclear $^{1}H$–$^{1}H$ correlation NMR studies of molecular interactions in membranes and membrane–protein complexes are appealing, as they benefit from the high proton abundance in biological systems and good signal strength. Proton nuclear Overhauser enhancement spectroscopy (NOESY) is a common solution state NMR technique that can be applied to membranes in solid-state NMR MAS experiments to estimate distance of pairs of protons between 2 to 5 Å (Holte and

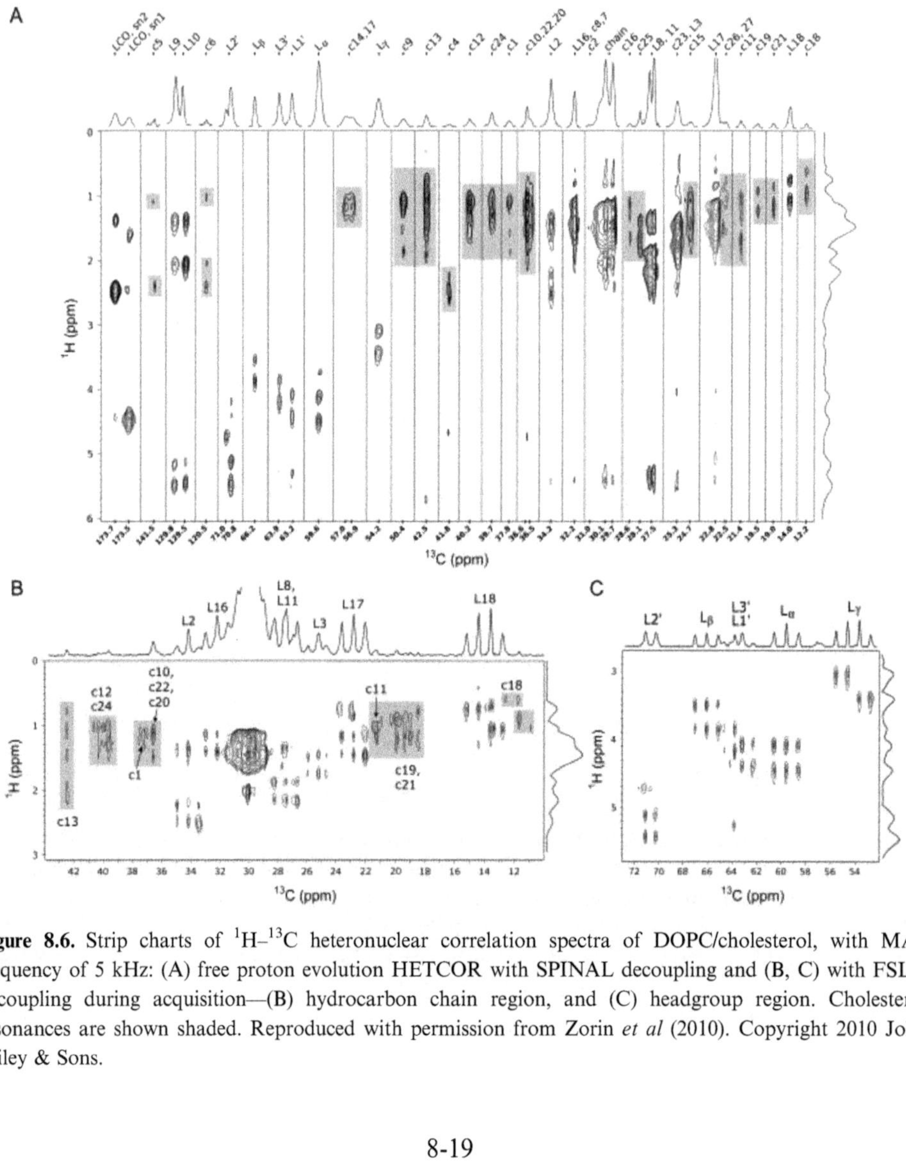


**Figure 8.6.** Strip charts of $^{1}H$–$^{13}C$ heteronuclear correlation spectra of DOPC/cholesterol, with MAS frequency of 5 kHz: (A) free proton evolution HETCOR with SPINAL decoupling and (B, C) with FSLG decoupling during acquisition—(B) hydrocarbon chain region, and (C) headgroup region. Cholesterol resonances are shown shaded. Reproduced with permission from Zorin *et al* (2010). Copyright 2010 John Wiley & Sons.

Gawrisch 1997). Solid-state $^{1}H$–$^{1}H$ NOESY MAS experiments have been used to study interaction between small molecules and membrane (Aucoin *et al* 2009, Davis *et al* 1995), determining the location of bound drugs and peptides (Le Guernevé and Seigneuret 1996, Ramamoorthy and Xu 2013, Zhang *et al* 2003), as well as the location of ethanol accumulation in the membrane (Holte and Gawrisch 1997). Well resolved $^{1}H$ NOESY spectra can be obtained in these samples due to high mobility of lipids within membrane, leading to effective reduction of $^{1}H$ line width by MAS.

Isotope-enrichment is normally required to observe homonuclear $^{13}C$–$^{13}C$ correlations in membrane molecular systems in the absence of DNP enhancement (see next section). Chemical shift correlations are often used to reveal connectivity networks and molecular interactions in membrane systems. A number of pulse sequences have been developed for homonuclear correlation including PDSD (Szeverenyi *et al* 1982), DARR (Takegoshi *et al* 2001) and RAD (Morcombe *et al* 2004). Variation in mixing times can vary the coherence exchange distance and can be used to differentiate between intramolecular structural couplings and intermolecular interactions. Carbon homonuclear correlation spectroscopy in solid-state NMR is commonly used to determined membrane protein structure in membrane (Das *et al* 2012, Mandala *et al* 2018, Park *et al* 2012, Wylie *et al* 2014), oligomeric conformation (Wang *et al* 2013) and interactions of multicomponent membrane protein complexes (Pinto *et al* 2017).

Double quantum (DQ) methods have been developed to observe selectively labelled spin pairs while suppressing natural abundance background, for example to resolve low molarity ligand or membrane protein from the background of abundant lipids. Examples of DQ excitation methods include POST-C7 (Hohwy *et al* 1998), R symmetry based R14 (figure 8.2) (Carravetta *et al* 2000) and five-fold symmetry based SPC5 (Hohwy *et al* 1999). Two-dimensional homonuclear correlation spectroscopy with double quantum filtering (DQF) can simplify the spectrum of a biomembrane system by removing background signals and long intramolecular correlations (Lopez *et al* 2008a). Homonuclear $^{13}C$–$^{13}C$ spectroscopy with a double quantum filter was employ to study the $^{13}C$ labelled TM helix of influenza M2 protein and yeast biosynthetic $^{13}C$ enriched cholesterol, allowing the identification of protein–cholesterol interaction (Elkins *et al* 2018). In addition to DQF shift correlation spectroscopy, coherence evolution within the DQ state can be used to deliver distance-dependent couplings and constraints. In a double quantum-single quantum (DQSQ) correlation spectroscopy, the indirect dimension is the sum of the coupled isotropic chemical shift. For example, a pair of nuclei, A with chemical shift $\delta_A$ and B with chemical shift $\delta_B$, will produce resonances at $\delta_{SQ} = \delta_A$ and $\delta_{SQ} = \delta_B$ in the SQ dimension with $\delta_{DQ} = \delta_A + \delta_B$ in the DQ dimension. Double quantum-single quantum correlation spectroscopy provides better resolution and cleaner spectra from spin pairs with a good chemical shift difference, as well as removing diagonal peaks, which are advantageous in a relatively complex system. Membrane protein in lipid membrane have been studied using DQSQ $^{13}C$–$^{13}C$ correlation to address the problem of large natural abundant background signals, and provide information such as chemical shift assignments and backbone structures (Lopez *et al* 2008b, Luca *et al* 2003).

## 8.9 Dynamic nuclear polarization

NMR, as a powerful and versatile tool of molecular characterization at the atomic level, is limited in its sensitivity by the comparatively low magnetic moments of atomic nuclei and the resulting low magnetic polarization. This leads to long acquisition times and limits in the scope of biological applications amenable to NMR characterization. A new 'old friend' has recently come to the fore bringing signal enhancement in solid-state MAS NMR spectroscopy in excess of hundreds of times through dynamic nuclear polarization (DNP) (Hall *et al* 1997). DNP can deliver NMR signal enhancement by coupling selectively the nuclear systems of interest to unpaired electrons on strategically introduced engineered biradicals (Song *et al* 2006). The efficiency in commonly used solid effect DNP-enhanced NMR experiments increases at low temperature and measurements are normally performed at 90–100 K to balance sensitivity gain with the increase in relaxation times at lower temperatures. Signal gain, however, usually comes at a resolution tradeoff under such conditions and due to proximity of paramagnetic radicals.

Increased availability of DNP MAS NMR instrumentation has driven a rapid increase in biological studies, such as membrane-associated peptides, membrane protein complexes, and larger macromolecular systems. DNP has had success in studies of amyloid fibrils (Bayro *et al* 2011) and their transient structures (Potapov *et al* 2015), intrinsically disordered proteins (Uluca *et al* 2018), viral capsids (Gupta *et al* 2016) and large protein complexes (Gelis *et al* 2013). DNP has also been applied in combination with restricted sample geometry and orientation to enhance spatial selectivity (Salnikov *et al* 2010).

Applications of DNP MAS NMR to membrane protein studies include crystalline membrane protein systems, such as bR (Bajaj *et al* 2009, Ni *et al* 2018) and channelrhodopsin (Becker-Baldus *et al* 2015) with reported 40 to 70-fold signal enhancement. Both proteins form ordered two-dimensional arrays of high-density protein assemble in the membrane, which may contribute to the observed high signal enhancement. Signal enhancement from DNP MAS NMR has made possible otherwise challenging studies of membrane proteins in proteoliposome (Elkins *et al* 2018, Koers *et al* 2014, Ong *et al* 2013) through significant reduction in experimental time. DNP also aid NMR investigation of biological systems in cells (Albert *et al* 2018, Takahashi *et al* 2013), cell lysate (Viennet *et al* 2016) and cell milieus (Frederick *et al* 2015), normally characterized by low molar presence of the molecular target and large contribution from complex background signal.

The signal enhancement of DNP MAS NMR depends largely on the nature of the sample, the preparation method and the type of biradical used. The commonly used biradicals are soluble TOTAPOL (Song *et al* 2006) and AMUPol (Sauvée *et al* 2013), and apolar TEKPol (Zagdoun *et al* 2013). Specialized DNP radicals designed for defined positions have been investigated, such as lipid anchored radicals (Fernández-de-Alba *et al* 2015), membrane proteins with site directed spin labelling (van der Cruijsen *et al* 2015) or radical bound peptides for specific targeted protein interaction (Viennet *et al* 2016) and structure of AMP assemblies *in situ* (Sani *et al* 2019).

The low temperatures required for typical DNP MAS NMR experiments necessitate the use of cryoprotectants to prevent segregation of sample molecules and radicals. The most common DNP cryoprotectant solution consists of a mixture of $^{12}C$-$d_6$-gylcerol, $D_2O$ and $H_2O$ at a molar ratio around 60:30:10. Experimental parameters, such as the type and concentration of radical and the exact composition of cryoprotectant and sample often need to be extensively optimized to achieve high enhancement factors.

In complex molecular systems containing bacterial lipopolysaccharide and phospholipid mixed membranes the coupling to the biradical spin system depends on the level of LPS hydration and in native membrane extracts signal enhancement between 10 and 30-fold is readily observed (figure 8.7).

Another advantage of DNP application to membrane studies is the preferential signal enhancement from surfaces over bulk (Lesage *et al* 2010). The use of DNP leads to significant reduction in experimental times that make possible complex experiments normally impractical on conventional MAS NMR systems, such as $^{31}P$ filtered $^{13}C$–$^{13}C$ correlations, double quantum experiments and three-dimensional solid-state spectroscopy. Despite the need for using low temperatures for signal gain in most experiments, DNP holds a great promise in revolutionizing molecular studies in membranes and opens new avenues for investigations of native membrane extracts, as well as membrane studies in whole bacteria and cells.

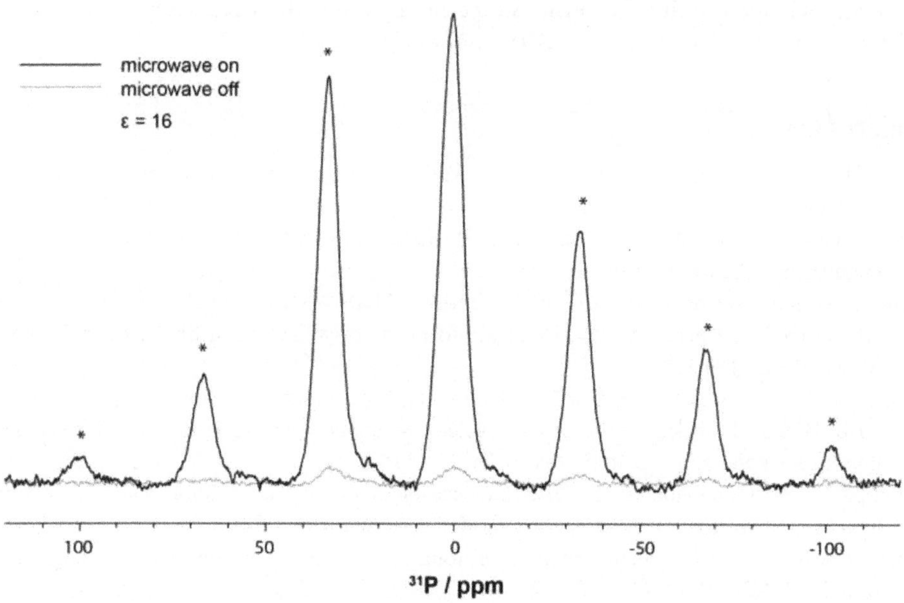


**Figure 8.7.** DNP enhanced $^{31}P$ CP MAS NMR of rough type LPS from *E. coli* reconstituted in DOPC membranes in the presence of AMUPol with (black) and without (grey) microwave irradiation at 8 kHz MAS. Asterisks indicate spinning sidebands.

## 8.10 Summary

Rapid technological and methodological developments have revolutionized the field of structural biology with modern crystallography and x-ray free-electron (XFEL) techniques, a recent boom in FEG cryoelectron microscopy, as well as solution NMR moving across the GHz boundary. Despite global databases showing a healthy growth in structural content from molecular systems in solution and recent breakthroughs in membrane protein structural analysis, understanding the molecular interactions with and within membranes in their organizational and functional complexity remains largely out of reach.

Solid-state NMR remains the method of choice for gaining high-resolution insights into the structure, organization and dynamics of membranes. Following decades of instrument and method development, the field is mature and offers a versatile toolkit for studying protein–protein, protein–lipid, protein–ligand or substrate interactions, as well as membrane protein function. Established and recent techniques follow through wideline, aligned and magic angle spinning methods, as well as comprehensive pulse sequence libraries that can provide sensitive measures of molecular dynamics and interatomic distances at sub-angstrom accuracy. Together with a revolution in DNP instrumentation, radical design and methodology, solid-state NMR is breaking the limitations of biophysical models and faces head on the challenges of studying native membranes and whole cells.

## Acknowledgements

We acknowledge funding form the Biotechnology and Biological Sciences Research Council grant BB/N010426/1 to BBB that supports VY.

## References

Agarwal V *et al* 2014 De novo 3D structure determination from sub-milligram protein samples by solid-state 100 kHz MAS NMR spectroscopy *Angew. Chem. Int. Ed.* **53** 12253–6

Akutsu H and Seelig J 1981 Interaction of metal ions with phosphatidylcholine bilayer membranes *Biochemistry* **20** 7366–73

Alam T M and Holland G P 2006 $^{1}$H–$^{13}$C INEPT MAS NMR correlation experiments with $^{1}$H–$^{1}$H mediated magnetization exchange to probe organization in lipid biomembranes *J. Magn. Reson.* **180** 210–21

Albert B J, Gao C, Sesti E L, Saliba E P, Alaniva N, Scott F J, Sigurdsson S T and Barnes A B 2018 Dynamic nuclear polarization nuclear magnetic resonance in human cells using fluorescent polarizing agents *Biochemistry* **57** 4741–6

Amin D N and Hazelbauer G L 2012 Influence of membrane lipid composition on a transmembrane bacterial chemoreceptor *J. Biol. Chem.* **287** 41697–705

Andreas L B *et al* 2016 Structure of fully protonated proteins by proton-detected magic-angle spinning NMR *PNAS* **113** 9187–92

Angelis A A D and Opella S J 2007 Bicelle samples for solid-state NMR of membrane proteins *Nat. Protoc.* **2** 2332–8

Arnold A A, Genard B, Zito F, Tremblay R, Warschawski D E and Marcotte I 2015 Identification of lipid and saccharide constituents of whole microalgal cells by $^{13}$C solid-state NMR *Biochim. Biophys. Acta (BBA)* **1848** 369–77

Aucoin D, Camenares D, Zhao X, Jung J, Sato T and Smith S O 2009 High-resolution $^{1}$H MAS RFDR NMR of biological membranes *J. Magn. Reson.* **197** 77–86

Bajaj V S, Mak-Jurkauskas M L, Belenky M, Herzfeld J and Griffin R G 2009 Functional and shunt states of bacteriorhodopsin resolved by 250 GHz dynamic nuclear polarization–enhanced solid-state NMR *PNAS* **106** 9244–9

Barbet-Massin E *et al* 2014 Rapid proton-detected nmr assignment for proteins with fast magic angle spinning *J. Am. Chem. Soc.* **136** 12489–97

Bayro M J, Debelouchina G T, Eddy M T, Birkett N R, MacPhee C E, Rosay M, Maas W E, Dobson C M and Griffin R G 2011 Intermolecular structure determination of amyloid fibrils with magic-angle spinning and dynamic nuclear polarization NMR *J. Am. Chem. Soc.* **133** 13967–74

Bechinger B, Zasloff M and Opella S J 1993 Structure and orientation of the antibiotic peptide magainin in membranes by solid-state nuclear magnetic resonance spectroscopy *Protein Sci.* **2** 2077–84

Becker-Baldus J *et al* 2015 Enlightening the photoactive site of channelrhodopsin-2 by DNP-enhanced solid-state NMR spectroscopy *PNAS* **112** 9896–901

Bennett A E, Rienstra C M, Auger M, Lakshmi K V and Griffin R G 1995 Heteronuclear decoupling in rotating solids *J. Chem. Phys.* **103** 6951–8

Bernat P, Paraszkiewicz K, Siewiera P, Moryl M, Płaza G and Chojniak J 2016 Lipid composition in a strain of *Bacillus subtilis*, a producer of iturin A lipopeptides that are active against uropathogenic bacteria *World J. Microbiol. Biotechnol.* **32** 157

Bielecki A, Kolbert A C and Levitt M H 1989 Frequency-switched pulse sequences: homonuclear decoupling and dilute spin NMR in solids *Chem. Phys. Lett.* **155** 341–6

Bonev B, Watts A, Bokvist M and Gröbner G 2001a Electrostatic peptide–lipid interactions of amyloid-β peptide and pentalysine with membrane surfaces monitored by $^{31}$P MAS NMR *Phys. Chem. Chem. Phys.* **3** 2904–10

Bonev B B 2013 High-resolution solid-state NMR of lipid membranes *Advances in Planar Lipid Bilayers and Liposomes* (Amsterdam: Elsevier) pp 299–329

Bonev B B, Breukink E, Swiezewska E, De Kruijff B and Watts A 2004 Targeting extracellular pyrophosphates underpins the high selectivity of nisin *FASEB J.* **18** 1862–9

Bonev B B, Chan W C, Bycroft B W, Roberts G C K and Watts A 2000 Interaction of the lantibiotic nisin with mixed lipid bilayers: A $^{31}$P and $^{2}$H NMR study *Biochemistry* **39** 11425–33

Bonev B B, Gilbert R J C, Andrew P W, Byron O and Watts A 2001b Structural analysis of the protein/lipid complexes associated with pore formation by the bacterial toxin pneumolysin *J. Biol. Chem.* **276** 5714–9

Bonev B B, Lam Y-H, Anderluh G, Watts A, Norton R S and Separovic F 2003 Effects of the eukaryotic pore-forming cytolysin equinatoxin II on lipid membranes and the role of sphingomyelin *Biophys. J.* **84** 2382–92

Bonev B B and Morrow M R 1997 Effect of pressure on the dimyristoylphosphatidylcholine bilayer main transition *Phys. Rev.* E **55** 5825–33

Breukink E, Wiedemann I, Kraaij C, van Kuipers O P, Sahl H-G and de Kruijff B 1999 Use of the cell wall precursor lipid II by a pore-forming peptide antibiotic *Science* **286** 2361–4

Brötz H, Bierbaum G, Leopold K, Reynolds P E and Sahl H-G 1998 The lantibiotic mersacidin inhibits peptidoglycan synthesis by targeting lipid II *Antimicrob. Agents Chemother.* **42** 154–60

Brown D A and London E 2000 Structure and function of sphingolipid- and cholesterol-rich membrane rafts *J. Biol. Chem.* **275** 17221–4

Brown M F and Seelig J 1978 Influence of cholesterol on the polar region of phosphatidylcholine and phosphatidylethanolamine bilayers *Biochemistry* **17** 381–4

Buttke T M and Ingram L O 1978 Mechanism of ethanol-induced changes in lipid composition of *Escherichia coli*: inhibition of saturated fatty acid synthesis *in vivo Biochemistry* **17** 637–44

Carravetta M, Edén M, Zhao X, Brinkmann A and Levitt M H 2000 Symmetry principles for the design of radiofrequency pulse sequences in the nuclear magnetic resonance of rotating solids *Chem. Phys. Lett.* **321** 205–15

Cegelski L 2013 REDOR NMR for drug discovery *Bioorg. Med. Chem. Lett.* **23** 5767–75

Ciesielski F, Davis B, Rittig M, Bonev B B and O'Shea P 2012 Receptor-independent interaction of bacterial lipopolysaccharide with lipid and lymphocyte membranes; the role of cholesterol *PLoS One* **7** e 38677

Ciesielski F, Griffin D C, Rittig M and Bonev B B 2009 High-resolution J-coupled $^{13}$C MAS NMR spectroscopy of lipid membranes *Chem. Phys. Lipids* **161** 77–85

Ciesielski F, Griffin D C, Rittig M, Moriyón I and Bonev B B 2013 Interactions of lipopolysaccharide with lipid membranes, raft models—a solid state NMR study *Biochim. Biophys. Acta (BBA)—Biomembr.* **1828** 1731–42

Clifton L A, Skoda M W A, Le Brun A P, Ciesielski F, Kuzmenko I, Holt S A and Lakey J H 2015 Effect of divalent cation removal on the structure of gram-negative bacterial outer membrane models *Langmuir* **31** 404–12

Cruijsen E A W, van der Prokofyev A V, Pongs O and Baldus M 2017 Probing conformational changes during the gating cycle of a potassium channel in lipid bilayers *Biophys. J.* **112** 99–108

Cullis P R and De Kruijff B 1978 Polymorphic phase behaviour of lipid mixtures as detected by $^{31}$P NMR. Evidence that cholesterol may destabilize bilayer structure in membrane systems containing phosphatidylethanolamine *Biochim. Biophys. Acta (BBA)—Biomembr.* **507** 207–18

Cullis P R and De Kruijff B 1979 Lipid polymorphism and the functional roles of lipids in biological membranes *Biochim. Biophys. Acta (BBA)* **559** 399–420

Das B B, Nothnagel H J, Lu G J, Son W S, Tian Y, Marassi F M and Opella S J 2012 Structure determination of a membrane protein in proteoliposomes *J. Am. Chem. Soc.* **134** 2047–56

Davis J H 1983 The description of membrane lipid conformation, order and dynamics by $^{2}$H-NMR *Biochim. Biophys. Acta (BBA)* **737** 117–71

Davis J H, Auger M and Hodges R S 1995 High resolution $^{1}$H nuclear magnetic resonance of a transmembrane peptide *Biophys. J.* **69** 1917–32

de Planque M R R, Bonev B B, Demmers J A A, Greathouse D V, Koeppe R E, Separovic F, Watts A and Killian J A 2003 Interfacial anchor properties of tryptophan residues in transmembrane peptides can dominate over hydrophobic matching effects in peptide–lipid interactions *Biochemistry* **42** 5341–8

De Smet M J, Kingma J and Witholt B 1978 The effect of toluene on the structure and permeability of the outer and cytoplasmic membranes of escherichia coli *Biochim. Biophys. Acta (BBA)—Biomembr.* **506** 64–80

den Kamp J A F O, Redai I and van Deenen L L M 1969 Phospholipid composition of *Bacillus subtilis J. Bacteriol.* **99** 298–303

Dertli E, Colquhoun I J, Gunning A P, Bongaerts R J, Gall G L, Bonev B B, Mayer M J and Narbad A 2013 Structure and biosynthesis of two exopolysaccharides produced by *Lactobacillus johnsonii* FI9785 *J. Biol. Chem.* M113.507418

Dombek K M and Ingram L O 1984 Effects of ethanol on the *Escherichia coli* plasma membrane *J. Bacteriol.* **157** 233–9

Dufourc E J, Mayer C, Stohrer J, Althoff G and Kothe G 1992 Dynamics of phosphate head groups in biomembranes. Comprehensive analysis using phosphorus-31 nuclear magnetic resonance lineshape and relaxation time measurements *Biophys. J.* **61** 42–57

Dunnick J K and O'Leary W M 1970 Correlation of bacterial lipid composition with antibiotic resistance *J. Bacteriol.* **101** 892–900

Eliasz A W, Chapman D and Ewing D F 1976 Phospholipid phase transitions. Effects of n-alcohols, n-monocarboxylic acids, phenylalkyl alcohols and quatenary ammonium compounds *Biochim. Biophys. Acta (BBA)—Biomembr.* **448** 220–33

Elkins M R, Sergeyev I V and Hong M 2018 Determining cholesterol binding to membrane proteins by cholesterol $^{13}$C labeling in yeast and dynamic nuclear polarization NMR *J. Am. Chem. Soc.* **140** 15437–49

Epand R F, Savage P B and Epand R M 2007 Bacterial lipid composition and the antimicrobial efficacy of cationic steroid compounds (Ceragenins) *Biochim. Biophys. Acta (BBA)—Biomembr.* **1768** 2500–9

Etzkorn M, Martell S, Andronesi O C, Seidel K, Engelhard M and Baldus M 2007 Secondary structure, dynamics, and topology of a seven-helix receptor in native membranes, studied by solid-state NMR spectroscopy *Angew. Chem. Int. Ed.* **46** 459–62 200602139

Fernández-de-Alba C, Takahashi H, Richard A, Chenavier Y, Dubois L, Maurel V, Lee D, Hediger S and De Paëpe G 2015 Matrix-free DNP-enhanced NMR spectroscopy of liposomes using a lipid-anchored biradical *Chem.—A Eur. J.* **21** 4512–7

Frederick K K, Michaelis V K, Corzilius B, Ong T-C, Jacavone A C, Griffin R G and Lindquist S 2015 Sensitivity-enhanced NMR reveals alterations in protein structure by cellular milieus *Cell* **163** 620–8

Fung B M, Khitrin A K and Ermolaev K 2000 An improved broadband decoupling sequence for liquid crystals and solids *J. Magn. Reson.* **142** 97–101

Garavito R M and Ferguson-Miller S 2001 Detergents as tools in membrane biochemistry *J. Biol. Chem.* **276** 32403–6

Gelis I, Vitzthum V, Dhimole N, Caporini M A, Schedlbauer A, Carnevale D, Connell S R, Fucini P and Bodenhausen G 2013 Solid-state NMR enhanced by dynamic nuclear polarization as a novel tool for ribosome structural biology *J. Biomol. NMR* **56** 85–93

Gidden J, Denson J, Liyanage R, Ivey D M and Lay J O 2009 Lipid compositions in *Escherichia coli* and *Bacillus subtilis* during growth as determined by MALDI-TOF and TOF/TOF mass spectrometry *Int. J. Mass Spectrom.* **283** 178–84

Glaubitz C and Watts A 1998 Magic angle-oriented sample spinning (MAOSS): a new approach toward biomembrane studies *J. Magn. Reson.* **130** 305–16

Good D B, Wang S, Ward M E, Struppe J, Brown L S, Lewandowski J R and Ladizhansky V 2014 Conformational dynamics of a seven transmembrane helical protein anabaena sensory rhodopsin probed by solid-state NMR *JACS* **136** 2833–42

Griffin R G 1976 Observation of the effect of water on the phosphorus-31 nuclear magnetic resonance spectra of dipalmitoyllecithin *J. Am. Chem. Soc.* **98** 851–3

Grigorieff N, Ceska T A, Downing K H, Baldwin J M and Henderson R 1996 Electron-crystallographic refinement of the structure of bacteriorhodopsin *J. Mol. Biol.* **259** 393–421
Grogan D W and Cronan J E 1997 Cyclopropane ring formation in membrane lipids of bacteria *Microbiol. Mol. Biol. Rev.* **61** 429–41
Gullion T and Schaefer J 1989 Rotational-echo double-resonance NMR *J. Magn. Reson.* **81** 196–200
Gupta R *et al* 2016 Dynamic nuclear polarization enhanced MAS NMR spectroscopy for structural analysis of HIV-1 protein assemblies *J. Phys. Chem.* B **120** 329–39
Hagn F, Etzkorn M, Raschle T and Wagner G 2013 Optimized phospholipid bilayer nanodiscs facilitate high-resolution structure determination of membrane proteins *J. Am. Chem. Soc.* **135** 1919–25
Hall D A, Maus D C, Gerfen G J, Inati S J, Becerra L R, Dahlquist F W and Griffin R G 1997 Polarization-enhanced NMR spectroscopy of biomolecules in frozen solution *Science* **276** 930–2
Hallock K J, Henzler Wildman K, Lee D-K and Ramamoorthy A 2002 An innovative procedure using a sublimable solid to align lipid bilayers for solid-state NMR studies *Biophys. J.* **82** 2499–503
Hartmann S R and Hahn E L 1962 Nuclear double resonance in the rotating frame *Phys. Rev.* **128** 2042–53
Heipieper H J and de Bont J A 1994 Adaptation of *Pseudomonas putida* S12 to ethanol and toluene at the level of fatty acid composition of membranes *Appl. Environ. Microbiol.* **60** 4440–4
Heipieper H J, Weber F J, Sikkema J, Keweloh H and de Bont J A M 1994 Mechanisms of resistance of whole cells to toxic organic solvents *Trends Biotechnol.* **12** 409–15
Heise H, Hoyer W, Becker S, Andronesi O C, Riedel D and Baldus M 2005 Molecular-level secondary structure, polymorphism, and dynamics of full-length α-synuclein fibrils studied by solid-state NMR *PNAS* **102** 15871–6
Hohwy M, Jakobsen H J, Eden M, Levitt M H and Nielsen N C 1998 Broadband dipolar recoupling in the nuclear magnetic resonance of rotating solids: a compensated C7 pulse sequence *J. Chem. Phys.* **108** 2686–94
Hohwy M, Rienstra C M, Jaroniec C P and Griffin R G 1999 Fivefold symmetric homonuclear dipolar recoupling in rotating solids: application to double quantum spectroscopy *J. Chem. Phys.* **110** 7983–92
Holland G P and Alam T M 2006 Multi-dimensional $^{1}H$–$^{13}C$ HETCOR and FSLG-HETCOR NMR study of sphingomyelin bilayers containing cholesterol in the gel and liquid crystalline states *J. Magn. Reson.* **181** 316–26
Holland G P and Alam T M 2008 Unique backbone-water interaction detected in sphingomyelin bilayers with $^{1}H/^{31}P$ and $^{1}H/^{13}C$ HETCOR MAS NMR spectroscopy *Biophys. J.* **95** 1189–98
Holland G P, McIntyre S K and Alam T M 2006 Distinguishing individual lipid headgroup mobility and phase transitions in raft-forming lipid mixtures with $^{31}P$ MAS NMR *Biophys. J.* **90** 4248–60
Holte L L and Gawrisch K 1997 Determining ethanol distribution in phospholipid multilayers with MAS–NOESY spectra *Biochemistry* **36** 4669–74
Hyde A J, Parisot J, McNichol A and Bonev B B 2006 Nisin-induced changes in bacillus morphology suggest a paradigm of antibiotic action *Proc. Natl Acad. Sci.* **103** 19896–901
Ingram L O 1976 Adaptation of membrane lipids to alcohols *J. Bacteriol.* **125** 670–8

Ingram L O 1986 Microbial tolerance to alcohols: role of the cell membrane *Trends Biotechnol.* **4** 40–4

Inoue A and Horikoshi K 1989 A pseudomonas thrives in high concentrations of toluene *Nature* **338** 264–6

Jia L, Liang S, Sackett K, Xie L, Ghosh U and Weliky D P 2015 REDOR solid-state NMR as a probe of the membrane locations of membrane-associated peptides and proteins *J. Magn. Reson.* **253** 154–65

Johnson C L, Ridley H, Marchetti R, Silipo A, Griffin D C, Crawford L, Bonev B, Molinaro A and Lakey J H 2014 The antibacterial toxin colicin N binds to the inner core of lipopolysaccharide and close to its translocator protein *Mol. Microbiol.* **92** 440–52

Johnson C L, Ridley H, Pengelly R J, Salleh M Z and Lakey J H 2013 The unstructured domain of colicin N kills *Escherichia coli Mol. Microbiol.* **89** 84–95

Kaneda T 1977 Fatty acids of the genus Bacillus: an example of branched-chain preference *Bacteriol. Rev.* **41** 391–418

Kaneda T 1991 Iso- and anteiso-fatty acids in bacteria: biosynthesis, function, and taxonomic significance *Microbiol. Mol. Biol. Rev.* **55** 288–302

Kim S J, Cegelski L, Studelska D R, O'Connor R D, Mehta A K and Schaefer J 2002 Rotational-echo double resonance characterization of vancomycin binding sites in *Staphylococcus aureus Biochemistry* **41** 6967–77

Kim S J, Chang J and Singh M 2015 Peptidoglycan architecture of Gram-positive bacteria by solid-state NMR *Biochim. Biophys. Acta (BBA)—Biomembr.* **1848** 350–62

Koers E J *et al* 2014 NMR-based structural biology enhanced by dynamic nuclear polarization at high magnetic field *J. Biomol. NMR* **60** 157–68

Koynova R and Caffrey M 1998 Phases and phase transitions of the phosphatidylcholines *Biochim. Biophys. Acta (BBA)—Rev. Biomembr.* **1376** 91–145

Ladizhansky V 2017 Applications of solid-state NMR to membrane proteins *Biochim. Biophys. Acta (BBA)—Proteins Proteom.* **1865** 1577–86

Lafleur M, Fine B, Sternin E, Cullis P R and Bloom M 1989 Smoothed orientational order profile of lipid bilayers by $^2$H-nuclear magnetic resonance *Biophys. J.* **56** 1037–41

Lanne A B M, Goode A, Prattley C, Kumari D, Drasbek M R, Williams P, Conde-Álvarez R, Moriyón I and Bonev B B 2019 Molecular recognition of lipopolysaccharide by the lantibiotic nisin *Biochim. Biophys. Acta (BBA)—Biomembr.* **1861** 83–92

Le Guernevé C and Seigneuret M 1996 High-resolution mono- and multidimensional magic angle spinning $^1$H nuclear magnetic resonance of membrane peptides in nondeuterated lipid membranes and $H_2O$ *Biophys. J.* **71** 2633–44

Le Maire M, Champeil P and Møller J V 2000 Interaction of membrane proteins and lipids with solubilizing detergents *Biochim. Biophys. Acta (BBA)—Biomembr.* **1508** 86–111

Leão C and Van Uden N 1984 Effects of ethanol and other alkanols on passive proton influx in the yeast *Saccharomyces cerevisiae Biochim. Biophys. Acta (BBA)—Biomembr.* **774** 43–8

Lebreton F and Cattoir V 2019 Resistance to glycopeptide antibiotics *Bacterial Resistance to Antibiotics—From Molecules to Man* (Hoboken, NJ: Wiley) pp 51–80

Lee A G 2004 How lipids affect the activities of integral membrane proteins *Biochim. Biophys. Acta (BBA)—Biomembr.* **1666** 62–87

Lee M and Goldburg W I 1965 Nuclear-magnetic-resonance line narrowing by a rotating rf field *Phys. Rev.* **140** A1261–71

Lee T-Y, Yeh V, Chuang J, Chung Chan J C, Chu L-K and Yu T-Y 2015 Tuning the photocycle kinetics of bacteriorhodopsin in lipid nanodiscs *Biophys. J.* **109** 1899–906

Lesage A *et al* 2010 Surface enhanced NMR spectroscopy by dynamic nuclear polarization *J. Am. Chem. Soc.* **132** 15459–61

Lingwood D and Simons K 2010 Lipid rafts as a membrane-organizing principle *Science* **327** 46–50

Loffhagen N, Härtig C, Geyer W, Voyevoda M and Harms H 2007 Competition between cis, trans and cyclopropane fatty acid formation and its impact on membrane fluidity *Eng. Life Sci.* **7** 67–74

Lopez J J, Kaiser C, Shastri S and Glaubitz C 2008a Double quantum filtering homonuclear MAS NMR correlation spectra: a tool for membrane protein studies *J. Biomol. NMR* **41** 97–104

Lopez J J, Shukla A K, Reinhart C, Schwalbe H, Michel H and Glaubitz C 2008b The structure of the neuropeptide bradykinin bound to the human G-protein coupled receptor bradykinin B2 as determined by solid-state NMR spectroscopy *Angew. Chem. Int. Ed.* **47** 1668–71

Lopez-Campistrous A, Semchuk P, Burke L, Palmer-Stone T, Brokx S J, Broderick G, Bottorff D, Bolch S, Weiner J H and Ellison M J 2005 Localization, annotation, and comparison of the *Escherichia coli* K-12 proteome under two states of growth *Mol. Cell. Proteom.* **4** 1205–9

Loudet C, Manet S, Gineste S, Oda R, Achard M-F and Dufourc E J 2007 Biphenyl bicelle disks align perpendicular to magnetic fields on large temperature scales: a study combining synthesis, solid-state NMR, TEM, and SAXS *Biophys. J.* **92** 3949–59

Luca S, White J F, Sohal A K, Filippov D V, Boom J H, van Grisshammer R and Baldus M 2003 The conformation of neurotensin bound to its G protein-coupled receptor *PNAS* **100** 10706–11

Ly H V and Longo M L 2004 The influence of short-chain alcohols on interfacial tension, mechanical properties, area/molecule, and permeability of fluid lipid bilayers *Biophys. J.* **87** 1013–33

Malcolm I C, Ross J C and Higinbotham J 2005 A study of the headgroup motion of sphingomyelin using 31P NMR and an analytically soluble model *Solid State Nucl. Magn. Reson.* **27** 247–56

Mandala V S, Williams J K and Hong M 2018 Structure and dynamics of membrane proteins from solid-state NMR *Annu. Rev. Biophys.* **47** 201–22

Marassi F M and Opella S J 2000 A solid-state NMR index of helical membrane protein structure and topology *J. Magn. Reson.* **144** 150–5

Medeiros-Silva J, Jekhmane S, Paioni A L, Gawarecka K, Baldus M, Swiezewska E, Breukink E and Weingarth M 2018 High-resolution NMR studies of antibiotics in cellular membranes *Nat. Commun.* **9** 3963

Mishra P and Prasad R 1988 Role of phospholipid head groups in ethanol tolerance of *Saccharomyces cerevisiae Microbiology* **134** 3205–11

Morcombe C R, Gaponenko V, Byrd R A and Zilm K W 2004 Diluting abundant spins by isotope edited radio frequency field assisted diffusion *J. Am. Chem. Soc.* **126** 7196–7

Moreau C, Le Floch M, Segalen J, Leray G, Metzinger L, de Certaines J D and Le Rumeur E 1999 Static and magic angle spinning $^{31}$P NMR spectroscopy of two natural plasma membranes *FEBS Lett.* **461** 258–62

Morein S, Andersson A-S, Rilfors L and Lindblom G 1996 Wild-type *Escherichia coli* cells regulate the membrane lipid composition in a window between gel and non-lamellar structures *J. Biol. Chem.* **271** 6801–9

Morris G A and Freeman R 1979 Enhancement of nuclear magnetic resonance signals by polarization transfer *J. Am. Chem. Soc.* **101** 760–2

Mulholland S, Turpin E R, Bonev B B and Hirst J D 2016 Docking and molecular dynamics simulations of the ternary complex nisin 2:lipid II *Sci. Rep.* 6

Ni Q Z, Can T V, Daviso E, Belenky M, Griffin R G and Herzfeld J 2018 Primary transfer step in the light-driven ion pump bacteriorhodopsin: an irreversible u-turn revealed by dynamic nuclear polarization-enhanced magic angle spinning NMR *J. Am. Chem. Soc.* **140** 4085–91

Nickels J D *et al* 2017 *Bacillus subtilis* lipid extract, a branched-chain fatty acid model membrane *J. Phys. Chem. Lett.* **8** 4214–7

Nomura K, Ferrat G, Nakajima T, Darbon H, Iwashita T and Corzo G 2005 Induction of morphological changes in model lipid membranes and the mechanism of membrane disruption by a large scorpion-derived pore-forming peptide *Biophys. J.* **89** 4067–80

Ong Y S, Lakatos A, Becker-Baldus J, Pos K M and Glaubitz C 2013 Detecting substrates bound to the secondary multidrug efflux pump EmrE by DNP-enhanced solid-state NMR *J. Am. Chem. Soc.* **135** 15754–62

Opella S J and Marassi F M 2004 Structure determination of membrane proteins by NMR spectroscopy *Chem. Rev.* **104** 3587–606

Opella S J, Marassi F M, Gesell J J, Valente A P, Kim Y, Oblatt-Montal M and Montal M 1999 Structures of the M2 channel-lining segments from nicotinic acetylcholine and NMDA receptors by NMR spectroscopy *Nat. Struct. Biol.* **6** 374–9

Oursel D, Loutelier-Bourhis C, Orange N, Chevalier S, Norris V and Lange C M 2007 Lipid composition of membranes of *Escherichia coli* by liquid chromatography/tandem mass spectrometry using negative electrospray ionization *Rapid Commun. Mass Spectrom.* **21** 1721–8

Overall S A, Zhu S, Hanssen E, Separovic F and Sani M-A 2019 *In situ* monitoring of bacteria under antimicrobial stress using $^{31}$P solid-state NMR *Int. J. Mol. Sci.* **20** 181

Pake G E 1948 Nuclear resonance absorption in hydrated crystals: fine structure of the proton line *J. Chem. Phys.* **16** 327–36

Parisot J, Carey S, Breukink E, Chan W C, Narbad A and Bonev B 2008 Molecular mechanism of target recognition by subtilin, a class I lanthionine antibiotic *Antimicrob. Agents Chemother.* **52** 612–8

Park S H *et al* 2012 Structure of the chemokine receptor CXCR1 in phospholipid bilayers *Nature* **491** 779–83

Park S H, Mrse A A, Nevzorov A A, Mesleh M F, Oblatt-Montal M, Montal M and Opella S J 2003 Three-dimensional structure of the channel-forming trans-membrane domain of virus protein 'u' (Vpu) from HIV-1 *J. Mol. Biol.* **333** 409–24

Park S H, Prytulla S, De Angelis A A, Brown J M, Kiefer H and Opella S J 2006 High-resolution NMR spectroscopy of a GPCR in aligned bicelles *J. Am. Chem. Soc.* **128** 7402–3

Pines A, Gibby M G and Waugh J S 1972 Proton-enhanced nuclear induction spectroscopy $^{13}$C chemical shielding anisotropy in some organic solids *Chem. Phys. Lett.* **15** 373–6

Pinheiro T J T and Watts A 1994 Resolution of individual lipids in mixed phospholipid membranes and specific lipid-cytochrome c interactions by magic-angle spinning solid-state phosphorus-31 NMR *Biochemistry* **33** 2459–67

Pinkart H C, Wolfram J W, Rogers R and White D C 1996 Cell envelope changes in solvent-tolerant and solvent-sensitive *Pseudomonas putida* strains following exposure to o-xylene *Appl. Environ. Microbiol.* **62** 1129–32

Pinto C, Mance D, Julien M, Daniels M, Weingarth M and Baldus M 2017 Studying assembly of the BAM complex in native membranes by cellular solid-state NMR spectroscopy *J. Struct. Biol.*

Potapov A, Yau W-M, Ghirlando R, Thurber K R and Tycko R 2015 Successive stages of amyloid-β self-assembly characterized by solid-state nuclear magnetic resonance with dynamic nuclear polarization *J. Am. Chem. Soc.* **137** 8294–307

Prosser R S, Evanics F, Kitevski J L and Al-Abdul-Wahid M S 2006 Current applications of bicelles in NMR studies of membrane-associated amphiphiles and proteins *Biochemistry* **45** 8453–65

Ramamoorthy A and Xu J 2013 2D $^{1}$H/$^{1}$H RFDR and NOESY NMR experiments on a membrane-bound antimicrobial peptide under magic angle spinning *J. Phys. Chem.* B **117** 6693–700

Ramos J L, Duque E, Rodríguez-Herva J-J, Godoy P, Haïdour A, Reyes F and Fernández-Barrero A 1997 Mechanisms for solvent tolerance in bacteria *J. Biol. Chem.* **272** 3887–90

Rappolt M, Hickel A, Bringezu F and Lohner K 2003 Mechanism of the lamellar/inverse hexagonal phase transition examined by high resolution x-ray diffraction *Biophys. J.* **84** 3111–22

Raschle T, Hiller S, Yu T-Y, Rice A J, Walz T and Wagner G 2009 Structural and functional characterization of the integral membrane protein VDAC-1 in lipid bilayer nanodiscs *J. Am. Chem. Soc.* **131** 17777–9

Reichhardt C, Ferreira J A G, Joubert L-M, Clemons K V, Stevens D A and Cegelski L 2015 Analysis of the *Aspergillus fumigatus* biofilm extracellular matrix by solid-state nuclear magnetic resonance spectroscopy *Eukaryot. Cell* **14** 1064–72

Renault M, Boxtel R T, Bos M P, Post J A, Tommassen J and Baldus M 2012 Cellular solid-state nuclear magnetic resonance spectroscopy *PNAS* **109** 4863–8

Retel J S *et al* 2017 Structure of outer membrane protein G in lipid bilayers *Nat. Commun.* **8** 2073

Romaniuk J A H and Cegelski L 2015 Bacterial cell wall composition and the influence of antibiotics by cell-wall and whole-cell NMR *Philos. Trans. R Soc. Lond. B Biol. Sci.* **370** 20150024

Rossjohn J, Feil S C, McKinstry W J, Tweten R K and Parker M W 1997 Structure of a cholesterol-binding, thiol-activated cytolysin and a model of its membrane form *Cell* **89** 685–92

Salnikov E, Rosay M, Pawsey S, Ouari O, Tordo P and Bechinger B 2010 Solid-state NMR spectroscopy of oriented membrane polypeptides at 100 K with signal enhancement by dynamic nuclear polarization *J. Am. Chem. Soc.* **132** 5940–1

Sanders C R and Landis G C 1995 Reconstitution of membrane proteins into lipid-rich bilayered mixed micelles for NMR studies *Biochemistry* **34** 4030–40

Sanders C R and Schwonek J P 1992 Characterization of magnetically orientable bilayers in mixtures of dihexanoylphosphatidylcholine and dimyristoylphosphatidylcholine by solid-state NMR *Biochemistry* **31** 8898–905

Sanghera N *et al* 2011 Deciphering the molecular details for the binding of the prion protein to main ganglioside GM1 of neuronal membranes *Chem. Biol.* **18** 1422–31

Sani M-A, Zhu S, Hofferek V and Separovic F 2019 Nitroxide spin–labeled peptides for DNP-NMR in-cell studies *FASEB J.* **33** 11021–7

Sardessai Y and Bhosle S 2002 Tolerance of bacteria to organic solvents *Res. Microbiol.* **153** 263–8

Sauvée C, Rosay M, Casano G, Aussenac F, Weber R T, Ouari O and Tordo P 2013 Highly efficient, water-soluble polarizing agents for dynamic nuclear polarization at high frequency *Angew. Chem. Int. Ed.* **52** 10858–61

Schmidt-Rohr K, Clauss J and Spiess H W 1992 Correlation of structure, mobility, and morphological information in heterogeneous polymer materials by two-dimensional wide-line-separation NMR spectroscopy *Macromolecules* **25** 3273–7

Schneider R, Ader C, Lange A, Giller K, Hornig S, Pongs O, Becker S and Baldus M 2008 Solid-state NMR spectroscopy applied to a chimeric potassium channel in lipid bilayers *J. Am. Chem. Soc.* **130** 7427–35

Seddon J M 1990 Structure of the inverted hexagonal (HII) phase, and non-lamellar phase transitions of lipids *Biochim. Biophys. Acta (BBA)—Biomembr.* **1031** 1–69

Seddon J M, Cevc G and Marsh D 1983 Calorimetric studies of the gel-fluid (L.beta.-L.alpha.) and lamellar-inverted hexagonal (L.alpha.-HII) phase transitions in dialkyl- and diacylphosphatidylethanolamines *Biochemistry* **22** 1280–9

Seelig J 1978 $^{31}$P nuclear magnetic resonance and the head group structure of phospholipids in membranes *Biochim. Biophys. Acta (BBA)—Biomembr.* **515** 105–40

Seelig J 1977 Deuterium magnetic resonance: theory and application to lipid membranes *Q. Rev. Biophys.* **10** 353–418

Shi L, Kawamura I, Jung K-H, Brown L S and Ladizhansky V 2011 Conformation of a seven-helical transmembrane photosensor in the lipid environment *Angew. Chem. Int. Ed.* **50** 1302–5

Shukla R *et al* 2020 Mode of action of teixobactins in cellular membranes *Nat. Commun.* **11** 2848

Sikkema J, de Bont J A and Poolman B 1995 Mechanisms of membrane toxicity of hydrocarbons *Microbiol. Rev.* **59** 201–22

Silhavy T J, Kahne D and Walker S 2010 The bacterial cell envelope *Cold Spring Harb. Perspect. Biol.* **2** a000414

Siliakus M F, van der Oost J and Kengen S W M 2017 Adaptations of archaeal and bacterial membranes to variations in temperature, pH and pressure *Extremophiles* **21** 651–70

Sinnige T, Houben K, Pritisanac I, Renault M, Boelens R and Baldus M 2015 Insight into the conformational stability of membrane-embedded BamA using a combined solution and solid-state NMR approach *J. Biomol. NMR* **61** 321–32

Sizun C and Bechinger B 2002 Bilayer sample for fast or slow magic angle oriented sample spinning solid-state NMR spectroscopy *J. Am. Chem. Soc.* **124** 1146–7

Skipski V P, Barclay M, Reichman E S and Good J J 1967 Separation of acidic phospholipids by one-dimensional thin-layer chromatography *Biochim. Biophys. Acta (BBA)—Lipids Lipid Metabol.* **137** 80–9

Slichter C P 1990 *Principles of Magnetic Resonance* 3rd edn (Berlin: Springer)

Smith R, Cornell B A, Keniry M A and Separovic F 1983 $^{31}$P nuclear magnetic resonance studies of the association of basic proteins with multilayers of diacyl phosphatidylserine *Biochim. Biophys. Acta (BBA)—Biomembr.* **732** 492–8

Song C, Hu K-N, Joo C-G, Swager T M and Griffin R G 2006 TOTAPOL: a biradical polarizing agent for dynamic nuclear polarization experiments in aqueous media *J. Am. Chem. Soc.* **128** 11385–90

Soubias O, Niu S-L, Mitchell D C and Gawrisch K 2008 Lipid–rhodopsin hydrophobic mismatch alters rhodopsin helical content *J. Am. Chem. Soc.* **130** 12465–71

Sullivan K H 1979 Alteration of the fatty acid composition of escherichia coli by growth in the presence of normal alcohols *J. Bacteriol.* **138** 6

Szeverenyi N M, Sullivan M J and Maciel G E 1982 Observation of spin exchange by two-dimensional Fourier transform $^{13}$C cross polarization-magic-angle spinning *J. Magn. Reson. (1969)* **47** 462–75

Takahashi H, Ayala I, Bardet M, De Paëpe G, Simorre J-P and Hediger S 2013 Solid-state NMR on bacterial cells: selective cell wall signal enhancement and resolution improvement using dynamic nuclear polarization *J. Am. Chem. Soc.* **135** 5105–10

Takegoshi K, Nakamura S and Terao T 2001 $^{13}C$–$^{1}H$ dipolar-assisted rotational resonance in magic-angle spinning NMR *Chem. Phys. Lett.* **344** 631–7

Tauskela J S and Thompson M 1992 A $^{31}P$-NMR spin-lattice relaxation and $^{31}P\{^{1}H\}$ nuclear overhauser effect study of sonicated small unilamellar phosphatidylcholine vesicles *Biochim. Biophys. Acta (BBA)—Biomembr.* **1104** 137–46

Treviño J, Bayón C, Ardá A, Marinelli F, Gandolfi R, Molinari F, Jimenez-Barbero J and Hernáiz M J 2014 New insights into glycopeptide antibiotic binding to cell wall precursors using SPR and NMR spectroscopy *Chem.–A Eur. J.* **20** 7363–72

Tycko R 2011 Solid-state NMR studies of amyloid fibril structure *Annu. Rev. Phys. Chem.* **62** 279–99

Uluca B, Viennet T, Petrović D, Shaykhalishahi H, Weirich F, Gönülalan A, Strodel B, Etzkorn M, Hoyer W and Heise H 2018 DNP-enhanced MAS NMR: a tool to snapshot conformational ensembles of α-synuclein in different states *Biophys. J.* **114** 1614–23

Uttlová P, Pinkas D, Bechyňková O, Fišer R, Svobodová J and Seydlová G 2016 Bacillus subtilis alters the proportion of major membrane phospholipids in response to surfactin exposure *Biochim. Biophys. Acta (BBA)—Biomembr.* **1858** 2965–71

Valiyaveetil F I, Zhou Y and MacKinnon R 2002 Lipids in the structure, folding, and function of the KcsA K+ channel *Biochemistry* **41** 10771–7

van der Cruijsen E A W, Koers E J, Sauvée C, Hulse R E, Weingarth M, Ouari O, Perozo E, Tordo P and Baldus M 2015 Biomolecular DNP-supported NMR spectroscopy using site-directed spin labeling *Chem.—A Eur. J.* **21** 12971–7

Viennet T, Viegas A, Kuepper A, Arens S, Gelev V, Petrov O, Grossmann T N, Heise H and Etzkorn M 2016 Selective protein hyperpolarization in cell lysates using targeted dynamic nuclear polarization *Angew. Chem. Int. Ed.* **55** 10746–50

Vinogradov E, Madhu P K and Vega S 1999 High-resolution proton solid-state NMR spectroscopy by phase-modulated Lee–Goldburg experiment *Chem. Phys. Lett.* **314** 443–50

Wang S, Munro R A, Shi L, Kawamura I, Okitsu T, Wada A, Kim S-Y, Jung K-H, Brown L S and Ladizhansky V 2013 Solid-state NMR spectroscopy structure determination of a lipid-embedded heptahelical membrane protein *Nat. Methods* **10** 1007–12

Warschawski D E, Arnold A A, Beaugrand M, Gravel A, Chartrand É and Marcotte I 2011 Choosing membrane mimetics for NMR structural studies of transmembrane proteins *Biochim. Biophys. Acta (BBA)—Biomembr.* **1808** 1957–74

Weber F J and de Bont J A M 1996 Adaptation mechanisms of microorganisms to the toxic effects of organic solvents on membranes *Biochim. Biophys. Acta (BBA)—Rev. Biomembr.* **1286** 225–45

Wu C H, Ramamoorthy A and Opella S J 1994 High-resolution heteronuclear dipolar solid-state NMR spectroscopy *J. Magn. Reson.* A **109** 270–2

Wylie B J, Bhate M P and McDermott A E 2014 Transmembrane allosteric coupling of the gates in a potassium channel *PNAS* **111** 185–90

Yang Y, Yao H and Hong M 2015 Distinguishing bicontinuous lipid cubic phases from isotropic membrane morphologies using $^{31}P$ solid-state NMR spectroscopy *J. Phys. Chem.* B **119** 4993–5001

Yeh V, Lee T-Y, Chen C-W, Kuo P-C, Shiue J, Chu L-K and Yu T-Y 2019 Highly efficient Transfer of 7TM membrane protein from native membrane to covalently circularized nanodisc *Sci. Rep.* **8** 13501–11

Yeh V, Goode A, Eastham G, Rambo R P, Inoue K, Doutch J and Bonev B B 2020 Membrane stability in the presence of methacrylate esters *Langmuir* **36** 9649–57

Yeh V, Hsin Y, Lee T-Y, Chan J C C, Yu T-Y and Chu L-K 2016 Lipids influence the proton pump activity of photosynthetic protein embedded in nanodiscs *RSC Adv.* **6** 88300–5

Zagdoun A *et al* 2013 Large molecular weight nitroxide biradicals providing efficient dynamic nuclear polarization at temperatures up to 200 K *J. Am. Chem. Soc.* **135** 12790–7

Zhang W, Crocker E, McLaughlin S and Smith S O 2003 Binding of peptides with basic and aromatic residues to bilayer membranes phenylalanine in the myristoylated alanine-rich c kinase substrate effector domain penetrates into the hydrophobic core of the bilayer *J. Biol. Chem.* **278** 21459–66

Zorin V, Ciesielski F, Griffin D C, Rittig M and Bonev B B 2010 Heteronuclear chemical shift correlation and J-resolved MAS NMR spectroscopy of lipid membranes *Magn. Reson. Chem.* **48** 925–34

# Chapter 9

## Mapping dynamics in membrane proteins with solid-state NMR: methods and examples

**Kaustubh R Mote and P K Madhu**

The ability to track large-amplitude changes in the orientation of transmembrane segments in membrane proteins is important to understand how they function. Solid-state NMR has a distinct advantage over other structural-biology techniques in this aspect, as the functionally important dynamics can be studied in near native conditions. The dependence of dynamics on important variables such as the lipid composition can be extensively ascertained, in particular, with magic-angle spinning (MAS) methods. The technique allows the detection of conformational changes, motional order parameters, as well as the orientation of transmembrane segments. With the availability of probes that can achieve MAS frequencies higher than 60 kHz, many of the relevant pulse sequences can be used in conjunction with $^{1}$H detection for better sensitivity, especially in mass-limited samples. Here, we review certain MAS-NMR and oriented-NMR methods that can be used to study dynamics in membrane proteins.

### 9.1 Introduction

Membrane proteins pose a wide variety of challenges to contemporary structural-biology techniques. The natural environment for these proteins is a lipid bilayer. Their functional preparation in such an environment can often be recalcitrant to high-resolution structural studies, suffering from disadvantages such as poor expression and a tendency to aggregate or misfold *in vitro*. Even when these issues are avoided, repeating secondary-structure elements and heterogeneity of the sample as well as a preponderance of hydrophobic amino acids can decrease the resolution in NMR studies. However, given their importance as pharmacological targets in a number of diseases, these are often highly valuable targets for structural studies (Overington *et al* 2006). Recent developments in x-ray crystallography and Cryo-EM techniques have given an impetus to the work on large membrane-protein

doi:10.1088/978-0-7503-2532-5ch9

complexes (Moraes *et al* 2014, Cheng 2018). On the other hand, the niche advantage of NMR, particularly solid-state NMR, is that these proteins can be studied in their native environment, under fully functional conditions. This gives direct insight into the dynamics that are responsible for the activity of these proteins. To date, NMR remains the only experimental technique that can give residue-specific information on dynamics in a fully functional state, under physiological conditions.

A major problem in understanding membrane-protein dynamics relates to the conformational changes that alter the transmembrane orientation of helices. These are not accompanied by large changes in the secondary structures, and, hence, can often have little impact on the isotropic chemical shifts. Detecting these by monitoring chemical-shift changes with traditional solution-state or magic-angle spinning (MAS) solid-state NMR is a difficult task. Yet, the most functionally relevant changes which occur in a wide array of membrane proteins belong to this class. For example, secondary active transporters undergo large transmembrane helix re-orientations, alternately opening up to opposite sites of the membrane (Forrest *et al* 2011). These changes, classified under the alternating access model (Jardetzky 1966), form the structural basis for the coupling of metabolite gradients across biological membranes. They can involve large-scale changes in transmembrane helix orientations, with changes in angle by as much as 15° being common (Holt and Killian 2010). Similarly, changes in the orientation of transmembrane segments in G-protein coupled receptors have been implicated in their functioning (Kobilka 2007). Monitoring these changes is thus critical and ideally suited for NMR-based studies.

In general, oriented NMR, under non-spinning conditions of the sample, is one of the few techniques that can directly map such conformational changes. Insights into how aromatic and basic residues anchor transmembrane segments and influence the tilt and dynamics of these segments have been obtained using oriented NMR techniques (Vostrikov *et al* 2010a, 2010b). These have been used to monitor conformational changes in secondary active transporters as well (Basting *et al* 2006). Oriented NMR methods, however, are normally performed with $^{15}N$ requiring tedious labeling schemes. Since oriented NMR experiments, such as polaristion inversion spin-exchange at magic-angle (PISEMA) (Wu *et al* 1994) and geometric analysis of labeled alanines (GALA) (van der Wel *et al* 2002), give information on the orientation of peptide planes with respect to the bilayer normal, they can be directly used to map these changes. Under MAS, however, such information is not directly available from the isotropic chemical shifts. Instead, experiments that determine anisotropic parameters such as the heteronuclear dipole–dipole (DD) coupling and chemical-shift anisotropy (CSA), which are sensitive to motion, must be used (De Paëpe 2012, Krushelnitsky *et al* 2013). Given the versatility of MAS NMR in mapping the backbone as well as side-chain resonances, and the flexibility that exists in preparing membrane-protein samples with arbitrary lipid composition, MAS NMR is expected to be a method of choice for many spectroscopists interested in studying membrane proteins. In particular, the development of probes that allow MAS frequencies, $\omega_r/2\pi$, greater than 60 kHz may revolutionise the spectroscopy of these proteins. However, the application of pulse sequences that are used to measure anisotropic parameters at slow-moderate MAS frequencies is not straightforward in the projected high MAS frequency regime. Here, we will look at some of the schemes in vogue to study dynamics under MAS as well as oriented NMR. In the case of MAS, we will concentrate on a couple of them in particular, namely, dipolar-coupling chemical-shift correlation (DIPSHIFT) (Munowitz *et al* 1981) and rotational-echo double-resonance (REDOR) (Gullion and Schaefer 1989), as they both may have a great potential to map dynamics and motional-order parameters in membrane proteins under a wide range of MAS frequencies.

## 9.2 NMR strategies to monitor dynamics

Membrane proteins, similar to most other biomolecules, are intrinsically dynamic. The time scales of the internal dynamics and exchange rates with other molecules vary significantly. The proteins may also sample a large number of transient and low-populated conformations. Recent work suggests that membrane-embedded proteins exhibit dynamics that are distinct from soluble proteins, and have a residual conformational entropy as a result of this. This is proposed as the driving factor in many of the membrane-protein functions (O'Brien *et al* 2020).

Nuclear spins are extremely sensitive to the frequencies of fluctuation of the internal spin interactions. These cause relaxation of the spins in particular and are responsible for the intrinsic dynamics in general. NMR has methods to probe this wide range of dynamics and also individually address the internal spin interactions depending upon the time scales and the strength of the concerned interactions. There are at least four classical cases depending on the relative magnitudes of the inverse of the time scale of dynamics, $k$, and the interaction strength, $\delta$: (i) when the motional time scale is faster than the time scale associated with the strength of the interaction, $k \gg \delta$: this leads to averaging of the interaction. From the resultant spectral line narrowing, information about the amplitude of motion and geometry can be ascertained. Some of the experiments used in this regard are DIPSHIFT (Munowitz *et al* 1981), wide-line separation (WISE) (Schmidt-Rohr *et al* 2012), Lee–Goldburg cross-polarisation (LG-CP) (van Rossum *et al* 2000), CSA recoupling (Tycko *et al* 1989), and $^2$H quadrupole echo (Vega *et al* 1989). (ii) When the motional time scale is slower than that of the concerned interaction, $k \ll \delta$: This corresponds to the so-called 'rigid limit' where one can observe contributions from individual species. This domain is covered by exchange spectroscopy (EXSY) (Schmidt-Rohr and Spiess 2012), stimulated echo (Marasinghe *et al* 2005), and centreband-only detection of exchange (CODEX) (de Azevedo *et al* 1999). (iii) The so-called slow-to-intermediate motion, $k < \delta$: This is best probed with spin-echo experiments, either Hahn echo for spin-$\frac{1}{2}$ systems or quadrupole echo for spin-1 systems. (iv) When the two time scales are comparable leading to an intermediate regime, $k \sim \delta$: this manifests in broadening of the spectral lines mapping out DD couplings, free-energy estimates, and transient states (Rothwell and Waugh 1981, de Azevedo *et al* 2008, Vallurupalli *et al* 2012). Figure 9.1 depicts a schematic representation of these regimes detailing the kind of NMR experiments useful for various time scales including a snap shot of the REDOR curves which show an

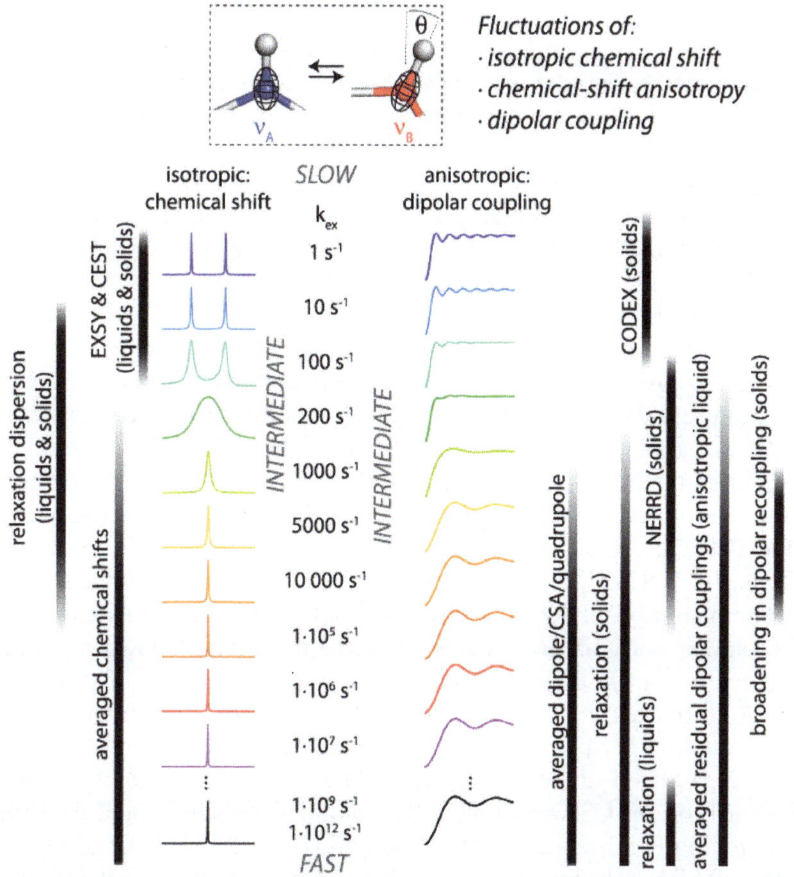


**Figure 9.1.** A schematic showing averaging of NMR interactions with various time scales leading to different spectral responses and the NMR tools to investigate each of them. From the top in counter-clockwise order: schematic of an *A*–*B* bond undergoing orientation, $\nu_A$ and $\nu_B$ being their isotropic chemical shifts; two-state isotropic chemical-shift averaging and resultant spectral response; averaging of heteronuclear DD interaction at a MAS frequency of 40 kHz, for various time scales, obtained with REDOR dephasing curves; list of NMR techniques to probe the dynamics at various time scales. CEST stands for chemical-exchange saturation transfer (van Zijl and Yadav 2011), and NERRD for near-rotary resonance relaxation dispersion (Marion *et al* 2018). Reprinted from Schanda (2019), copyright (2019), with permission from Elsevier.

increasing trend in the DD couplings as the spins traverse from fast to intermediate to slow time-scale regimes (Schanda 2019).

The intrinsic dynamics and MAS average out the relevant anisotropic interactions in membrane proteins, as in other samples. The pertinent ones of interest with regard to geometry and dynamics are heteronuclear DD couplings and CSA as we will focus only on methods involving $^{1}H$, $^{13}C$, and $^{15}N$ spins. The experiments usually employed in the study of dynamics map out the heteronuclear DD couplings or the CSA or correlate them with isotropic chemical shifts. They belong to the separated-local-field (SLF) class (Hester *et al* 1976). Literally, these are experiments which separate out local fields in action, chemical shifts and DD couplings in the current case. In doing so, the experiment provides a correlation between these two interactions and can be used for site-specific analysis (Krushelnitsky *et al* 2013).

SLF experiments employ strategies based on pulse methods manipulating heteronuclear interactions. One strategy is CP-based. These include WISE (Schmidt-Rohr and Mao 2002, Yao *et al* 2004, Schmidt-Rohr *et al* 2012), CP with polarisation inversion (CPPI) (Wu and Zilm 1993), PISEMA (Wu *et al* 1994), and LG-CP (van Rossum *et al* 2000). All these correlate heteronuclear DD interactions with high-resolution isotropic chemical-shift information, mostly of the rare spins, and ideally try to remove the $^{1}$H–$^{1}$H DD coupling with homonuclear decoupling schemes (Mote *et al* 2016).

The second class of schemes recouples the heteronuclear interactions with radio-frequency (RF) pulses, either 180° or 360° or their composite versions. Such methods may also suppress the homonuclear couplings among $^{1}$H in some cases, and they are, in general, applicable at higher MAS frequencies. There are commonly three ways the pulsed methods are implemented: (i) employing 180° pulses for recoupling either the heteronuclear DD couplings or the CSA. DIPSHIFT (Munowitz *et al* 1981) and REDOR (Gullion and Schaefer 1989) are the classic examples of the former type and 2DCSA (Tycko *et al* 1989, Orr and Duer 2006) of the latter. (ii) For a better compensation of RF imperfections, but recoupling CSA, employing 360° pulses is another way, named as separation of undistorted powder patterns with effortless recoupling (SUPER) (Liu *et al* 2002). (iii) Making use of schemes such as $CN_n^{\nu}$ and $RN_n^{\nu}$ that exploit the symmetry of the internal spin interactions with a selective recoupling of specific interactions and removing the unwanted couplings (Levitt 2002). One of these is the recoupling of the chemical shift anisotropy (ROCSA) method for CSA recoupling (Chan and Tycko 2003) and the other is the phase-alternating R-symmetry (PARS) method (Hou *et al* 2014) and its windowed derivative *w*PARS scheme (Lu *et al* 2016) for recoupling heteronuclear DD interactions. The recoupling pulses prevent the near-complete averaging out of the anisotropic interactions and recouple a fraction of them, the fraction being theoretically tractable.

Besides the above, to probe slow motions, on the order of milliseconds, exchange spectroscopy (Marasinghe *et al* 2005, Schmidt-Rohr and Spiess 2012), stimulated echo schemes (Marasinghe *et al* 2005), and stimulated echo under MAS with recoupling, such as CODEX, (de Azevedo *et al* 1999), may be used.

This chapter will address mainly some of these SLF schemes. In general, all such methods should have the following properties (Dvinskikh and Sandström 2005): (i) tunable scaling factors for recoupling of the spin interactions so that both strong and weak couplings can be measured. (ii) Suppression of $^{1}$H–$^{1}$H DD couplings if heteronuclear DD coupling involving $^{1}$H or the $^{1}$H CSA is being measured. (iii) The relevant mixing block should have sufficiently short cycle times enabling measurement of stronger couplings without aliasing. (iv) Less dependence on RF imperfections. (v) The schemes should be applicable over a wide range of RF field strengths and MAS frequencies, noting that at higher MAS frequencies, the RF conditions should be still acceptable so that sample heating and duty-cycle effects are under control.

## 9.3 Monitoring dynamics with CP-based methods

CP-based methods generally can track fast motions and are effective in providing a quick and qualitative assessment of the motions present in various systems. Whilst these may have to be combined with relaxation measurements to get more quantitative results, here, we sketch out the principles and applications of a select few methods. It may be noted that CP by itself selects out immobile species and direct polarisation (of the rare spins) selects both mobile and immobile species. Intermediate time scale motion often interferes with CP intensity. Very large-amplitude motion may be also tracked by *J*-coupling based methods, such as INEPT, as they can selectively target mobile segments (Andronesi *et al* 2005). An example of CP- and INEPT-based dynamics will be discussed later.

### 9.3.1 WISE

2D wide-line separation, WISE, method gives an assessment of the $^1$H–$^1$H couplings and motional effect on them by correlating $^1$H resonances, in the $t_1$ dimension, to the high-resolution $^{13}$C or $^{15}$N resonances in the $t_2$ dimension (Schmidt-Rohr and Mao 2002, Yao *et al* 2004, Schmidt-Rohr *et al* 2012). After the evolution of the $^1$H resonances, selective transfer to the dipole-bonded $^{13}$C spins is normally implemented. For this LG-CP (Lee and Goldburg 1965, Huster *et al* 2001) is used. Figure 9.2(a) shows the WISE pulse scheme. The LG-CP contact time is usually 50–500 μs to ensure selective polarisation transfer. Figure 9.2(b) shows the 2D WISE spectra of an elastin–mimetic protein, poly(Lys-25), with a repeat sequence of VPGVG as a function of hydration. Figure 9.2(c) shows the $^1$H cross-sections for two residues corresponding to the backbone site as a function of hydration. These provide first-hand information on the degree of motional narrowing. For instance, hydration till 20% does not change the $^1$H line shape very much, but there is a drastic

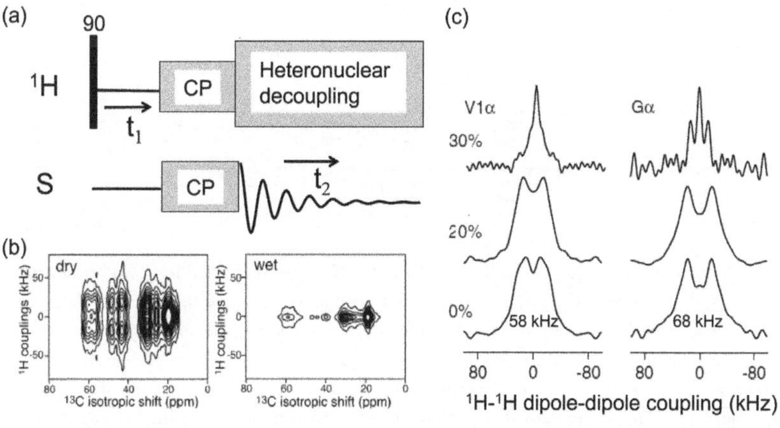


**Figure 9.2.** (a) The WISE pulse scheme. (b) 2D spectra of the polyLys-25 elastin–mimetic protein at hydration levels of 0% and 30%. (c) $^1$H cross-sections of indicated residues at hydration levels of 0%, 20%, and 30%. The MAS frequency was 4.0 kHz and for the CP transfer, LG-CP was used with a mixing time of 100–200 μs. (b) and (c) Adapted with permission from Yao *et al* (2004), copyright (2004), Wiley.

effect at 30%. Here, the quality of the spectrum is also relatively worse as CP becomes inefficient due to motion (Hong *et al* 2002). Such motions of large amplitudes, causing conformational distribution, can be then related to the function of the corresponding biomolecular systems.

### 9.3.2 CPPI

CPPI is the standard CP scheme combined with polarisation inversion, figure 9.3(a) (Wu and Zilm 1993). The scheme essentially maintains the heteronuclear DD coupling by matching the two RF fields (on the $^1$H and *S* channel) to the first-order spinning sidebands. This scheme has most of the desirable features for extracting heteronuclear DD couplings listed earlier. In the original version, it was used for spectral editing to distinguish $CH_3$, $CH_2$ and CH groups (Wu and Zilm 1993). It was used for the extraction of couplings, and thereby distances and order parameters later (Dvinskikh *et al* 2003, Dvinskikh and Sandström 2005, Chevelkov *et al* 2009, Asami and Reif 2017). One of the applications of this scheme was in the elucidation

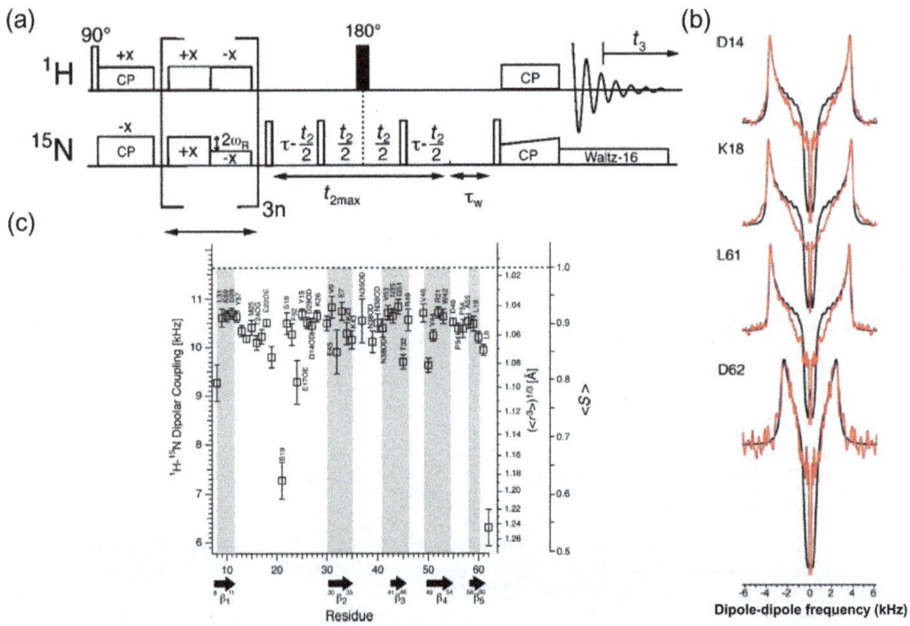


**Figure 9.3.** (a) A correlation experiment scheme with the CPPI block enclosed in square brackets. The overall pulse scheme, hence, correlates the heteronuclear DD couplings, in $t_1$, with $^{15}$N evolution, in $t_2$, and $^1$H evolution, in $t_3$. The experiment, unless done at high MAS frequencies, needs perdeuterated samples as the detection is on the $^1$H channel or requires homonuclear DD decoupling. (b) $^{15}$N–$^1$H line shapes, simulations in red and experiments in black, for certain residues of the protein system $\alpha$-spectrin SH3 domain. (c) Experimental $^1$H–$^{15}$N DD couplings for the protein system as a function of residue. Assignments correspond to the carbonyl hydrogen bonding acceptor of the respective amide proton. On the right-hand side, the vertical axis shows the hypothetical values for the average $^{15}$N–$^1$H bond length, assuming no motion, and the order parameter, *S*. The MAS frequency was 20 kHz. Reprinted with permission from Chevelkov *et al* (2009), copyright (2009), American Chemical Society.

of the $^{1}H$–$^{15}N$ couplings in the $\alpha$-spectrin SH3 domain. Figure 9.3(b) shows the experimental and simulated heteronuclear DD line shapes of certain residues of the proteins system. Figure 9.3(c) shows the estimated coupling strengths as a function of the residue for the protein system along with the order parameter. CPPI has been implemented at MAS frequencies of 20–60 kHz (Asami and Reif 2017).

### 9.3.3 LG-CP

LG-CP method, figure 9.4(a), implemented in a 2D way, is useful to get an idea of fast motions. It yields a good estimate of the heteronuclear distances from the CP build-up curves. An LG (Lee–Goldburg) irradiation is applied on the $^{1}H$-channel during the normal spin-lock pulse on the heteronuclear spin. This ensures suppression of $^{1}H$–$^{1}H$ couplings paving the way for a reliable estimate of heteronuclear coupling constants. The transient oscillations during the CP contact time, which is $t_1$ evolution, for each of the resolved $^{13}C$ or $^{15}N$ sites in $t_2$ evolution can be analysed for distance elucidation (Hong *et al* 2002).

LG-CP has found applications in the study of membrane-protein motions (Huster *et al* 2001, Hong *et al* 2002). It can provide information on the amplitude and geometry of the segmental motions. The recoupled $^{13}C$–$^{1}H$ and $^{15}N$–$^{1}H$ line shapes are good indicators of the geometry of various segments. Figure 9.4(b) shows 2D spectrum of colicin 1a channel domain protein correlating the $^{13}C$–$^{1}H$ recoupled heteronuclear DD line shapes with the resolved $^{13}C$ isotropic shifts. Figure 9.4(c) shows extracted line shapes of certain functional groups of four residues in the protein. These give a good idea of the geometry of the membrane-bound protein system and insight into the functional aspects (Hong *et al* 2002).

### 9.3.4 Dynamics in oriented NMR

In oriented systems, PISEMA experiments are generally used to determine heteronuclear DD couplings and CSA (Wu *et al* 1994). These give direct information on the orientation of the peptide plane with respect to the bilayer normal. Since these interactions can be substantially averaged by motion, PISEMA can be directly used to map the whole-molecule dynamics as well as rotational diffusion in membrane

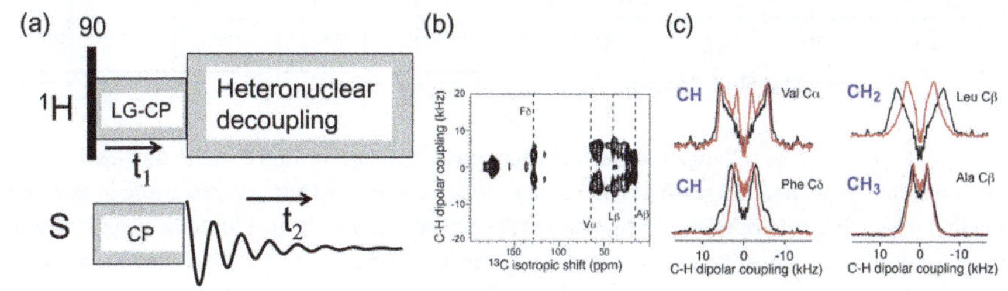


**Figure 9.4.** (a) Schematic of the LG-CP pulse scheme. (b) 2D spectrum of the colicin 1a channel protein system acquired with the LG-CP scheme correlating $^{13}C$–$^{1}H$ couplings with $^{13}C$ chemical shifts. (c) Extracted $^{13}C$–$^{1}H$ heteronuclear DD line shapes of the indicated functional groups and residues of the protein system. The MAS frequency was 12–13 kHz. Adapted with permission from Hong *et al* (2002), copyright (2002), American Chemical Society.

proteins. Several authors have shown how the appearance of the polar index slant angle (PISA) wheel pattern of peaks in alpha-helical transmembrane segments changes with different types of motion (Page *et al* 2008, Esteban-Martín *et al* 2009, Nevzorov 2011). The PISEMA experiment can also be used to directly map the changes in conformation in membrane proteins. For example, figure 9.5 shows the inter-conversion of EmrE, a secondary active transporter, between two conformations which give distinct resonances in a PISEMA spectrum (Gayen *et al* 2013). An advantage of such an approach is that it directly gives the rate of exchange (using exchange-spectroscopy) as well as the exact orientation of the peptides in the two conformations, which is otherwise not possible to obtain. The PISEMA experiment has been adapted to work under MAS to measure heteronuclear DD couplings as well (Yamamoto *et al* 2005).

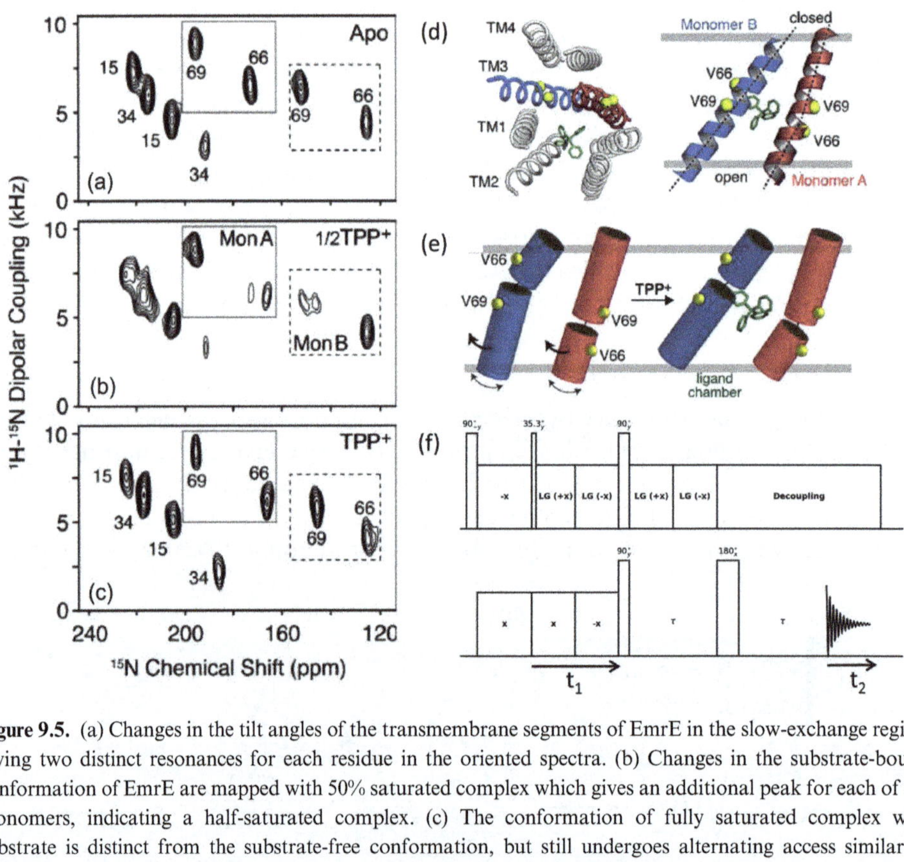


**Figure 9.5.** (a) Changes in the tilt angles of the transmembrane segments of EmrE in the slow-exchange regime giving two distinct resonances for each residue in the oriented spectra. (b) Changes in the substrate-bound conformation of EmrE are mapped with 50% saturated complex which gives an additional peak for each of the monomers, indicating a half-saturated complex. (c) The conformation of fully saturated complex with substrate is distinct from the substrate-free conformation, but still undergoes alternating access similar to former. (d) The anti-parallel EmrE dimer has two distinct orientations for each transmembrane helix, one coming from each of the monomers. (e) The changes in transmembrane orientation upon binding TPP+ are mapped using the orientation restraints from panels (a)–(c). (f) Pulse sequence of the sensitivity-enhanced version of the PISEMA experiment (Gopinath and Veglia 2009) which was used to acquire spectra in panels (a)–(c). (a)–(e) Adapted with permission from Gayen *et al* (2013), copyright (2013), Wiley.

Another class of experiments that can be used to determine transmembrane protein orientation and dynamics is the geometric analysis of labeled alanines (GALA) method (van der Wel *et al* 2002, Strandberg *et al* 2004). This method relies on mapping the averaging of the $^2H$ quadrupolar interaction of the $CD_3$ group in suitably labeled alanine residues in transmembrane segments. Since methyl groups undergo fast rotation, the angle which the $C_\alpha$–$C_\beta$ bond vector makes with the external magnetic field in an oriented system determines the extent of averaging. Since this bond vector has a simple relationship with the peptide-plane orientation, the orientation of the peptide planes can be determined. Given the large strength of the quadrupole interaction, and its high sensitivity to motional averaging, this is a powerful method to map orientation and dynamics of transmembrane peptides (Vostrikov *et al* 2011). However, it requires selective labeling of alanines, and hence, can be applied to segments that can only be synthesised using solid-phase peptide synthesis. Nevertheless, a wealth of information about the dependence of peptide-plane orientation and dynamics of aromatic and basic residues has been obtained using this method (Vostrikov *et al* 2010a). Resonances seen in a PISEMA spectrum are sensitive to the tilt and azimuthal angle changes in transmembrane helices, and can be used to directly track them, figures 9.6(a), (b). The appearance of the PISA wheel is also sensitive to rotational diffusion around the bilayer normal, where the normal is perpendicular to the magnetic field direction, and can be used to directly monitor these changes, figures 9.6(c)–(e).

## 9.4 Monitoring dynamics with heteronuclear DD recoupling-based methods

As stated earlier, there are two interactions for spin-$\frac{1}{2}$ systems, namely heteronuclear DD couplings, and CSA, that yield dynamics, orientation, and order parameter information. In fact, both of these are rank-1 in spin terms. Hence, any pulse scheme, which manipulates the spin part of the Hamiltonian, would recouple both of these. A differentiation may be done by placing 180° pulses to refocus the chemical-shift interaction whilst recoupling the DD couplings. Similarly, appropriate heteronuclear spin decoupling may be implemented whilst recoupling CSA or combine with high-enough MAS frequencies. The RF nutation frequency required for the decoupling may have to be 2–3 times higher than that applied for the CSA recoupling pulses.

### 9.4.1 DIPSHIFT

DIPSHIFT can be considered as the first of the 'recoupling' schemes in solid-state NMR, although perhaps it was not so classified. DIPSHIFT belongs to the classical SLF experiments under MAS (Munowitz *et al* 1981). This is used mostly for torsion-angle and order-parameter elucidation (Hong *et al* 1997, Ivanir-Dabora *et al* 2015), but DIPSHIFT results can also be used to obtain internuclear distances. This experiment monitors the signal intensity for each chemical-shift resolved site as a function of the recoupled heteronuclear DD couplings. Recoupling is accomplished

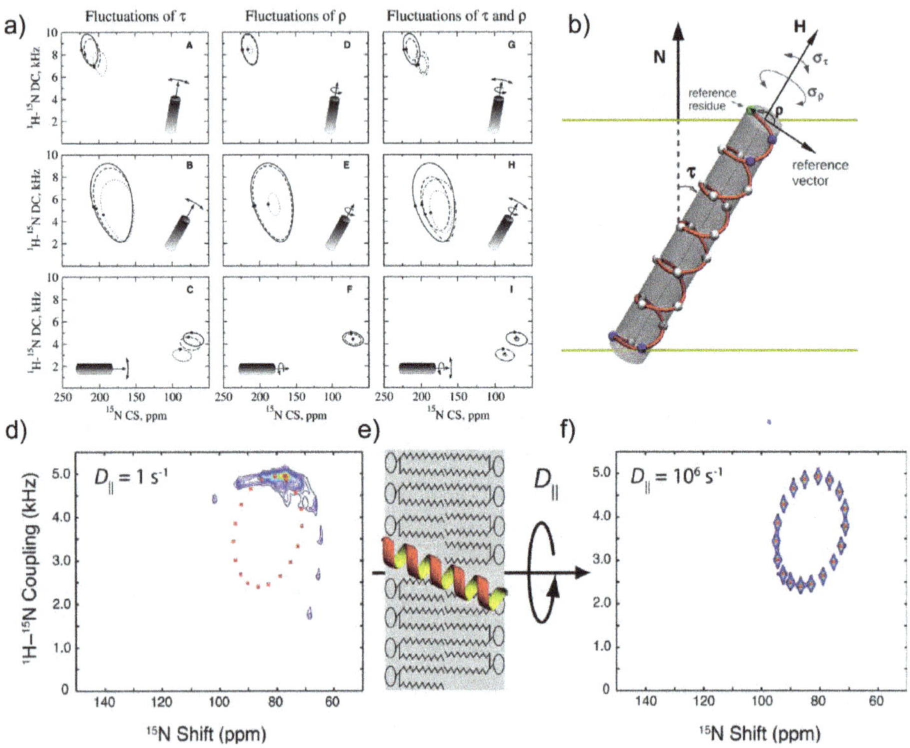


**Figure 9.6.** (a) The influence of dynamics represented by changes in tilt ($\theta$) and rotation angle ($\rho$) of a transmembrane segment, as defined by the representation in panel (b). (c, e) Changes in the appearance of PISEMA spectrum with different rotational-diffusion rates about an axis parallel to the bilayer normal (d). Resonances that usually appear at the lower DD couplings disappear as they are broadened out when rotational diffusion is restricted (d), but are seen when fast rotational diffusion is present (e). Panels (a), (b) are reprinted from Esteban-Martín *et al* (2009), copyright (2009), with permission from Elsevier, and panels (c)–(e) are adapted with permission from Nevzorov (2011), copyright (2011), American Chemical Society.

by 180° pulses in a rotor period ($\tau_r$) preceding which the $t_1$ evolution takes place. During the $t_1$ period, the $S$ magnetisation ($^{13}$C or $^{15}$N) evolves under heteronuclear DD coupling, the influence of which is detected in say $t_2$ period, monitoring the chemical shift of the $S$ spin. During the $t_1$ period, homonuclear DD decoupling may be implemented so that the dephasing of the signal is maximal with $^{13}$C/$^{15}$N–$^{1}$H DD recoupling.

The original DIPSHIFT (Munowitz *et al* 1981) employed only one 180° pulse for recoupling, figure 9.7(a), leading to dephasing or DIPSHIFT curves as shown in figure 9.7(c). The recoupling could be amplified by placing multiple 180° pulses as shown in figure 9.7(b) leading to a deeper modulation of the dephasing curves, an example of which is shown in figure 9.7(d). An advantage of the amplification is the possibility of performing the experiments at higher MAS frequencies that can lead to better chemical-shift resolution as well as the measurement of smaller DD couplings (Hong *et al* 1997, Cobo *et al* 2012). The DIPSHIFT curves can be fitted with standard expressions to obtain order parameter and also distances.

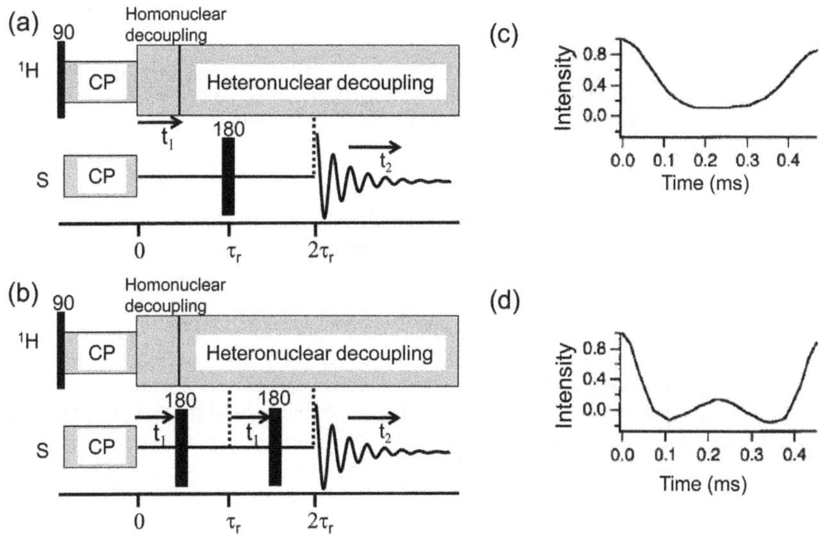


**Figure 9.7.** (a) The original DIPSHIFT pulse scheme employing one 180° pulse for heteronuclear DD recoupling. (b) The DIPSHIFT pulse scheme, making amplification possible for the recoupled heteronuclear DD couplings, in this case twice compared to that in (a). The DIPSHIFT curves with (c) one 180° pulse and (d) with two 180° pulses for recoupling. The sample was NAV and MAS frequency was 2137 ± 5 Hz. (c) and (d) are reprinted from Hong *et al* (1997), copyright (1997), with permission from Elsevier.

### 9.4.2 REDOR

REDOR is one of the robust recoupling schemes used in solid-state NMR to measure internuclear distances (Gullion and Schaefer 1989, Gullion 1998). The method involves, in the simplest case for the measurement of $^{13}C$–$^{15}N$ couplings ($I$–$S$), application of 180° pulses at a rotor period, $\tau_r$, for chemical-shift refocusing during which the heteronuclear DD couplings behave as under MAS, i.e. they are averaged out to the first order. The observed signals are referred to as $S_0$ in this case. This is followed by another experiment which has in addition 180° pulses in a rotor period, on the $^{15}N$ ($S$) channel. This prevents the averaging of the $^{15}N$–$^{13}C$ DD coupling, and the acquired $^{13}C$ signal, compared with the first case (no recoupling pulse), will be attenuated. The difference in the intensity, $\Delta S = S_0 - S$, which is $T_2'$ corrected, is a measure of the recoupled DD interaction. The schemes are repeated for increasing duration, $\tau$, and the difference (or even the ratio of the absolute values of the signal with and without the recoupling pulses) may be plotted as a function of the mixing time. Such a REDOR dephasing curve may be fitted using numerical simulations or analytical expression based on Bessel functions to get distances and order parameters (Mueller 1995, Gehman *et al* 2007).

Figure 9.8(a) shows a schematic of the pulse blocks used to acquire $S_0$ (MAS) and $S$ (REDOR). Here, the scheme has the basic element in four rotor periods and the two recoupling pulses are separated by $0.5\tau_r$. This implementation may vary and different versions exist in the literature (Dusold and Sebald 2000). Figure 9.8(a) also shows the time evolution of the DD Hamiltonian, $\mathcal{H}^{m=\pm1}_{D,\beta=\frac{\pi}{4},\gamma=0}$ and $\mathcal{H}^{m=\pm2}_{D,\beta=\frac{\pi}{4},\gamma=0}$, which

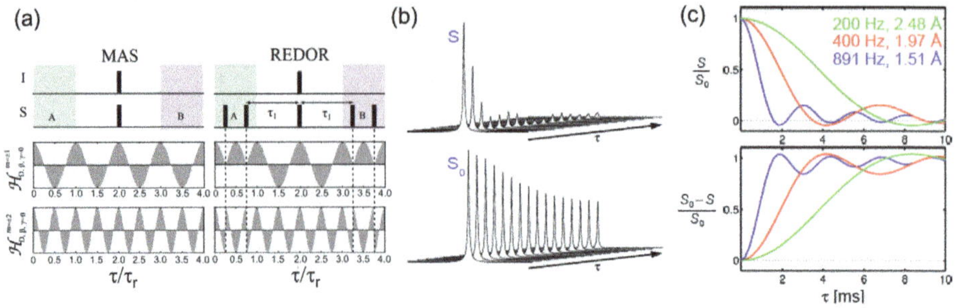


**Figure 9.8.** (a) REDOR scheme in the simplest version. Both the control and the recoupled versions are shown yielding signals often notated as $S_0$ and $S$, respectively. (b) Simulated spectra as a function of the mixing time for the schemes in (a). (c) The REDOR dephasing curves for three coupling constants. (a) also has the time evolution of the Fourier components of the DD Hamiltonian showing the averaging effect under MAS and recoupling effect with the 180° pulses during rotor periods. The important aspect to note here is the absence of the $m = \pm 2$ Fourier component which is a consequence of the duration between the recoupling pulses. (a) Adapted from Jain *et al* (2019a), with the permission of AIP publishing.

are the first and second Fourier components of the DD Hamiltonian. Here, $\beta$ and $\gamma$ are the Euler angles for a particular crystallite. For a reliable estimate of the distances, it is imperative not to have both of these at the same time, and keeping the duration between the two recoupling pulses as $0.5\tau_r$ meets this requirement (Jain *et al* 2019a). Figure 9.8(b) shows simulated spectral response at a MAS frequency of 10kHz for a coupling constant of 891 Hz. Figure 9.8(c) shows two ways in which the REDOR dephasing curves may be plotted, in general $\frac{\Delta S}{S_0}$ is the one that is used. It is already clear that the stronger the coupling, the faster the dephasing, and this is an issue which will be addressed later.

REDOR, due to its simplicity and robustness to various experimental parameters, has found applications in a variety of systems (Dusold and Sebald 2000). In biological systems, it has been used to measure selective distances for the confirmation of salt bridges, and related examples (Jaroniec *et al* 2001, Chandra *et al* 2017). Similarly, it has been used extensively in membrane-protein systems leading to insights on dynamics and conformational heterogeneity targeting nuclei like $^2$H and $^{31}$P (Jia *et al* 2015, Yang *et al* 2018).

### 9.4.3 Symmetry-based schemes

The schemes designed on the basis of the symmetry of the internal spin interactions, such as, $CN_n^\nu$ and $RN_n^\nu$ (Levitt 2002), may also be used to understand the motion in membrane-protein systems. An attempt in this regard was made in the context of model molecules making use of phase-alternating $R$ symmetry (PARS) and its windowed version, *w*PARS (Hou *et al* 2014, Lu *et al* 2016). All the schemes mentioned so far would also recouple CSA of the observed spins as the spin rank of heteronuclear DD and CSA interactions is the same, 1. By phase alternating the $RN$ blocks, it is possible to suppress the CSA contribution. However, $RN$ schemes are known to be sensitive to RF imperfections, such as RF-field inhomogeneity. To

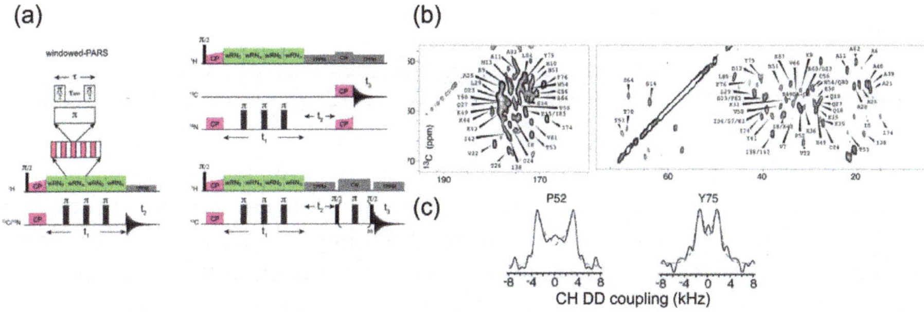


**Figure 9.9.** (a) Various multi-dimensional schemes employing *w*PARS as the heteronuclear DD recoupling block. The figure comprises $^{1}$H–S (either $^{13}$C or $^{15}$N) 2D scheme, $^{1}$H–$^{15}$N with $^{13}$C detection 3D scheme, and $^{1}$H–$^{13}$C with $^{13}$C detection 3D scheme. (b) Aromatic region of the $^{13}$C–$^{13}$C correlation spectra of DIC8 protein obtained with the third of the pulse scheme in (a). (c) Dipole–dipole line shapes of two residues of DLC8 extracted from the 3D $^{13}$C–$^{13}$C correlation spectrum along with the fits. This gives the DD couplings and the motional information from the order-parameter estimates. $R10_1^3$ was the symmetry element with a MAS frequency of 20 kHz. Adapted with permission from Lu *et al* (2016), copyright (2016), published by the PCCP Owner Societies.

circumvent this, *w*PARS was introduced which consists of well-designed windows in the basic constituent of the *R* element, that is in the 180° pulse, resulting in the basic element being two 90° pulses separated by a window. The window duration will additionally dictate the RF field strength for a given MAS frequency.

The schematic of *w*PARS is shown in figure 9.9(a). PARS would be the same without the windows. Results from *w*PARS are shown in figures 9.9(b) and (c). These clearly show the potential of PARS kind of schemes to extract the hetero-nuclear DD couplings in a reliable fashion and with site-specific resolution by incorporating into three-dimensional experiments.

*w*PARS has the following pros and cons: (i) CSA recoupling is largely avoided. This will be useful for certain functional groups with large CSA in membrane-protein systems and also for experiments in high static magnetic fields where there are the additional benefits of dispersion, sensitivity, and resolution. (ii) The sequence is tolerant to imperfections with regard to RF compared with other schemes, except REDOR. (iii) The effect of homonuclear $^{1}$H–$^{1}$H DD couplings is minimal in PARS, which is not the case in REDOR or DIPSHIFT. (iv) Higher scaling factor compared with PARS and other schemes due to the splitting of the 180° pulses. This helps in the measureme(nt of weaker couplings. (v) Need for higher RF irradiation amplitudes than PARS for the same *RN* symmetry scheme due to the windows. As both PARS and *w*PARS employ rotor-synchronised recoupling blocks, they will require 'prohibitive' RF amplitudes for MAS frequencies, say, >40 kHz.

## 9.5 Monitoring dynamics with CSA recoupling methods

CSA patterns also reveal structural information. They give a reliable estimate of the CSA parameters, such as, anisotropy and asymmetry. They can be analysed to elucidate the dynamics in comparison with the rigid-limit value.

### 9.5.1 2DCSA and SUPER

One of the initial chemical-shift recoupling schemes employed under magic-angle spinning, notably 2DCSA, is shown in figure 9.10(a) (Tycko *et al* 1989, Orr and Duer 2006). This involves, again, judiciously placed 180° pulses for CSA recoupling, figure 9.10(b). The heteronuclear spin decoupling on the $^1H$ channel ensures only CSA recoupling. The RF amplitude on the $^1H$ channel during the 180° pulses should be such that any kind of Hartmann–Hahn matching is avoided. This experiment maps out the CSA pattern in the indirect dimension for each of the resolved $^{13}C$ sites in the direct dimension. A careful optimisation, cogwheel phase cycling (Levitt *et al* 2002), and a constant-time implementation of the original scheme was shown to be better in giving undistorted line shapes and making the scheme tolerant to residual DD couplings (Orr and Duer 2006).

The original version of the 2DCSA (Tycko *et al* 1989) was improved with the use of 360° pulses in the recoupling block, figure 9.10(c). Such a scheme, separation of undistorted power patters with effortless recoupling (SUPER), gave better line shapes and compensation against various RF artefacts (Liu *et al* 2002). SUPER suffers from the presence of homonuclear $^1H$–$^1H$ couplings. However, the heteronuclear spin decoupling RF amplitude may be mismatched with respect to the

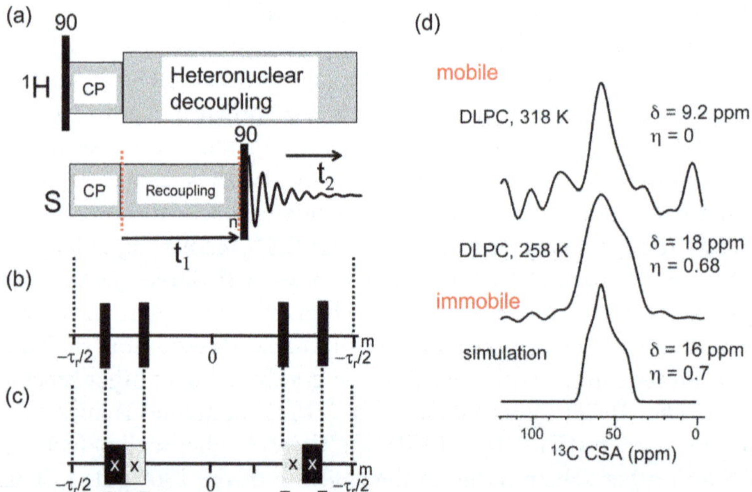


**Figure 9.10.** (a) General schematic of the 2D pulse scheme for CSA recoupling. The 90° pulse, before $t_2$ evolution, is used to get both cosine- and sine-modulated date sets for pure-absorption mode line shapes. The recoupling block could be a scheme with (b) 180° pulses, 2DCSA, or with (c) 360° pulses, SUPER. The scaling factor for the CSA recoupling will depend on the values of $T$. Both $n$ and $m$ may have integer values. $n$ determines how many rotor periods within which 180° pulses are applied. $m$ determines how many times the recoupling block is repeated in a given $t_1$ increment. (d) SUPER line shapes for the $C_\alpha$-resonance of the L40 residue of the M2 peptide of the influenza virus in DLPC membrane at two temperatures showing the mobility setting in and the variation from the rigid limit from simulation. The CSA, $\delta$, and asymmetry, $\eta$, values for each of the line shapes are also indicated in the figure. (b) Adapted with permission from Cady *et al* (2007), copyright (2007), American Chemical Society.

recoupling pulses by a factor of two only as against a factor of 3 in the case of 2DCSA. Figure 9.10(d) shows SUPER line shapes for the $C_\alpha$-resonance of the L40 residue of the M2 peptide of the influenza virus in DLPC phospholipid membrane at two temperatures showing the influence of mobility.

### 9.5.2 Symmetry-based schemes for CSA recoupling

Both *C* and *R* schemes with appropriate symmetry numbers can be used for CSA recoupling. These schemes also allow for efficient recoupling at higher MAS frequencies, although, the RF amplitude requirements may increase depending upon the values of *N* and *n*. Two pertinent schemes are ROCSA (Chan and Tycko 2003), based on *C* symmetry, and RNCSA (Hou *et al* 2012), based on *R* symmetry.

ROCSA uses, for simplicity, $Cn_n^1$ type of symmetry. The phase shifts between each *C*-block could be in multiples of 90°. POST-*C* elements are used to compensate for RF inhomogeneity. The entire pulse schematic along with the positioning of the *C* elements with a suggestion for the POST-*C* element is shown in figure 9.11(a). An advantage of the ROCSA scheme is the removal of the homonuclear $^1$H–$^1$H couplings to first order and yielding a CSA pattern very close to or even the same as the static CSA pattern. The scheme, however, requires an RF nutation frequency of 4.283 times the MAS frequency for the recoupling block and may be inefficient at high static magnetic fields, that is, in the presence of large CSA. Figure 9.11(b) shows an application of the ROCSA scheme (Su *et al* 2008).

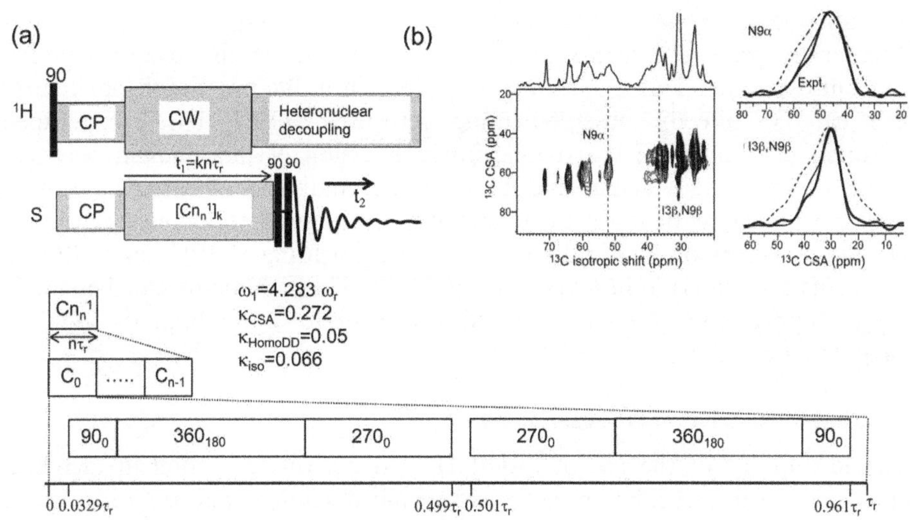


**Figure 9.11.** (a) Schematic of the ROCSA pulse scheme with the recoupling element sketched out in detail. $\kappa$ corresponds to the scaling factor of the respective interaction, CSA, homonuclear DD couplings, and isotropic chemical shift. (b) 2D-ROCSA spectrum and CSA line shapes of two residues of U–$^{15}$N,$^{13}$C–I3, N9 labeled pentratin in DMPS/DMPG phospholipid bilayers. The thick, thin, and dashed lines correspond to experimental, best fit, and numerical simulations respectively. The MAS frequency was 8 kHz. (b) Reprinted from Su *et al* (2008), copyright (2008), with permission from Elsevier.

The RNCSA scheme has a similar schematic as the SUPER, figure 9.10(a), except that the recoupling block is an *R*-symmetry scheme. Instead of the 90° pulse in figure 9.10(a), the schematic has the *R*-block flanked by two 90° pulses, as *R*-schemes are effective *z*-rotation methods. $\gamma$-encoded *R*-schemes are used and, hence, they are also effective in the presence of large CSA. The RF requirements for the recoupling block, again, could become prohibitively large at higher MAS frequencies.

## 9.6 Examples of studying dynamics in membrane proteins

A few representative examples are highlighted here to emphasise the role of some of the discussed methods in the study of dynamics in membrane-protein systems.

### 9.6.1 Cross-polarisation based filtering

Transmembrane segments in membrane proteins are often rigid, and $^{13}$C and $^{15}$N nuclei can be observed with cross-polarisation from $^{1}$H. The rigidity of the system is, however, lower than that seen in microcrystalline proteins (O'Brien *et al* 2020), and this can be readily observed as changes in both the resolution and sensitivity with variations in temperature. In the membrane system EmrE, it was shown that the sensitivity and resolution change significantly when temperature changes from below to above the lipid phase-transition value (Banigan *et al* 2014). As expected, the sensitivity is the highest below the phase-transition temperature, where the dynamics in the system is expected to be quenched. The best resolution was also observed below the phase-transition temperature, where the fluidity of the bilayer corresponded to ~5%.

Membrane protein domains that are exposed outside the bilayer can have significant dynamics. This was shown for a domain in the protein phospholamban which can be directly observed using refocused-INEPT based experiments (Andronesi *et al* 2005). The dynamic interconversion of this domain between a free R-state and a rigid T-state and its role in the inhibition of Sarco(endo)pasmic reticulum Ca-ATPase have been determined in detail using techniques that look at either the dynamics or the rigid part of the protein using INEPT- and CP-based experiments alternately (Gustavsson *et al* 2013). These domains can be simultaneously mapped using cleverly designed sequences to detect both the rigid and flexible domains (Gopinath *et al* 2017).

### 9.6.2 Rotational-echo based experiments

A classic example of the use of DIPSHIFT to determine motion in membrane proteins was in the elucidation of the side-chain dynamics that are critical in the transport of protons via the M2 channel (Hu *et al* 2010). REDOR between His-$N_{\delta}$1 and His-$C_{\alpha}$ was used to tease out the pH-dependent changes in the $\chi_2$ torsion angle, and DIPSHIFT was used to map out the pH-dependent changes in the order parameters of the imidazole side chain. The changes in the relative orientation of Trp41 and His37 as a function of pH, a critical step in the conductance of protons, was mapped directly using $^{13}$C–$^{19}$F REDOR experiments using $^{13}$C-His and a singly

$^{19}F$-labeled Trp residue, as shown in figure 9.12 (Williams *et al* 2013). DIPSHIFT was also used, along with other experiments, on Anabena sensory rhodopsin, a 7-TM protein, to show that this protein undergoes slow, small-amplitude domain motions, which gave insight into the changes induced due to the binding of a transducer (Good *et al* 2017). These experiments have also been used to extract backbone order parameters for Y2R, a G-protein coupled receptor and delineate differences in dynamics in different lipid environments (Schrottke *et al* 2017).

The TEDOR (Hing *et al* 1992) experiment (which uses the REDOR recoupling block to achieve heteronuclear polarisation transfer) has been exploited to determine inter-protomer contacts in oligomeric membrane proteins, aided by dynamic nuclear polarisation (Maciejko *et al* 2015). TEDOR not only reports on the residues that make contacts to stabilise the oligomers, but the TEDOR build-up curves also

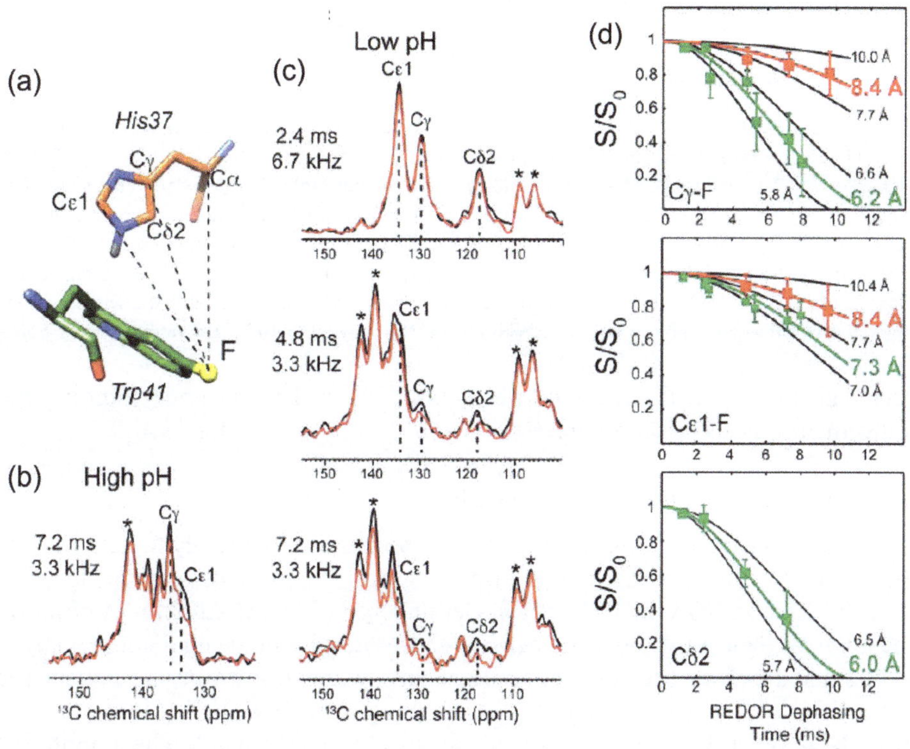


**Figure 9.12.** The changes in dynamics and orientation of a tryptophan side chain in M2 channel is deciphered by monitoring its interaction with the side chain nuclei in a histidine residue using $^{13}C$–$^{19}F$ REDOR experiment. (a) Structural-schematic of the interaction between Trp41 and His37. (b) $^{13}C$ CP spectra with (red) and without (black) REDOR dephasing pulses on $^{19}F$ channel at high pH. (c) $^{13}C$ CP spectra with (red) and without (black) $^{19}F$ REDOR pulses at low pH, at three different spinning frequencies. (d) REDOR dephasing profiles under high (red) and low (green) pH conditions for different residues indicate difference in the orientation and dynamics of these residues. Black curves in (d) indicate theoretical estimates for the dephasing curves at the indicated distances. Reprinted from Williams *et al* (2013), copyright (2013), with permission from Elsevier.

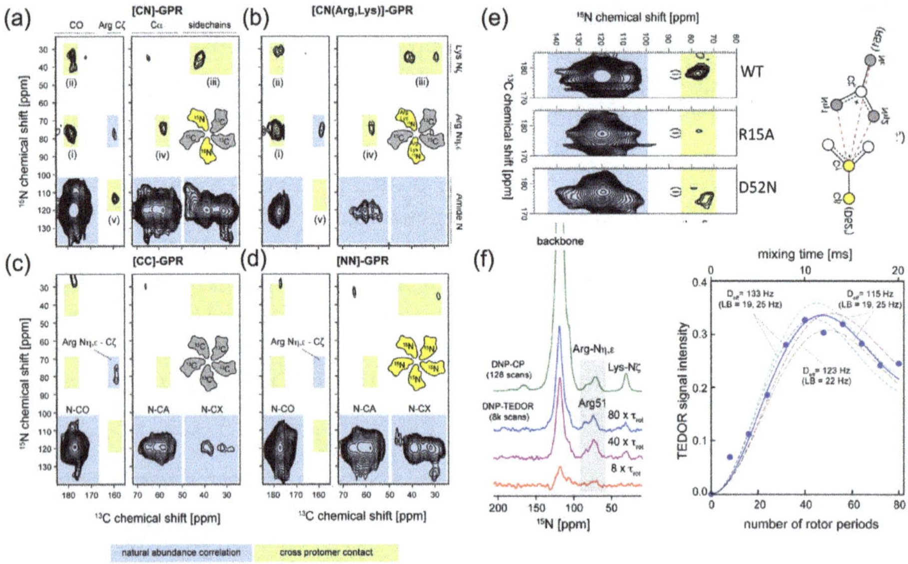

**Figure 9.13.** Use of TEDOR (which is a REDOR-based experiment) to determine the interprotomer contacts in a multimeric membrane protein. TEDOR spectra from (a) sample consisting of uniformly $^{13}$C-labeled protomers mixed with uniformly $^{15}$N-labeled protomers, and (b) uniformly $^{13}$C-labeled protomers mixed with Arg and Lys $^{15}$N-labeled protomers showing distinct $^{15}$N–$^{13}$C cross peaks that are absent in control samples of either (c) uniformly $^{13}$C- or (d) uniformly $^{15}$N-labeled samples. (e) These contacts are absent or substantially decreased in R51A and D52N mutants revealing their specific nature. (f) The presence of the salt bridge is also confirmed using a TEDOR build-up curve between R51 and D52, consistent with the structure shown alongside. Adapted with permission from Maciejko *et al* (2015), copyright (2015), American Chemical Society.

directly give accurate distances between these, which help in understanding their arrangements, as shown in figure 9.13.

### 9.6.3 Protein structures using SLF experiments

The *R* class of symmetry-based sequences has been used extensively to determine the topology of transmembrane helices in proteins. When proteins undergo fast rotational diffusion in bilayers, different nuclei undergo different amplitudes of motion depending on their position in the helix, with this amplitude being identical only for residues that are along the same face of the helix. When anisotropic parameters, such as the CSA or the heteronuclear DD coupling, are measured, these vary in a sinusoidal manner due to the helicity of transmembrane segments. The amplitude of the helix has a direct bearing on the tilt of this segment with respect to the bilayer normal, whereas the phase has a direct bearing on the azimuthal angle of the helix. This information, together with distance restraints from homonuclear correlation experiments, can be directly used to determine the structure and topology of transmembrane segments (Park *et al* 2010, Das *et al* 2012, Lu *et al* 2013), as shown in figure 9.14. Such experiments can also be done using DIPSHIFT and were used to determine structures of membrane proteins and map how these change in different conditions (Cady *et al* 2007, Mandala *et al* 2020).

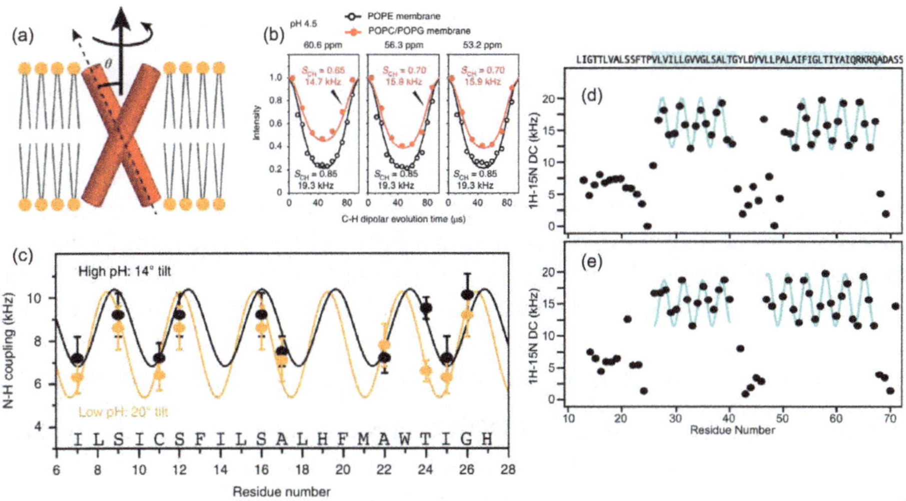


**Figure 9.14.** (a) Membrane proteins undergo rapid rotational motion along the bilayer normal, which gives each $^{15}N$–$^{1}H$ and $^{13}C$–$^{1}H$ bond vectors different amplitude of motion when the tilt angle $\theta$ is $>0°$, and consequently, different order parameters, as shown here for M2 channel. (b) These order parameters can be determined for each residue (three of which are shown), and under different conditions. (c) When the order parameters, or the effective DD couplings, are plotted against the residue number, they show characteristic oscillatory pattern in helical transmembrane domains. Further, the changes in the tilt angle as a function of pH can be directly mapped from these experiments. The oscillatory patterns are equivalent to those obtained using oriented NMR (d), but can show differences in helix length (e), most likely due to the different environments used under oriented and MAS conditions. Results in (d) and (e) are from the membrane protein MerFT. (a)–(c) are reprinted by permission from Springer Nature, (Mandala *et al* 2020), copyright (2020), and (d), (e) are adapted with permission from Das *et al* (2012), copyright (2012), American Chemical Society.

## 9.7 Schemes for higher MAS and stronger couplings

Most of the schemes, explained so far, deal with couplings between rare spins for distance and order-parameter elucidation. The main reason for bypassing $^{1}H$ here is the stronger couplings between $^{1}H$–$^{15}N$ or $^{1}H$–$^{13}C$. These lead to faster dephasing and inadequate oscillations or data points that can be collected in an experiment. It is, hence, necessary, to have schemes that scale down the coupling constants. Another important aspect is that most of the schemes seen up until now have been applied at MAS frequencies of not more than 10–20 kHz, except CPPI, REDOR, and PARS family of methods. PARS may involve high RF, on the order of 150 kHz or more, depending on the combination of the symmetry numbers. The same is also true for DIPSHIFT, REDOR, and CPPI, where at higher MAS frequencies, RF nutation frequencies scale up beyond reasonable numbers. Hence, another vital need is to have schemes that can be applied at higher MAS frequencies, >40 kHz, with RF nutation frequencies on the order of the MAS frequency or, in fact, lower. This gives the advantage of resolution along with sample and probe stability. Whilst looking for schemes to satisfy both the conditions, with regard to dynamics elucidation it is safe to rely on schemes like DIPSHIFT and REDOR as they have been proven to be robust, easy to implement and analyse, and shown to be effective for a wide range of nuclear spins.

### 9.7.1 REDOR-schemes with dipole–dipole coupling scaling

The original REDOR work (Gullion and Schaefer 1989) itself suggests scaling of the couplings by shifting the 180° pulses from their designated positions, called shifted-REDOR. A more rigorous analysis of such a protocol was made recently where the aim was to understand the scaling properties and removal of unwanted Fourier coefficients in the heteronuclear DD Hamiltonian for a clean dephasing curve (Jain *et al* 2017, 2019a). Figure 9.15(a) shows a typical REDOR dephasing curve against $\lambda$, the universal constant. Here $b_{IS} = -\frac{\mu_0}{4\pi}\frac{\gamma_I\gamma_S\hbar}{r_{IS}^3}$. $I$ is typically the observed spin and $S$ is normally the spin on which the REDOR pulses are applied. From the dephasing curve it is clear that the number of points, $n$, which a REDOR experiment can collect in the most important initial part of the build-up curve, is given by

$$n = \frac{1.7}{2b_{IS}\tau_r} \tag{9.1}$$

Figure 9.15(b) shows the $n$ value for a range of spin couplings as a function of the MAS frequency. It is clear, here, that stronger the couplings, lesser the number of points. This can lead to a mis-interpretation of the REDOR data (the same applies

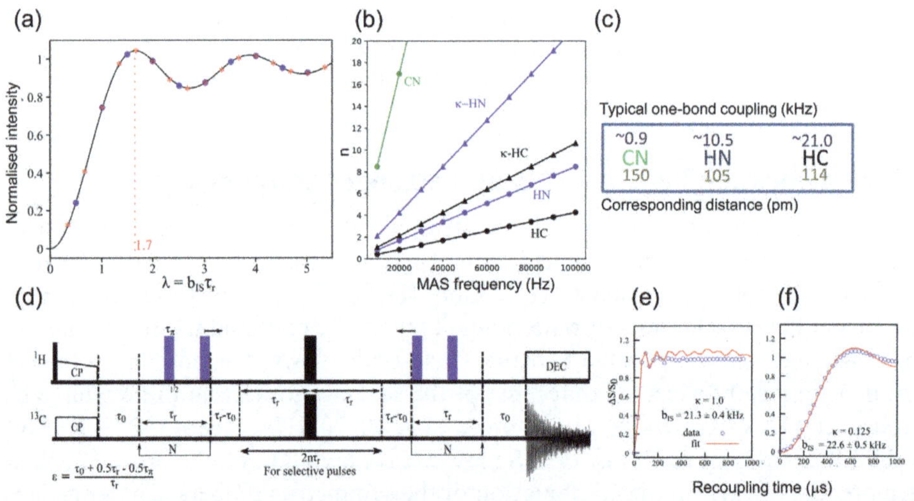


**Figure 9.15.** (a) Typical REDOR dephasing curve plotted against the so-called universal constant, $\lambda$. (b) Number of data points, $n$, that can be acquired in a REDOR experiment for various coupling constant regimes as a function of the MAS frequency with and without scaling, $\kappa$. (c) Typical heteronuclear DD coupling constants encountered in solid-state NMR. (d) General REDOR $mR_{\epsilon_1}^{\epsilon_2}$, scheme. REDOR dephasing curves for $C_\alpha$ $H\alpha$ pair on a sample of deuterated alanine obtained with (e) $mR_{0.25}^{0.75}$, $\kappa = 1$ and (f) $mR_{0.02}^{0.52}$, $\kappa = 0.125$. The MAS frequency was 62.5 kHz with the nutation frequency of the recoupling pulses being around 125 kHz. The advantage of scaling is quite evident in (f) with better oscillations due to larger number of data points. (d)–(f) Adapted from Jain *et al* (2019a), with the permission of AIP Publishing.

to the DIPSHIFT curves) and wrong distances. Ideally, it is desired to have as high $n$ as possible for a very reliable fit, this means, either higher MAS frequencies (lower $\tau_r$) or lower couplings. In the case of stronger couplings, the situation can be made better by introducing a scaling factor, $\kappa$, for the recoupled interaction, so that

$$n = \frac{1.7}{\kappa(2b_{IS}\tau_r)} \tag{9.2}$$

A ruler for the relevant DD couplings is shown in figure 9.15(c).

Considering the placement of the recoupling pulses with respect to the central 180° pulse, REDOR may have symmetrical and asymmetrical versions; here only the symmetrical version of REDOR application is considered. In both cases there are two kinds of scaling: one coming from the placement of the recoupling 180° pulses, and the other from the finite width of the respective pulses, denoted as $\phi = \frac{2\tau_p}{\tau_r}$. Having REDOR implementation with two 180° pulses in a rotor period, the classical REDOR with symmetrical implementation, may be notated as $mR_{\epsilon_1}^{\epsilon_2}$, where $\epsilon = \frac{1}{\tau_r}$. Here, $m$ refers to mirror-symmetric version and $\epsilon_1$ and $\epsilon_2$ determine the position of the recoupling pulses with respect to the central 180° pulse.

The best REDOR performance (suppressing the second Fourier component of the DD Hamiltonian and only retaining the first component—this way, the oscillations are neatly preserved as there is no mixing of components (Jain *et al* 2019a)) is obtained with $\epsilon_1 - \epsilon_2 = 0.5$ as in the original REDOR implementation. However, the best scaling is obtained when both the pulses are shifted, simultaneously, keeping the separation always equal of $\frac{\tau_r}{2}$, unlike in the case of S-REDOR (Gullion and Schaefer 1989).

A schematic of the revisited REDOR scheme is shown in figure 9.15(d). Although this is for $^{13}C$–$^{1}H$ coupling measurements, it could be modified to three channels or for any other nuclear spin. Here, the pulses coloured in blue are shifted in tandem so that the separation between them is always $\frac{\tau_r}{2}$. The scheme also allows for ample space to place any frequency-selective pulses, if need arises. Figures 9.15(e) and (f) show the results from the two-pulse shifted REDOR on a sample of deuterated alanine monitoring $C_\alpha$–$H_\alpha$ couplings. The inadequacy of the standard REDOR, $\kappa = 1$, is clear with the experiment only resulting in three data points, whilst the scheme with $\kappa = 0.125$ gives much better dephasing curve, and also an excellent fit. This is also one of the attempts of pushing REDOR to high MAS, here 62.5 kHz, with moderate RF nutation frequencies of around 120–130 kHz. A similar attempt was done in case of REDOR and CPPI with low-enough RF nutation frequencies at MAS frequencies of 60–100 kHz (Asami and Reif 2017).

### 9.7.2 Equivalence of DIPSHIIT and REDOR pulse schemes

A careful look at figure 9.15(a) brings in an interesting aspect. Figure 9.16(a) is an adaptation of the REDOR scheme shown in figure 9.15(a). It is clear that the pulse scheme can be implemented in two ways: one by varying $n$, for a fixed $\epsilon$, and the other by varying $\epsilon$, for a fixed $n$. The former is clearly the REDOR implementation

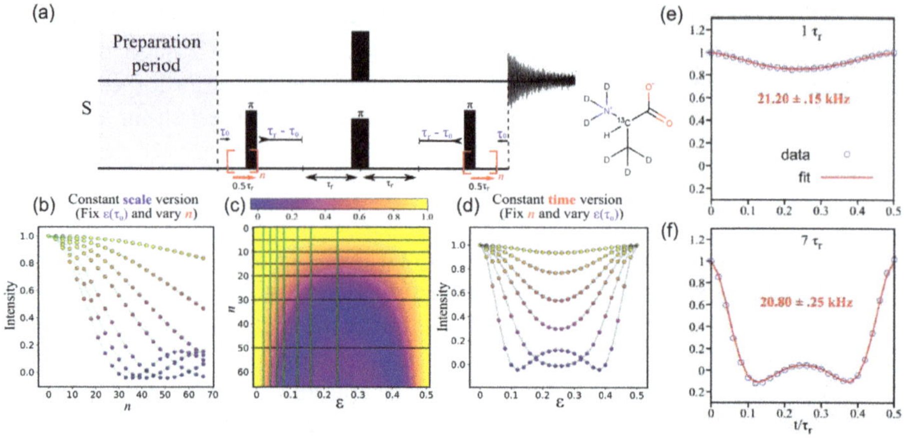


**Figure 9.16.** (a) Pulse schematic for the measurement of heteronuclear DD couplings. Keeping $\epsilon$ constant and varying $n$, REDOR curves result and the other way, DIPSHIFT curves result. (b) REDOR dephasing curves for various scaling factors. (c) Two-dimensional contour plot depicting the intensity of the recoupled signal as a function of $\epsilon$ and $n$. (d) DIPSHIFT curves for various values of $n$, the larger the $n$, the better is the dephasing, and more reliable the estimate of the geometry parameters. DIPSHIFT curves from (a) for the CH pair in deuterated alanine. The curves are shown for two $n$ values indicating amplification in the observed curves. The MAS frequency was 62.5 kHz and the recoupling pulses were with a nutation frequency of 200 kHz on $^1$H and 100 kHz on $^{13}$C. Adapted from Jain *et al* (2019b), copyright (2019), with permission from Elsevier.

and the latter the DIPSHIFT implementation. Hence, both DIPSHIFT and REDOR can be treated as equivalent and may be visualised within a single pulse scheme (Jain *et al* 2019b). It is only a certain experimental parameter that is varied in either case, the number of repetitions of the recoupling block in REDOR, and the position of the recoupling 180° pulse in DIPSHIFT. In fact, theoretically, both are identical as both have exactly the same effective Hamiltonian.

Figure 9.16(b) shows simulated REDOR dephasing curves with the scheme in (a). Figure 9.16(c) shows a simulated two-dimensional contour plot showing the intensity of the recoupled signal as a function of the two relevant experimental parameters, $n$ and $\epsilon$. Figure 9.16(d) shows simulated DIPSHIFT curves for the scheme in (a). Both the REDOR and DIPSHIFT curves can be obtained from figure 9.16(b) by taking the slices corresponding to constant $\epsilon$ or constant $n$. $\epsilon$ in this case is related to $\tau_0$, figure 9.15(d).

Figures 9.16(e) and (f) show DIPSHIFT curves with the scheme, figure 9.16(a). The CH curves are from a sample of deuterated alanine. It is clear that the DIPSHIFT curves give a very reliable estimate of the coupling-related parameters by repeating $n$ several times. The repetition amplifies the DIPSHIFT signal.

The equivalence between REDOR and DIPSHIFT brings in certain other features as well. DIPSHIFT could become very well the experiment of choice for the measurement of both weak and strong DD couplings. It is less time consuming as the control experiment is in-built unlike REDOR. Also, DIPSHIFT may be done to get only half of the so-called ‘W’ curve as that has all the relevant information for the extraction of couplings. Such a curve may be obtained with more precision with high repetitions of $n$. The limiting factors are the $T_2'$ relaxation of the observed resonance, very large number of pulses/repetitions required for the measurement of weak couplings, and high nutation frequencies for the pulses as MAS frequencies become higher. It is important to consider losses in the signal-to-noise ratio for DIPSHIFT due to $T_2'$ than REDOR, and, hence, REDOR may be better for the measurement of weak couplings.

### 9.7.3 DIPSHIFT and REDOR at higher MAS frequencies

How can REDOR and DIPSHIFT be redesigned so that they can be effective with low RF nutation frequencies at MAS frequencies >40 kHz? The first-order average Hamiltonian for the REDOR scheme may be written as:

$$\mathcal{H}^{(1)}_{\mathrm{DD,REDOR}} = \sin(\epsilon\omega_r)\left\{\frac{\cos\frac{\pi\phi}{2}}{1-\phi^2}\right\}[2\sqrt{2}\,\delta_D\cos(\gamma)\sin(2\beta)2I_zS_z] \tag{9.3}$$

where $\delta_D = 2b_{IS}$ and $\beta$, $\gamma$ being the Euler angles of transformation from the principal-axis system to the laboratory frame. The term in curly braces denotes the scaling of the DD interactions with REDOR due to the finite duration of the recoupling pulses. This has been exploited earlier with two 180° pulses occupying one rotor period (Jaroniec *et al* 2000). Here, $\phi = 1$, leading to a scaling factor of 0.78. However, in the case of the classical REDOR, $mR^{0.75}_{0.25}$, the second pulse can be shifted to $1.75\tau_r$ and the whole scheme can be shifted further by $0.5\tau_r$ without affecting the effective Hamiltonian. What is accomplished here is having two 180° pulses in three rotor periods which allows RF amplitudes as low as $\frac{1}{3}$ of $\omega_r$ to be used. The scaling factor goes to zero as $\phi$ tends to 3. However, it is possible to have practically $\phi$ between 1.4 and 2 leading to a scaling factor of 0.62 to 0.33 and minimum RF nutation frequencies of $0.72\omega_r$ to $0.5\omega_r$. For the regular REDOR, the minimum RF nutation frequency is $\omega_r$.

The scheme, with such a deferral of echo in practice, is called deferred rotational-echo double resonance, DEDOR (Jain *et al* 2020). The schematic of DEDOR is shown in figure 9.17(a). Substantially lower RF nutation frequencies can be then used with this practice of deferring the echo to later rotor periods.

The experimental verification of the DEDOR scheme is shown in figure 9.17(b) where CN curves are shown for a sample of glycine. In figure 9.17(b) panels (a)–(e) show the dephasing curves for various scaling factors, $\phi$, leading to RF nutation frequencies of 62.5 to 31.25 kHz for a MAS frequency of 62.5 kHz. Similar results have been obtained for DIPSHIFT and other schemes making use of echo pulses, for instance, TEDOR (Hing *et al* 1992, Jain *et al* 2020). The panel (f) shows this scaling, $\kappa_\phi$, as a function of $\phi$, making it clear the region of practical $\phi$ so that $\kappa \neq 0$ and where the RF nutation frequencies are lower than the MAS frequencies.

The criteria for DIPSHIFT and REDOR kind of schemes for efficient recoupling are the following:

- Scale down the DD couplings for the measurement of strong couplings.
- Arbitrary scaling factor possibilities without increasing RF requirements.

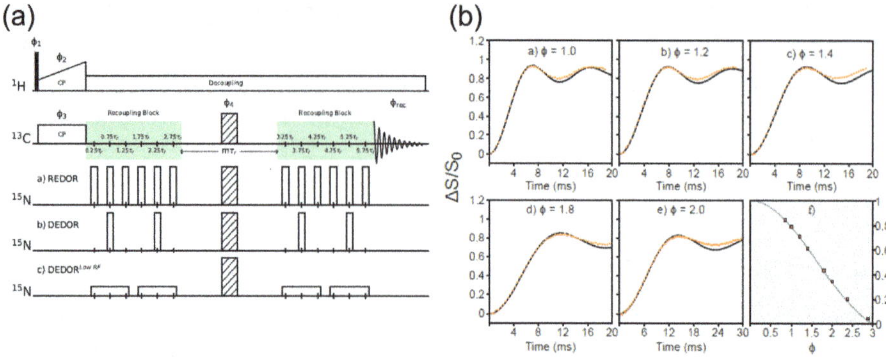

**Figure 9.17.** (a) DEDOR pulse scheme showing the standard REDOR and DEDOR with two recoupling 180° pulses in three rotor periods, and a similar one but with the recoupling 180° pulses having a duration of more than a rotor period leading to lower RF nutation frequencies. (b) $^{13}C$–$^{15}N$ DEDOR dephasing curves for a sample of 2-$^{13}C$–$^{15}N$ glycine for various $\phi$ with (a)–(e) having the recoupling pulses at an RF nutation frequency of 62.5, 52.08, 44.64, 34.72, and 31.25 kHz, respectively, for a MAS frequency of 62.5 kHz. Panel (f) compares the theoretical scaling factor, $\kappa$, as a function of $\phi$ (solid line) and the experimentally calculated scaling factor (points). This indicates the working region of DEDOR with regard to practical scaling factors. Adapted with permission from Jain *et al* (2020), copyright (2009), American Chemical Society.

- Non-mixing of the first- and second-Fourier components of the DD Hamiltonian.
- Work with non-prohibitive RF amplitudes under all possible MAS frequencies.
- Tolerant to experimental issues, such as, finite pulse widths, CSA, and RF inhomogeneity.
- Feasibility of placing selective pulse blocks for further spin manipulation.
- Feasibility of placing blocks, such as, PMLG (Madhu 2009, Mote *et al* 2016), for the removal of homonuclear DD interactions involving $^1H$ spins.

DEDOR satisfies these to a large extent.

## 9.8 Conclusions

This review has outlined various methods, both under MAS and static conditions, that may be used for the elucidation of dynamics in membrane-protein systems, with some representative examples. The methods essentially target heteronuclear DD couplings and CSA interaction. New developments are also hinted at although they have been tested only on model compounds. They are the schemes ideally suited for high MAS frequencies with the possibility of applying recoupling pulses at low RF amplitudes. This will be advantageous for membrane proteins, the advantages being, higher resolution, less sample, and heat-stability of the samples.

## Acknowledgements

We thank the National Facility for High-Field NMR, TIFR Hyderabad, where some of the experiments were carried out. Some of the work mentioned in this review was funded by intramural funds at TIFR-Hyderabad from the Department of Atomic Energy (DAE), India. We acknowledge Dr G Rajalakshmi for a careful reading of the manuscript.

## References

Andronesi O C, Becker S, Seidel K, Heise H, Young H S and Baldus M 2005 Determination of membrane protein structure and dynamics by magic-angle-spinning solid-state NMR spectroscopy *J. Am. Chem. Soc.* **127** 12965–74

Asami S and Reif B 2017 Comparative study of REDOR and CPPI derived order parameters by $^{1}$H-detected MAS NMR and MD simulations *J. Phys. Chem.* B **121** 8719–30

Banigan J R, Gayen A and Traaseth N J 2014 Correlating lipid bilayer fluidity with sensitivity and resolution of polytopic membrane protein spectra by solid-state NMR spectroscopy *Biochim. Biophys. Acta Biomembr.* **1848** 334–41

Basting D, Lehner I, Lorch M and Glaubitz C 2006 Investigating transport proteins by solid state NMR *Naunyn. Schmiedebergs Arch. Pharmacol.* **372** 451–64

Cady S D, Goodman C, Tatko C D, DeGrado W F and Hong M 2007 Determining the orientation of uniaxially rotating membrane proteins using unoriented samples: a $^{2}$H, $^{13}$C, and $^{15}$N solid-state NMR investigation of the dynamics and orientation of a transmembrane helical bundle *J. Am. Chem. Soc.* **129** 5719–29

Chan J C C and Tycko R 2003 Recoupling of chemical shift anisotropies in solid-state NMR under high-speed magic-angle spinning and in uniformly $^{13}$C-labeled systems *J. Chem. Phys.* **118** 8378–89

Chandra B, Bhowmik D, Maity B K, Mote K R, Dhara D, Venkatramani R, Maiti S and Madhu P K 2017 Major reaction coordinates linking transient amyloid-$\beta$ oligomers to fibrils measured at atomic level *Biophys. J.* **113** 805–16

Cheng Y 2018 Membrane protein structural biology in the era of single particle cryo-em *Curr. Opin. Struct. Biol.* **52** 58–63

Chevelkov V, Fink U and Reif B 2009 Accurate determination of order parameters from $^{1}$H, $^{15}$N dipolar couplings in MAS solid-state NMR experiments *J. Am. Chem. Soc.* **131** 14018–22

Cobo M F, Achilles A, Reichert D, Deazevedo E R and Saalwächter K 2012 Recoupled separated-local-field experiments and applications to study intermediate-regime molecular motions *J. Magn. Reson.* **221** 85–96

Das B B, Nothnagel H J, Lu G J, Son W S, Tian Y, Marassi F M and Opella S J 2012 Structure determination of a membrane protein in proteoliposomes *J. Am. Chem. Soc.* **134** 2047–56

de Azevedo E R, Hu W G, Bonagamba T J and Schmidt-Rohr K 1999 Centerband-only detection of exchange: efficient analysis of dynamics in solids by NMR *J. Am. Chem. Soc.* **121** 8411–2

de Azevedo E R, Saalwachter K, Pascui O, de Souza A A, Bonagamba T J and Reichert D 2008 Intermediate motions as studied by solid-state separated local field NMR experiments *J. Chem. Phys.* **128** 104505

De Paëpe G 2012 Dipolar recoupling in magic angle spinning solid-state nuclear magnetic resonance *Annu. Rev. Phys. Chem.* **63** 661–84

Dusold S and Sebald A 2000 Dipolar recoupling under magic-angle spinning conditions *Annu. Rep. NMR Spectrosc.* **41** 185–264

Dvinskikh S V and Sandström D 2005 Frequency offset refocused PISEMA-type sequences *J. Magn. Reson.* **175** 163–9

Dvinskikh S V, Zimmermann H, Maliniak A and Sandström D 2003 Separated local field spectroscopy of columnar and nematic liquid crystals *J. Magn. Reson.* **163** 46–55

Esteban-Martín S, Strandberg E, Fuertes G, Ulrich A S and Salgado J 2009 Influence of whole-body dynamics on $^{15}$N PISEMA NMR spectra of membrane proteins: a theoretical analysis *Biophys. J.* **96** 3233–41

Forrest L R, Krämer R and Ziegler C 2011 The structural basis of secondary active transport mechanisms *Biochim. Biophys. Acta* **1807** 167–88

Gayen A, Banigan J R and Traaseth N J 2013 Ligand-induced conformational changes of the multidrug resistance transporter EmrE probed by oriented solid-state NMR spectroscopy *Angew. Chem. Int. Ed.* **52** 10321–4

Gehman J D, Separovic F, Lu K and Mehta A K 2007 Boltzmann statistics rotational-echo double resonance-analysis *J. Phys. Chem.* B **111** 7802–11

Good D, Pham C, Jagas J, Lewandowski J R and Ladizhansky V 2017 Solid-state NMR provides evidence for small-amplitude slow domain motions in a multi-spanning transmembrane $\alpha$-helical protein *J. Am. Chem. Soc.* **139** 9246–58

Gopinath T, Nelson S E and Veglia G 2017 $^1$H-detected MAS solid-state NMR experiments enable the simultaneous mapping of rigid and dynamic domains of membrane proteins *J. Magn. Reson.* **285** 101–7

Gopinath T and Veglia G 2009 Sensitivity enhancement in static solid-state NMR experiments via single and multiple-quantum dipolar coherences *J. Am. Chem. Soc.* **131** 5754–6

Gullion T 1998 Introduction to rotational-echo, double-resonance NMR *Concepts Magn. Reson.* **10** 277–89

Gullion T and Schaefer J 1989 Rotational-echo double-resonance NMR *J. Magn. Reson.* **81** 196–200

Gustavsson M, Verardi R, Mullen D G, Mote K R, Traaseth N J, Gopinath T and Veglia G 2013 Allosteric regulation of SERCA by phosphorylation-mediated conformational shift of phospholamban *Proc. Natl Acad. Sci.* **110** 17338–43

Hester R K, Ackerman J L, Neff B L and Waugh J S 1976 Separated local field spectra in NMR: determination of structure of solids *Phys. Rev. Lett.* **36** 1081–3

Hing A W, Vega S and Schaefer J 1992 Transferred-echo double-resonance NMR *J. Magn. Reson.* **96** 205–9

Holt A and Killian J A 2010 Orientation and dynamics of transmembrane peptides: the power of simple models *Eur. Biophys. J.* **39** 609–21

Hong M, Gross J D, Rienstra C M, Griffin R G, Kumashiro K K and Schmidt-Rohr K 1997 Coupling amplification in 2D MAS NMR and its application to torsion angle determination in peptides *J. Magn. Reson.* **129** 85–92

Hong M, Yao X, Jakes K and Huster D 2002 Investigation of molecular motions by Lee–Goldburg cross-polarization NMR spectroscopy *J. Phys. Chem.* B **106** 7355–64

Hou G, Byeon I-J, Ahn J, Gronenborn A M and Polenova T 2012 Recoupling of chemical shift anisotropy by r-symmetry sequences in magic angle spinning NMR spectroscopy *J. Chem. Phys.* **137** 134201

Hou G, Lu X, Vega A J and Polenova T 2014 Accurate measurement of heteronuclear dipolar couplings by phase-alternating r-symmetry (PARS) sequences in magic angle spinning NMR spectroscopy *J. Chem. Phys.* **141** 104202

Hu F, Luo W and Hong M 2010 Mechanisms of proton conduction and gating in Influenza M2 proton channels from solid-state NMR *Science* **330** 505–8

Huster D, Xiao L and Hong M 2001 Solid-state nmr investigation of the dynamics of the soluble and membrane-bound colicin ia channel-forming domain† *Biochemistry* **40** 7662–74

Ivanir-Dabora H, Nimerovsky E, Madhu P K and Goldbourt A 2015 Site-resolved backbone and side-chain intermediate dynamics in a carbohydrate-binding module protein studied by magic-angle spinning NMR spectroscopy *Chemistry* **21** 10778–85

Jain M G, Mote K R, Hellwagner J, Rajalakshmi G, Ernst M, Madhu P K and Agarwal V 2019a Measuring strong one-bond dipolar couplings using REDOR in magic-angle spinning solid-state NMR *J. Chem. Phys.* **150** 134201

Jain M G, Rajalakshmi G, Agarwal V, Madhu P K and Mote K R 2019b On the direct relation between REDOR and DIPSHIFT experiments in solid-state NMR *J. Magn. Reson.* **308** 106563

Jain M G, Rajalakshmi G, Equbal A, Mote K R, Agarwal V and Madhu P K 2017 Sine-squared shifted pulses for recoupling interactions in solid-state NMR *J. Chem. Phys.* **146** 244201

Jain M G, Rajalakshmi G, Madhu P K, Agarwal V and Mote K R 2020 Overcoming prohibitively large radiofrequency demands for the measurement of internuclear distances with solid-state NMR under fast magic-angle spinning *J. Phys. Chem.* B **124** 1444–51

Jardetzky O 1966 Simple allosteric model for membrane pumps *Nature* **211** 969–70

Jaroniec C P, Tounge B A, Herzfeld J and Griffin R G 2001 Frequency selective heteronuclear dipolar recoupling in rotating solids: accurate $^{13}$C–$^{15}$N distance measurements in uniformly $^{13}$C, $^{15}$N-labeled peptides *J. Am. Chem. Soc.* **123** 3507–19

Jaroniec C P, Tounge B A, Rienstra C M, Herzfeld J and Griffin R G 2000 Recoupling of heteronuclear dipolar interactions with rotational-echo double-resonance at high magic-angle spinning frequencies *J. Magn. Reson.* **146** 132–9

Jia L, Liang S, Sackett K, Xie L, Ghosh U and Weliky D P 2015 REDOR solid-state NMR as a probe of the membrane locations of membrane-associated peptides and proteins *J. Magn. Reson.* **253** 154–65

Kobilka B K 2007 G protein coupled receptor structure and activation *Biochim. Biophys. Acta (BBA)-Biomembr.* **1768** 794–807

Krushelnitsky A, Reichert D and Saalwachter K 2013 Solid-state nmr approaches to internal dynamics of proteins: from picoseconds to microseconds and seconds *Acc. Chem. Res.* **46** 2028–36

Lee M and Goldburg W I 1965 Nuclear magnetic resonance line narrowing by a rotating RF field *Phys. Rev.* A **140** 1261–71

Levitt M H 2002 *Symmetry-based Pulse Sequences in Magic-angle Spinning Solid-state NMR* (New York: Wiley) pp 165–96

Levitt M H, Madhu P K and Hughes C E 2002 Cogwheel phase cycling *J. Magn. Reson.* **155** 300–6

Liu S-F, Mao J-D and Schmidt-Rohr K 2002 A robust technique for two-dimensional separation of undistorted chemical-shift anisotropy powder patterns in magic-angle-spinning NMR *J. Magn. Reson.* **155** 15–28

Lu G J, Tian Y, Vora N, Marassi F M and Opella S J 2013 The structure of the Mercury transporter MerF in phospholipid bilayers: a large conformational rearrangement results from N-terminal truncation *J. Am. Chem. Soc.* **135** 9299–302

Lu X, Zhang H, Lu M, Vega A J, Hou G and Polenova T 2016 Improving dipolar recoupling for site-specific structural and dynamics studies in biosolids NMR: windowed RN-symmetry sequences *Phys. Chem. Chem. Phys.* **18** 4035–44

Maciejko J, Mehler M, Kaur J, Lieblein T, Morgner N, Ouari O, Tordo P, Becker-Baldus J and Glaubitz C 2015 Visualizing specific cross-protomer interactions in the homo-oligomeric membrane protein proteorhodopsin by dynamic-nuclear-polarization-enhanced solid-state NMR *J. Am. Chem. Soc.* **137** 9032–43

Madhu P 2009 High-resolution solid-state NMR spectroscopy of protons with homonuclear dipolar decoupling schemes under magic-angle spinning *Solid State Nucl. Magn. Reson.* **35** 2–11

Mandala V S, Loftis A R, Shcherbakov A A, Pentelute B L and Hong M 2020 Atomic structures of closed and open influenza B M2 proton channel reveal the conduction mechanism *Nat. Struc. Mol. Biol.* **27** 160–7

Marasinghe P A B, Buffy J J, Schmidt-Rohr K and Hong M 2005 Membrane curvature change induced by an antimicrobial peptide detected by $^{31}P$ exchange NMR *J. Phys. Chem.* B **109** 22036–44

Marion D, Gauto D F, Ayala I, Giandoreggio-Barranco K and Schanda P 2018 Microsecond protein dynamics from combined Bloch–McConnell and near-rotary-resonance $R_{1\rho}$ relaxation-dispersion MAS NMR *Chem. Phys. Chem.* **20** 276–84

Moraes I, Evans G, Sanchez-Weatherby J, Newstead S and Stewart P D S 2014 Membrane protein structure determination—the next generation *Biochim. Biophys. Acta-Biomembr.* **1838** 78–87

Mote K R, Agarwal V and Madhu P K 2016 Five decades of homonuclear dipolar decoupling in solid-state NMR: status and outlook *Prog. Nucl. Magn. Reson. Spectrosc.* **97** 1–39

Mueller K 1995 Analytic solutions for the time evolution of dipolar-dephasing NMR signals *J. Magn. Reson. Ser.* A **113** 81–93

Munowitz M G, Griffin R G, Bodenhausen G and Huang T H 1981 Two-dimensional rotational spin-echo nuclear magnetic resonance in solids: correlation of chemical shift and dipolar interactions *J. Am. Chem. Soc.* **103** 2529–33

Nevzorov A A 2011 Orientational and motional narrowing of solid-state NMR lineshapes of uniaxially aligned membrane proteins *J. Phys. Chem.* B **115** 15406–14

O'Brien E S, Fuglestad B, Lessen H J, Stetz M A, Lin D W, Marques B S, Gupta K, Fleming K G and Wand J J 2020 Membrane proteins have distinct fast internal motion and residual conformational entropy *Angew. Chem.* **59** 11108–14

Orr R M and Duer M J 2006 Recoupling of chemical-shift anisotropy powder patterns in MAS NMR *J. Magn. Reson.* **181** 1–8

Overington J P, Al-Lazikani B and Hopkins A L 2006 How many drug targets are there? *Nat. Rev. Drug Discov.* **5** 993–6

Page R C, Kim S and Cross T A 2008 Transmembrane helix uniformity examined by spectral mapping of torsion angles *Structure* **16** 787–97

Park S H, Das B B, De Angelis A A, Scrima M and Opella S J 2010 Mechanically, magnetically, and 'rotationally aligned' membrane proteins in phospholipid bilayers give equivalent angular constraints for NMR structure determination *J. Phys. Chem.* B **114** 13995–4003

Rothwell W P and Waugh J S 1981 Transverse relaxation of dipolar coupled spin systems under rf irradiation: detecting motions in solids *J. Chem. Phys.* **74** 2721–32

Schanda P 2019 Relaxing with liquids and solids–a perspective on biomolecular dynamics *J. Magn. Reson.* **306** 180–6

Schmidt-Rohr K, Fritzsching K J, Liao S Y and Hong M 2012 Spectral editing of two-dimensional magic-angle-spinning solid-state NMR spectra for protein resonance assignment and structure determination *J. Biomol. NMR* **54** 343–53

Schmidt-Rohr K and Mao J-D 2002 Efficient CH-group selection and identification in $^{13}$C solid-state NMR by dipolar DEPT and $^{1}$H chemical-shift filtering *J. Am. Chem. Soc.* **124** 13938–48

Schmidt-Rohr K and Spiess H 2012 *Multidimensional Solid-State NMR and Polymers* (Amsterdam: Elsevier)

Schrottke S, Kaiser A, Vortmeier G, Els-Heindl S, Worm D, Bosse M, Schmidt P, Scheidt H A, Beck-Sickinger A G and Huster D 2017 Expression, functional characterization, and solid-state NMR investigation of the g protein-coupled GHS receptor in bilayer membranes *Sci. Rep.* **7** 46128

Strandberg E, Ozdirekcan S, Rijkers D T S, van der Wel P C a, Koeppe R E, Liskamp R M J and Killian J A 2004 Tilt angles of transmembrane model peptides in oriented and non-oriented lipid bilayers as determined by $^{2}$H solid-state NMR *Biophys. J.* **86** 3709–21

Su Y, Mani R, Doherty T, Waring A J and Hong M 2008 Reversible sheet-turn conformational change of a cell-penetrating peptide in lipid bilayers studied by solid-state NMR *J. Mol. Biol.* **381** 1133–44

Tycko R, Dabbagh G and Mirau P A 1989 Determination of chemical-shift-anisotropy lineshapes in a two-dimensional magic-angle-spinning NMR experiment *J. Magn. Reson.* **85** 265–74

Vallurupalli P, Bouvignies G and Kay L E 2012 Studying 'invisible' excited protein states in slow exchange with a major state conformation *J. Am. Chem. Soc.* **134** 8148–61

van der Wel P C A, Strandberg E, Killian J A and Koeppe R E 2002 Geometry and intrinsic tilt of a tryptophan-anchored transmembrane alpha-helix determined by $^{2}$H NMR *Biophys. J.* **83** 1479–88

van Rossum B J, de Groot C P, Ladizhansky V, Vega S and de Groot H J M 2000 A method for measuring heteronuclear $^{1}$H–$^{13}$C distances in high speed MAS NMR *J. Am. Chem. Soc.* **122** 3465–72

van Zijl P C M and Yadav N N 2011 Chemical exchange saturation transfer (CEST): what is in a name and what isn't? *Magn. Reson. Med.* **65** 927–48

Vega A, Poupko R and Luz Z 1989 Dynamic $^{2}$H NMR by quadrupole echo train in the fast-exchange limit. Ring inversion of cyclohexane-$d_{12}$ in liquid crystalline solvents *J. Magn. Reson.* **83** 111–27

Vostrikov V V, Daily A E, Greathouse D V and Koeppe R E 2010a Charged or aromatic anchor residue dependence of transmembrane peptide tilt *J. Biol. Chem.* **285** 31723–30

Vostrikov V V, Hall B A, Greathouse D V, Koeppe R E and Sansom M S P 2010b Changes in transmembrane helix alignment by arginine residues revealed by solid-state NMR experiments and coarse-grained MD simulations *J. Am. Chem. Soc.* **132** 5803–11

Vostrikov V V, Grant C V, Opella S J and Koeppe R E 2011 On the combined analysis of $^{2}$H and $^{15}$N/$^{1}$H solid-state NMR data for determination of transmembrane peptide orientation and dynamics *Biophys. J.* **101** 2939–47

Williams J K, Zhang Y, Schmidt-Rohr K and Hong M 2013 pH-dependent conformation, dynamics, and aromatic interaction of the gating tryptophan residue of the influenza M2 proton channel from solid-state NMR *Biophys. J.* **104** 1698–708

Wu C, Ramamoorthy A and Opella S 1994 High-resolution heteronuclear dipolar solid-state NMR spectroscopy *J. Magn. Reson. Ser.* A **109** 270–2

Wu X and Zilm 1993 Complete spectral editing in CPMAS NMR *J. Magn. Reson. Ser.* A **102** 205–13

Yamamoto K, Ermakov V L, Lee D K and Ramamoorthy A 2005 PITANSEMA-MAS, a solid-state NMR method to measure heteronuclear dipolar couplings under MAS *Chem. Phys. Lett.* **408** 118–22

Yang H, Staveness D, Ryckbosch S M, Axtman A D, Loy B A, Barnes A B, Pande V S, Schaefer J, Wender P A and Cegelski L 2018 REDOR NMR reveals multiple conformers for a protein kinase C ligand in a membrane environment *ACS Cent. Sci.* **4** 89–96
Yao X L, Conticello V P and Hong M 2004 Investigation of the dynamics of an elastin-mimetic polypeptide using solid-state NMR *Magn. Reson. Chem.* **42** 267–25

# Chapter 10

# Solid-state NMR studies of peripherally membrane-associated proteins: dealing with dynamics, disorder and dilute conditions

**Patrick C A van der Wel**

## Abstract

Proteins that associate with cellular membranes via an interaction with the membrane surface are known as peripheral, amphitropic or conditional membrane proteins. These interactions depend on the membrane lipid composition, facilitating lipid-dependent targeting to specific (sub)cellular membranes. Multidomain proteins may contain one or more specialized lipid binding domains. Despite their important roles, limited information is available on their molecular structure in the membrane-bound state. Solid-state NMR offers the potential to reveal the missing information but faces some practical challenges. This chapter reviews applications of solid-state NMR to three peripheral membrane proteins: cardiolipin-bound cytochrome c, phosphatidylinositol-binding Pleckstrin homology domain and membrane-associated myelin basic protein. We describe solutions to common challenges, such as the dynamics of the membrane-bound protein and the preparation of suitable samples with biological relevance. Modern NMR hardware, molecular biology and sample preparation methods pave the way for effective studies of many interesting peripheral membrane proteins by solid-state NMR.

## 10.1 Peripheral and conditional membrane proteins

### 10.1.1 What are peripheral membrane proteins?

Proteins are often classified as membrane-associated or 'soluble' proteins, but this classification can be imperfect or ambiguous. A notable class of proteins is able to bind lipid membranes via a surface-mediated interaction and does so in a lipid- or condition-dependent fashion. What this indicates is that these proteins are in principle soluble but attain a membrane-bound state when exposed to specific kinds

doi:10.1088/978-0-7503-2532-5ch10

of membranes. These proteins have been described as peripheral membrane proteins, amphitropic or 'conditional' membrane proteins (Johnson and Cornell 1999, Moravcevic *et al* 2012, Monje-Galvan and Klauda 2016). Understanding the structure and function of these peripheral membrane proteins is important as they are involved in many cellular processes (Johnson and Cornell 1999, Brunecky *et al* 2005, Hurley 2006, 2010, Lemmon 2008, Halskau *et al* 2009, p. 200; Moravcevic *et al* 2012, Basso *et al* 2016, Van Doren *et al* 2017, Chen *et al* 2019, Ruskamo *et al* 2020).

These proteins are functionally variable and structurally diverse, meaning that they deploy a variety of molecular interactions in their selective membrane binding. Some traditional examples of peripheral membrane proteins are globular single-domain proteins, but many more are multidomain proteins outfitted with lipid-binding domains that bind lipids via a peripheral interaction. Lipid-specific binding domains are categorized as families. For instance, phosphoinositol (PI) lipid variants are recognized and bound by such domains such as Pleckstrin homology (PH), Phox (PX), FYVE, and FERM (four point one, ezrin, radixin and moesin) domains (figure 10.1) (Lemmon 2003, Brunecky *et al* 2005, Seet and Hong 2006, Pemberton and Balla 2019). The lipid binding affinity varies greatly, with some multidomain proteins relying on the combination of multiple weakly binding domains to attain proper selectivity and function. Other proteins feature lipid-specific binding motifs that are less well understood (Planas-Iglesias *et al* 2015, Stepanyants *et al* 2015). Interactions with lipids are not just used to direct proteins to specific subcellular locations, but also affect their structure and function (Zhou and Cross 2013). This is perhaps particularly true for (peripherally binding) proteins that interact with the membrane interface (Kinnunen *et al* 1994, Watts 1997). In some cases, lipid binding leads to structural changes in a folded domain; in others a gain or loss of structural order between a folded and disordered structural state (figure 10.1(E)). However, to fully understand these structural changes, we need to know the structure of the protein in both its unbound (soluble) state and its membrane-bound state. The latter, especially, is a challenge.

### 10.1.2 Structural studies of peripheral membrane proteins.

Numerous structures of lipid-binding domains have been determined in the absence of a lipid membrane. Both x-ray crystallography and solution NMR have proved effective in revealing the protein without substrate or bound to a soluble substrate (or mimic) (figures 10.1(A–E)) (Ferguson *et al* 1995, Lomize *et al* 2007, Song *et al* 2007, Guerrero-Valero *et al* 2009, Halskau *et al* 2009, Moravcevic *et al* 2012, Chinthalapudi *et al* 2018). However, the experimentally determined structures do not reflect the membrane-bound state. In solution NMR, the membrane-protein complex becomes too large for in-depth study due to slow tumbling and enhanced relaxation rates. The membrane-surface bound state is not compatible with crystallization, prohibiting the use of x-ray diffraction. In light of the lack of experimental tools, some efforts have been toward deploying computational modeling methods to the question of how proteins bind the membrane surface

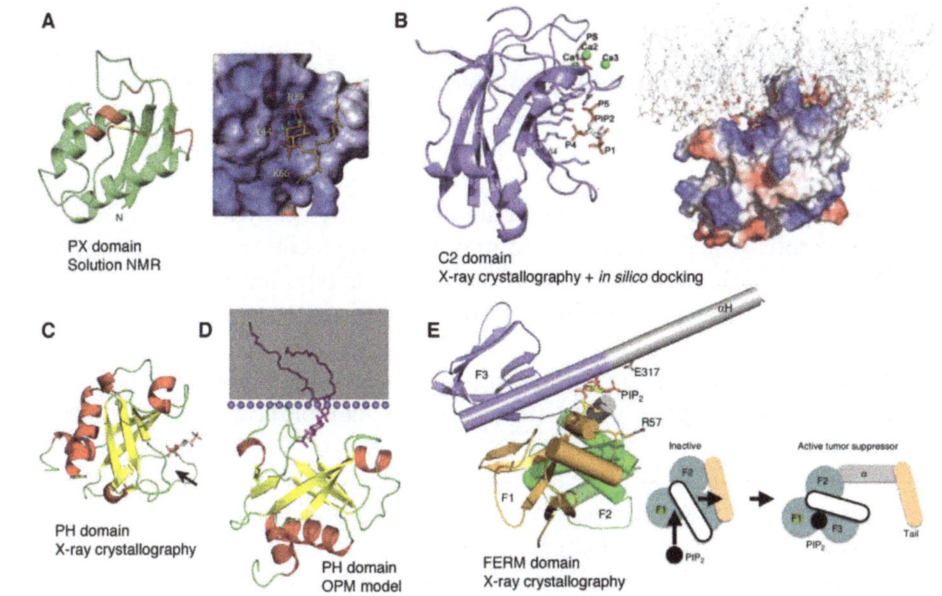


**Figure 10.1.** Selected structures of peripheral membrane proteins. (A) Solution NMR revealed the structure and binding site of sorting nexin 2 PX domain (Song *et al* 2007). Figure adapted with permission from Song *et al* (2007), copyright John Wiley & Sons. (B) X-ray crystallographic structure (left) and membrane-bound docking model (right) of PKCα-C2 domain. Adapted from Guerrero-Valero *et al* (2009) with permission from PNAS. (C) X-ray crystallography structure of the PH domain of PLC-δ1, with a bound soluble substrate mimic (arrow) (Ferguson *et al* 1995). (D) Model for the same structure bound to a membrane surface, from the OPM database (Lomize *et al* 2006, 2007). The top of the figure shows the predicted membrane location along with lipid acyl chains back-added to the soluble substrate in the original crystal structure. (E) Lipid binding to its FERM domain causes structural changes in tumor suppressor neurofibromin 2 (Chinthalapudi *et al* 2018). Panel E is adapted from Chinthalapudi *et al* (2018), under its Creative Commons Attribution 4.0 International License.

(Lomize *et al* 2007, Monje-Galvan and Klauda 2016). These approaches start with a known unbound structure and evaluate how it most likely binds to and 'sits' on a lipid bilayer surface. Some studies (e.g. figure 10.1(B)) apply docking methods to the protein structure determined in the absence of a lipid membrane. Alternatively, the 'orientations of proteins in membranes' (OPM) database (Lomize *et al* 2006, 2007) presents modeled structures for many proteins present in the protein databank (PDB). Figure 10.1(D) includes an example of the proposed binding mode for the PH domain of a phospholipase (PLC-δ1), based on its known crystal structure (figure 10.1(C)).

## 10.2 Solid-state NMR studies of peripheral membrane proteins

Recent studies have started to show avenues for using solid-state NMR (ssNMR) spectroscopy to examine how peripheral membrane proteins bind membranes and how membrane interactions modulate their structure and dynamics. Although based on the same principles as liquid-state NMR, in ssNMR one probes samples that lack the rapid isotropic tumbling required for the former. Thanks to ongoing technological and methodological improvements, ssNMR is now able to dissect the molecular features of many different kinds of biological 'solids', including protein aggregates, virus capsids, membrane proteins and large protein complexes (Tang *et al* 2013, Abramov *et al* 2015, Van der Wel 2018). Samples used in ssNMR are characterized by a lack of rapid tumbling of individual molecules, but do not need to be dry or rigid (Mandal *et al* 2017). Hydrated and semi-solid samples are routinely studied and permit the observation of important dynamic features of biological assemblies, proteins and membrane components. To facilitate the preparation of such samples, specialized 'packing tools' have become popular, in which the sample of interest is pelleted or sedimented straight into the ssNMR sample holder (Böckmann *et al* 2009, Mandal *et al* 2017). This approach (figure 10.2(A)) is both time efficient and ensures a high concentration of the fully hydrated material in the sample, and is also of great utility for peripheral membrane protein samples (see below).

Biomolecular ssNMR studies of membrane-associated proteins and peptides sometimes rely on the preparation of oriented membranes (Bechinger *et al* 2011, Van der Wel 2014, Hansen *et al* 2015). Such experiments permit the measurement of orientations relative to the lipid bilayer of transmembrane helices or amino acid side chains (Cornell *et al* 1988, Ketchem *et al* 1993, Marassi 2002, Van der Wel *et al* 2002, 2007). However, for the study of peripheral membrane proteins most recent successful studies combine ssNMR with magic angle spinning (MAS). In this technique, the entire NMR sample is spun rapidly at a fixed angle relative to the magnetic field, in an attempt to mimic the signal-narrowing effects of rapid isotropic tumbling in liquid-state NMR. MAS NMR does not require the time-consuming sample alignment process and detects the same isotropic NMR chemical shifts as are

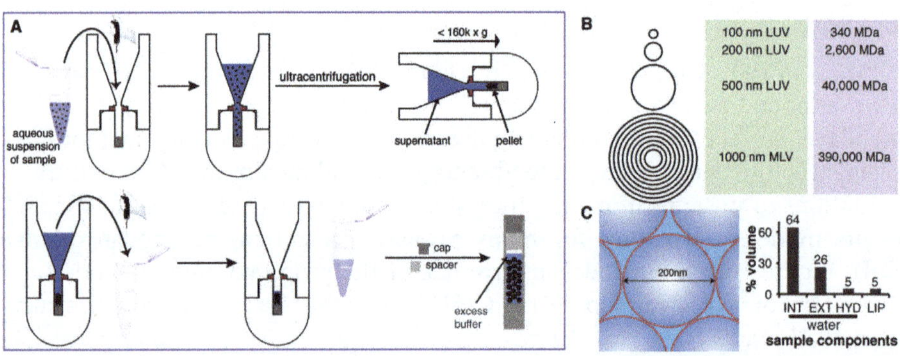


**Figure 10.2.** Membrane samples and sample preparation for membrane protein ssNMR. (A) Procedure for pelleting membrane samples into an MAS rotor sample holder, while maintaining full hydration. (B) Comparative sizes and approximate masses of different types of vesicle-based membrane mimics, showing unilamellar and multilamellar vesicles. Mandal *et al* (2017) with permission of Springer (C) Calculations of water content and sample composition in densely packed LUVs shows that a large fraction of the sample will be water, limiting the ability to pack large amounts of protein for ssNMR. INT = internal water; EXT = external (bulk) water; HYD = membrane-associated hydration water; LIP = lipids. Adapted with permission from Mandal and Van der Wel (2016), with permission for Springer.

seen in solution NMR. One implication of this direct analogy is that spectra and data obtained by MAS ssNMR are directly comparable to corresponding solution NMR results. Changes in chemical shifts of a protein in its soluble and immobilized form (e.g. membrane-bound, aggregated or crystallized) are indicative of structural and motional changes (Frericks Schmidt *et al* 2007, Boatz *et al* 2017, Li *et al* 2019). Going far beyond chemical shift analysis, modern MAS ssNMR offers a large and ever growing toolkit of atomic-level structure and dynamics measurements (Andronesi *et al* 2005, McDermott 2009, Lewandowski 2013, Tang *et al* 2013).

Thus, ssNMR offers in principle the ability to probe immobilized proteins bound to unilamellar and multilamellar vesicles (figures 10.1(B–C)), unsuitable for solution NMR. It may seem like a perfect tool to understand the features of peripheral membrane proteins and protein-binding domains, especially as it can be directly compared to results from liquid-state NMR on the free protein. Next we examine three sets of published case studies, in which we discuss practical challenges that complicate the real-world application of ssNMR, along with approaches to overcome some of these challenges.

## 10.3 Case studies of ssNMR on peripheral membrane proteins

### 10.3.1 Lipid-bound cytochrome c as a pro-apoptotic lipid peroxidase

Mitochondrial cytochrome c (cyt-c) is a heme protein found in the mitochondrial intermembrane space (IMS), where it plays an essential role in electron transport. In this role the protein binds to the IMS-facing interfaces of integral membrane proteins (figure 10.3(A); left). These protein–protein interactions have been extensively studied by various structural biology techniques, including NMR (Lange and Hunte 2002, Solmaz and Hunte 2008, Gentry *et al* 2017, Shimada *et al* 2017, Freeman *et al* 2018). Cyt-c can also directly bind negatively charged lipid membranes, facilitated at least in part by its high positive net charge (figure 10.3(A); middle). As a consequence, cyt-c is often presented as a paradigm for peripheral binding to acidic membranes (Kinnunen *et al* 1994, Rytömaa and Kinnunen 1995, Salamon and Tollin 1996). A renewed interest in this topic arose upon the discovery of a novel role for cyt-c in mitochondrial apoptosis: the catalytic peroxidation of polyunsaturated mitochondrial lipids and in particular cardiolipin (CL), generating pro-apoptotic signals (figure 10.3(A,C–D)) (Kagan *et al* 2020). This enzymatic function is activated by the formation of a CL-cyt-c complex at the interface of the inner mitochondrial membrane (IMM) with the IMS, and catalyzes the formation of specific peroxidized lipid derivatives that act as proapoptotic signals (Kagan *et al* 2020).

NMR in various forms has been applied to cyt-c in solution and interacting with detergents, lipids and protein partners, with solution and solid-state NMR studies going back to the 1970s and 1980s (Oldfield and Allerhand 1973, Brown and Wuthrich 1977, Waltham, Cornell and Smith 1986, Roder *et al* 1988). A busy period of *solid-state* NMR studies of cyt-c lipid interactions followed in the 1990s, coming from the Watts group (Spooner and Watts 1991a, 1991b, 1992, Spooner *et al* 1993, Pinheiro and Watts 1994b, Pinheiro and Watts 1994a, Pinheiro *et al* 1997,

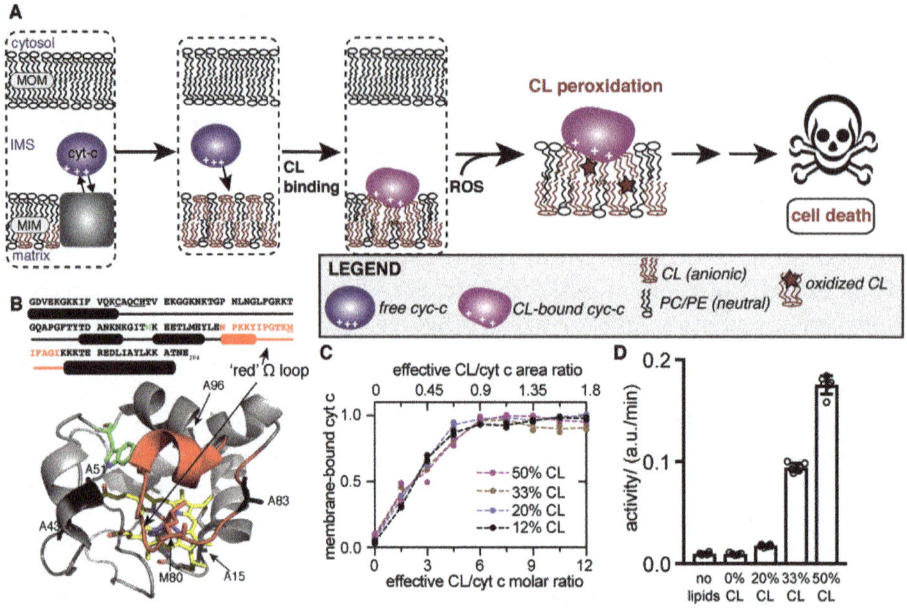


**Figure 10.3.** Cytochrome c as a peripheral membrane protein in mitochondrial apoptosis. (A) CL in the mitochondrial inner membrane (MIM) is to a large extent engaged in protein interactions and normally enriched in the inner leaflet. Increases in unbound CL in the IMS-facing leaflet enables formation of a peripheral cyt-c/CL complex that catalyzes lipid peroxidation by reactive oxygen species (ROS). Oxidized products then act as pro-apoptotic signals. (B) Sequence and structure of cyt-c in its unbound lipid-free state (Bushnell *et al* 1990). (C) Cyt-c binding to CL-containing LUVs determined by a co-sedimentation assay that mimics the ssNMR sample packing procedure from figure 10.2. (D) Reconstitution of CL-dependent peroxidase activity *in vitro*, based on fluorescent activity assays. The observed activity is strongly dependent on the fraction of CL in the vesicular membrane. Panels C–D adapted from (Mandal *et al* 2015, Li *et al* 2019) with permission from Elsevier.

Pinheiro *et al* 1997, Watts 1997). Several different groups have examined the interaction between CL lipids and cyt-c by solution NMR (Brown and Wuthrich 1977, O'Brien *et al* 2015, Kobayashi *et al* 2016, Mohammadyani *et al* 2018). However, even when employing rapidly tumbling bicelles, small vesicles or nanodiscs, complete binding to the lipid bilayer renders the membrane-associated protein invisible in solution NMR spectra. Under certain conditions one observes a partial disappearance of peaks in such spectra, reflecting partial or transient binding. There is little to no evidence of chemical shift changes in the still-visible peaks, further preventing an identification or characterization of membrane-induced changes in structure or dynamics. As the bound protein is invisible, solution NMR does not permit the observation of structural changes upon stable association of the protein with the membrane.

A natural question is then whether ssNMR (which does not require rapidly tumbling) may reveal the structural and dynamic changes of the membrane-bound protein. And indeed, both traditional 'static' and MAS ssNMR methods have both been deployed to address this question. Static ssNMR has examined the lipid side, mostly via $^{31}P$ and $^{2}H$ ssNMR experiments (de Kruijff and Cullis 1980,

Rietveld *et al* 1983, Devaux *et al* 1986, Waltham *et al* 1986, Spooner and Watts 1991a, 1991b, Spooner *et al* 1993, Pinheiro and Watts 1994b, Pinheiro *et al* 1997). These studies provided important and valuable information not accessible by other spectroscopic methods. Nonetheless, they yield limited insight into the protein itself. Such data only became available through the combined use of MAS NMR experiments and stable-isotope (primarily $^{13}$C) labeling of the protein. Such studies on peripherally bound cyt-c were performed in the early 1990s (Spooner and Watts 1992), based on the introduction of $^{13}$C methyl groups into the methionine side chain of (natural abundance) cyt-c (figure 10.4(A)). The labeled Met residues are of particular interest, since one of them acts as a ligand to the heme iron, with implications for the peroxidase activity of the protein (figure 10.3(B)).

A decade later optimized routines for uniform $^{13}$C,$^{15}$N-labeling of cyt-c were first reported (Szabo *et al* 2000, Rumbley *et al* 2002). Informed by these protocols, and the solution NMR studies that they enabled (Liu *et al* 2003, Kapralov *et al* 2011, Volkov *et al* 2012), we initiated multidimensional MAS NMR studies of membrane-bound

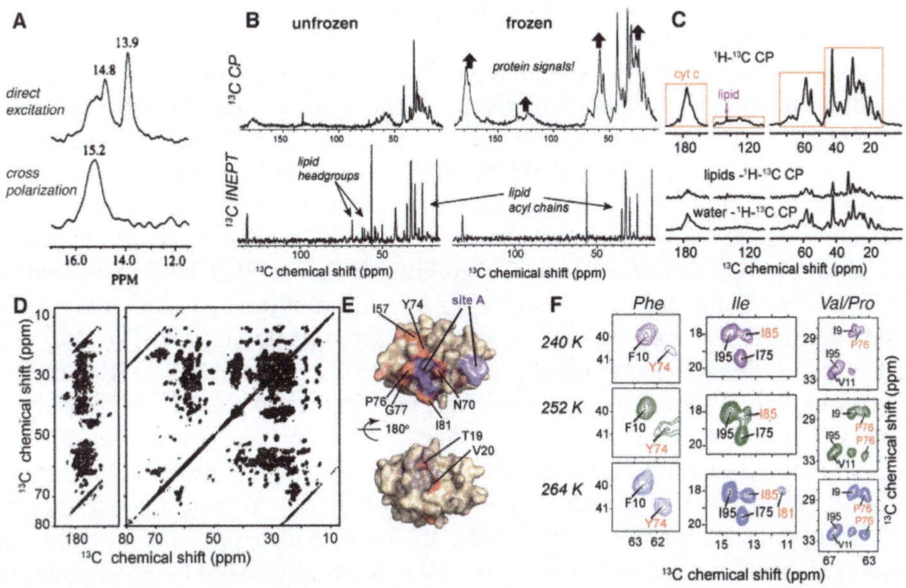

**Figure 10.4.** SSNMR studies of membrane-bound cyt-c. (A) $^{13}$C MAS NMR spectral region showing the $^{13}$C methyl signals of specifically labeled Met in the membrane-bound cyt-c, comparing CP (below) and direct $^{13}$C excitation (top). Adapted with permission from Spooner and Watts (1992), copyright 1992 American Chemical Society. (B) Left: above the freezing point, virtually no signal is seen from the $^{13}$C-labeled LUV-bound cyt-c, whether detected by CP or INEPT MAS NMR. Right: partly freezing the sample boosts the protein signal in CP, while the lipid membrane core remains fluid. (C) Optimal sample conditions for CP ssNMR (top) occur at temperatures where most water is frozen, but the protein is surrounded by fluid water and liquid-crystalline lipids. This is shown in 1D slices from a $^{1}$H–$^{13}$C CP 2D spectrum, featuring $^{1}$H–$^{1}$H transfer from lipid (mid) or water protons (bottom) to the protein. (D) 2D $^{13}$C–$^{13}$C CP/DARR spectrum on the membrane-bound protein. (E) Location of prominent chemical shift changes induced by membrane binding. (F) Temperature-dependent peak width changes affect residues facing the membrane interface. Panels B-F adapted with permission from Mandal *et al* (2015), Li *et al* (2019) with permission from Elsevier.

cyt-c (Mandal *et al* 2015, 2017, Li *et al* 2019). Our ssNMR studies were designed to probe the structural features of the CL-bound cyt-c in context of its proapoptotic peroxidase function (figures 10.3(A,C–D)) (Kagan *et al* 2020). *A priori*, several questions were of interest, based on the existing literature. Does the protein largely unfold upon membrane binding (Muga *et al* 1991, Spooner and Watts 1991a, 1991b, de Jongh *et al* 1995, Hong *et al* 2012)? Is the protein on the membrane surface or embedded in the membrane core? Does cyt-c binding disrupt the membrane's bilayer structure (de Kruijff and Cullis 1980)? What is the (residual) fold of cyt-c in its bound state?

#### *10.3.1.1 Preparing samples for MAS ssNMR*

Small membrane-bound peptides are sometimes amenable to protocols in which peptide and lipids are mixed in organic solvents, dried, and only then hydrated with aqueous buffer (Li *et al* 2007, Van der Wel *et al* 2007, Gehman and Separovic 2011, Bechinger and Salnikov 2012, Hoop *et al* 2012). However, this protocol is not suitable for more complex proteins, including most peripheral proteins. For such proteins, one generally reconstitutes the hydrated membranes of the desired lipid composition first, as mimics of the natural membrane. This implies the presence of the corresponding target lipid species (e.g. CL, PI or phosphatidylserine (PS) lipids), depending on the protein's lipid specificity. Then, a separately prepared (aqueous) solution of the protein is added to the membrane, permitting the subsequent binding interaction to take place, to form the desired protein–lipid complex.

MAS NMR studies commonly use multilamellar vesicles (MLVs) or large unilamellar vesicles (LUVs) as membrane mimics, as these most closely match the lipid bilayer properties of biological membranes (figure 10.2(B)). Given that the peripheral membrane protein is added after formation of the hydrated vesicles, they cannot access lipids that are not exposed on the outside surface. For MLVs in particular, this limits the amount of protein that can fit into a given NMR sample. It also can be hard to know the 'effective' protein:lipid ratio for MLVs, as only the exposed lipids are relevant. Third, the protein–lipid interactions become even harder to detect in the presence of large numbers of 'internal' lipids if those do not (or only partially) engage with the externally added protein. These concerns are reduced, though not fully eliminated, when using LUVs in which only the internal lipid monolayer would not be accessible. A tempting alternative would use non-spherical lipid bilayer models, such as bicelles and nanodiscs (Warschawski *et al* 2011, Durr *et al* 2013, Dörr *et al* 2014, Ding *et al* 2015). However, peripherally binding proteins could engage in non-native interactions with the curved non-bilayer parts of these lipid assemblies and distort the lipid bilayer structure itself. Thus, they may be less suitable for peripheral membrane proteins, unless careful control and comparison studies are done.

Once formed, protein–lipid complexes are recovered by lyophilization or, preferably, (ultra)centrifugation. The resulting pellet or rehydrated powder can then be studied by ssNMR. Increasingly, researchers eschew lyophilization due to the risk of irrevocable structural changes and associated broadening of the ssNMR spectra (Gregory *et al* 1993, Martin and Zilm 2003, Hisao *et al* 2016). Instead, a popular approach is to perform ultracentrifugation (or sedimentation) to transfer membrane samples into the MAS rotor sample holder (figure 10.3(C)) (Böckmann *et al* 2009, Mandal *et al* 2017), which avoids damage due to sample dehydration or freezing. The use of LUVs (rather than MLVs) for preparing membrane ssNMR samples does affect this process. Small-diameter LUVs resist pelleting, requiring increasingly high centrifugal forces ('sedimentation' conditions) (Mandal *et al* 2017).

*10.3.1.2 Balancing sample size, concentration and biological relevance*

One challenge in designing the ssNMR sample conditions is that the protein itself can constitute only a fraction of the total sample, especially if one aspires to biological relevance. For maximum ssNMR signal, one favors the highest amount of protein, at the expense of other sample components. This would lead to high protein-to-lipid (P/L) ratios and minimal sample hydration. However, biological relevance limits both parameters (Mandal *et al* 2015, 2017, Li *et al* 2019). In the case of cyt-c, and presumably also other peripheral membrane proteins, the interaction and structure change as the P/L molar ratio gets (non-physiologically) high (Oellerich *et al* 2002). Note also that the recognized 'target' lipid (i.e., CL in the case of cyt-c) is often only a minor component of the relevant cellular membrane. Fortunately, progress in NMR instrumentation and methods make it ever more feasible to work at lower and more relevant P/L ratios, compared to earlier ssNMR studies. Biological membranes need proper hydration to behave in a 'natural' way, with 'proper' hydration requiring that a membrane/water mixture with at least 40%–50% by weight of hydration water (Bechinger and Seelig 1991, Dvinskikh *et al* 2005). The geometric characteristics of water-filled LUVs mean that they inherently contain very high amounts of water relative to the amount of lipids present (figure 10.2(C)) (Mandal and Van der Wel 2016, Mandal *et al* 2017). Altogether, it is clear that each protein in the sample requires quite a large number of other sample components, which all take up precious space in the sample rotor. A fruitful approach may be to perform in-depth structural studies at higher P/L ratios, while checking for P/L ratio dependence with functional assays and 1D ssNMR (figures 10.3(C–D)).

*10.3.1.3 Optimizing the MAS ssNMR experimental conditions*

In modern biomolecular ssNMR, cross-polarization (CP)-based methods are the most commonly used. Yet, our initial CP MAS NMR studies on the cyt-c/LUV samples yielded very little signal, even though the protein was fully $^{13}C$,$^{15}N$-labeled (figure 10.4(B)). The presence of the protein was apparent not only from the red color of the sample, but also the strong $^{13}C$ signals seen in $^{13}C$ direct excitation experiments. In some cases, dynamic protein regions can be visualized by INEPT-based MAS NMR, which is highly effective for flexible and disordered protein regions (Andronesi *et al* 2005, Matlahov and Van der Wel 2018). Yet, for these cyt-c samples, very little INEPT-based protein signal was detected, suggesting that the membrane-bound protein lacked extensive flexible disordered segments. Low-temperature experiments yield relatively narrow peaks, indicative of a largely folded rather than frozen-out disordered structure (figure 10.4(D)). These data argued against the presence of a largely unfolded protein on the membrane, in contrast to prior reports of large scale unfolding being behind cyt-c peroxidase activity (Mandal *et al* 2015). Our initial results suffered from weak signal strengths, limited by our ability to pack sufficient labeled protein and maximize the signal from the available protein. In subsequent work, we optimized sample conditions to enable analysis of the binding mode and comparison of different lipid compositions. This was based in part on increasing the P/L ratio and optimizing experimental ssNMR conditions. Of especial importance in the latter was the careful tweaking of the sample temperature for optimal CP performance and linewidths. These were found to be best, not at the lowest temperatures, but rather at the point where the bulk sample water was frozen but the protein was still surrounded by fluid lipids and fluid membranes, as detected via 1D and 2D $^{1}H/^{13}C$ MAS ssNMR (figures 10.4(B–C)). In this context it is important to note the relevance of our use of mono- and polyunsaturated lipid membranes, which do not enter the gel transition until well below the freezing point of water (Mandal and Van der Wel 2016). Varying the temperature to lower temperatures actually resulted in residue-specific signal losses and line broadening (figure 10.4(F)), which we attributed to mobility of a crucial loop of cyt-c, present at the protein's membrane interacting face (figures 10.3(B), 10.4(D)). One way to dramatically increase the signal-to-noise (and work at even lower P/L ratios) would be to deploy dynamic nuclear polarization, which can offer orders of magnitude in signal enhancement (van der Wel *et al* 2006, Barnes *et al* 2008). However, this requires low temperatures that suppress biologically interesting dynamics and often cause extensive line broadening that limit the ability to determine site-specific information unless targeted labeling is applied.

Altogether, ssNMR studies allowed for unique insights into the membrane-bound state of cyt-c and are being used to better understand the molecular origins of the biologically important peroxidase activity of the CL-cyt-c proteolipid complex.

### 10.3.2 Myelin basic protein

Myelin basic protein (MBP) is a peripheral membrane protein of 18.5 kDa size involved in myelin maintenance. The protein interacts with various proteins and membranes, facilitating the clustering of membranes by interacting with multiple in parallel (figure 10.5(A)). In its unbound soluble state, it is an intrinsically disordered protein (IDP), which gains increased secondary structure upon membrane binding. The effect of MBP binding on the membrane itself was studied by monitoring the lipid by ssNMR (Pointer-Keenan *et al* 2004). More recently, the Ladizhansky group has spent considerable effort applying ssNMR to MBP interaction with membranes (and in other contexts) (Zhong *et al* 2007, Ahmed *et al* 2009, 2010, Harauz *et al* 2009, Libich *et al* 2010, Ward *et al* 2015). In the preparation of LUV-bound protein samples, the authors benefited from a natural tendency for the protein to 'aggregate' vesicles into more 'pellet-able' assemblies. Through 2D and 3D INEPT-based MAS NMR techniques, the most flexible segments of the membrane-bound $^{13}C$,$^{15}N$-labeled protein were identified, assigned and studied (Zhong *et al* 2007). Figure 10.5(B) shows an example of a 2D projection spectrum of a 3D $^{13}C$-detected

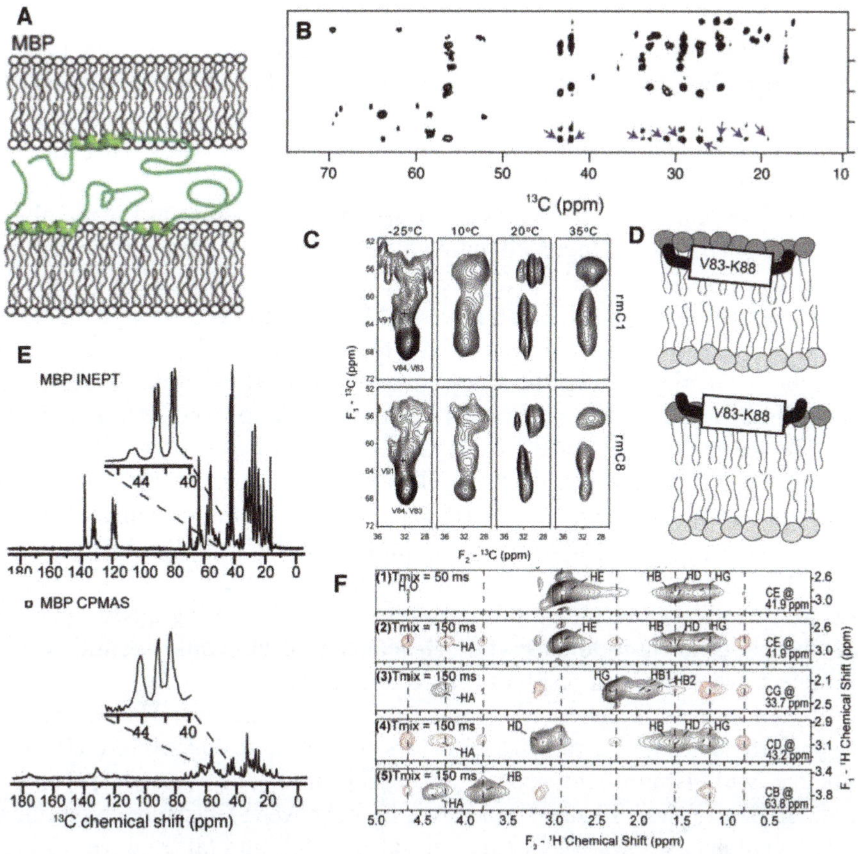

**Figure 10.5.** MAS ssNMR of membrane-bound MBP. (A) Schematic representation of MBP and how it brings together multiple membranes. (B) 2D projection of $^{13}C$-detected HHCC NOE/INEPT/TOBSY 3D MAS NMR of the flexible regions of membrane-bound MBP. (C) CP-based MAS NMR of membrane-bound MBP variants rmC1 and rmC8. (D) Models showing different interactions for rmC1 and rmC8 charge variants. (E) Comparison of INEPT and CP-based ssNMR peak heights of membrane-bound MBP. (F) $^{1}H$-detection enabled 3D INEPT MAS ssNMR, showing intra- and inter-residue cross peaks. Adapted with permission from Zhong *et al* (2007), Ahmed *et al* (2010), Ward *et al* (2015).

NOE/TOBSY-based HHCC experiment that demonstrates that these flexible residues interact strongly with water. Significant parts of the membrane-bound protein remained undetected, pointing to an immobilization of the central region (K68-K102) (Harauz *et al* 2009). An important strategy deployed next was residue-specific (Val/Asn) isotopic labeling, taking advantage of the fact that these residues are present in this core region (Ahmed *et al* 2010). Through 2D CP-based ssNMR, INEPT-based ssNMR and variable temperature studies, a short α-helical segment was observed to be immobilized in the membrane (and be visible by CP ssNMR), yet have a high level of mobility (figures 10.5(C–D)). Of note, in the CP-based ssNMR experiments the authors added a pre-CP 90° pulse on the $^{13}C$ channel to gain additional signal from more mobile parts of the sample (Rienstra *et al* 2002,

Wylie *et al* 2005, Li, Wylie and Rienstra 2006). Differences in mobility were observed as temperature-dependent changes in the relative intensities of CP-based cross-peaks for the three Val residues in the membrane-interacting region (figure 10.5(C)). These measurements allowed the comparison of how two different charge variants (rmC1 and rmC8) interact in different ways with the lipid bilayer membrane (figure 10.5(D)).

The membrane-bound protein may gain in secondary structure at the membrane interface but nonetheless is highly dynamic. This was visualized in comparative INEPT versus CP ssNMR analysis (figure 10.5(E)). The CP-based signal is much lower than that attained by INEPT-based ssNMR, making a thorough structural analysis by CP-based multidimensional MAS NMR difficult. For this reason, the authors focused on maximizing the information from INEPT approaches (Ward *et al* 2015). They turned to $^{1}H$ detection for dramatically improved sensitivity in 1D, 2D and 3D spectroscopy, aided by $^{2}H_2O$ solvent exchange (e.g. 3D data in figure 10.5(F)). The high solvent accessibility and dynamics of the peripherally bound protein facilitated rapid $^{1}H$–$^{2}H$ (H/D) exchange of the exchangeable sites. Notably, the exchange was done with the membrane/protein samples already packed inside the MAS rotor. This is an important observation as LUVs tend to float in $D_2O$ (or $D_2O$ buffers), which greatly complicates the pelleting-based protocols discussed earlier. The multidimensional $^{1}H$-detected spectra data identified additional mobile solvent-exposed residues (Ward *et al* 2015).

A number of aspects of the experimental details are of interest. The experiments were all performed using 3.2 mm MAS rotors on 600 and 800 MHz ($^{1}H$) NMR spectrometers, with 4 mg of the protein (alongside lipids and water). The authors employed 'moderate' MAS rates (typically ~12 kHz MAS) because a significant decrease in signal intensity was observed at elevated MAS rates (at or above 15 kHz), which the authors attribute to phase separation and local dehydration (Gawrisch and Gaede 2007, Zhong *et al* 2007, Ward *et al* 2015).

### 10.3.3 The Pleckstrin homology domain of PLC-δ1

Among the lipid-specific binding modules that many proteins employ to conditionally bind specific cellular membranes, the PH domains are found in many different kinds of proteins (Lemmon and Ferguson 2001). Preceding much of the above work, a series of papers in the 2000s has applied ssNMR to the study of the PH domain of phospholipase C-δ1 (PLC-δ1) (Tuzi *et al* 2003, Uekama *et al* 2007, 2009). The PLC-δ1 PH domain features ~120 amino acids and is responsible for localizing the protein to the plasma membrane, by binding the headgroup of phosphatidylinositol 4,5-bisphosphate ($PIP_2$, or $PI(4,5)P_2$) lipids. PLC-δ1 then can hydrolyze this membrane-embedded substrate. The PH domain features a mixed α-helical/β-sheet architecture with loops of a β-sandwich forming the stereospecific binding site (figure 10.1) (Ferguson *et al* 1995, Lemmon and Ferguson 2001). While the binding of soluble model substrates has been studied, how this interaction takes place on a real membrane surface has remained opaque. Some simulation groups have looked at this, with figure 10.1(B) showing the model from the OPM database (Lomize *et al* 2007).

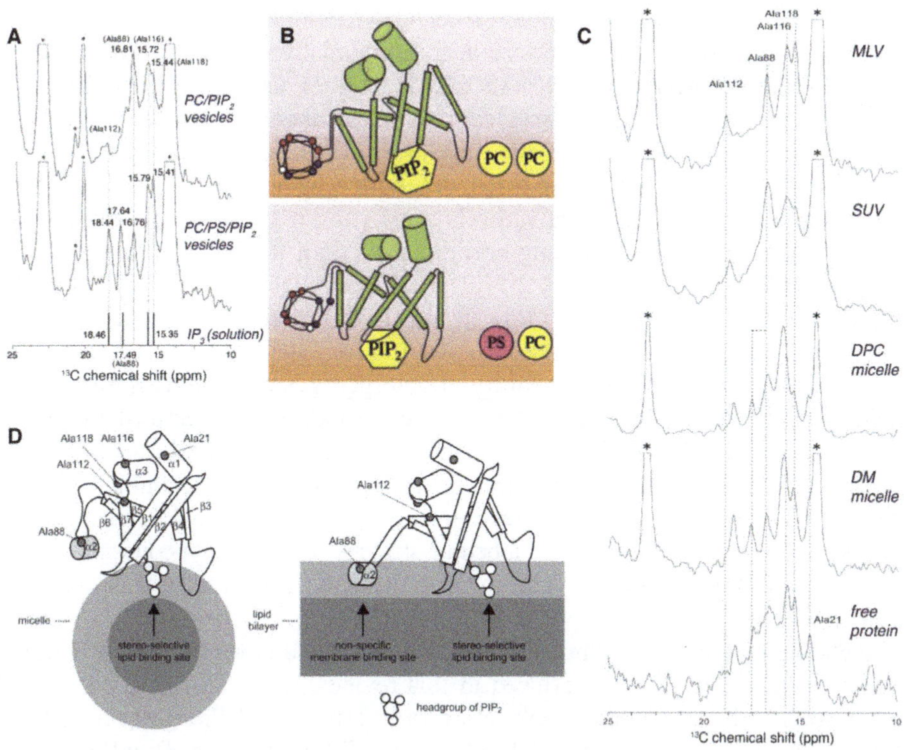


**Figure 10.6.** SSNMR studies of lipid-bound PH domain of PLC-δ1. (A) Direct excitation $^{13}$C MAS NMR spectra of $^{13}CH_3$-Ala-labeled PLC-δ1 PH domain bound to phosphatidylcholine (PC)/PIP$_2$ (top) or PC/PS/PIP$_2$ vesicles (bottom). Peaks from the lipids are indicated with asterisks (*). Bars at the bottom show the chemical shifts for IP$_3$-bound PH in solution. (B) Schematic models of lipid-dependent structural changes in the PH domain. (C) $^{13}$C MAS NMR spectra of PLC-δ1 PH bound to PIP$_2$-containing membrane mimics of different curvature, showing notable changes in the Ala88 residue for highly curved micelles. (D) SSNMR-based model showing different binding modes to biologically relevant lipid bilayers (right) and highly curved detergent micelles (left). Adapted with permission from Uekama *et al* (2007 copyright John Wiley & Sons, 2009, copyright Elsevier).

The ssNMR studies provided insights into the structural differences between membrane-bound PLC-δ1 PH and its solution-state or micelle-bound conformations. A key strategy here was also the use of residue-specific isotope labeling: $^{13}$C-methyl-labeling of the alanines, which are distributed throughout the protein (figure 10.6(D)). This enabled the effective use of 1D rather than multidimensional ssNMR. Assignment of specific alanines was facilitated by site-directed mutagenesis of Ala to Gly or Leu residues (Tuzi *et al* 2003). With this approach it was shown that binding to membranes containing PIP$_2$ resulted in changes in some Ala residues relative to solution NMR of the protein bound to soluble IP$_3$ (figure 10.6(A)). The residues undergoing structural change were identified as Ala-88 and Ala-112. The authors concluded that the protein has a unique conformation and dynamics at the lipid bilayer surface, distinct from the solution state (Tuzi *et al* 2003). Analogous experiments then examined the role of different lipid interactions such as the presence of anionic PS lipids (figure 10.6(A)) (Uekama *et al* 2007). Differences in measured membrane affinity were accompanied by structural changes in the protein (figure 10.6(B)). Next, MAS NMR studies using different membrane mimics, all containing $PIP_2$, dissected the importance of membrane curvature (figure 10.6(C)). Small unilamellar vesicles (SUVs), large unilamellar vesicles (LUVs) and multilamellar vesicles (MLVs), which have a relatively small degree of membrane curvature, all behaved relatively similarly. There is a difference in the linewidth with the SUVs giving broader lines than the MLVs (figure 10.6(C)). Likely this reflects at least in part the different dynamics of, and within, the SUVs compared to the MLVs. Much more dramatic differences between micelles and proper lipid bilayers were interpreted to stem from the differences in curvature, raising concerns about the applicability of detergents or curved membrane mimics for studying these kinds of proteins. The corresponding structural models are shown in figure 10.6(D). The figure shows how the Ala residues (which were selectively $^{13}CH_3$ labeled) are distributed throughout the key secondary structure elements, enabling them to act as effective sensors of functionally relevant structural changes. Thus, the residue-specific labeling strategy with methyl-containing amino acids has proved an important strategy in each of the case studies analyzed in this review article.

We end here with a few technical and methodological notes on these PH domain ssNMR studies. The data reproduced in this review article show $^{13}C$ MAS NMR spectra measured with $^{13}C$ direct excitation rather than $^{13}C$ CP. However, CP MAS NMR was performed (Tuzi *et al* 2003) to detect different levels of mobility for the Ala residues but the CP signal was generally weak due to the high mobility of the protein bound to the liquid crystalline membrane. The presented data were acquired using a 400 MHz ($^1H$ frequency) NMR instrument, with 5 mm MAS rotors spinning at relatively slow MAS frequencies of 2.6 to 4 kHz. Note that each 1D spectrum required 20 to 40 thousand acquisitions for a reasonable signal-to-noise (i.e., 1–2 days per spectrum).

## 10.4 Lessons learned

Looking at the case studies, we see many commonalities between the challenges faced and that both shared and complementary solutions were described. All three membrane-bound peripheral **proteins experience extensive dynamics**, which reduced the effectiveness of canonical CP-based ssNMR. Indeed, the mobility can be such that CP experiments yield very little signal. Relaxation-induced signal losses and inefficient dipolar-based transfers largely prohibit multidimensional CP MAS NMR under such conditions. This feature is shared by all three proteins examined above and is not just a hallmark of one of the IDP-like proteins. In our perspective, the underlying mobility stems from a combination of factors:

- *Intramolecular mobility*: partial unfolding or flexible disorder within the protein, induced in part by destabilizing interactions with the membrane surface.

- The protein is not binding to a rigid surface but rather a *fluid and flexible membrane*. By being bound to freely moving lipids within the bilayer, the protein is subject to high overall mobility (diffusion and wobbling motions (Dufourc *et al* 1992)).

These dynamics present a big challenge as they prohibit the application of much of CP-based multidimensional ssNMR. The case studies show some ways of tackling this challenge.

- Turning to either *direct excitation* methods or *INEPT-based experiments*. It is important to note that the latter only work well for truly flexible regions and does not allow the observation of the peripheral proteins' folded domains.
- Relying partly on *1D ssNMR* rather than multidimensional ssNMR, and simply acquiring a lot of scans to attain acceptable signal/noise ratios.
- In 1D spectroscopy full isotopic labeling becomes less effective. Instead it is ideally combined with *site- or residue-specific labeling methods*, where the labeled residues are either distributed throughout the protein or in particularly interesting locations. Given their modest costs and beneficial NMR characteristics, methyl-containing amino acids have been especially popular (e.g. Ala, Met, Val).

The above approaches are functional in the presence of dynamics. A complementary approach is to modulate the problematic mobility. Here, different approaches have been discussed or can be considered:

- *Reduce the sample temperature* to suppress mobility to make CP-based ssNMR (more) effective.
- Freeze out mobility at a temperature where the lipids and/or solvent are in the *gel state* or frozen phase, respectively. A caveat with this approach is the need to verify biological relevance. Another concern is that this recovers the rigidity needed to perform CP-based ssNMR but can result in extensive line broadening that can nonetheless prohibit a detailed analysis. Also here, specific isotopic labeling may be beneficial.
- Recently, we have reported on a compromise where part of the solvent is frozen (suppressing whole-membrane/protein mobility) while the lipid membrane (with unsaturated lipids) and membrane-proximal waters are not (Li *et al* 2019). The surface-bound protein remains surrounded by fluid water and excessive line broadening is avoided. It remains to be seen whether similar results can be obtained with other proteins.

Several studies have reported membrane sample spectral changes at MAS rates exceeding 10–15 kHz (using 3.2 and 4 mm MAS rotors) (Gawrisch and Gaede 2007, Zhong *et al* 2007). For now, the primary approach to tackle this is based on sticking to MAS rates below 15 kHz. Since these data were acquired with 3.2 mm rotors, the critical MAS rate of 15 kHz would be equivalent to approximately 1 million times the g-force being experienced by the sample (Mandal *et al* 2017). Smaller or larger rotors can, respectively, be spun faster or slower than 15 kHz if one wants to stay below the critical g-force. It has previously been suggested that density-matched buffers (with an optimized $H_2O/D_2O$ ratio) may allow work also at faster MAS rates (Gawrisch and Gaede 2007). However, to the best of our knowledge this has not yet been tested in practice for peripheral membrane proteins.

The proper preparation of LUV-bound peripheral membrane protein samples limits the amount of (labeled) protein that can be fit into the μL-sized volumes of MAS rotors. This fraction gets 'worse' the larger the LUV diameter is, suggesting smaller LUVs are better from that perspective. However, small LUVs bring their own problems. They are difficult to pellet, making the efficient packing of MAS rotors challenging. The sedimentation, or ultracentrifugation-based, approach is a great help (Böckmann *et al* 2009, Mandal *et al* 2017). In small LUVs overall tumbling is faster and the membrane components rapidly traverse the spherical shape of the vesicle, leading to worse line broadening in certain ssNMR experiments (Burnell *et al* 1980). Thus, likely a compromise in terms of vesicle size is optimal, with the best choice dependent on the system at hand.

## 10.5 Conclusions

Peripheral membrane proteins present a number of challenges when it comes to in-depth studies by ssNMR. Dynamics, disorder and relatively dilute sample conditions conspire to make many standard methods of modern biomolecular ssNMR challenging to implement. However, the reviewed approaches help overcome some of these challenges. Similar challenges (and solutions) have been described for smaller membrane-surface-binding peptides, and interested readers may also refer to this literature (Bechinger and Salnikov 2012, Porcelli *et al* 2013, Su *et al* 2013). With the aid of these approaches, ssNMR is poised to provide crucial insights into the wider family of peripheral membrane proteins as part of the ever more integrated toolbox of structural biology techniques.

## Acknowledgements

Our work on peripheral membrane protein studies by ssNMR has been supported by grants from the US National Institutes of Health grants R01 GM113908 and R01 GM121583, and instrument grant S10 1S10RR033477.

## References

Abramov G, Morag O and Goldbourt A 2015 Magic-angle spinning NMR of intact bacteriophages: insights into the capsid, DNA and their interface *J. Magn. Reson.* **253** 80–90

Ahmed M A M *et al* 2009 Induced secondary structure and polymorphism in an intrinsically disordered structural linker of the CNS: solid-state NMR and FTIR spectroscopy of myelin basic protein bound to actin *Biophys. J.* **96** 180–91

Ahmed M A M *et al* 2010 Solid-state NMR spectroscopy of membrane-associated myelin basic protein—conformation and dynamics of an immunodominant epitope *Biophys. J.* **99** 1247–55

Andronesi O C *et al* 2005 Determination of membrane protein structure and dynamics by magic-angle-spinning solid-state NMR spectroscopy *J. Am. Chem. Soc.* **127** 12965–74

Barnes A B *et al* 2008 High-field dynamic nuclear polarization for solid and solution biological NMR *Appl. Magn. Reson.* **34** 237–63

Basso L G M, Mendes L F S and Costa-Filho A J 2016 The two sides of a lipid-protein story *Biophys. Rev.* **8** 179–91

Bechinger B, Resende J M and Aisenbrey C 2011 The structural and topological analysis of membrane-associated polypeptides by oriented solid-state NMR spectroscopy: established concepts and novel developments *Biophys. Chem.* **153** 115–25

Bechinger B and Salnikov E S 2012 The membrane interactions of antimicrobial peptides revealed by solid-state NMR spectroscopy *Chem. Phys. Lipids* **165** 282–301

Bechinger B and Seelig J 1991 Conformational changes of the phosphatidylcholine headgroup due to membrane dehydration. A $^2$H-NMR study *Chem. Phys. Lipids* **58** 1–5

Boatz J C *et al* 2017 Cataract-associated P23T γD-crystallin retains a native-like fold in amorphous-looking aggregates formed at physiological pH *Nat. Commun.* **8** 15137

Böckmann A *et al* 2009 Characterization of different water pools in solid-state NMR protein samples *J. Biomol. NMR* **45** 319–27

Brown L R and Wuthrich K 1977 NMR and ESR studies of the interactions of cytochrome c with mixed cardiolipin-phosphatidylcholine vesicles *Biochim. Biophys. Acta* **468** 389–410

Brunecky R *et al* 2005 Investigation of the binding geometry of a peripheral membrane protein *Biochemistry* **44** 16064–71

Burnell E E, Cullis P R and de Kruijff B 1980 Effects of tumbling and lateral diffusion on phosphatidylcholine model membrane $^{31}$P-NMR lineshapes *Biochim. Biophys. Acta* **603** 63–9

Bushnell G W, Louie G V and Brayer G D 1990 High-resolution three-dimensional structure of horse heart cytochrome c *J. Mol. Biol.* **214** 585–95

Chen K E, Healy M D and Collins B M 2019 Towards a molecular understanding of endosomal trafficking by Retromer and Retriever *Traffic* **20** 465–78

Chinthalapudi K *et al* 2018 Lipid binding promotes the open conformation and tumor-suppressive activity of neurofibromin 2 *Nat. Commun.* **9** 1338

Cornell B A *et al* 1988 Conformation and orientation of gramicidin a in oriented phospholipid bilayers measured by solid state carbon-13 NMR *Biophys. J.* **53** 67–76

de Jongh H H, Ritsema T and Killian J A 1995 Lipid specificity for membrane mediated partial unfolding of Cytochrome C *FEBS Lett.* **360** 255–60

de Kruijff B and Cullis P R 1980 Cytochrome c specifically induces non-bilayer structures in cardiolipin-containing model membranes *Biochim. Biophys. Acta* **602** 477–90

Devaux P F *et al* 1986 Interaction of cytochrome c with mixed dimyristoylphosphatidylcholine-dimyristoylphosphatidylserine bilayers: a deuterium nuclear magnetic resonance study *Biochemistry* **25** 3804–12

Ding Y *et al* 2015 Solid-state NMR of the *Yersinia pestis* outer membrane protein Ail in lipid bilayer nanodiscs sedimented by ultracentrifugation *J. Biomol. NMR* **61** 275–86

Dörr J M *et al* 2014 Detergent-free isolation, characterization, and functional reconstitution of a tetrameric K+ channel: the power of native nanodiscs *Proc. Natl Acad. Sci. USA* **111** 18607–12

Dufourc E J *et al* 1992 Dynamics of phosphate head groups in biomembranes. Comprehensive analysis using phosphorus-31 nuclear magnetic resonance lineshape and relaxation time measurements *Biophys. J.* **61** 42–57

Durr U H N, Soong R and Ramamoorthy A 2013 When detergent meets bilayer: birth and coming of age of lipid bicelles *Prog. Nucl. Magn. Reson. Spectrosc.* **69** 1–22

Dvinskikh S V, Castro V and Sandström D 2005 Probing segmental order in lipid bilayers at variable hydration levels by amplitude- and phase-modulated cross-polarization NMR *Phys. Chem. Chem. Phys.* **7** 3255–7

Ferguson K M *et al* 1995 Structure of the high affinity complex of inositol trisphosphate with a phospholipase C pleckstrin homology domain *Cell* **83** 1037–46

Freeman S L *et al* 2018 Solution structure of the cytochrome P450 reductase–cytochrome ccomplex determined by neutron scattering *J. Biol. Chem.* **293** 5210–9

Frericks Schmidt H L *et al* 2007 Crystal polymorphism of protein GB1 examined by solid-state NMR spectroscopy and x-ray diffraction *J. Phys. Chem.* B **111** 14362–9

Gawrisch K and Gaede H C 2007 Measurement of lateral diffusion rates in membranes by pulsed magnetic field gradient, magic angle spinning-proton nuclear magnetic resonance *Methods Mol. Biol.* **400** 257–65

Gehman J D and Separovic F 2011 Solid-state NMR of amyloid membrane interactions *Methods Mol. Biol.* **752** 165–77

Gentry K A *et al* 2017 Kinetic and structural characterization of the effects of membrane on the complex of cytochrome B 5 and cytochrome C *Sci. Rep.* **7** 7793

Gregory R B *et al* 1993 The influence of hydration on the conformation of lysozyme studied by solid-state $^{13}$C-NMR spectroscopy *Biopolymers* **33** 513–9

Guerrero-Valero M *et al* 2009 Structural and mechanistic insights into the association of PKCα-C2 domain to PtdIns(4,5)$P_2$ *Proc. Natl Acad. Sci.* **106** 6603–7

Halskau Ø, Muga A and Martinez A 2009 Linking new paradigms in protein chemistry to reversible membrane-protein interactions *Curr. Protein Pept. Sci.* **10** 339–59

Hansen S K *et al* 2015 Solid-state NMR methods for oriented membrane proteins *Prog. Nucl. Magn. Reson. Spectrosc.* **88–89** 48–85

Harauz G, Ladizhansky V and Boggs J M 2009 Structural polymorphism and multifunctionality of myelin basic protein *Biochemistry* **48** 8094–104

Hisao G S *et al* 2016 An efficient method and device for transfer of semisolid materials into solid-state NMR spectroscopy rotors *J. Magn. Reson.* **265** 172–6

Hong Y *et al* 2012 Origin of the conformational heterogeneity of cardiolipin-bound cytochrome C *J. Am. Chem. Soc.* **134** 18713–23

Hoop C L *et al* 2012 Structural characterization of the caveolin scaffolding domain in association with cholesterol-rich membranes *Biochemistry* **51** 90–9

Hurley J H 2006 Membrane binding domains *Biochim. Biophys. Acta* **1761** 805–11

Hurley J H 2010 The ESCRT complexes *Crit. Rev. Biochem. Mol. Biol.* **45** 463–87

Johnson J E and Cornell R B 1999 Amphitropic proteins: regulation by reversible membrane interactions (review) *Mol. Membr. Biol.* **16** 217–35

Kagan V E *et al* 2020 Redox phospholipidomics of enzymatically generated oxygenated phospholipids as specific signals of programmed cell death *Free Radic. Biol. Med.* **147** 231–41

Kapralov A A *et al* 2011 Topography of tyrosine residues and their involvement in peroxidation of polyunsaturated cardiolipin in cytochrome C/cardiolipin peroxidase complexes *Biochim. Biophys. Acta* **1808** 2147–55

Ketchem R R, Hu W and Cross T A 1993 High-resolution conformation of gramicidin A in a lipid bilayer by solid-state NMR *Science* **261** 1457–60

Kinnunen P K J *et al* 1994 Lipid dynamics and peripheral interactions of proteins with membrane surfaces *Chem. Phys. Lipids* **73** 181–207

Kobayashi H, Nagao S and Hirota S 2016 Characterization of the cytochrome c membrane-binding site using cardiolipin-containing bicelles with NMR *Angew. Chem.* **128** 14225–8

Lange C and Hunte C 2002 Crystal structure of the yeast cytochrome bc1 complex with its bound substrate cytochrome c *Proc. Natl Acad. Sci. USA* **99** 2800–5

Lemmon M A 2003 Phosphoinositide recognition domains *Traffic* **4** 201–13

Lemmon M A 2008 Membrane recognition by phospholipid-binding domains *Nat. Rev. Mol. Cell Biol.* **9** 99–111

Lemmon M A and Ferguson K M 2001 Molecular determinants in pleckstrin homology domains that allow specific recognition of phosphoinositides *Biochem. Soc. Trans.* **29** 377–84

Lewandowski J R 2013 Advances in solid-state relaxation methodology for probing site-specific protein dynamics *Acc. Chem. Res.* **46** 2018–27

Li C *et al* 2007 Solid-state NMR characterization of conformational plasticity within the transmembrane domain of the influenza A M2 proton channel *Biochim. Biophys. Acta* **1768** 3162–70

Li M *et al* 2019 Surface-binding to cardiolipin nanodomains triggers cytochrome c pro-apoptotic peroxidase activity via localized dynamics *Structure* **27** 806–15

Li Y, Wylie B J and Rienstra C M 2006 Selective refocusing pulses in magic-angle spinning NMR: characterization and applications to multi-dimensional protein spectroscopy *J. Magn. Reson.* **179** 206–16

Libich D S *et al* 2010 Fuzzy complexes of myelin basic protein: NMR spectroscopic investigations of a polymorphic organizational linker of the central nervous system *Biochem. Cell Biol.* **88** 143–55

Liu W *et al* 2003 Backbone and side-chain heteronuclear resonance assignments and hyperfine NMR shifts in horse cytochrome C *Protein Sci.* **12** 2104–8

Lomize A L *et al* 2007 The role of hydrophobic interactions in positioning of peripheral proteins in membranes *BMC Struct. Biol.* **7** 44

Lomize M A *et al* 2006 OPM: orientations of proteins in membranes database *Bioinformatics* **22** 623–5

Mandal A *et al* 2015 Structural changes and proapoptotic peroxidase activity of cardiolipin-bound mitochondrial cytochrome C *Biophys. J.* **109** 1873–84

Mandal A *et al* 2017 On the use of ultracentrifugal devices for routine sample preparation in biomolecular magic-angle-spinning NMR *J. Biomol. NMR* **67** 165–78

Mandal A and Van der Wel P C A 2016 MAS $^{1}$H NMR probes freezing point depression of water and liquid-gel phase transitions in liposomes *Biophys. J.* **111** 1965–73

Marassi F M 2002 NMR of peptides and proteins in oriented membranes *Concept Magn. Res* **14** 212–24

Martin R W and Zilm K W 2003 Preparation of protein nanocrystals and their characterization by solid state NMR *J. Magn. Reson.* **165** 162–74

Matlahov I and Van der Wel P C A 2018 Hidden motions and motion-induced invisibility: dynamics-based spectral editing in solid-state NMR *Methods* **148** 123–35

McDermott A 2009 Structure and dynamics of membrane proteins by magic angle spinning solid-state NMR *Annu. Rev. Biophys.* **38** 385–403

Mohammadyani D *et al* 2018 Structural characterization of cardiolipin-driven activation of cytochrome c into a peroxidase and membrane perturbation *Biochim. Biophys. Acta* **1860** 1057–68

Monje-Galvan V and Klauda J B 2016 Peripheral membrane proteins: tying the knot between experiment and computation *Biochim. Biophys. Acta* **1858** 1584–93

Moravcevic K, Oxley C L and Lemmon M A 2012 Conditional peripheral membrane proteins: facing up to limited specificity *Structure* **20** 15–27

Muga A, Mantsch H H and Surewicz W K 1991 Membrane binding induces destabilization of cytochrome c structure *Biochemistry* **30** 7219–24

O'Brien E S *et al* 2015 Defining the apoptotic trigger: the interaction of cytochrome c and cardiolipin *J. Biol. Chem.* **290** 30879–87

Oellerich S, Wackerbarth H and Hildebrandt P 2002 Spectroscopic characterization of nonnative conformational states of cytochrome C *J. Phys. Chem.* B **106** 6566–80

Oldfield E and Allerhand A 1973 Cytochrome C: observation of numerous single-carbon sites of the reduced and oxidized species by means of natural-abundance $^{13}$C nuclear magnetic resonance spectroscopy *Proc. Natl Acad. Sci. USA* **70** 3531–5

Pemberton J G and Balla T 2019 Polyphosphoinositide-binding domains: insights from peripheral membrane and lipid-transfer proteins *Adv. Exp. Med. Biol.* **1111** 77–137

Pinheiro T J *et al* 1997 Structural and kinetic description of cytochrome c unfolding induced by the interaction with lipid vesicles *Biochemistry* **36** 13122–32

Pinheiro T J T, Duer M J and Watts A 1997 Phospholipid headgroup dynamics in DOPG-$d_5$-cytochrome c complexes as revealed by $^2$H and $^{31}$P NMR: the effects of a peripheral protein on collective lipid fluctuations *Solid State Nucl. Magn. Reson.* **8** 55–64

Pinheiro T J T and Watts A 1994a Lipid specificity in the interaction of cytochrome C with anionic phospholipid bilayers revealed by solid-state $^{31}$P NMR *Biochemistry* **33** 2451–8

Pinheiro T J and Watts A 1994b Resolution of individual lipids in mixed phospholipid membranes and specific lipid-cytochrome c interactions by magic-angle spinning solid-state phosphorus-31 NMR *Biochemistry* **33** 2459–67

Planas-Iglesias J *et al* 2015 Cardiolipin interactions with proteins *Biophys. J.* **109** 1282–94

Pointer-Keenan C D *et al* 2004 Investigation of the interaction of myelin basic protein with phospholipid bilayers using solid-state NMR spectroscopy *Chem. Phys. Lipids* **132** 47–54

Porcelli F *et al* 2013 On the role of NMR spectroscopy for characterization of antimicrobial peptides *Methods Mol. Biol.* **1063** 159–80

Rienstra C M *et al* 2002 Determination of multiple torsion-angle constraints in U-$^{13}$C,$^{15}$N-labeled peptides: 3D $^1$H–$^{15}$N–$^{13}$C–$^1$H dipolar chemical shift NMR spectroscopy in rotating solids *J. Am. Chem. Soc.* **124** 11908–22

Rietveld A *et al* 1983 Interaction of cytochrome C and its precursor apocytochrome c with various phospholipids *EMBO J.* **2** 907–13

Roder H, Elöve G A and Englander S W 1988 Structural characterization of folding intermediates in cytochrome-C by H-exchange labeling and proton NMR *Nature* **335** 700–4

Rumbley J N, Hoang L and Englander S W 2002 Recombinant equine cytochrome C in *Escherichia coli*: high-level expression, characterization, and folding and assembly mutants *Biochemistry* **41** 13894–901

Ruskamo S *et al* 2020 Cryo-EM, x-ray diffraction, and atomistic simulations reveal determinants for the formation of a supramolecular myelin-like proteolipid lattice *J. Biol. Chem.* **295** 8692–705

Rytömaa M and Kinnunen P K 1995 Reversibility of the binding of cytochrome C to liposomes. Implications for lipid-protein interactions *J. Biol. Chem.* **270** 3197–202

Salamon Z and Tollin G 1996 Surface plasmon resonance studies of complex formation between cytochrome c and bovine cytochrome c oxidase incorporated into a supported planar lipid bilayer. II. Binding of cytochrome C to oxidase-containing cardiolipin/phosphatidylcholine membranes *Biophys. J.* **71** 858–67

Seet L-F and Hong W 2006 The Phox (PX) domain proteins and membrane traffic *Biochim. Biophys. Acta* **1761** 878–96

Shimada S *et al* 2017 Complex structure of cytochrome C–cytochrome C oxidase reveals a novel protein–protein interaction mode *EMBO J.* **36** 291–300

Solmaz S R N and Hunte C 2008 Structure of complex III with bound cytochrome C in reduced state and definition of a minimal core interface for electron transfer *J. Biol. Chem.* **283** 17542–9

Song J *et al* 2007 Solution structure of human sorting nexin 22 *Protein Sci.* **16** 807–14

Spooner P J R *et al* 1993 Dynamics in a protein-lipid complex: nuclear magnetic resonance measurements on the headgroup of cardiolipin when bound to cytochrome c *Biophys. J.* **65** 106–12

Spooner P J R and Watts A 1991a Reversible unfolding of cytochrome c upon interaction with cardiolipin bilayers. 1. Evidence from deuterium NMR measurements *Biochemistry* **30** 3871–9

Spooner P J R and Watts A 1991b Reversible unfolding of cytochrome C upon interaction with cardiolipin bilayers. II. Evidence from phosphorus-31 NMR measurements *Biochemistry* **30** 3880–5

Spooner P J R and Watts A 1992 Cytochrome C interactions with cardiolipin in bilayers: a multinuclear magic-angle spinning NMR study *Biochemistry* **31** 10129–38

Stepanyants N *et al* 2015 Cardiolipin's propensity for phase transition and its reorganization by dynamin-related protein 1 form a basis for mitochondrial membrane fission *Mol. Biol. Cell* ed P Bassereau **26** 3104–16

Su Y, Li S and Hong M 2013 Cationic membrane peptides: atomic-level insight of structure-activity relationships from solid-state NMR *Amino Acids* **44** 821–33

Szabo C M *et al* 2000 Expression of doubly labeled *Saccharomyces cerevisiae* iso-1 ferricytochrome c and $^{1}$H, $^{13}$C and $^{15}$N chemical shift assignments by multidimensional NMR *FEBS Lett.* **482** 25–30

Tang M, Comellas G and Rienstra C M 2013 Advanced solid-state NMR approaches for structure determination of membrane proteins and amyloid fibrils *Acc. Chem. Res.* **46** 2080–8

Tuzi S *et al* 2003 Structure and dynamics of the phospholipase C-δ1 pleckstrin homology domain located at the lipid bilayer surface *J. Biol. Chem.* **278** 28019–25

Uekama N *et al* 2007 Phosphatidylserine induces functional and structural alterations of the membrane-associated pleckstrin homology domain of phospholipase C-δ1 *FEBS J.* **274** 177–87

Uekama N *et al* 2009 Influence of membrane curvature on the structure of the membrane-associated pleckstrin homology domain of phospholipase C-δ1 *Biochim. Biophys. Acta* **1788** 2575–83

Van der Wel P C A *et al* 2002 Geometry and intrinsic tilt of a tryptophan-anchored transmembrane α-helix determined by $^{2}$H NMR *Biophys. J.* **83** 1479–88

van der Wel P C A *et al* 2006 Dynamic nuclear polarization of amyloidogenic peptide nanocrystals: GNNQQNY, a core segment of the yeast prion protein Sup35p *J. Am. Chem. Soc.* **128** 10840–6

Van der Wel P C A *et al* 2007 Orientation and motion of tryptophan interfacial anchors in membrane-spanning peptides *Biochemistry* **46** 7514–24

Van der Wel P C A 2014 Lipid dynamics and protein-lipid interactions in integral membrane proteins: insights from solid-state NMR *eMagRes* **3** 111–8

Van der Wel P C A 2018 New applications of solid-state NMR in structural biology *Emerg. Top. Life Sci.* **2** 57–67

Van Doren S R *et al* 2017 Peripheral membrane associations of matrix metalloproteinases *Biochim. Biophys. Acta Mol. Cell. Res.* **1864** 1964–73

Volkov A N *et al* 2012 Redox-dependent conformational changes in eukaryotic cytochromes revealed by paramagnetic NMR spectroscopy *J. Biomol. NMR* **52** 245–56

Waltham M C, Cornell B A and Smith R 1986 Association of ferri- and ferro-cytochrome c with lipid multilayers: a $^{31}$P solid-state NMR study *Biochim. Biophys. Acta* **862** 451–6

Ward M E *et al* 2015 Proton detection for signal enhancement in solid-state NMR experiments on mobile species in membrane proteins *J. Biomol. NMR* **63** 375–88

Warschawski D E *et al* 2011 Choosing membrane mimetics for NMR structural studies of transmembrane proteins *Biochim. Biophys. Acta* **1808** 1957–74

Watts A 1997 Protein conformational distortions induced by membrane interfacial interactions *Biochem. Soc.* T **25** 1119–24

Wylie B J *et al* 2005 Site-specific $^{13}$C chemical shift anisotropy measurements in a uniformly 15N,13C-labeled microcrystalline protein by 3D magic-angle spinning NMR spectroscopy *J. Am. Chem. Soc.* **127** 11946–7

Zhong L *et al* 2007 Solid-state NMR spectroscopy of 18.5 kDa myelin basic protein reconstituted with lipid vesicles: spectroscopic characterisation and spectral assignments of solvent-exposed protein fragments *Biochim. Biophys. Acta* **1768** 3193–205

Zhou H-X and Cross T A 2013 Influences of membrane mimetic environments on membrane protein structures *Annu. Rev. Biophys.* **42** 361–92

**IOP** Publishing

Solid-State NMR

Applications in biomembrane structure

**Frances Separovic and Marc-Antoine Sani**

# Chapter 11

# Structural dynamics of G protein-coupled receptors in lipid membranes investigated by solid-state NMR spectroscopy

**Daniel Huster**

## Overview/Abstract

G protein-coupled receptors (GPCRs) belong to the most important signaling molecules in the human body. These proteins transmit an extracellular signal, triggered by absorption of a photon or binding of an ion, a small molecule, a peptide or a protein into a cascade of structural changes leading to binding of intracellular effector proteins. In spite of the great success in the structure determination of GPCRs by crystallography and cryo electron microscopy, for the understanding of the biological function of these pharmacologically highly important proteins, spectroscopic data that particularly addresses their molecular dynamics is essential. To this end, magnetic resonance spectroscopy has been very successful describing the dynamic conformational changes GPCRs undergo in their course of action. Nuclear magnetic resonance (NMR) spectroscopy has particularly contributed as it allows investigation of protein dynamics over a very broad window of time scales. Here, we summarize the procedures of preparing GPCRs for NMR investigation in sufficient quantity, with the necessary isotopic labeling and their reconstitution into membrane mimetics and true lipid membranes for optimal stabilization of the very sensitive proteins. The most common NMR methods are summarized to achieve information on the dynamic conformational equilibria of GPCRs before recent examples from the literature are discussed. The examples introduced in this chapter will focus on investigations of GPCRs reconstituted into planar lipid bilayers investigated by solid-state NMR methods.

doi:10.1088/978-0-7503-2532-5ch11

## 11.1 Introduction

Ever since the first crystal structures of GPCRs were published (Cherezov *et al* 2007, Palczewski *et al* 2000) it has been increasingly recognized that these molecules are highly dynamic in their natural membrane environment (Latorraca *et al* 2017). GPCRs are heptahelical membrane proteins that convert chemical signals arriving at the extracellular side of a GPCR in the form of an ion, a small molecule, a peptide or a protein into a biological response. This process is mediated by structural transitions within the protein leading to intracellular G-protein binding and initiation of an intracellular signal transduction cascade and finally desensitization by β-arrestin. GPCRs constitute the largest family of signaling molecules in biology. Thanks to very intense research over the last 15 years that led to significant advances in protein engineering and stabilization, along with sophisticated methods of membrane protein crystallization and methodological developments in x-ray and cryo-electron microscopy, we now have access to the structures of about 70 unique GPCRs (Shimada *et al* 2019), some of which are in the activated and/or G-protein bound states (García-Nafría and Tate 2019, Hilger *et al* 2018) and also most recently structures in complex with β-arrestins were published (Huang *et al* 2020, Staus *et al* 2020). However, as often noted previously, crystal structures represent static snapshots of highly mobile molecules and the mode of action of these receptors can only been deciphered with the help of spectroscopic tools to allow directly addressing the structural dynamics of GPCRs in response to activation by agonist binding and subsequent interaction with intracellular G-proteins and arrestins. In this endeavor, NMR spectroscopy carried out in solution has contributed various important insights revealing details of the dynamic transitions between the conformational states of these molecules (Bostock *et al* 2019). In spite of this great success, solution NMR methods require the proteins of interest to tumble quickly in solution rendering investigation of GPCRs in planar membranes, which represent the natural environment of these molecules, impossible. Rather, for solution NMR applications, GPCRs are solubilized in small size membrane mimetics, which usually insufficiently represent the properties of true planar lipid membranes. An alternative is solid-state NMR spectroscopy, which can be applied to GPCRs reconstituted into planar lipid membranes of relatively native composition. Furthermore, it is important to realize that the conformational ensemble of a GPCR in the lipid bilayer is distinctly different from that in detergent micelles (van Eps *et al* 2017), necessitating the investigation of GPCR conformational equilibria in true membranes. Though often inferior in resolution, solid-state NMR in combination with specific isotope labeling has contributed to our understanding of GPCR function and structural dynamics. Here, we will discuss some recent solid-state NMR results of GPCRs in their native membrane environment.

The molecular basis for the pronounced dynamics of GPCRs is described by a complex energy landscape in which these molecules exist that is characterized by shallow minima allowing for the GPCRs to overcome low energy barriers in thermal equilibrium and higher barriers that need to be crossed as the receptors go through the circle of activation by agonist binding, G-protein interaction and desensitizing by arrestin binding (Deupi and Kobilka 2010, Manglik and Kobilka 2014). Many recent NMR studies have elegantly shown that GPCRs in the apo or ligand bound states exist in an equilibrium of several conformations (Bostock *et al* 2019, Manglik *et al* 2015, Shimada *et al* 2019, van Eps *et al* 2017). This has the important consequence that the assumption of an on/off switch between the inactive and active conformations is too simplistic to describe the biological function of GPCRs. Rather, the activation of these molecules follows a very complex pattern of highly dynamic structural changes between several receptor sub-states each characterized by a specific conformation. These conformational transitions during activation and interaction with intracellular effectors can only be studied in atomistic detail using spectroscopic tools (Bostock *et al* 2019, Weis and Kobilka 2018).

This calls for classical biophysical techniques that allow investigation of reconstituted receptor molecules in lipid bilayers, such as Fourier transform infrared, fluorescence, EPR or solid-state NMR spectroscopy (Briddon and Kilpatrick 2018, Chavez-Abiega *et al* 2019, van Eps *et al* 2015, Wingler *et al* 2019). Results of the latter technique will be discussed in the current chapter. Solid-state NMR was used to describe the structure of the CXCR1 receptor in phospholipid membranes (Park *et al* 2012b). It also holds great potential for the quantitative description of the molecular dynamics of membrane reconstituted GPCRs over a broad window of correlation times (Palmer *et al* 1996). In the following, we will summarize a few general aspects of GPCR production, labeling and reconstitution for a broad range of NMR investigations but discuss exclusively recent studies that have used solid-state NMR methods to characterize GPCR dynamics. For a comprehensive overview on the investigation of the conformational equilibria of GPCRs by solution NMR, the reader is referred to recently published reviews (Bostock *et al* 2019, Shimada *et al* 2019).

## 11.2 Preparative options for studying GPCRs by NMR spectroscopic tools

As with most membrane proteins, preparation of high quality samples of GPCRs in a stabilizing environment is of utmost importance but remains a big challenge (Das *et al* 2015). The problems start already at the expression level. Protein engineering has helped to increase expression yield and prepare more stable GPCR variants. Several expression systems have been utilized, strategies for selective isotope labeling have been developed and purification protocols optimized. Along with the biochemical optimization, several molecular assemblies have been developed to keep GPCRs in a stable membrane mimicking a true membrane environment. A summary of the most important procedures is given in figure 11.1.

### 11.2.1 Construct optimization and expression systems

Preparation of recombinant GPCRs for successful crystallization trials requires careful protein engineering to obtain stabilized receptors by thermostabilization (Vaidehi *et al* 2016), loop modification (Rosenbaum *et al* 2007), mutagenesis to reduce the number of free cysteins (Witte *et al* 2013), and/or attachment of tags for

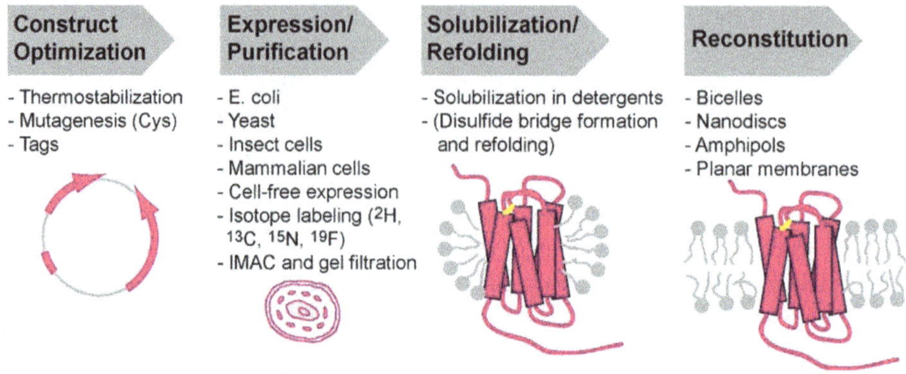


**Figure 11.1.** Summary of common procedures/systems required for production and preparation of stable GPCRs for structural and dynamic characterization by NMR spectroscopy.

increased expression and/or purification (Henrich *et al* 2015, Vandemoortele and Gevaert 2019). Careful optimization of each receptor plasmid prior to expression and protein reconstitution is key to successful biophysical characterization of the molecules.

Various microorganisms are available to express GPCRs in various quantities and qualities (Sarramegna *et al* 2006). As NMR is a technique of low sensitivity, usually high yield expression is preferred. However, the functionality of the proteins produced in expression systems that provide high protein quantities is typically low. While GPCRs are not native to the most commonly used expression host *Echerichia coli* (*E. coli*), special tricks have to be applied to induce expression of these molecules and insertion into the inner membrane of the host bacterium (Grisshammer *et al* 2005, Krepkiy *et al* 2006, Schlinkmann *et al* 2012, Yeliseev *et al* 2005). Under these conditions, a relatively high percentage of correctly folded receptors at rather low yields are obtained. Alternatively, GPCRs can be expressed in good yield in *E. coli* as inclusion bodies (Park *et al* 2006, Schmidt *et al* 2009) but, under these circumstances, the proteins are aggregated and have to be renatured following often very extensive and cumbersome refolding protocols (Baneres *et al* 2003, Das *et al* 2015, Kiefer *et al* 1999, Park *et al* 2012a, Schmidt *et al* 2009, Schmidt *et al* 2018). In most cases, these procedures yield a mix of misfolded and correctly folded GPCRs that have to be separated, for instance by affinity chromatography (Bosse *et al* 2011). Nevertheless, for NMR applications, *E. coli* expression represents a good option as large quantities of receptors can be produced and isotopic enrichment is straightforward at a relatively low cost.

In contrast to the prokaryotic expression in *E. coli*, the methylotrophic yeast *Pichia pastoris* (*P. pastoris*) represents an eukaryotic host for expression of the eukaryotic GPCRs (Clark *et al* 2018). *P. pastoris* also provides attractive options for isotopic labeling. Quite intriguing are options for specific labeling of isoleucine methyl groups on a perdeuterated protein background providing ideal conditions for methyl-TROSY experiments (Clark *et al* 2017). Also U-$^{15}$N and selective amino acid labeling has been reported (Eddy *et al* 2018b, Eddy *et al* 2018a). Straightforward purification protocols yield high levels of folded and functional receptor for structural studies (Clark *et al* 2017, Eddy *et al* 2018b, Eddy *et al* 2018a, Mizumura *et al* 2020).

A large number of studies in the field have reported results on receptors that were expressed in infected insect cells from *Spodoptera frugiperda* (Sf9) (Massotte 2003), but also other strains, which provide highly functional and high quality protein at the expense of low expression yields and expensively large expression volumes. For isotopic labeling, cells have to be fed with fully isotope labeled amino acids from supplemented media representing the most expensive option as cultures have to be large for sufficient yields. However, recent advances have led to the development of relatively inexpensive extracts from yeast and algae for isotopic labeling which represent an attractive alternative (Franke *et al* 2018).

Mammalian cell lines such as HEK293 certainly provide the most native environment for GPCRs but also provide the lowest expression yields. In GPCR structural biology, this method has only been successfully established and optimized for rhodopsin and $^{15}$N-Trp (Werner *et al* 2007) and even $^{13}$C labeling of quite a few residue types in rhodopsin has been described (Ahuja *et al* 2009).

Finally, cell free expression techniques have become very advanced such that they provide sizable yields of protein in very small volumes (achievable yields are on the order of 1 mg/ml), which is attractive for specific isotopic labeling of GPCRs (Henrich *et al* 2015). The most commonly used protocols involve *E. coli* extracts that are prepared from bacterial cultures but extracts from other microorganisms have also been described (Bernhard and Tozawa 2013, Kögler *et al* 2019). As cell free expression is carried out as an open process, all extracts can be supplemented with isotopically labeled precursors or amino acids which will not be metabolized (Hoffmann *et al* 2018). However, the problem of refolding as reported for *E. coli* expression in inclusion bodies remains. To circumvent this problem, co-translational folding of GPCRs in the presence of membrane mimetics such as detergent micelles or nanodiscs have been described in cell-free expression strategies so that the expressed protein is refolded and reconstituted within the cell-free expression batch (Rues *et al* 2016). Alternatively, renaturation protocols as applied for preparation of GPCRs from inclusion bodies have to be used (Krug *et al* 2020).

### 11.2.2 Isotopic labeling

Isotopic labeling is essential to all NMR studies of membrane protein. The simplicity of uniform protein labeling with $^{2}$H, $^{13}$C and $^{15}$N from basic molecular precursors such as $D_2O$, $^{13}$C-glucose, and $^{15}$N-ammonium salts in *E. coli* culture represents the basis of the large success in NMR studies of soluble proteins. However, due to increased NMR line width and reduced signal dispersion due to the predominantly $\alpha$-helical bundle or $\beta$-barrel secondary structure of membrane proteins, uniform labeling often proves difficult for such applications. In NMR studies of membrane proteins and of GPCRs in particular, usually more specific labeling schemes have to be applied.

Fully labeling represents still the best way to study (membrane) proteins provided the line width is sufficiently low and the resolution good. These prerequisites are fulfilled for quite a few proteins but GPCRs, due to their intrinsic dynamics and structural heterogeneity, tend to provide broader NMR lines. In such cases, it can be helpful to apply more specifically labeled glucose or glycerol precursors for more fractional labeling (Castellani *et al* 2002, Hong 1999, Hong and Jakes 1999, Loquet *et al* 2011, Lundström *et al* 2007) instead of using U-$^{13}$C-glucose as the carbon source. Such a strategy has led to the first and so far only characterization of a GPCR structure by solid-state NMR (Park *et al* 2012b). Relatively high backbone labeling is advantageous for obtaining sequential assignments along the protein backbone but this is usually hampered in GPCRs by the comparably large NMR linewidths caused by intermediate timescale motion and structural heterogeneity.

Thus, specific labeling often represents the only alternative. By this approach, only a few isotopic labels are introduced either as native residues or by modification or replacement of natural residues by artificial amino acids or their sidechains. All these specific labeling strategies are advantageous because of the high selectivity and the relatively small number of NMR signals obtained from these isolated sites. However, as sequential connectivities are typically not available for the isolated labels, signal assignment of these sites is often very cumbersome by preparing receptor variants in site-directed mutagenesis lacking certain sites (Imai *et al* 2020).

For their favorable NMR properties, which include highest sensitivity and long relaxation times, $^{13}$C-labeled methyl groups have become very popular probes in GPCR research. One of the first studies in that regard used $^{13}$C-formaldehyde to modify lysine sidechains with $^{13}$C-methyl groups (Bokoch *et al* 2010). By covalently attaching one or two methyl groups, the positive charge of the lysine sidechain remains intact (Larda *et al* 2012) and a highly sensitive NMR label is introduced. Further strategies for perturbation free methyl labeling were developed in ongoing research resulting in protocols that allow for specific $^{13}CH_3$-ε-methionine labeling (Kofuku *et al* 2012, Nygaard *et al* 2013). Another useful alternative option is $^{15}$N-labeling of tryptophan residues, which can be used as both intrinsic (Eddy *et al* 2018b) or extrinsic probes (Eddy *et al* 2018a). Tryptophan sidechains also represent good NMR probes with relatively unique chemical shifts far from the amide signals of the receptor and the natural abundance background of the aliphatic signals of detergents and lipids. However, backbone labeling of tryptophan in cell culture is difficult as the enzyme tryptophanase cleaves off the Trp sidechain leaving the backbone subject to isotope scrambling (Berger *et al* 2013).

For the backbone labeling of GPCRs, $^{13}$C-glycine labeling of the $A_{2A}$ adenosine receptor in yeast culture (Eddy *et al* 2018b), $^{15}$N-valine labeling of the β1 adrenergic receptor in Sf9 cells (Isogai *et al* 2016) and $^{13}$C-labeling of various residues in rhodopsin in HEK293 cells has been reported (Ahuja *et al* 2009, Kimata *et al* 2016).

An alternative to isotopic labeling of native residues in GPCRs is the introduction of artificial amino acids or sidechain modification of natural or extrinsic cysteine residues. In this regard, $^{19}$F especially as one of the most sensitive NMR nuclei has been used. The most common strategy is modification of engineered cysteine residues with 2,2,2 trifluoroethanethiol (TET) (Frei *et al* 2020, Liu *et al* 2012), trifluoromethyl (–$COCF_3$) (Chung *et al* 2012) or trifluoromethyl moiety 2-bromo-*N*-(4-(trifluoromethyl)phenyl) acetamide (BTFMA) (Ye *et al* 2016, Ye *et al* 2018).

### 11.2.3 Reconstitution into membranes and membrane mimetics

Many membrane mimetic assemblies of molecules have been described that can stabilize membrane proteins in solution, many of which are amenable to solution NMR techniques (Klopfer and Hagn 2019, Laguerre *et al* 2016). These systems consist of polymers, detergents, micelles or bicelles, lipid/protein and lipid/polymer. The choice of the right membrane mimetic is crucial as it can significantly influence the rate of exchange between individual receptor conformations (Chung *et al* 2012, van Eps *et al* 2017) and thus the quality of the observed NMR spectra. Nevertheless, reconstitution into planar membranes remains the gold standard, as any membrane mimetic often poorly represents essential properties such as intrinsic curvature, lateral stress profile and the specificity of headgroup chemistry, chain length and degree of unsaturation, and specific lipids such as cholesterol. Consequently, the conformational ensemble of a GPCR in the lipid bilayer is distinctly different from that in detergent micelles (van Eps *et al* 2017). Various systems to solubilize GPCRs for solution NMR investigations have been reviewed recently (Klopfer and Hagn 2019, Puthenveetil 2019, Ravula *et al* 2019, Seddon *et al* 2004).

Key requirements of good membrane mimetics for solution NMR spectroscopy include a small molecular weight of the assembly to allow for fast isotropic rotational diffusion of the protein/polymer or protein/detergent complex as well as sufficient stability to prevent intermediate exchange between receptor states as it broadens NMR lines, along with sufficient stability of the GPCR/membrane mimetic complex to avoid intermolecular aggregation at the high protein concentrations required for NMR studies, which is in the ~100 μM range. To this end, specific detergents for GPCR stabilization such as LMNG have been developed (Chae *et al* 2010, Eddy *et al* 2018b, Frei *et al* 2020, Umai *et al* 2020). But also classical detergents such as maltosides (Isogai *et al* 2016, Mizumura *et al* 2020, Nygaard *et al* 2013) enjoy high popularity. Also the cholesterol analogue cholesteryl hemisuccinate has been identified as an important stabilizing additive (Grisshammer *et al* 2005).

Bicelles represent more membrane-like discoidal lipid assemblies that were introduced as membrane mimetics of various sizes consisting of a bilayer preferring phospholipid and a short chain detergent (CHAPSO or DHPC) that forms the rim of the lipid disc. Bicelles can be prepared in different sizes amenable for both solution and solid-state NMR spectroscopy as determined by the molar ratio of phospholipid to detergent, the q ratio (Ram and Prestegard 1988). A further advantageous property is the alignment of large bicelles in the magnetic field of the NMR spectrometer (Prosser *et al* 1996, Sanders and Prestegard 1990), which has also been utilized in GPCR research (Park *et al* 2006). Thus, bicelles of various sizes can be used in combination with solution and solid-state NMR studies (Park *et al* 2011a). Bicelles can be prepared from a variety of different lipids to model a natural environment for GPCRs (Schmidt *et al* 2018, Thomas *et al* 2015). Especially high q bicelles are very well suited for solid-state NMR experiments as they resemble planar membranes with small holes and pores (Sternin *et al* 2001, Triba *et al* 2005).

Somewhat more stable membrane mimetics are nanodiscs, approximately 10 nm diameter disk-shaped lipid bilayer assemblies held together by a protein ring of the membrane scaffolding protein from high density lipoprotein (Bayburt *et al* 2002). These systems have also been prepared using covalently circularized membrane scaffold protein providing very homogeneous preparations especially since empty nanodiscs can easily be removed in a $Ni^{2+}$ affinity chromatography step (Nasr *et al* 2017). An advantage of nanodiscs is that specific lipids known to be essential for GPCR function in biology such as polyunsaturated phospholipids can be incorporated, providing a relatively native environment for GPCRs, which has been shown to influence the conformational equilibrium of the adenosine $A_{2A}$ receptor (Mizumura *et al* 2020). However, GPCRs loaded nanodiscs approach a molecular weight of 200 kDa restricting the choice of useful NMR applications to mostly methyl-TROSY experiments.

Without doubt, the best environment for investigation of GPCRs is planar lipid membranes with composition which resembles that of the natural membrane as closely as possible (Oates and Watts 2011). This includes lipids of different chain lengths and degrees of unsaturation, headgroup structure as well as sterols such as cholesterol. These environments can only be satisfactorily provided in planar membranes, preferably at a low protein-to-lipid molar ratio. There is also tremendous research interest in elucidating the specificity of interaction of membrane proteins with specific lipid molecules (Huster 2014, Oates and Watts 2011). The most investigated GPCR in that regard is bovine rhodopsin, for which the importance of chemical parameters such as lipid chain length and headgroup structure, degree of unsaturation and cholesterol content (Soubias *et al* 2006, Soubias *et al* 2008, Soubias *et al* 2015, Soubias and Gawrisch 2005, Soubias and Gawrisch 2013) as well as physical parameters such as intrinsic curvature, lipid order and lateral stress profile, and membrane elasticity (Botelho *et al* 2002, Gibson and Brown 1993, Soubias *et al* 2010) have been studied in detail. This way, the membrane and specific lipids can be viewed as important allosteric regulators of GPCR function (Soubias *et al* 2015).

The general reconstitution protocol for membrane proteins involves solubilization of the protein in the best suited detergent, mixing with solubilized lipids or lipid bicelles, and removal of the detergent by dialysis or biobeads and methyl-β-cyclodextrin (Das *et al* 2015, Park *et al* 2012a, Skrzypek *et al* 2018). Although allegedly simple, however, for optimal reconstitution the general protocol needs to be optimized for each system including the choice of detergents for solubilization, the choice of lipid for reconstitutions and the buffer conditions, which requires time and effort. Certainly, careful activity tests should be commenced to ensure full protein activity.

## 11.3 Methodological options for studying dynamics of GPCRs by solid-state NMR

Solid-state NMR is an ideal tool to comprehensively investigate the molecular dynamics of GPCRs, addressing both the time scale and the amplitude of a respective molecular motion (Huster 2005, Palmer *et al* 1996, Schanda and Ernst 2016). As solid-state NMR spectroscopy does not require isotropic rotational diffusion of the protein of interest, molecular dynamics can be observed directly and is not masked by the overall tumbling of the GPCR and the respective membrane mimetic in solution. Solid-state NMR can be carried out under stationary conditions yielding NMR powder spectra that contain information on the geometry of motion or under magic angle spinning (MAS) conditions, providing higher resolution for analysis of conformation dependent isotropic chemical shifts and relaxation rates (Das *et al* 2015, Elkins and Hong 2019, Huster 2005, Ladizhansky 2017). Information from anisotropic NMR interactions that is lost in MAS applications can be reintroduced into the MAS spectra by recoupling (Baldus 2002, Dusold and Sebald 2000) using separated local field experiments (Waugh 1976). A summary of the four most common experimental parameters that report dynamic properties and structural transitions in proteins by solid-state NMR are illustrated in figure 11.2.

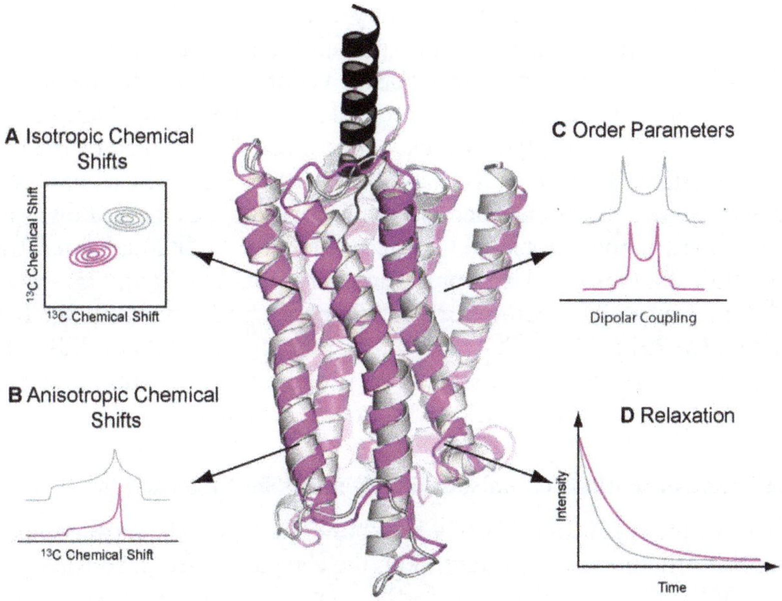


**Figure 11.2.** Superposition of models of the neuropeptide Y receptor type 2 in the apo (gray) and the agonist bound state (magenta) with the agonist, neuropeptide Y (NPY), shown in black (Kaiser *et al* 2015) and methodological principles for studying GPCR dynamics by solid-state NMR. Essential NMR parameters reporting the conformational dynamics of the GPCR are depicted: (A) Conformational transitions between structurally different receptor states are observed by isotropic chemical shift changes. Motional averaging leads to characteristic changes of (B) NMR powder spectra or MAS sideband patterns, which can be converted into (C) molecular order parameters. (D) Stochastic fluctuations of molecular bond vectors drive NMR relaxation allowing one to obtain motional correlation times and information on the geometry of the molecular motion.

### 11.3.1 Isotropic chemical shift changes

Isotropic chemical shifts are the simplest parameters detected by NMR that are related to local protein structure. Structural transitions of a protein usually cause alterations in the chemical shift. As the chemical shift is determined by the local magnetic environment of a given nucleus—it reflects the influence of the electron cloud around a nucleus of interest—which is related to the backbone and/or sidechain structure of an amino acid in a protein (Wishart and Sykes 1994). Isotropic chemical shifts are also related to the environment of a nucleus and especially ring current effects induced by aromatic sidechains are useful to detect tertiary structure changes (Tang and Lam 2019). As GPCRs are characterized by an ensemble of conformations, each individual structure gives rise to a specific isotropic chemical shift of a particular backbone site. Thus, this simple parameter represents an ideal tool to study the conformations of a GPCR and transitions between them (Bostock *et al* 2019). Gradual chemical shift changes upon GPCR interaction with partial and full agonists determined by solution NMR techniques have been reported (Eddy *et al* 2018b, Imai *et al* 2020, Manglik *et al* 2015, Solt *et al* 2017).

Difficulties can arise when a GPCR exists in many conformations with little structural differences as it can lead to very broad NMR lines causing losses in spectral resolution, which introduces the difficulty that the individual states can no longer be distinguished. If the exchange between conformations occurs on the fast or slow timescale limit, further information on the rate of exchange can be obtained. Intermediate timescale exchange with exchange times in the range of the inverse of the frequency difference of the individual NMR signals will cause signal broadening and intensity losses. Isotropic chemical shift analysis has been most successful in solution NMR applications of GPCRs where the individual conformational ensembles, and changes in their populations in response to ligand binding and G-protein interaction, have been described in detail (Eddy *et al* 2018b, Imai *et al* 2020, Isogai *et al* 2016, Manglik *et al* 2015, Solt *et al* 2017). Also solid-state NMR chemical shift changes of rhodopsin reconstituted into planar bilayer membranes have been reported (Ahuja *et al* 2009, Kimata *et al* 2016).

### 11.3.2 Influence of motions on anisotropic chemical shift interactions

While the isotropic chemical shifts of a backbone site report the local structure of an individual conformation of a given protein, the chemical shift anisotropy giving rise to broad NMR powder spectra can further report details about the geometry of the molecular motion of a respective site. NMR powder patterns dominated by the chemical shift anisotropy reflect the screening of the external magnetic field by the magnetic environment of a nucleus that is described by a second rank tensor. This chemical shielding tensor is related to the local electronic environment and assumes a specific orientation with respect to the molecular frame. Molecular motions of a given amplitude can scale down the shielding tensor and change its orientation with respect to the motional axis, which leads to characteristic chemical shift changes (Lewis *et al* 1985) that are manifested either in alterations of the NMR powder spectra detected under static conditions or the intensity distribution of the MAS sideband pattern. Thus, anisotropic chemical shifts report on the geometry of a molecular motion provided the correlation time of motion is faster than the inverse of the interaction strength, which depends on the magnetic field strength of the spectrometer and the chemical shift anisotropy of the given nucleus but usually amounts to correlation times faster than a few or a few tens of microseconds. While molecular motions of small and moderate amplitudes scale down the NMR powder or MAS sideband spectra, motions of large amplitude lead to very narrow and even isotropic lines. Thus, the specific spectral shape of an NMR powder pattern and detection of isotropic lines instead of a powder pattern can provide a lot of information on the mobility of a GPCR in the lipid membrane. However, NMR powder spectra of $^{13}$C or $^{15}$N nuclei are typically acquired using cross-polarization from $^{1}$H nuclei. Cross polarization (CP) depends on the dipolar coupling between the heteronucleus to its attached proton(s) for the polarization transfer, and the magnitude of the dipolar coupling is also subject to motional averaging. Thus, a dynamic bias applies when exciting NMR coherences by CP methods (Schmidt *et al* 2014). Direct detection provides a true intensity distribution but such methods are hardly used as quantification requires long repetition times due to very long relaxation times and, especially for the low gyromagnetic ratio nucleus $^{15}$N, direct detection is very inefficient (Thomas *et al* 2015).

### 11.3.3 Molecular order parameters

Molecular order parameters which report the motional amplitude of a given bond vector, typically $^{1}$H–$^{13}$C or $^{2}$H–C, have long been used in NMR biophysical studies to characterize the dynamics of biological systems (Davis 1983, Seelig 1977). For a rigid bond vector, the $^{1}$H–$^{13}$C dipolar coupling or $^{2}$H–C quadrupolar coupling has a maximal value which corresponds to an order parameter of 1. As the molecular amplitude increases, the dipolar or quadrupolar interactions are successively scaled down, leading to motionally averaged interaction strengths that give rise to order parameters between 1 and 0. The ratio of this motionally averaged interaction by its rigid limit value defines the molecular order parameter. If the motional amplitude is 90° the order parameter is 0. Mathematical models can be employed to convert the molecular order parameter into amplitude information of the respective bond vector (Brüschweiler and Wright 1994). Molecular order parameters can be determined very conveniently from static $^{2}$H NMR using direct polarization which results in a Pake doublet for a single deuterium site or a superposition of such Pake doublets if several sites are labeled. The deconvolution of such complicated $^{2}$H NMR spectra is often not straightforward as the assignments are usually not known (Vogel *et al* 2016). More attractive are $^{13}$C-detected separated local field experiments that measure a motionally averaged dipolar coupling in the indirect dimension of a two-dimensional NMR experiment for each resolved signal in an MAS spectrum for which ideally the assignment is known. Numerous NMR pulse sequences are available to measure motionally averaged $^{1}$H–$^{13}$C dipolar couplings (Baldus 2002, Dusold and Sebald 2000). Provided the resolution is reasonable and NMR signals can be detected with sufficient resolution, such experiments can provide a very complete picture of the motions of a protein (Aharoni and Goldbourt 2018).

### 11.3.4 Relaxation rates

Finally, relaxation rates provide a very comprehensive means to characterize the molecular motions of a membrane protein. Relaxation of NMR coherences is caused by stochastic fluctuations of the interaction fields: thus the relaxation rates determined in an NMR experiment can report on the molecular dynamics of a given bond vector of the protein undergoing such dynamics (Ladizhansky 2017, Schanda and Ernst 2016). For all relaxation rates, the exact dependence on spectral density functions that provide insight into the fluctuating interaction fields is known (Ladizhansky 2017, Palmer *et al* 1996, Schanda and Ernst 2016). The spectral density functions are the Fourier transforms of the correlation function of a specific motion and various molecular models can be applied to model relaxation times providing information on the correlation time and amplitude of motion of a membrane protein (Good *et al* 2014, Reuther *et al* 2006). As the correlation times of motion can also very conveniently be determined in molecular dynamics (MD) simulations, the combination of NMR and MD has also been shown to be highly advantageous to help interpret experimental results (Smith *et al* 2019, Vogel *et al* 2007, Vogel *et al* 2010). Although relaxation rates are highly useful experimental quantities, very little work has been reported so far with regard to GPCRs. The few examples published so far include the dynamics of the retinal in rhodopsin which has been characterized by $^{2}H$ NMR relaxation in the solid state (Struts *et al* 2011), the dynamics of the peptides dynorphin bound to the human κ opoid receptor (O'Connor *et al* 2015) and ghrelin bound to the GHS receptor (Ferré *et al* 2019), and the dynamics of $^{19}F$ probes on the β2 adrenoreceptor (Chung *et al* 2012). Since NMR relaxometry has been very successful to characterize the molecular dynamics of heptahelical bacterial proteins (Good *et al* 2014), it is expected that the method will potentially play a major role in characterizing the molecular dynamics of GPCRs as well.

## 11.4 Examples

Solid-state NMR techniques have been used to determine a structural model of the CXCR1 receptor (Park *et al* 2012b) and provide quite a few models for complexes of peptide ligand receptors with their agonists. These include the ligands of the neurotensin receptor (Luca *et al* 2003), the GHS receptor (Bender *et al* 2019, Ferré *et al* 2019), the neuropeptide Y receptor type 1 (Yang *et al* 2018) and 2 (Kaiser *et al* 2015), and the bradykinin B1 (Joedicke *et al* 2018) and B2 (Lopez *et al* 2008) receptors. However, explicit dynamics studies of GPCRs are limited. Most data on the dynamics of a GPCR are available for rhodopsin, which was subject to numerous reviews (Mertz *et al* 2012, Pope *et al* 2013, Struts *et al* 2015). Here, we will highlight a few examples of peptide ligand GPCRs

### 11.4.1 Chemokine receptor CXCR1

Chemokine receptors are involved in many biological processes, in particular inflammation, metastasis, embryogenesis, host defense against infection and innate immunity. The ligands of chemokine receptors are small molecules such as the protein IL-8 (Fernandez and Lolis 2002). The chemokine receptor CXCR1 expressed in *E. coli* as inclusion bodies, refolded and reconstituted into lipid bilayers (Park *et al* 2012a) was the first receptor for which comprehensive solid-state NMR experiments yielded a structural model of the molecule reconstituted in lipid bilayers determined solely by solid-state NMR spectroscopy (Park *et al* 2012b).

In addition to the structural studies, also the molecular dynamics of the CXCR1 was investigated by solid-state NMR (Park *et al* 2011b). Using CXCR1 reconstituted into small isotropically tumbling bicelles, Opella and coworkers detected only a small number of mobile backbone sites stemming from the N- and C-termini of the receptor that would provide signals under solution NMR conditions. No evidence for any backbone motions for the rest of the receptor molecule was found. For addressing the motions of the receptor as a whole in the membrane, the motional averaging of the receptor carbonyl and amide sites were investigated (Park *et al* 2011b) using stationary $^{13}$C and $^{15}$N NMR powder patterns as first reported for bacteriorhodopsin (Lewis *et al* 1985). From characteristic narrowing and lineshapes of the NMR powder patterns indicative of an axially symmetric chemical shift anisotropy tensor it was concluded that the receptor was undergoing axially symmetric rotational diffusion about the membrane normal but only when the membrane is in the liquid crystalline phase state (Park *et al* 2011b). The same conclusion could also be obtained from $^{13}$C MAS NMR at low rotational frequencies (figure 11.3). While isotropic $^{13}$CO signals without any rotational sideband intensity were detected at 30 °C, the $^{13}$CO region of the spectra showed the sideband pattern characteristic of the full chemical shift tensor of the carbonyl sites. The rate of axially symmetric motions of the receptor was estimated to be >30 kHz. A somewhat slower timescale was probed by the $^{15}$N NMR experiments of

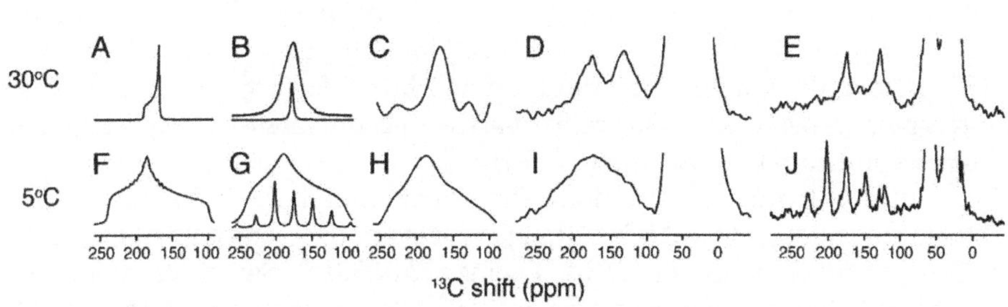


**Figure 11.3.** Simulated and experimental $^{13}$CO solid-state NMR spectra of CXCR1 35% $^{13}$C enriched in DMPC/POPC membranes at 30 °C (top row) and 5 °C (bottom row): (A, F) Simulated $^{13}$C NMR powder spectra, and (B, G) $^{13}$C MAS NMR spectra illustrating the effect of axially symmetric motions about the membrane normal; (C, H) experimental $^{13}$C CSA recoupling experiment; (D, I) $^{13}$C CP NMR powder spectra of a stationary sample; and (E, J) $^{13}$C NMR spectrum obtained under MAS conditions with a slow spinning frequency of 5 kHz. Reprinted with permission from (Park *et al* 2011b). Copyright 2011 American Chemical Society.

the amide labeled CXCR1 but, as the $^{15}$N chemical shift tensor is almost aligned with the $^{15}$N–$^{1}$H bond, these spectra are generally not so sensitive. Furthermore, the maximal contact time in the $^{15}$N CP NMR spectra was 1 ms, more mobile sites could not be detected due to their weak dipolar couplings (*vide infra*). This might have been the reason why it was stated that no evidence for motions in the loops of CXCR1 was found (Park *et al* 2011b). Other than the axially symmetric rotational diffusion and the mobility of a few sites in the termini, the CXCR1 receptor was found to be highly rigid.

### 11.4.2 Neuropeptide Y receptor type 2

There are four neuropeptide Y (NPY) receptors in humans: a crystal structure is currently only available for the Y1 receptor (Yang *et al* 2018). Our group has been interested in studying the molecular dynamics of the Y2 receptor (Y2R) prepared in *E. coli* and refolded from inclusion bodies reconstituted into planer lipid bilayers (Schmidt *et al* 2018, Witte *et al* 2013). The ligand of the Y2R is the 36 amino acid peptide amide neuropeptide Y (NPY). To study the dynamics of the receptor in bilayer membranes, the molecule was uniformly $^{13}$C/$^{15}$N labeled and static $^{15}$N NMR spectra showed a powder pattern dominated by an axially symmetric chemical shift anisotropy tensor suggesting that the receptor undergoes axially symmetric rotational diffusion about the membrane normal. However, superimposed on the powder pattern, a significant proportion of isotropic signal intensity from the $^{15}$N amide sites was also detected (Schmidt *et al* 2014). As the CP transfer from $^{1}$H to $^{15}$N is a significantly biased towards rigid structures, $^{15}$N CP NMR spectra were recorded as a function of contact time. As the CP contact time increased, more mobile sites featuring lower dipolar coupling rare polarized and consequently an increase in the intensity of these near isotropic NMR signals was observed as the contact time increased (figure 11.4). Analysis of the signal intensity of the isotropic and anisotropic spectral components recorded as a function of CP contact time revealed that the narrow NMR signals accounted for about 40% of the signal intensity, which was confirmed by directly polarized $^{15}$N NMR spectra as shown in figure 11.4D (Thomas *et al* 2015). Similarly, static $^{2}$H NMR spectra of U-$^{2}$H labeled Y2R showed a superposition of a broad Pake spectrum indicative of rigid receptor segments as well as rather narrow spectral intensity, indicating highly mobile receptor segments (Schmidt *et al* 2014).

In an attempt to quantify the molecular dynamics in more detail and to achieve better site resolution, $^{13}$C MAS NMR experiments of the reconstituted U-$^{13}$C-Y2R were carried out (Schmidt *et al* 2014, Thomas *et al* 2015). Signal resolution of the highly labeled molecule, however, was only sufficient to distinguish $CH_3$, $CH_2$, and CH groups. In the protein backbone, only the Gly C$\alpha$ signals could be detected separately. Separated local field experiments using the DipShift pulse sequence (Munowitz *et al* 1981) allowed determining C–H order parameters for the backbone and the sidechain carbons (figures 11.4(E, F)). These order parameters trivially decrease from the backbone towards the sidechain but more interestingly depended also on the CP contact times. Highest order parameters were detected at short CP

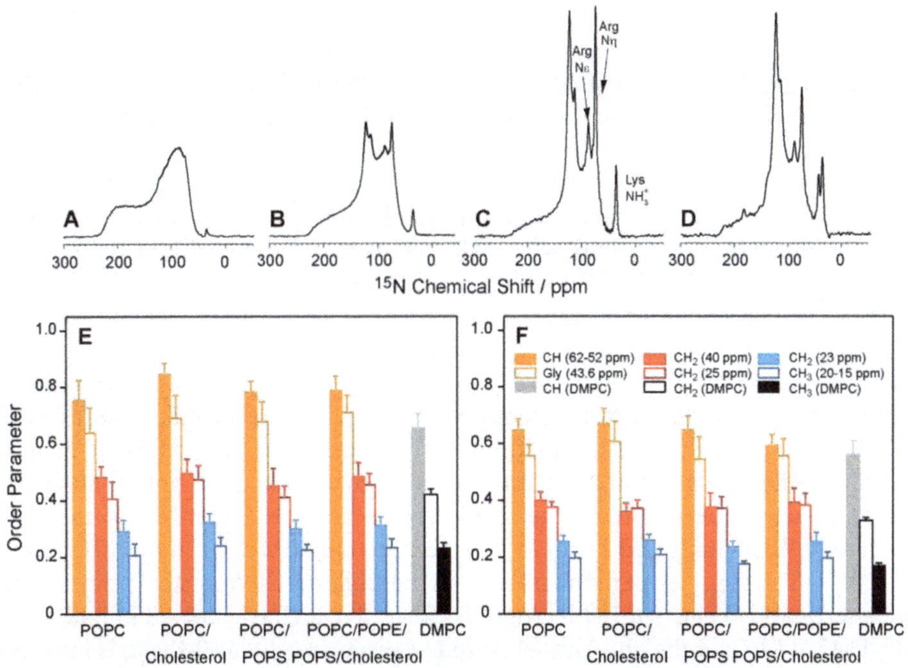


**Figure 11.4.** $^{15}$N NMR spectra from a stationary sample of Y2R reconstituted into POPC membranes and $^{1}$H–$^{13}$C order parameters when reconstituted into membranes of different composition determined from DipShift experiments recorded at a temperature of 37 °C: (A-D) $^{15}$N NMR spectra were acquired using CP with contact times of (A) 70 μs, (B) 2 ms, (C) 8 ms, and (D) direct polarization. (E–F) $^{1}$H–$^{13}$C order parameters of the Y2R in various membranes were determined from DipShift experiments using (E) CP with contact time of 0.7 ms, and (F) direct polarization. Data adapted from Thomas *et al* (2015) with permission. Copyright 2015, Springer Nature.

contact times as only highly rigid sites with large dipolar couplings are polarized. These order parameters measured in DipShift experiments with short CP contact times were assigned to the central transmembrane segments of Y2R. At longer contact times, the residues in more mobile sites such as the helix ends and the loop and tail structures with weaker dipolar couplings also were excited and the detected order parameters dropped to lower values. Lowest order parameters were detected when DipShift experiments were conducted with direct polarization of the $^{13}$C nuclei, where no dynamic bias existed (figure 11.4(F)). In the directly polarized DipShift experiments, order parameters of 0.55 for the backbone CH, 0.33 for the sidechain $CH_2$, and 0.17 for the $CH_3$ groups were detected for Y2R reconstituted into DMPC membranes, corresponding to motional amplitudes greater than 50° (Schmidt *et al* 2014). In more physiological phospholipids (POPC) and lipid mixtures (POPC/cholesterol, POPC/POPS, and POPC/POPE/POPS/cholesterol) the detected order parameters were a bit higher, likely due to the slightly larger hydrophobic thickness of these membranes (figures 11.4(E, F)) (Thomas *et al* 2015). These results confirmed that GPCRs in liquid crystalline membranes are much more mobile that other membrane proteins of comparable size (Good *et al* 2014, Huster *et al* 2001, Liang *et al* 2010) and also than the CXCR1 receptor (Park *et al* 2011b).

While generally interesting, these studies on the mobility of the Y2R suffered from lack of site resolution in the NMR spectra. Therefore, we prepared specifically $^{13}$C-Trp labeled Y2R by cell-free synthesis. Y2R contains six native Trp residues, which were labeled and provided relatively well resolved $^{13}$C NMR spectra (figure 11.5). The NMR signals were assigned using $^{13}$C–$^{15}$N correlation experiments in receptor preparations where all Trp were $^{13}$C labeled and the successor amino acid was $^{15}$N labeled. As each Trp had a unique amino acid type successor, the $^{13}$C–$^{15}$N detected under dynamic nuclear polarization (DNP) provide the unique chemical shift for the respective Trp residue (Krug *et al* 2020). $^{13}$C NMR spectra were acquired for the apo state, Y2R in the presence of the ligand NPY, and in complex with NPY and arrestin (Krug *et al* 2020). Unique chemical shift changes indicated the conformational equilibria of the receptor (figure 11.5). It could be shown that the activation of Y2R was mediated by well-known molecular switches of GPCR activation involving the toggle switch residue Trp281$^{6.48}$ of the highly conserved SWLP motif and Trp327$^{7.55}$ adjacent to the NPxxY$^{7.53}$ motif (Traskowski *et al* 2012). The superscript numbers refer to the Ballesteros/Weinstein nomenclature (Ballesteros and Weinstein 1995). Also, Trp207$^{5.26}$, which is involved in interactions with TM5, TM6 and the antiparallel strand of the extracellular β-sheet, is influenced by binding of the agonist expressed by a pronounced chemical shift change upon ligand binding. Furthermore, a conformationally preserved 'cysteine lock'-Trp116$^{23.50}$ was identified. This represents the first example where the conformational equilibrium of a GPCR in lipid membranes has been comprehensively investigated by solid-state NMR spectroscopy (Krug *et al* 2020).

While these isotropic chemical shift changes detected in the $^{13}$C MAS NMR spectra intriguingly describe the structural dynamics of the Y2R in DPMC membranes, DipShift order parameters detected for the six resolved Trp sites did not indicate very large differences (figure 11.5). This suggests that the functional structural dynamics of the Y2R occurs on a slower timescale than the single digit microseconds to which DipShift experiments are sensitive. Furthermore, Trp residues are relatively rigid sites as the indole ring with its large surface area is frequently involved in hydrogen bonds, π stacking, cation-π and van der Waals interactions (Hulme 2013).

### 11.4.3 Growth hormone secretagogue receptor

The human growth hormone secretagogue (GHS) receptor plays an important role in the regulation of appetite and food intake. GHS has been identified as an important target for pharmacological treatment of obesity and eating disorders (Wiedmer *et al* 2007). The 28 amino acid peptide hormone ghrelin is the ligand of the GHS receptor, which is lipid modified by an octanoyl group at residue Ser3 (Kojima *et al* 1999). Similar to the Y2R, the dynamics of the GHS receptor also has been investigated for the molecule reconstituted into DMPC or POPC membranes (Schrottke *et al* 2017). Static $^{15}$N NMR spectra showed a similar superposition of an

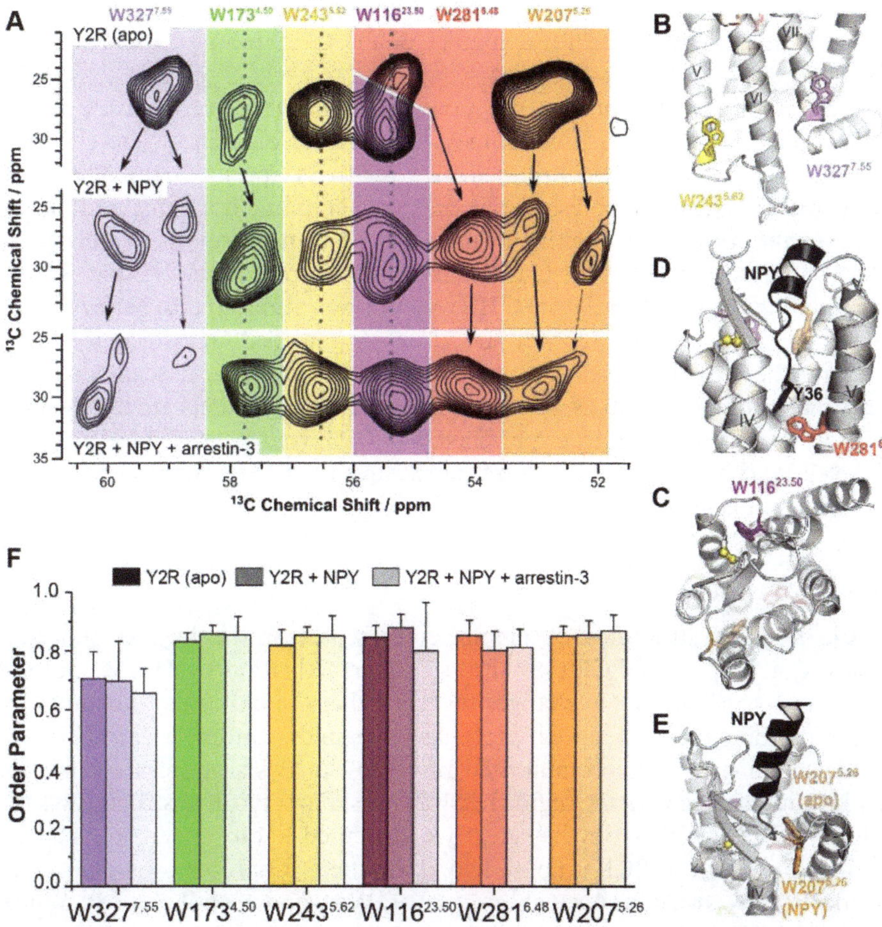


**Figure 11.5.** Conformational dynamics of the Y2R reconstituted into DMPC membranes: (A) Cα/Cβ region of $^{13}C$–$^{13}C$ DARR correlation experiments of $^{13}C$-Trp-labeled Y2R in DMPC membranes at −30 °C in different functional states, top row: apo state; middle row: NPY-bound state; bottom row: NPY- and arrestin-3 bound state. $^{13}C$ MAS NMR experiments were carried using a 600 MHz NMR spectrometer at a MAS frequency of 11 to 12 kHz. (B–E) Structural model of the Y2R illustrating the interactions of various Trp residues. Models were taken from Kaiser *et al* (2015). (B) The sidechains of Trp243$^{5.62}$ and Trp327$^{7.55}$ localized on the intracellular side facing the membrane; (C) Trp116$^{23.50}$ forms a 'cysteine lock' with the indole ring stacked against the conserved disulfide bridge; (D) interaction of the toggle switch residue Trp281$^{6.48}$ with the C-terminal residue Tyr36 of the agonist NPY (black); (E) Conformation of Trp207$^{5.26}$ in the apo and the NPY-bound states indicating a putative conformational change between these states. (F) Motional order parameters of the $^{1}$Hα–$^{13}$Cα bond vectors of the 6 Trp residues of Y2R. Order parameters were determined at a temperature of 5 °C. The numbers in superscript refer to the Ballesteros/Weinstein nomenclature (Ballesteros and Weinstein 1995). Figure modified from Krug *et al* (2020) with permission, copyright John Wiley & Sons.

axially symmetric powder pattern with quite a few isotropic narrow backbone signals of the GHS receptor, which account for approximately 40% highly mobile sites. Explicit dynamics measurements using DipShift and r-PDLF (Dvinskikh *et al* 2004) pulse sequences showed order parameters for the GHS receptor that were a little bit higher than for the Y2R. Agonist or inverse agonist binding did not significantly alter the equilibrium dynamics of the GHS receptor (Schrottke *et al* 2017).

To achieve higher resolution, the GHS receptor was prepared by cell-free synthesis (Pacull *et al* 2020). Methionines, arginines or histidines were chosen for $^{13}C/^{15}N$-labeling as they are distributed over relatively specific areas of the GHS receptor. Methionines are representative of transmembrane domains, arginines are enriched in the loops and flanking regions of the transmembrane α-helices, and histidines are concentrated in the C-terminus of the receptor in the His-tag as well as transmembrane region. Overall, the GHS receptor was also found to be very mobile. Among the labeled sites, U-$^{13}$C-Met and U-$^{13}$C-Arg labeled GHS receptor showed order parameters of 0.45 and 0.48, respectively, for the helical segments, which suggests that both the central as well as the flanking helix elements are highly mobile in the molecule. Even lower order parameters of 0.23 were measured for the U-$^{13}$C-His labeled GHS receptor, which were mostly from the C-terminal His-tag and is highly flexible (Pacull *et al* 2020).

## 11.5 Conclusions

Atomistic resolved structures of biomolecules represent the key to understand biological function. As GPCRs go through their activation cycle, the molecules undergo rapid conformational changes revealing that the biological function of the receptors is strongly related to their molecular dynamics (Latorraca *et al* 2017). For full understanding of these complicated processes, analysis of spectroscopic data ideally in combination with the highly resolved crystal or cryo-EM structures is most promising. NMR spectroscopy has greatly contributed to our understanding of the activation circles of GPCRs upon ligand binding, G-protein interaction and desensitization by arrestin (Bostock *et al* 2019, Weis and Kobilka 2018). While it is not the method of choice for a full structure determination of large membrane proteins and protein complexes, application of specific isotope labeling allows focus on hotspots of GPCR activation such as the well conserved molecular switches. Solution NMR, which is usually superior in the achieved resolution, can be carried out in physiological buffers, and allows for titration studies of the ligand and intracellular effectors as both sides of the receptor are accessible. However, the technique is limited by the requirement of isotropic tumbling membrane mimetics that insufficiently represent true lipid bilayers. Solid-state NMR techniques are carried out in planer bilayer membranes with physiologically relevant phospholipids and cholesterol in physiological buffers. Thus, the specificity of receptor–lipid interactions also can be investigated quantitatively (Huster 2014, Soubias *et al* 2006, Soubias and Gawrisch 2005). A limitation of the technique is that the multilamellar membrane preparations prevent the titration of agonists or antagonists or intracellular effectors. Thus, each state of a receptor has to be prepared in an individual sample rendering solid-state NMR approaches more time consuming than the solution NMR counterparts. However, unique advantages of solid-state NMR are the comprehensive investigation of receptor structural dynamics in the absence of the overall tumbling of the molecule in solution. Although currently relatively underexplored, the technique will likely contribute more to our understanding of the intriguing structural dynamics of GPCRs. From the body of data available so far, a picture emerges portraying GPCRs as highly dynamic molecules in liquid crystalline membranes undergoing dynamic rapid structural transitions upon activation, signal transduction and desensitization as well as pronounced equilibrium dynamics with relatively large amplitudes. Although crystal and cryo-EM structures have become available for approximately 70 unique GPCRs, this impressive number still accounts for less than 10% of the over 800 GPCR species in humans. Though highly desirable, it is clear that not all pharmacologically important receptors will become the target for atomistic structure determination in each relevant state and in complex with all intracellular proteins in the near future. Thus, alternative approaches must be pursued to reveal the structural dynamics of these important molecules. We believe that the solid-state NMR techniques discussed here will represent important contributions towards this endeavour.

## Acknowledgement

Funding from the Deutsche Forschungsgemeinschaft (DFG, German Research Foundation) through CRC 1423, project number 421152132, subprojects A02 and A04 is acknowledged. The author would like to thank all colleagues at the Department of Chemical Sciences, Tata Institute of Fundamental Research in Mumbai, India, for hosting and supporting him during the lockdown due to the Corona virus outbreak in India. Dr Ulrike Krug is acknowledged for help preparing the figures for this chapter.

## References

Aharoni T and Goldbourt A 2018 Dynamics and rigidity of an intact filamentous bacteriophage virus probed by magic angle spinning NMR *Chemistry* **24** 8737–41

Ahuja S *et al* 2009 Helix movement is coupled to displacement of the second extracellular loop in rhodopsin activation *Nat. Struct. Mol. Biol.* **16** 168–75

Baldus M 2002 Correlation experiments for assignment and structure elucidation of immobilized polypeptides under magic angle spinning *Prog. Nucl. Magn. Reson. Spectrosc.* **41** 1–47

Ballesteros J A and Weinstein H 1995 Integrated methods for the construction of three dimensional models and computational probing of structure function relations in G protein-coupled receptors *Methods Neuroscience* 25 edn ed S C Sealfon and P M Conn (San Diego, CA: Academic) pp 366–428

Baneres J L, Martin A, Hullot P, Girard J P, Rossi J C and Parello J 2003 Structure-based analysis of GPCR function: conformational adaptation of both agonist and receptor upon leukotriene B-4 binding to recombinant BLT1 *J. Mol. Biol.* **329** 801–14

Bayburt T H, Grinkova Y V and Sligar S G 2002 Self-assembly of discoidal phospholipid bilayer nanoparticles with membrane scaffold proteins *Nano Lett.* **2** 853–6

Bender B J, Vortmeier G, Ernicke S, Bosse M, Kaiser A, Els-Heindl S, Krug U, Beck-Sickinger A G, Meiler J and Huster D 2019 Structural model of ghrelin bound to its G protein-coupled receptor *Structure* **27** 537–44

Berger C, Berndt S, Pichert A, Theisgen S and Huster D 2013 Efficient isotopic tryptophan labeling of membrane proteins by an indole controlled process conduct *Biotechnol. Bioeng.* **110** 1681–90

Bernhard F and Tozawa Y 2013 Cell-free expression—making a mark *Curr. Opin. Struct. Biol.* **23** 374–80

Bokoch M P *et al* 2010 Ligand-specific regulation of the extracellular surface of a G-protein-coupled receptor *Nature* **463** 108–12

Bosse M, Thomas L, Hassert R, Beck-Sickinger A G, Huster D and Schmidt P 2011 Assessment of a fully active class A G protein-coupled receptor isolated from *in vitro* folding *Biochemistry.* **50** 9817–25

Bostock M J, Solt A S and Nietlispach D 2019 The role of NMR spectroscopy in mapping the conformational landscape of GPCRs *Curr. Opin. Struct. Biol.* **57** 145–56

Botelho A V, Gibson N J, Thurmond R L, Wang Y and Brown M F 2002 Conformational energetics of rhodopsin modulated by nonlamellar-forming lipids *Biochemistry.* **41** 6354–68

Briddon S J and Kilpatrick L E 2018 Studying GPCR pharmacology in membrane microdomains: fluorescence correlation spectroscopy comes of age *Trends Pharmacol. Sci.* **39** 158–74

Brüschweiler R and Wright P E 1994 NMR order parameters of biomolecules: a new analytical representation and application to the Gaussian axial fluctuation model *J. Am. Chem. Soc.* **116** 8426–7

Castellani F, van Rossum B, Diehl A, Schubert M, Rehbein K and Oschkinat H 2002 Structure of a protein determined by solid-state magic-angle-spinning NMR spectroscopy *Nature* **420** 98–102

Chae P S *et al* 2010 Maltose-neopentyl glycol (MNG) amphiphiles for solubilization, stabilization and crystallization of membrane proteins *Nat. Methods* **7** 1003–8

Chavez-Abiega S, Goedhart J and Bruggeman F J 2019 Physical biology of GPCR signalling dynamics inferred from fluorescence spectroscopy and imaging *Curr. Opin. Struct. Biol.* **55** 204–11

Cherezov V *et al* 2007 High-resolution crystal structure of an engineered human $\beta_2$-adrenergic G protein-coupled receptor *Science* **318** 1258–65

Chung K Y, Kim T H, Manglik A, Alvares R, Kobilka B K and Prosser R S 2012 Role of detergents in conformational exchange of a G protein-coupled receptor *J. Biol. Chem.* **287** 36305–11

Clark L, Dikiy I, Rosenbaum D M and Gardner K H 2018 On the use of *Pichia pastoris* for isotopic labeling of human GPCRs for NMR studies *J. Biomol. NMR* **71** 203–11

Clark L D, Dikiy I, Chapman K R K E, Aramini J, LeVine M V, Khelashvili G, Rasmussen S G, Gardner K H and Rosenbaum D M 2017 Ligand modulation of sidechain dynamics in a wild-type human GPCR *eLife* **6** e28505

Das B B, Park S H and Opella S J 2015 Membrane protein structure from rotational diffusion *Biochim. Biophys. Acta* **1848** 229–45

Davis J H 1983 The description of membrane lipid conformation, order and dynamics by $^2$H NMR *Biochim. Biophys. Acta* **737** 117–71

Deupi X and Kobilka B K 2010 Energy landscapes as a tool to integrate GPCR structure, dynamics, and function *Physiology* **25** 293–303

Dusold S and Sebald A 2000 Dipolar recoupling under magic-angle spinning conditions *Ann. Rep. NMR Spectrosc.* **41** 185–264

Dvinskikh S V, Zimmermann H, Maliniak A and Sandstrom D 2004 Measurements of motionally averaged heteronuclear dipolar couplings in MAS NMR using R-type recoupling *J. Magn. Reson.* **168** 194–201

Eddy M T, Gao Z G, Mannes P, Patel N, Jacobsen K A, Katritch V, Stevens R C and Wüthrich K 2018a Extrinsic tryptophans as NMR probes of allosteric coupling in membrane proteins: application to the A2A adenosine receptor *J. Am. Chem. Soc.* **140** 8228–35

Eddy M T *et al* 2018b Allosteric coupling of drug binding and intracellular signaling in the A2A adenosine receptor *Cell* **172** 68–80

Elkins M R and Hong M 2019 Elucidating ligand-bound structures of membrane proteins using solid-state NMR spectroscopy *Curr. Opin. Struct. Biol.* **57** 103–9

Fernandez E J and Lolis E 2002 Structure, function, and inhibition of chemokines *Annu. Rev. Pharmacol. Toxicol.* **42** 469–99

Ferré G *et al* 2019 Structure and dynamics of G protein-coupled receptor-bound ghrelin reveal the critical role of the octanoyl chain *Proc. Natl Acad. Sci. USA* **116** 17525–30

Franke B, Opitz C, Isogai S, Grahl A, Delgado L, Gossert A D and Grzesiek S 2018 Production of isotope-labeled proteins in insect cells for NMR *J. Biomol. NMR* **71** 173–84

Frei J N, Broadhurst R W, Bostock M J, Solt A, Jones A, Gabriel F, Tandale A, Shrestha B and Nietlispach D 2020 Conformational plasticity of ligand-bound and ternary GPCR complexes studied by $^{19}$F NMR of the $\beta_1$-adrenergic receptor *Nat. Commun.* **11** 669

García-Nafría J and Tate C G 2019 Cryo-EM structures of GPCRs coupled to $G_s$, $G_i$ and $G_o$ *Mol. Cell. Endocrinol.* **488** 1–13

Gibson N J and Brown M F 1993 Lipid headgroups and acyl chain composition modulate the MI-MII equilibrium of rhodopsin in recombinant membranes *Biochemistry* **32** 2438–54

Good D B, Wang S, Ward M E, Struppe J, Brown L S, Lewandowski J R and Ladizhansky V 2014 Conformational dynamics of a seven transmembrane helical protein anabaena sensory rhodopsin probed by solid-state NMR *J. Am. Chem. Soc.* **136** 2833–42

Grisshammer R, White J F, Trinh L B and Shiloach J 2005 Large-scale expression and purification of a G-protein-coupled receptor for structure determination—an overview *J. Struct. Funct. Genomics* **6** 159–63

Henrich E, Hein C, Dötsch V and Bernhard F 2015 Membrane protein production in *Escherichia coli* cell-free lysates *FEBS Lett.* **589** 1713–22

Hilger D, Masureel M and Kobilka B K 2018 Structure and dynamics of GPCR signaling complexes *Nat. Struct. Mol. Biol.* **25** 4–12

Hoffmann B, Löhr F, Laguerre A, Bernhard F and Dötsch V 2018 Protein labeling strategies for liquid-state NMR spectroscopy using cell-free synthesis *Prog. Nucl. Magn. Reson. Spectrosc.* **105** 1–22

Hong M 1999 Determination of multiple F-torsion angles in proteins by selective and extensive $^{13}$C labeling and two-dimensional solid-state NMR. *J. Magn. Reson.* **139** 389–401

Hong M and Jakes K S 1999 Selective and extensive $^{13}$C labeling of a membran protein for solid-state NMR investigation *J. Biomol. NMR.* **14** 71–4

Huang W, Masureel M, Qu Q, Janetzko J, Inoue A, Kato H E, Robertson M J, Nguyen K C, Skiniotis G and Kobilka B K 2020 Structure of the neurotensin receptor 1 in complex with ß-arrestin 1. *Nature* **579** 303–8

Hulme E C 2013 GPCR activation: a mutagenic spotlight on crystal structures *Trends Pharmacol. Sci.* **34** 67–84

Huster D 2005 Investigations of the structure and dynamics of membrane-associated peptides by magic angle spinning NMR *Prog. Nucl. Magn. Reson. Spectrosc.* **46** 79–107

Huster D 2014 Solid-state NMR spectroscopy to study protein-lipid interactions *Biochim. Biophys. Acta* **1841** 1146–60

Huster D, Xiao L and Hong M 2001 Solid-state NMR investigation of the dynamics of soluble and membrane-bound colicin Ia channel-forming domain *Biochemistry* **40** 7662–74

Imai S, Yokomizo T, Kofuku Y, Shiraishi Y, Ueda T and Shimada I 2020 Structural equilibrium underlying ligand-dependent activation of $\beta_2$-adrenoreceptor *Nat. Chem. Biol.* **16** 430–9

Isogai S, Deupi X, Opitz C, Heydenreich F M, Tsai C J, Brueckner F, Schertler G F, Veprintsev D B and Grzesiek S 2016 Backbone NMR reveals allosteric signal transduction networks in the ß1-adrenergic receptor *Nature* **530** 237–41

Joedicke L *et al* 2018 The molecular basis of subtype selectivity of human kinin G-protein-coupled receptors *Nat. Chem. Biol.* **14** 284–90

Kaiser A *et al* 2015 Unwinding of the C-terminal residues of neuropeptide Y is critical for Y(2) receptor binding and activation *Angew. Chem. Int. Ed. Engl.* **54** 7446–9

Kiefer H, Maier K and Vogel R 1999 Refolding of G-protein-coupled receptors from inclusion bodies produced in *Escherichia coli Biochem. Soc. Trans.* **27** 908–12

Kimata N, Pope A, Sanchez-Reyes O B, Eilers M, Opefi C A, Ziliox M, Reeves P J and Smith S O 2016 Free backbone carbonyls mediate rhodopsin activation *Nat. Struct. Mol. Biol.* **23** 738–43

Klopfer K and Hagn F 2019 Beyond detergent micelles: the advantages and applications of non-micellar and lipid-based membrane mimetics for solution-state NMR *Prog. Nucl. Magn. Reson. Spectrosc.* **114** 271–83

Kofuku Y, Ueda T, Okude J, Shiraishi Y, Kondo K, Maeda M, Tsujishita H and Shimada I 2012 Efficacy of the $\beta_2$-adrenergic receptor is determined by conformational equilibrium in the transmembrane region *Nat. Commun.* **3** 1045

Kögler L M, Stichel J and Beck-Sickinger A G 2019 Structural investigations of cell-free expressed G protein-coupled receptors *Biol. Chem.* **401** 97–116

Kojima M, Hosoda H, Date Y, Nakazato M, Matsuo H and Kangawa K 1999 Ghrelin is a growth-hormone-releasing acylated peptide from stomach *Nature* **402** 656–60

Krepkiy D, Wong K, Gawrisch K and Yeliseev A 2006 Bacterial expression of functional, biotinylated peripheral cannabinoid receptor CB2 *Protein Expr. Purif.* **49** 60–70

Krug U *et al* 2020 The conformational equilibrium of the neuropeptide Y2 receptor in bilayer membranes *Angew. Chem. Int. Ed.* https://onlinelibrary.wiley.com/doi/10.1002/anie.202006075

Ladizhansky V 2017 Applications of solid-state NMR to membrane proteins *Biochim. Biophys. Acta* **1865** 1577–86

Laguerre A, Löhr F, Heinrich E, Hoffmann B, Abdul-Manan N, Connolly P J, Perozo E, Moore J M, Bernhard F and Dötsch V 2016 From nanodiscs to isotropic bicelles: a procedure for solution nuclear magnetic resonance studies of detergent-sensitive integral membrane proteins *Structure* **24** 1830–41

Larda S T, Bokoch M P, Evanics F and Prosser R S 2012 Lysine methylation strategies for characterizing protein conformations by NMR *J. Biomol. NMR* **54** 199–209

Latorraca N R, Venkatakrishnan A J and Dror R O 2017 GPCR dynamics: structures in motion *Chem. Rev.* **117** 139–55

Lewis B A, Harbison G S, Herzfeld J and Griffin R G 1985 NMR structural analysis of a membrane protein: bacteriorhodopsin peptide backbone orientation and motion *Biochemistry* **24** 4671–9

Liang B, Arora A and Tamm L K 2010 Fast-time scale dynamics of outer membrane protein A by extended model-free analysis of NMR relaxation data *Biochim. Biophys. Acta* **1798** 68–76

Liu J J, Horst R, Katritch V, Stevens R C and Wüthrich K 2012 Biased signaling pathways in $\beta_2$-adrenergic receptor characterized by $^{19}$F NMR *Science* **335** 1106–10

Lopez J J, Shukla A K, Reinhard C, Schwalbe H, Michel H and Glaubitz C 2008 The structure of the neuropeptide bradykinin bound to the human G-protein coupled receptor bradykinin B2 as determined by solid-state NMR spectroscopy *Angew. Chem. Int. Ed. Engl.* **47** 1668–71

Loquet A, Lv G, Giller K, Becker S and Lange A 2011 $^{13}$C spin dilution for simplified and complete solid-state NMR resonance assignment of insoluble biological assemblies *J. Am. Chem. Soc.* **133** 4722–5

Luca S, White J F, Sohal A K, Filippov D V, van Boom J H, Grisshammer R and Baldus M 2003 The conformation of neurotensin bound to its G protein-coupled receptor *Proc. Natl Acad. Sci. USA* **100** 10706–11

Lundström P, Teilum K, Carstensen T, Bezsonova I, Wiesner S, Hansen D F, Religa T L, Akke M and Kay L E 2007 Fractional $^{13}$C enrichment of isolated carbons using [1-$^{13}$C]- or [2- $^{13}$C]-glucose facilitates the accurate measurement of dynamics at backbone C$^{\alpha}$ and side-chain methyl positions in proteins *J. Biomol. NMR* **38** 199–212

Manglik A and Kobilka B 2014 The role of protein dynamics in GPCR function: insights from the $\beta_2$AR and rhodopsin *Curr. Opin. Cell Biol.* **27** 136–43

Manglik A *et al* 2015 Structural insights into the dynamic process of $\beta_2$-adrenergic receptor signaling *Cell.* **161** 1101–11

Massotte D 2003 G protein-coupled receptor overexpression with the baculovirus-insect cell system: a tool for structural and functional studies *Biochim. Biophys. Acta* **1610** 77–89

Mertz B, Struts A V, Feller S E and Brown M F 2012 Molecular simulations and solid-state NMR investigate dynamical structure in rhodopsin activation *Biochim. Biophys. Acta* **1818** 241–51

Mizumura T, Kondo K, Kurita M, Kofuku Y, Natsume M, Imai S, Shiraishi Y, Ueda T and Shimada I 2020 Activation of adenosine A2A receptor by lipids from docosahexaenoic acid revealed by NMR *Sci. Adv.* **6** eaay8544

Munowitz M G, Griffin R G, Bodenhausen G and Huang T H 1981 Two-dimensional rotational spin-echo nuclear magnetic resonance in solids: correlation of chemical shift and dipolar interactions *J. Am. Chem. Soc.* **103** 2529–33

Nasr M L *et al* 2017 Covalently circularized nanodiscs for studying membrane proteins and viral entry *Nat. Methods* **14** 49–52

Nygaard R *et al* 2013 The dynamic process of $\beta_2$-adrenergic receptor activation *Cell* **152** 532–42

O'Connor C, White K L, Doncescu N, Didenko T, Roth B L, Czaplicki G, Stevens R C, Wüthrich K and Milon A 2015 NMR structure and dynamics of the agonist dynorphin peptide bound to the human kappa opioid receptor *Proc. Natl Acad. Sci. USA* **112** 11852–7

Oates J and Watts A 2011 Uncovering the intimate relationship between lipids, cholesterol and GPCR activation *Curr. Opin. Struct. Biol.* **21** 802–7

Pacull E, Sendker F, Bernhard F, Scheidt H A, Schmidt P, Huster D and Krug U 2020 Integration of cell-free expression and solid-state NMR to show the dynamic properties of different sites of growth hormone secretagogue receptor *Front. Pharmacol.* **11** 562113

Palczewski K *et al* 2000 Crystal structure of rhodopsin: a G protein-coupled receptor *Science* **289** 739–45

Palmer A G III, Williams J and McDermott A 1996 Nuclear magnetic resonance studies of biopolymer dynamics *J. Phys. Chem.* **100** 13293–310

Park S H, Casagrande F, Cho L, Albrecht L and Opella S J 2011a Interactions of interleukin-8 with the human chemokine receptor CXCR1 in phospholipid bilayers by NMR spectroscopy *J. Mol. Biol.* **414** 194–203

Park S H, Casagrande F, Chu M, Maier K, Kiefer H and Opella S J 2012a Optimization of purification and refolding of the human chemokine receptor CXCR1 improves the stability of proteoliposomes for structure determination *Biochim. Biophys. Acta* **1818** 584–91

Park S H, Casagrande F, Das B B, Albrecht L, Chu M and Opella S J 2011b Local and global dynamics of the G protein-coupled receptor CXCR1 *Biochemistry.* **50** 2371–80

Park S H *et al* 2012b Structure of the chemokine receptor CXCR1 in phospholipid bilayers *Nature* **491** 779–83

Park S H, Prytulla S, De Angelis A A, Brown J M, Kiefer H and Opella S J 2006 High-resolution NMR spectroscopy of a GPCR in aligned bicelles *J. Am. Chem. Soc.* **128** 7402–3

Pope A, Eilers M, Reeves P J and Smith S O 2013 Amino acid conservation and interactions in rhodopsin: probing receptor activation by NMR spectroscopy *Biochim. Biophys. Acta* **1837** 683–93

Prosser R S, Hunt S A, DiNatale J A and Vold R R 1996 Magnetically aligned membrane model systems with positive order parameter: switching the sign of $S_{zz}$ with paramagnetic ions *J. Am. Chem. Soc.* **118** 269–70

Puthenveetil R V O 2019 Solution NMR: a powerful tool for structural and functional studies of membrane proteins in reconstituted environments *J. Biol. Chem.* **294** 15914–31

Ram P and Prestegard J H 1988 Magnetic field induced ordering of bile salt/phospholipid micelles: new media for NMR structural investigations *Biochim. Biophys. Acta* **940** 289–94

Ravula T, Hardin N Z and Ramamoorthy A 2019 Polymer nanodiscs: advantages and limitations *Chem. Phys. Lipids* **219** 45–9

Reuther G, Tan K-T, Vogel A, Nowak C, Kuhlmann J, Waldmann H and Huster D 2006 The lipidated membrane anchor of the N-ras protein shows an extensive dynamics as revealed by solid-state NMR *J. Am. Chem. Soc.* **128** 13840–6

Rosenbaum D M *et al* 2007 GPCR engineering yields high-resolution structural insights into $\beta_2$-adrenergic receptor function *Science* **318** 1266–73

Rues R B, Dötsch V and Bernhard F 2016 Co-translational formation and pharmacological characterization of beta1-adrenergic receptor/nanodisc complexes with different lipid environments *Biochim. Biophys. Acta* **1858** 1306–16

Sanders C R and Prestegard J H 1990 Magnetically orientable phospholipid bilayers containing small amounts of bile salt analogue, CHAPSO *Biophys. J.* **58** 447–60

Sarramegna V, Muller I, Milon A and Talmont F 2006 Recombinant G protein-coupled receptors from expression to renaturation: a challenge towards structure *Cell. Mol. Life Sci.* **63** 1149–64

Schanda P and Ernst M 2016 Studying dynamics by magic-angle spinning solid-state NMR spectroscopy: principles and applications to biomolecules *Prog. Nucl. Magn. Reson. Spectrosc.* **96** 1–46

Schlinkmann K M, Honegger A, Türeci E, Robinson K E, Lipovsek D and Plückthun A 2012 Critical features for biosynthesis, stability, and functionality of a G protein-coupled receptor uncovered by all-versus-all mutations *Proc. Natl Acad. Sci. USA* **109** 9810–5

Schmidt P, Bender B J, Kaiser A, Gulati K, Scheidt H A, Hamm H E, Meiler J, Beck-Sickinger A G and Huster D 2018 Improved *in vitro* folding of the Y2 G protein-coupled receptor into bicelles *Front Mol. Biosci.* **4** 100

Schmidt P, Lindner D, Montag C, Berndt S, Beck-Sickinger A G, Rudolph R and Huster D 2009 Prokaryotic expression, *in vitro* folding, and molecular pharmacological characterization of the neuropeptide Y receptor type 2 *Biotechnol. Prog.* **25** 1732–9

Schmidt P, Thomas L, Müller P, Scheidt H A and Huster D 2014 The G protein-coupled neuropeptide Y receptor type 2 is highly dynamic in lipid membranes as revealed by solid-state NMR spectroscopy *Chemistry* **20** 4986–92

Schrottke S, Kaiser A, Vortmeier G, Els-Heindl S, Worm D, Bosse M, Schmidt P, Scheidt H A, Beck-Sickinger A G and Huster D 2017 Expression, functional characterization, and solid-state NMR investigation of the G protein-coupled GHS receptor in bilayer membranes *Sci. Rep.* **7** 46128

Seddon A M, Curnow P and Booth P J 2004 Membrane proteins, lipids and detergents: not just a soap opera *Biochim. Biophys. Acta* **1666** 105–17

Seelig J 1977 Deuterium magnetic resonance: theory and application to lipid membranes *Q. Rev. Biophys.* **10** 353–418

Shimada I, Ueda T, Kofuku Y, Eddy M T and Wüthrich K 2019 GPCR drug discovery: integrating solution NMR data with crystal and cryo-EM structures *Nat. Rev. Drug Discov.* **18** 59–82

Skrzypek R, Iqbal S and Callaghan R 2018 Methods of reconstitution to investigate membrane protein function *Methods* **147** 126–41

Smith A A, Ernst M, Riniker S and Meier B H 2019 Localized and collective motions in HET-s (218-289) fibrils from combined NMR relaxation and MD simulation *Angew. Chem. Int. Ed.* **58** 9383–8

Solt A S, Bostock M J, Shrestha B, Kumar P, Warne T, Tate C G and Nietlispach D 2017 Insight into partial agonism by observing multiple equilibria for ligand-bound and Gs-mimetic nanobody-bound $\beta_1$-adrenergic receptor *Nat. Commun.* **8** 1795

Soubias O and Gawrisch K 2005 Probing specific lipid-protein interaction by saturation transfer difference NMR spectroscopy *J. Am. Chem. Soc.* **127** 13110–1

Soubias O and Gawrisch K 2013 Rhodopsin-lipid interactions studied by NMR *Methods Enzymol.* **522** 209–27

Soubias O, Niu S L, Mitchell D C and Gawrisch K 2008 Lipid-rhodopsin hydrophobic mismatch alters rhodopsin helical content *J. Am. Chem. Soc.* **130** 12465–71

Soubias O, Teague W E and Gawrisch K 2006 Evidence for specificity in lipid-rhodopsin interactions *J. Biol. Chem.* **281** 33233–41

Soubias O, Teague W E Jr., Hines K G and Gawrisch K 2015 Rhodopsin/lipid hydrophobic matching-rhodopsin oligomerization and function *Biophys. J.* **108** 1125–32

Soubias O, Teague W E Jr., Hines K G, Mitchell D C and Gawrisch K 2010 Contribution of membrane elastic energy to rhodopsin function *Biophys. J.* **99** 817–24

Staus D P, Hu H, Robertson M J, Kleinhenz A, Wingler L M, Capel W D, Latorraca N R, Lefkowitz R J and Skiniotis G 2020 Structure of the M2 muscarinic receptor-β-arrestin complex in a lipid nanodisc *Nature* **579** 297–302

Sternin E, Nizza D and Gawrisch K 2001 Temperature dependence of DMPC/DHPC mixing in a bicellar solution and its structural implications *Langmuir* **17** 2610–6

Struts A V, Chawla U, Suchithranga M D C and Brown M F 2015 Investigation of rhodopsin dynamics in its signaling state by solid-state deuterium NMR spectroscopy *Methods Mol. Biol.* **1271** 133–58

Struts A V, Salgado G F, Martinez-Mayorga K and Brown M F 2011 Retinal dynamics underlie its switch from inverse agonist to agonist during rhodopsin activation *Nat. Struct. Mol. Biol.* **18** 392–4

Tang M and Lam D 2019 Paramagnetic solid-state NMR of proteins *Solid State Nucl. Magn. Reson.* **103** 9–16

Thomas L, Kahr J, Schmidt P, Krug U, Scheidt H A and Huster D 2015 The molecular dynamics of the G protein-coupled neuropeptide Y2 receptor in monounsaturated membranes investigated by solid-state NMR spectroscopy *J. Biomol. NMR* **61** 347–59

Traskowski B, Latek D, Yuan S, Ghoshdastider U, Debinski A and Filipek S 2012 Action of molecular switches in GPCRs--theoretical and experimental studies *Curr. Med. Chem.* **19** 1090–109

Triba M N, Warschawski D E and Devaux P F 2005 Reinvestigation by phosphorus NMR of lipid distribution in bicelles *Biophys. J.* **88** 1887–901

Umai S, Yokomizo T, Shiraishi Y, Ueda T and Shimada I 2020 Structural equilibrium underlying ligand-dependent activation of $\beta_2$-adrenoreceptor *Nat. Chem. Biol.* **16** 430–9

Vaidehi N, Grisshammer R and Tate C G 2016 How can mutations thermostabilize G-protein-coupled receptors? *Trends Pharmacol. Sci.* **37** 37–46

van Eps N, Caro L N, Morizumi T and Ernst O P 2015 Characterizing rhodopsin signaling by EPR spectroscopy: from structure to dynamics *Photochem. Photobiol.* **14** 1586–97

van Eps N, Caro L N, Morizumi T, Kusnetzow A K, Suczepek M, Hofmann K P, Bayburt T H, Sligar S G, Ernst O P and Hubbell W L 2017 Conformational equilibria of light-activated rhodopsin in nanodiscs *Proc. Natl. Acad. Sci. USA* **114** E3268–75

Vandemoortele G E S and Gevaert K 2019 Pick a tag and explore the functions of your pet protein *Trends Biotechnol.* **37** 1078–90

Vogel A, Reuther G, Roark M B, Tan K T, Waldmann H, Feller S E and Huster D 2010 Backbone conformational flexibility of the lipid modified membrane anchor of the human N-Ras protein investigated by solid-state NMR and molecular dynamics simulation *Biochim. Biophys. Acta* **1798** 275–85

Vogel A, Scheidt H A, Baek D J, Bittman R and Huster D 2016 Structure and dynamics of the aliphatic cholesterol side chain in membranes as studied by $^2$H NMR spectroscopy and molecular dynamics simulation *Phys. Chem. Chem. Phys.* **18** 3730–8

Vogel A, Tan K-T, Waldmann H, Feller S E, Brown M F and Huster D 2007 Flexibility of ras lipid modifications studied by $^2$H solid-state NMR and molecular dynamics simulations *Biophys. J.* **93** 2697–712

Waugh J S 1976 Uncoupling of local field spectra in nuclear magnetic resonance: determination of atomic positions in solids *Proc. Natl Acad. Sci. USA* **73** 1394–7

Weis W I and Kobilka B K 2018 The molecular basis of G protein-coupled receptor activation *Annu. Rev. Biochem.* **87** 897–919

Werner K, Lehner I, Dhiman H K, Richter C, Glaubitz C, Schwalbe H, Klein-Seetharaman J and Khorana H G 2007 Combined solid state and solution NMR studies of alpha,epsilon-N-15 labeled bovine rhodopsin *J. Biomol. NMR* **37** 303–12

Wiedmer P, Nogueiras R, Broglio F, D'Alessio D and Tschop M H 2007 Ghrelin, obesity and diabetes *Nat. Clin. Pract. Endocrinol. Metab.* **3** 705–12

Wingler L M, Elgeti M, Hilger D, Latorraca N R, Lerch M T, Staus D P, Dror R O, Kobilka B K, Hubbell W L and Lefkowitz R J 2019 Angiotensin analogs with divergent bias stabilize distinct receptor conformations *Cell* **176** 468–78

Wishart D S and Sykes B D 1994 Chemical shifts as a tool for structure determination *Methods Enzymol.* **239** 363–92

Witte K, Kaiser A, Schmidt P, Splith V, Thomas L, Berndt S, Huster D and Beck-Sickinger A G 2013 Oxidative *in vitro* folding of a cysteine deficient variant of the G protein-coupled neuropeptide Y receptor type 2 improves stability at high concentration *Biol. Chem.* **394** 1045–56

Yang Z *et al* 2018 Structural basis of ligand binding modes at the neuropeptide Y Y1 receptor *Nature* **556** 520–4

Ye L *et al* 2018 Mechanistic insights into allosteric regulation of the A2A adenosine G protein-coupled receptor by physiological cations *Nat. Commun.* **9** 1372

Ye L, van Eps N, Zimmer M, Ernst O P and Prosser R S 2016 Activation of the A2A adenosine G-protein-coupled receptor by conformational selection *Nature* **533** 265–8

Yeliseev A A, Wong K K, Soubias O and Gawrisch K 2005 Expression of human peripheral cannabinoid receptor for structural studies *Protein Sci.* **14** 2638–53

**IOP** Publishing

Solid-State NMR
Applications in biomembrane structure
**Frances Separovic and Marc-Antoine Sani**

# Chapter 12

# Hybridizing isotropic and anisotropic solid-state NMR restraints for membrane protein structure determination

**Daniel K Weber, Erik K Larsen, Tata Gopinath and Gianluigi Veglia**

## Abstract

Solid-state NMR (ssNMR) techniques for membrane proteins are categorized into two classes: oriented sample (OS) and magic angle spinning (MAS). While these approaches require different sample preparations and pulse sequences, they provide complementary information for the analysis of the structure and topology of membrane proteins embedded in their native lipid bilayer environments. The main difference between the two is that MAS-ssNMR produces highly resolved isotropic spectra by averaging out chemical shift anisotropy (CSA) and dipolar coupling (DC) interactions. In contrast, OS-ssNMR retains this lost information in highly resolved anisotropic spectra by reconstituting membrane proteins into mechanically or magnetically aligned lipid bilayers. Anisotropic parameters from OS-ssNMR report directly on the structural topology of membrane proteins. This feature is especially crucial to describe the membrane architecture and structural transitions of receptors, channels, pumps, and transporters, whose functional dynamics involve topological changes (bend, tilt, rotation, and piston-like motions) of transmembrane domains from inactive to active states. However, OS-ssNMR data are rarely integrated with those from MAS, despite their synergistic nature that would confer a functional relevance to these structures. In this chapter, we illustrate how these techniques can be integrated, providing a mathematical and diagrammatic explanation of anisotropic restraints and a short guide on how to hybridize them with isotropic restraints for routine structure determination.

doi:10.1088/978-0-7503-2532-5ch12  

## 12.1 Oriented-sample solid-state NMR (OS-ssNMR) structure determination of membrane proteins

At the heart of OS-ssNMR is the separated local field (SLF) experiment, which resolves anisotropic chemical shifts (CSA) and dipolar couplings (DC) of chemical groups such as $^{13}C\alpha$–$^{1}H\alpha$ and $^{15}N$–$^{1}H_N$ (Hester *et al* 1976). Due to conspicuous line-broadening arising from strong $^{1}H$–$^{1}H$ dipolar interactions, initial SLF experiments lacked site-specific resolution necessary to resolve the crowded spectra of membrane proteins. Opella and co-workers introduced PISEMA, a rotating frame SLF experiment that dramatically improves the sensitivity and resolution (Wu *et al* 1994) (figure 12.1(a)). PISEMA starts with a cross-polarization (CP) period followed by a 35° tilt angle pulse to create a zero quantum (ZQ) polarization-inversion (PI) state that doubles the intensity of the initial state. The ZQ state is then evolved in $t_1$ under the effect of heteronuclear DC obtained from the spin-exchange at the magic angle (SEMA). The SEMA period suppresses the homonuclear $^{1}H$–$^{1}H$ DC via a frequency-switched Lee–Goldburg (FSLG) sequence on the $^{1}H$ channel, while simultaneously recoupling heteronuclear DC by applying on-resonance $^{15}N$ spin-lock pulses. Hartmann–Hahn matching is obtained during the SEMA period by setting the RF amplitude of $^{15}N$ spin-lock pulses equal to the effective RF of the $^{1}H$ FSLG sequence. Compared to laboratory frame SLF experiments, PISEMA has a

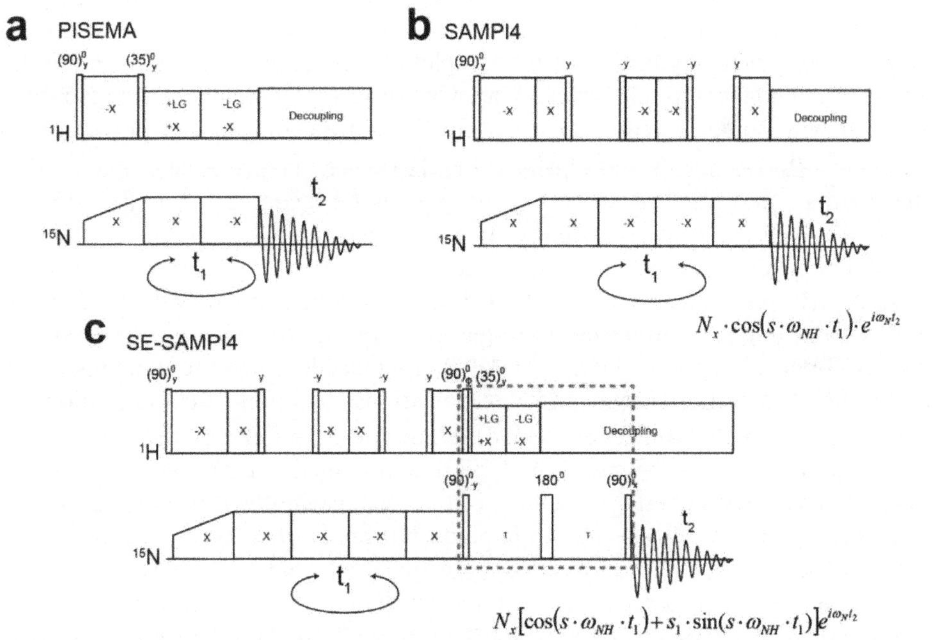


**Figure 12.1.** Example SLF pulse sequences used for OS-ssNMR, including the (a). PISEMA, (b). SAMPI4, and (c). the SE-SAMPI4 with the signal enhancement (SE) outlined in broken red lines. The $t_1$ DC evolution scaling factor ($s$) for SAMPI4 and SE-SAMPI4, corresponds to 0.67. The scaling factor of the sine component ($s_1$) is given by $\sin[\cos(\theta_m)\omega_{NH}\tau]$ (Gopinath and Veglia 2009).

higher dipolar scaling factor, with a resolution in the DC dimension up to ten times greater (Wu *et al* 1994). Other variants of PISEMA are SAMPI4 (figure 12.1(b)) and HIMSELF pulse sequences (Nevzorov and Opella 2007, Dvinskikh *et al* 2006b). These experiments ameliorate the problematic $^1$H-offset dependence in the DC dimension by using a magic sandwich or WIM24 recoupling pulses, respectively. The sensitivity of these pulse sequences can be significantly improved by introducing a sensitivity enhancement (SE) element (figure 12.1(c)), which increases the signal-to-noise ratio (S/N) up to 40% (Gopinath and Veglia 2009, Gopinath *et al* 2010). This simple modification consists of a spin-echo element prior to the $t_2$ acquisition period that combines both sine- and cosine-modulated coherences, one of which is discarded in the traditional PISEMA, SAMPI4, and HIMSELF experiments, resulting in a net enhancement of the S/N by $\sqrt{2}$. A more recent modification is the heat-compensated version, which eliminates temperature fluctuations resulting from the variable $t_1$ period with high-power spin-lock pulses. We found that this modification reduces lineshape distortions significantly (Wang *et al* 2019). Other noteworthy SLF sequences include the proton-evolved local field (PELF), and corresponding SE version, i.e., Hadamard-encoded (HE)-PELF (Gopinath *et al* 2011, Dvinskikh *et al* 2006a). These variants are suited for measuring weaker dipolar couplings in dynamic domains of membrane proteins.

To obtain high-resolution SLF spectra, membrane proteins must be reconstituted into a lipid bilayer that is homogenously aligned relative to the static magnetic field ($\vec{B}_0$). Classical sample preparations require the deposition of liposome/protein mixtures onto thin glass plates (mechanically aligned systems). Iterative cycles of de-hydration and re-hydration of the samples enable the orientation of the proteins and the lipid bilayers to take place. A stacking of several glass plates are placed in an RF coil with squared geometry orienting the normal of the lipid membrane ($\vec{n}$) parallel to $\vec{B}_0$ (Bechinger and Opella 1991). However, the presence of glass slides in the sample, and the necessity of using squared geometry for the RF coils dramatically reduces the sensitivity of these experiments (Veglia *et al* 2012). The introduction of bicelles has improved the sensitivity and resolution of the OS-ssNMR spectra (Dürr *et al* 2012). Using a defined molar ratio of detergents and lipids (e.g., 1,2-dihexanoyl-*sn*-glycero-3-phosphocholine:1,2-dimyristoyl-*sn*-glycero-3-phosphocholine, DHPC:DMPC), lipid bicelles form spontaneously within a specific temperature range and concentrations, assuming an orientation of $\vec{n}$ perpendicular to $\vec{B}_0$ (Sanders *et al* 1994). Under these conditions, the breadth of the CSA and DC is approximately half of that obtained for mechanically oriented systems. To achieve a better dispersion of the chemical shifts, it is possible to *flip* the orientation of the bicelles using lanthanides that bind either to adventitious sites on the lipid headgroups or are chelated by engineered lipid molecules (Prosser *et al* 1998). The magnetic susceptibility of the lanthanide induces an orientation of $\vec{n}$ parallel to $\vec{B}_0$ (Prosser *et al* 1996). Bicellar systems are fully hydrated and can be prepared with *ad hoc* protocols that preserve the function of larger membrane proteins and enzymes. Another advantage of bicellar preparations is the favorable relaxation properties of the membrane proteins, which enable the acquisition of multidimensional experiments (Mote *et al* 2011). The longer transverse relaxation times give rise to an increased resolution for uniformly labeled recombinant proteins. Several examples of membrane proteins oriented in bicelles are reported in figure 12.2(a–d). A suitable alternative to bicelles are macrodiscs, although, at this time, there are only a few examples available (Park *et al* 2011, Radoicic *et al* 2018). Both bicelles and macrodiscs promise to boost the resolution of OS-ssNMR spectra and enable the application of this technique to multi-pass membrane proteins (Weber and Veglia 2019).

## 12.2 Chemical shift anisotropy

The chemical shift ($\delta$) observed for an NMR-active nucleus ($^1$H, $^{13}$C, $^{15}$N, $^{19}$F, etc) depends on the shielding effect arising from the local electronic environment (Saitô *et al* 2010). In liquid-state NMR of proteins, secondary structure, protein–protein interactions, ligand–protein interactions, solvent environment, hydrogen bonding, etc, perturb this electronic environment causing the dispersion of their isotropic chemical shifts ($\delta_{\text{iso}}$). This is also true for membrane proteins reconstituted into detergent micelles and nanodiscs. However, the larger size of these systems and their associated slow molecular correlation times cause significant line broadening, hampering the application of multidimensional NMR experiments and sequential resonance assignments. Molecular size does not affect MAS-ssNMR experiments; however, resonance overlap, incomplete averaging of $^1$H–$^1$H DC and CSA, conformational exchange, and sample heterogeneity hinder the resonance assignments for membrane proteins. For α-helical membrane proteins, which constitute a significant class of therapeutic targets, this problem is exacerbated by the redundancy of their primary sequences and the limited amino-acid diversity of transmembrane segments.

In the absence of dynamic averaging, the electronic shielding environment surrounding a nucleus is anisotropic, i.e., each nucleus possesses a unique chemical

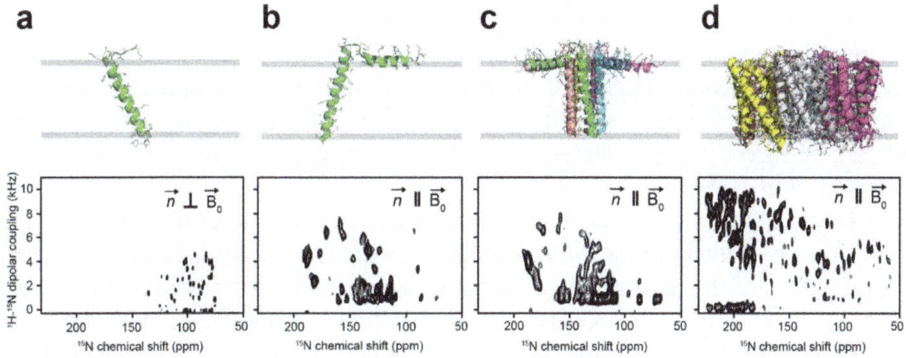


**Figure 12.2.** Examples of SLF spectra for (a). sarcolipin in *unflipped* bicelles (4 kDa; PDB 1JDM (Mascioni *et al* 2002)), (b). monomeric (6 kDa; PDB 4KB7 (Traaseth *et al* 2009)), and (c). pentameric phospholamban in *flipped* bicelles (5 x 6 kDa; PDB 2KYV (Verardi *et al* 2011)), and (d). the hexameric SatP acetate transporter in *flipped* bicelles (6 x 20 kDa; PDB 5ZUG (Sun *et al* 2018)).

shift that depends on its orientation relative to $\vec{B}_0$. A second rank tensor describes this anisotropy diagonalized into three principal components, which in Haeberlen notation (Harris *et al* 2008, Haeberlen 1976), i.e., ordered from most upfield (lowest ppm) to most downfield (highest ppm), are $\delta_{xx}$, $\delta_{yy}$ and $\delta_{zz}$. The fast molecular tumbling in liquids or mechanical spinning in MAS of solids average the tensor components ($\delta_{iso}$):

$$\delta_{iso} = (\delta_{xx} + \delta_{yy} + \delta_{zz})/3. \tag{12.1}$$

If the molecular reorientation is slow, then the signal is smeared into a broad line (i.e., powder pattern), which reflects all possible molecular orientations present in the sample. While CSA hampers the spectral resolution of techniques that rely on isotropic NMR parameters (e.g., liquid-state NMR and MAS), it is harnessed by OS-ssNMR, which necessitates homogenously aligned samples relative to $\vec{B}_0$ to resolve nuclear spin resonances based on their orientation ($\delta_{oriented}$). The latter has two main properties. First, chemical shift dispersion (i.e., spectral resolution) increases proportionally to the shift anisotropy ($\Delta\delta$):

$$\Delta\delta = \delta_{zz} - (\delta_{xx} + \delta_{yy})/2. \tag{12.2}$$

For example, the $^{15}$N shift tensor of backbone amides has the following components $\delta_{xx} = 57.3$ ppm, $\delta_{yy} = 81.2$ ppm, and $\delta_{zz} = 228.1$ ppm (Murray *et al* 2014), giving rise to $\Delta\delta$ ~160 ppm, which is an order of magnitude greater than the typical dispersion observed for $\delta_{iso}$ (~30 ppm). The second property is related to the nature of $\delta_{oriented}$, which depends on the orientation of the chemical shift tensor components ($\vec{\delta}_{xx}$, $\vec{\delta}_{yy}$ and $\vec{\delta}_{zz}$) relative to $\vec{B}_0$:

$$\delta_{oriented} = S \cdot [(\delta_{xx}\cos^2\theta_{xx} + \delta_{yy}\cos^2\theta_{yy} + \delta_{zz}\cos^2\theta_{zz}) - \delta_{iso}] + \delta_{iso}, \tag{12.3}$$

where $\theta_{xx}$, $\theta_{yy}$ and $\theta_{zz}$ are the polar angles of $\vec{\delta}_{xx}$, $\vec{\delta}_{yy}$ and $\vec{\delta}_{zz}$ relative to $\vec{B}_0$, and $S$ is the generalized order parameter for the protein and the alignment medium (section 12.4). Essentially, the scaling of the $\delta_{xx}$, $\delta_{yy}$ and $\delta_{zz}$ values are maximum if they are perfectly aligned and ordered along $\vec{B}_0$ (i.e., $\theta = 0°$, and $S = 1$), and become zero if they are perpendicular. Therefore, unlike $\delta_{iso}$, which requires empirical conversions to generate dihedral angle restraints, $\delta_{oriented}$ has a well-defined angular dependence and can be directly implemented as a restraint in structure calculation protocols (Ketchem *et al* 1994).

To solve equation (12.3) and implement $\delta_{oriented}$ as a structural restraint, $\delta_{xx}$, $\delta_{yy}$ and $\delta_{zz}$ must be defined in the molecular frame, which is accomplished by referencing the orientation of the tensor relative to bond vectors and to surrounding atoms using Euler rotation angles (α, β, and γ). Geometric parameters, along with their magnitudes, are derived using model compounds or proteins using ssNMR on single crystals, oriented liquid crystals, and powdered polycrystalline samples. Alternatively, it is possible to measure the magnitude of these tensor components via solution-state relaxation, sideband analysis under slow-spinning MAS conditions, CSA recoupling methods under fast MAS, and *ab initio* calculations (Saitô *et al* 2010). Using the $^{15}$N backbone amide as an example (Wu *et al* 1995), the tensor is approximated by directing $\vec{\delta}_{zz}$ along the N–H bond, $\vec{\delta}_{yy}$ perpendicular to the amide plane and using the cross product of $\vec{\delta}_{yy}$ and $\vec{\delta}_{zz}$ to set $\vec{\delta}_{xx}$. The tensor orientation is then finalized by a rotation around $\vec{\delta}_{yy}$ so that $\vec{\delta}_{zz}$ moves ~17° ($\beta$) towards the N–C′ bond (figure 12.3). The sign of the rotation is negative because of the right-hand rule. A similar approach is used to orient the tensors in structure calculation software such as XPLOR-NIH (Schwieters *et al* 2006) and CNS (Bertram *et al* 2000), as well as algorithms to back-calculate $\delta_{\text{oriented}}$ from molecular dynamics trajectories (Shi *et al* 2009a, Weber and Veglia 2019, Cheng *et al* 2013).

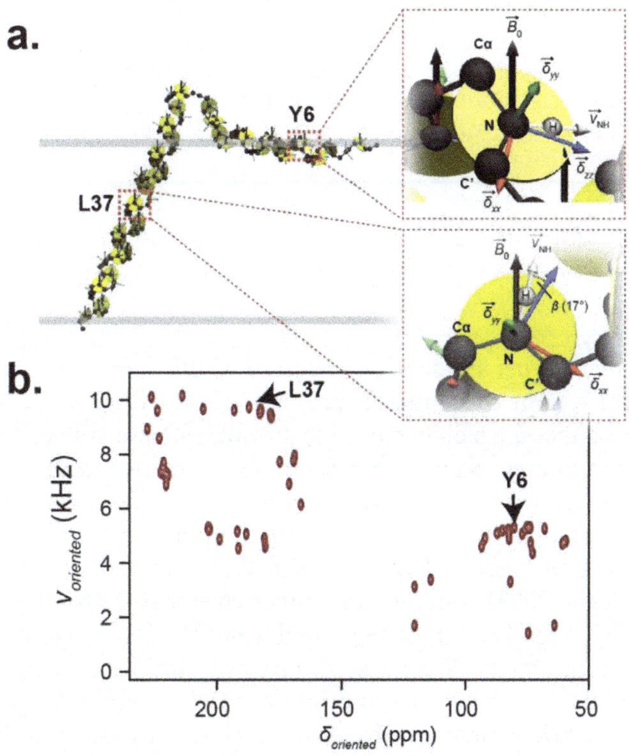


**Figure 12.3.** (a). Components of the $^{15}$N shift ($\vec{\delta}_{xx}$, $\vec{\delta}_{yy}$, and $\vec{\delta}_{zz}$) and $^{1}$H–$^{15}$N dipolar coupling ($\vec{\nu}_{NH}$) tensors projected onto the amide planes (yellow discs) of phospholamban (PDB 2KB7 (Traaseth *et al* 2009)). Residues L37 and Y6 are expanded to exemplify tensor orientations of residues located within the transmembrane and surface-bound helical segments. (b). An SLF spectrum back-calculated from PDB 2KB7. Peak FWHM linewidths were arbitrarily set to 100 Hz and 300 Hz in the $\delta_{\text{oriented}}$ and $\nu_{\text{oriented}}$ dimensions, respectively (60.8 MHz $^{15}$N Larmor frequency). Peak positions were predicted using a VMD (Humphrey *et al* 1996) Tcl script described elsewhere (Weber and Veglia 2019). Example calculations of $\delta_{\text{oriented}}$ and $\nu_{\text{oriented}}$ L37 and Y6 are shown in Box 12.1.

**Box 12.1: Determination of chemical shifts and dipolar coupling from tensors**

Figure 12.3 features two residues that lie in either the transmembrane (L37) or the surface bound (Y6) helix of phospholamban. The chemical shift ($\delta_{oriented}$) is determined by a summation of the contributions made by each of the components $\vec{\delta}_{xx}$, $\vec{\delta}_{yy}$ and $\vec{\delta}_{zz}$, which is a function of their polar angles ($\theta_{xx}$, $\theta_{yy}$ and $\theta_{zz}$) away from $\vec{B}_0$ (i.e., from the $z$-axis). Substituting these angles into equation 12.3, with the maximum contributions of each component (i.e., $\delta_{xx} = 57.3$ ppm, $\delta_{yy} = 81.2$ ppm, and $\delta_{zz} = 228.1$ ppm for backbone $^{15}$N nuclei), and neglecting $S$, the values of $\delta_{oriented}$ are found by:

| L37 | | Y6 | |
|---|---|---|---|
| $57.3\ \text{ppm} \times \cos^2 119.5°$ | (14 ppm) | $57.3\ \text{ppm} \times \cos^2 143.4°$ | (37 ppm) |
| $+\,81.2\ \text{ppm} \times \cos^2 80.8°$ | (2 ppm) | $+\,81.2\ \text{ppm} \times \cos^2 59.5°$ | (21 ppm) |
| $+\,228.1\ \text{ppm} \times \cos^2 31.2°$ | (167 ppm) | $+\,228.1\ \text{ppm} \times \cos^2 108.2°$ | (22 ppm) |
| $\delta_{oriented} = 183$ ppm | | $\delta_{oriented} = 80$ ppm | |

From these equations, it clear that the contributions of each component vary depending on the orientation of the helical segment. Similarly, the dipolar coupling ($\nu_{oriented}$) is determined from the polar angle of the dipolar vector ($\vec{\nu}_{NH}$) away from $\vec{B}_0$ ($\theta_{NH}$). Substituting $\theta_{NH}$, with $\nu_{max}$ set to 10.735 kHz, the values for $\nu_{oriented}$ are determined from equation 12.5 by:

**L37**

$$\frac{10.735\text{kHz}}{2} \cdot (3\cos^2 15.8° - 1)\nu_{oriented} = 9.5\ \text{kHz}$$

**Y6**

$$\frac{10.735\ \text{kHz}}{2} \cdot (3\cos^2 93.7° - 1)|\ \nu_{oriented}\ | = 5.3\ \text{kHz}$$

Anisotropic backbone amide $^{15}$N chemical shifts have been widely used in OS-ssNMR spectroscopy of membrane proteins primarily due to the high-resolution of the amide resonances, which behave like an isolated spin system and do not need homonuclear decoupling. Since the CSA reflects on the local geometry of the secondary structures of proteins, SLF spectra of helices often display regular periodic patterns that can be interpreted using the Polar Index Slant Angle method (Denny *et al* 2001, Marassi and Opella 2000, Weber *et al* 2020, Wang *et al* 2000, Mascioni and Veglia 2003). Additionally, other chemical shifts related to backbone nuclei such as $^{13}$C′ and $^{1}$H$_N$ can be measured using HETCOR (Ramamoorthy *et al* 1999) experiments, and provide complementary restraints for structural calculations. Oriented $^{13}$C and $^{15}$N shifts can also offer insights into sidechain orientation (Aisenbrey *et al* 2009, Ramamoorthy *et al* 1997, Separovic *et al* 1994, Hu *et al* 1995, Hu *et al* 1993). It is important to reiterate that the magnitude and orientation of chemical shift tensors are established based on model compounds and therefore assumed to be representative of all residues and uniform throughout protein samples. However, the latter is a rough approximation since $\delta_{iso}$ (equation 12.1) assumes residue- and structure-specific values that are critical to resolve individual residues by solution NMR or MAS-ssNMR. These variations, however, constitute a minor component of $\delta_{oriented}$, which introduces negligible errors in the determination of the overall protein topology (Shi *et al* 2009b). Glycine residues are the only exception and are often treated under the assumption of a unique tensor (Straus *et al* 2003).

## 12.3 Dipolar coupling

Like CSA, internuclear DC can be implemented as a direct restraint in membrane protein structure calculations (Ketchem *et al* 1993). DC depends on the internuclear distance, which is assumed to be fixed, and an angular term, which defines the orientation relative to $\vec{B}_0$. The internuclear distance ($r$) sets the maximum value of the DC ($\nu_{\max}$) and is a function of the gyromagnetic ratios of two nuclei ($\gamma_I$ and $\gamma_S$):

$$\nu_{\max} = \frac{-\mu_0 h \gamma_I \gamma_S}{8\pi^3 r^3}, \tag{12.4}$$

where $\mu_0$ is the vacuum permeability, and $h$ is Planck's constant. For example, the $\nu_{\max}$ for a $^1$H–$^{15}$N dipolar coupling is ~21.5 kHz. Since the SLF spectra are symmetric, in the OS-ssNMR field (figure 12.4), only half of the full DC is often considered (10.735 kHz). The observable DC ($\nu_{\text{oriented}}$) relates to $\nu_{\max}$ by:

$$\nu_{\text{oriented}} = S\frac{\nu_{\max}}{2}(3\cos^2\theta_{IS} - 1), \tag{12.5}$$

where $\theta_{IS}$ is the polar angle of the bond vector away from $\vec{B}_0$, and $S$ is the general order parameter used in equation (12.3). Assuming a rigid system, the $^1$H–$^{15}$N dipolar coupling varies from 10.735 kHz ($\theta = 0°$) to −5.37 kHz ($\theta = 90°$) in which the sign cannot be determined experimentally. However, ambiguity from the sign degeneracy for structure calculations is reduced when DC values are correlated with CSA (Bertram *et al* 2003). More recently, Nevzorov and co-workers reduced

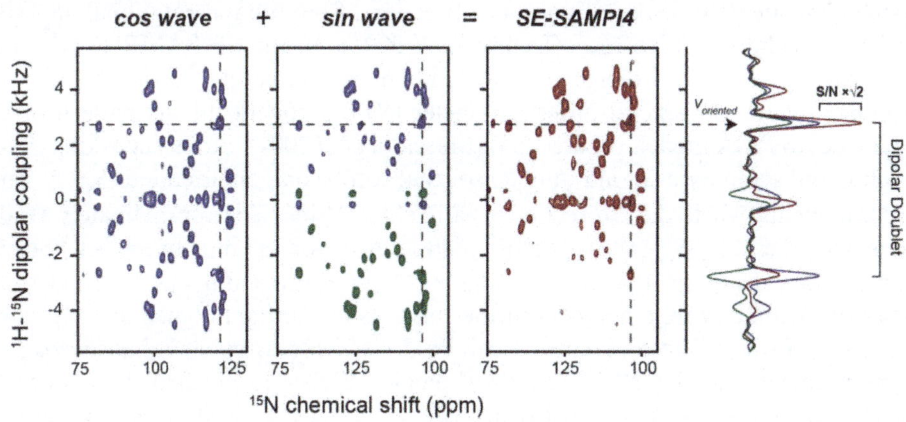


**Figure 12.4.** SE-SAMPI4 spectrum of sarcolipin in *unflipped* bicelles. The SE-SAMPI4 pulse sequence sums both the cosine and sine-modulated dipolar coherences to boost signal-to-noise up to $\sqrt{2}$ times greater compared to the standard SAMPI4 sequence, which only detects the cosine component (see figure 12.1). The splitting of the dipolar doublets seen in the cosine spectrum is the true $^1$H–$^{15}$N dipolar coupling value, although $\nu_{\text{oriented}}$ restraints are traditionally reported from one half of the spectrum.

this ambiguity by introducing a triple-resonance SAMPI4[2] sequence that measures $^{1}H\alpha$–$^{13}C\alpha$ and $^{1}H$–$^{15}N$ dipolar couplings in addition to $^{15}N$ chemical shifts, providing three orientational restraints per amide plane (Awosanya *et al* 2020).

Compared to residual dipolar couplings (RDCs) from liquid-state NMR, DCs are considerably more straightforward to implement as structural restraints. This is especially true for membrane proteins since they are embedded into a mechanically or magnetically aligned medium with known alignment tensor. Also, the alignment tensor is axially symmetric, meaning that it only has a magnitude in the direction of the membrane normal ($\vec{n}$) and no rhombic component. This compares favorably to RDCs, in which weak association of proteins with an alignment medium preserves only 0.1–1 % of the dipolar interaction (Chen and Tjandra 2012), with magnitudes and rhombicities of the alignment tensors being highly variable from system to system.

## 12.4 Scaling of orientational restraints due to conformational and topological dynamics

In early studies, the impact of motion on $\delta_{\text{oriented}}$ and $\nu_{\text{oriented}}$ was essentially neglected, which was partly due to the use of mechanically aligned lipid bilayers with low levels of hydration where membrane proteins have limited degrees of freedom. Under these conditions, the linewidths are significantly broad, ranging from ~5 to 10 ppm (Page *et al* 2008). With higher levels of hydration of magnetically aligned systems, membrane proteins display significantly narrower linewidths (less than 3 ppm), but the increased conformational and topological dynamics must be accounted for in structure calculations (De Angelis *et al* 2006, Park *et al* 2006). The dynamic scaling of the anisotropic parameters becomes apparent when fitting the SLF spectra of α-helical proteins using PISA wheel models (Marassi and Opella 2000, Wang *et al* 2000). Such models simulate SLF peak positions as a function of the tilt and rotational angles of idealized α-helices relative to $\vec{n}$ (Denny *et al* 2001). In the SE-SAMPI4 spectra of single-pass membrane proteins phospholamban and sarcolipin, it is necessary to introduce a variable general order parameter ($S$) to properly fit the pattern of the resonances to PISA models (figure 12.5) (Mascioni *et al* 2004). Strictly speaking, $S$ is a traceless and symmetric second rank tensor that can be diagonalized into $S_{xx}$, $S_{yy}$ and $S_{zz}$ components, where the $Z$ direction is parallel to $\vec{n}$. Since $S$ is approximately axially symmetric (i.e., $S_{xx}$-$S_{yy}$ = 0), values determined for $S$ mostly reflect the $S_{zz}$ contribution. However, rigorous analysis of $S$ for Gramicidin has revealed small values of asymmetry that can yield further insights into protein dynamics (Separovic *et al* 1993). Fitting $S$ to SLF spectra can be done easily using an exhaustive search function coded into the *PISA-SPARKY* plugin (Weber *et al* 2020) built into the *NMRFAM Sparky* software package (Lee *et al* 2015). Following an iterative procedure, it is possible to use PISA fitting for estimating protein dynamics, which in bicelles scales the $\delta_{\text{oriented}}$ (minus $\delta_{\text{iso}}$) and $\nu_{\text{oriented}}$ values by approximately 20%. For non-helical protein domains (i.e., β-barrels), the estimate of $S$ is not straightforward, and it might require specific assumptions (Mahalakshmi and Marassi 2008).

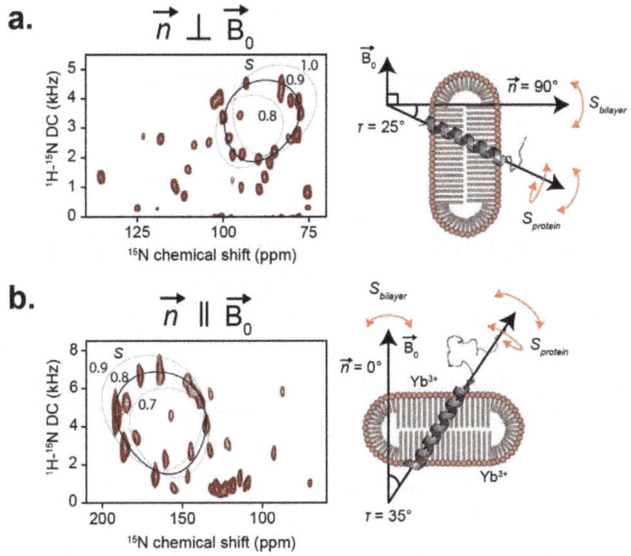


**Figure 12.5.** SE-SAMPI4 spectra of a. sarcolipin (PDB 1JDM (Mascioni *et al* 2002)), and b. an excited state of monomeric phospholamban (PDB 2LPF (De Simone *et al* 2013)) reconstituted into *unflipped* and *flipped* bicelles, respectively. Solid lines are PISA fits to spectra, and faded lines are the same fits recalculated with order parameters ± 0.1 ($S$); helix tilt angle kept the same). Right-side panels are schematics of each system illustrating the bicelle alignment order ($S_{\text{bilayer}}$), helix tilt angle ($\tau$), and rigid-body helix tilt and rotational fluctuations responsible for scaling SLF spectra ($S_{\text{protein}}$). $S_{\text{protein}}$ also includes localized backbone internal dynamics. From equation equation (12.6), the $S$ values fitted to sarcolipin (0.9) and phospholamban (0.8) suggest that dynamic scaling of SLF spectra are due to time-averaged overall uniaxial fluctuations ($\langle\vartheta\rangle$) of ~15° and ~20°, respectively.

An alternate approach is to allow for the alignment tensor to vary during structure calculation (Tian *et al* 2015).

The mode of motions that scale DC and CSA values in the SLF spectra are not entirely understood. As a first approximation, one should consider any dynamic process that results in fast-timescale angular excursions from the average orientation of the tensorial interaction. At slower timescales, these motions manifest as broad lineshapes (Quine *et al* 2006, Nevzorov *et al* 1999). Fast timescale dynamics, on the other hand, correspond to motions faster than the inverse of the reduced CSA ($\delta_{\text{oriented}} - \delta_{\text{iso}}$) and dipolar coupling ($\nu_{\text{oriented}}$). While the former is field-dependent, the latter is field-independent. The average tensorial interactions are the average values of $\theta_{xx}$, $\theta_{yy}$, $\theta_{zz}$ and $\theta_{IS}$ over the NMR timescale (equations 12.3 and 12.5). The order parameter takes the general form:

$$S = \left\langle \frac{3\cos^2\vartheta - 1}{2} \right\rangle \tag{12.6}$$

where $\vartheta$ is the instantaneous angular fluctuation of the tensorial interaction perpendicular to $\vec{n}$ (i.e., considering only the $S_{zz}$ component of $S$). The brackets denote that this is a time-averaged quantity, meaning that the equation is evaluated for every instantaneous angle over a set timescale. Using the PISA pattern scaling in figure 12.5 as an example, the generalized order parameters for sarcolipin ($S = 0.9$) and phospholamban ($S = 0.8$) could be explained by fluctuation amplitudes of ~15° and ~20°, respectively. These fluctuations are a product of order parameters of the protein ($S_{\text{protein}}$) and alignment medium ($S_{\text{bilayer}}$; i.e., $S = S_{\text{protein}} \cdot S_{\text{bilayer}}$). While, molecular dynamics simulations have shown that rigid body tilt angle *wobbling* accounts for the majority of the dynamics captured in SLF spectra (Esteban-Martín *et al* 2009, Straus *et al* 2003, Holt *et al* 2010, Perrin *et alet al* 2014), order parameters of 0.85 to 0.95 have also been applied for refinements of structures using RDCs and residual CSAs (RCSAs) in liquid-state NMR. These order parameters factor in the backbone internal dynamics of structural segments or domains (Tjandra *et al* 1997, Yao *et al* 2010). Lack of consideration of these motions can lead to overestimation of orientational restraints and ill-defined structures (Shi *et al* 2011, Chipot *et al* 2018). Uniaxial rotational diffusion about the bilayer normal constitutes another significant motion, but it can be discarded if the membrane $\vec{n}$ is parallel to $\vec{B}_0$, as in the case of *flipped* bicelles, since the $\theta_{xx}$, $\theta_{yy}$, $\theta_{zz}$, and $\theta_{IS}$ angles are not affected. For *unflipped* bicelles, however, the rotational diffusion does affect these angles, and the interpretation of SLF spectra is more complicated, especially at slow motional timescales (Tesch *et al* 2018). Since rotational diffusion of membrane proteins is generally fast (>$10^6$ Hz) even for sizeable multipass membrane proteins (Opella 2013, Park *et al* 2012, Lewis *et al* 1985), such cases are rare. A further complicating factor, especially for bicelles, is the bilayer disorder ($S_{\text{bilayer}}$) due to imperfect magnetic alignment (De Angelis *et al* 2006, Park *et al* 2006). This phenomenon becomes apparent from the analysis of $^2$H quadrupolar couplings and $^{31}$P CSA spectra of bicelles, which are reduced compared to supported lipid bilayers or vesicular membranes (Sanders *et al* 1994). However, lipid order parameters may also be affected by detergent-induced packing defects. Although deconvoluting the relative contributions from $S_{\text{protein}}$ and $S_{\text{bilayer}}$ is difficult, their overall effect on SLF spectra is identical and must be accounted for before embarking into the quasi-static structure calculation. The experimental values of $\delta_{\text{oriented}}$ are rescaled according to:

$$\delta_{\text{restraint}} = \frac{\delta_{\text{oriented}} - \delta_{\text{iso}}}{S}. \tag{12.7}$$

Notice that $\delta_{\text{restraint}}$ is represented in its reduced form relative to the isotropic chemical shift. Similarly, the experimental dipolar coupling ($\nu_{\text{oriented}}$) is scaled by:

$$\nu_{\text{restraint}} = \frac{\nu_{\text{oriented}}}{S}. \tag{12.8}$$

Additionally, equation (12.6) also handles the scaling effect due to the orientation of the alignment medium (i.e., $\vartheta = \vec{n}$), which factors into the $S_{\text{bilayer}}$ component of $S$. To account for this in the restraints, the order parameter $S$ in equations (12.7) and (12.8) must be multiplied by either 1.0 for *flipped* bicelles (Prosser *et al* 1996) or mechanically aligned systems ($\vec{n} = 0°$), or −0.5 for *unflipped* bicelles ($\vec{n} = 90°$).

Lastly, we should also assume that the scaling is uniform as $S_{\text{protein}}$ can be considered constant for all residues. A similar assumption is made for RDCs in structure calculations protocols. Of course, if residues or domains are very dynamic ($S < 0.6$) the refinement protocol using RDCs introduces significant errors (Tjandra *et al* 1997). The latter is not an issue for OS-ssNMR as dynamic residues are often not observable using CP-base pulse sequences.

## 12.5 Hybridization of isotropic and anisotropic restraints

The high-resolution structure determination of polypeptides by NMR spectroscopy requires a minimum number of non-redundant structural restraints (Russell *et al* 2019). For liquid-state NMR, there are different sources of structural restraints that can be implemented simultaneously (i.e., NOE, chemical shifts, J-couplings, RDCs, RCSAs, paramagnetic distances, and H-bonds). In contrast, membrane proteins in lipid membranes do not offer the same luxury. Indeed, DC and CSA encode considerable structural information, but they are insufficient to achieve a global resolution of less than 1–2 Å. Therefore, it is vital to supplement orientational restraints with isotropic restraints from either solution NMR or MAS-ssNMR. Isotropic chemical shifts provide additional torsion angle and distance restraints to help determine secondary, tertiary, and quaternary structures. For this reason, we developed the hybrid method that combines OS-ssNMR and solution NMR restraints into a simulated annealing protocol (Shi *et al* 2009b, Veglia *et al* 2012). We used this approach to determine the structures of monomeric and pentameric forms of phospholamban (Traaseth *et al* 2009, Verardi *et al* 2011, Vostrikov *et al* 2013). Other examples of structures determined with a similar strategy include the membrane-associated domain of huntingtin (Michalek *et al* 2013), influenza M2 proton channel (Cady *et al* 2010, Sharma *et al* 2010), bacterial mercury transporter MerF (Tian *et al* 2014a), bacteriophage major coat protein Pf1 (Park *et al* 2010), mitochondrial membrane protein stannin (Buck-Koehntop *et al* 2005), and microsomal cytochrome b5 (Ahuja *et al* 2013).

There are several different approaches to implement OS-ssNMR restraints in membrane protein structure calculations. For instance, the *structural fitting* methods developed by Marassi (Marassi and Opella 2003, Tian *et al* 2012) and Nevzorov (Lapin and Nevzorov 2019, Nevzorov and Opella 2003) groups, which generally set out to generate and fit atomic structural models and can achieve simultaneous assignment and high-resolution structure determination from SLF data alone. Another example is the *shiftless ssNMR* method, which computes protein structures solely from heteronuclear dipolar couplings (Yin and Nevzorov 2011, Wu and Opella 2008).

In the following section, we focus predominantly on the hybridization of isotropic and anisotropic restraints for simulated annealing calculations. The hybrid target function ($E_{\text{TOTAL}}$) for our calculations takes the form:

$$E_{\text{TOTAL}} = E_{\text{CHEM}} + E_{\text{ANI}} + E_{\text{ISO}}, \tag{12.9}$$

The total energy to minimize consists of a linear combination of geometric force field potentials ($E_{CHEM}$) and restraining potentials to enforce experimental isotropic ($E_{ISO}$) and anisotropic ($E_{ANI}$) NMR observables. In XPLOR-NIH (Schwieters *et al* 2006, Schwieters *et al* 2003, Schwieters *et al* 2018), $E_{CHEM}$ includes bonded terms for bond lengths ($E_{BON}$), bond angles ($E_{ANG}$), improper dihedral angles ($E_{IMPR}$), statistically determined proper dihedral angles ($E_{torsionDB}$) (Bermejo *et al* 2012), and a repulsive van der Waals non-bonded term ($E_{VDW-REPEL}$). More recently, Marassi and co-workers introduced the Effective Energy Function for XPLOR-NIH (EEFx) (Tian *et al* 2014b), which includes additional nonbonded terms including attractive van der Waals ($E_{VDW}$), electrostatic ($E_{ELEC}$), and solvation ($E_{SOLV}$) potentials. This force field was further developed by introducing an Implicit Membrane Model (IMMx), where $E_{ELEC}$ and $E_{SOLV}$ are modulated by a variable dielectric constant as a function of atomic $z$-coordinates from the membrane center ($z = 0$) (Tian *et al* 2015, Tian *et al* 2017). Therefore, with EEFx and IMMx, the overall form of $E_{CHEM}$ for membrane proteins, including associated weighting factors ($w$), is:

$$\begin{aligned} E_{CHEM} = {} & w_{BON}E_{BON} + w_{ANG}E_{ANG} + w_{IMPR}E_{IMPR} + w_{torsionDB}E_{torsionDB} \\ & + w_{VDW}E_{VDW} + w_{ELEC}^{IMMx}E_{ELEC}^{IMMx} + w_{SOLV}^{IMMx}E_{SOLV}^{IMMx}. \end{aligned} \tag{12.10}$$

The inclusion of EEFx and IMMx potentials has several advantages. First, the inclusion of electrostatic and solvation effects allows the force field to independently determine hydrogen bonds, salt bridges, and van der Waals interactions. Second, the inclusion of IMMx during folding and refinement substitutes for an additional rigid-body refinement stage using a virtual membrane, such as the empirical $E_z$ potential (Senes *et al* 2007) we have used previously to set the depth of insertion (Shi *et al* 2009b). Last, the physical constraints provided by the IMMx, including a user-specified bilayer thickness, reinforces the topological information encoded in experimental OS-ssNMR restraints by forcing the protein to satisfy both *hydrophobic matching* (Mouritsen and Bloom 1984) and *snorkeling* (Segrest *et al* 1990). These improvements, when applied to the single transmembrane helix of the HIV-1 *V*irus *P*rotein '*u*' (Vpu), yielded a folded and correctly oriented structure with only oriented $^{15}N$ chemical shifts applied, whereas additional torsion restraints were also necessary if $E_{VDW-REPEL}$ was the sole non-bonded term of $E_{CHEM}$ (Tian *et al* 2015).

The $E_{ISO}$ term of the target function enforces isotropic restraints by:

$$E_{ISO} = w_{NOE}E_{NOE} + w_{CDIH}E_{CDIH}, \tag{12.11}$$

where $E_{NOE}$ are distance potentials, which are most often NOE distances measured by solution NMR but can also be adapted for distances determined by MAS-ssNMR dipolar correlation experiments (Yan *et al* 2013). $E_{CDIH}$ includes dihedral restraints usually determined empirically from chemical shifts (Cornilescu *et al* 1999). While not as common, other isotropic restraints can be added, such as from paramagnetic relaxation enhancements (PREs), solvent PREs, relaxation rates, and H/D exchange experiments. Furthermore, if no solution NMR or MAS-ssNMR data are available, torsion angle restraints may be derived from SLF cross-peaks to build initial structural models (Achuthan *et al* 2008, Asbury *et al* 2006).

While $E_{\text{CHEM}}$ and $E_{\text{ISO}}$ terms are applied throughout the entire structural calculation, including a high-temperature torsion-angle dynamics step and the following simulated annealing, $E_{\text{ANI}}$ terms are generally applied only during the simulated annealing step, and their force constants are increased geometrically as the system is cooled. The $E_{\text{ANI}}$ potential includes the terms for anisotropic chemical shifts ($E_{\text{CSA}}$) and dipolar couplings ($E_{\text{DC}}$) determined by OS-ssNMR:

$$E_{\text{ANI}} = w_{\text{CSA}}E_{\text{CSA}} + w_{\text{DC}}E_{\text{DC}} = w_{\text{SLF}}(E_{\text{CSA}} + w_r E_{\text{DC}}), \tag{12.12}$$

with the relative weights against $E_{\text{ISO}}$ terms ($w_{\text{SLF}}$) and the relative weight of $E_{\text{DC}}$ to $E_{\text{CSA}}$ ($w_r$) optimized by cross-validation to prevent overfitting (Kim *et al* 2001, Shi *et al* 2009b). For cross-validation, $w_{\text{SLF}}$ and $w_r$ are systematically increased to optimize an $R$-factor (Kim *et al* 2001) defined by:

$$R = \frac{1}{N}\sum_{i=1}^{N}\left(\frac{|\text{Obs}_i^{\text{Exp}} - \text{Obs}_i^{\text{Calc}}|}{\text{Err}_i^{\text{Exp}}}\right)^2, \tag{12.13}$$

where $\text{Obs}_i^{\text{Exp}}$, $\text{Obs}_i^{\text{Calc}}$ and $\text{Err}_i^{\text{Exp}}$ are the experimental, calculated, and errors of the NMR observables (*i.e.*, $\delta_{\text{restraint}}$ and $\nu_{\text{restraint}}$) associated with a given residue $i$ normalized over $N$ residues. This is done both for an applied set of restraints ($R_{\text{work}}$) and a random set (~20 %) that have been purposely excluded ($R_{\text{free}}$) but still back-calculated from the structures. Optimal $w_{\text{SLF}}$ and $w_r$ values, which we have previously found as 5 and 3 (Shi *et al* 2009b), respectively (without EEFx terms), are determined when the $R$ values have minimized without penalizing the $E_{\text{CHEM}}$ and $E_{\text{ISO}}$ terms. Similarly, $R_{\text{work}}$ and $R_{\text{free}}$ should also be the same to provide additional assurance that $E_{\text{ANI}}$ terms are not competing with the chemical knowledge of the system.

While simulated annealing protocols are efficient for finding a global minimum energy from an unfolded starting configuration (figure 12.6), the resulting structures are devoid of the dynamic information encoded in the NMR observables. These quasi-static structures are assumed to be representative of a homogenous conformational ensemble. However, it is often that NMR-observables from dynamic proteins are averaged over a heterogeneous structural landscape that is important for function. Using MD simulations as an additional stage of structural refinement is, therefore, becoming increasingly common as technological improvements have made computational resources more accessible. MD refinement in explicit solvent and membranes also provides insights into protein–lipid interactions and allows protein dynamics to be correlated with order parameters determined from SLF experiments. However, force field deficiencies and insufficient sampling can introduce artifactual conformations and convergence issues if experimental restraints are not applied. This is true even for single-pass membrane proteins that exhibit a significant degree of topological freedom that cannot be sampled over conventional simulation timescales. CSA and DC restraints can be used to bias simulations towards realistic energy landscapes (De Simone *et al* 2014, Cheng *et al* 2013). With

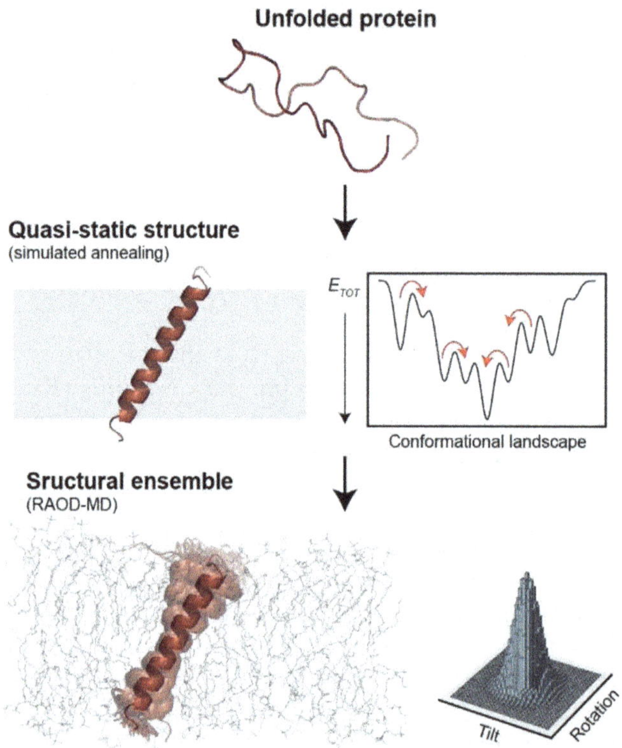


**Figure 12.6.** Simplified structure determination workflow. Simulated annealing is effective for finding the global energy minimum of a conformational landscape when starting from an unfolded state. The final structures are considered quasi-static since the dynamics underpinning an NMR observable cannot be portrayed through energy minimized structures. These missing dynamics can be modeled by RAOD-MD refinement, which can be tilt and rotational angle distributions in the case of a single helix.

the De Simone group, we have recently applied ensemble restraints for OS-ssNMR spectra, which we refer to as *Replica-Averaged Orientational Restrained MD* (RAOR-MD), for phospholamban and sarcolipin (De Simone *et al* 2014, Sanz-Hernández *et al* 2016). In this approach, CSA and DC are applied over multiple independent simulations, i.e., replicas, computed simultaneously. This approach allows heterogeneous conformational ensembles to be generated over picosecond-nanosecond timescales with strict adherence to OS-ssNMR observables.

## 12.6 Summary

We provided a brief survey on the use of anisotropic restraints from SLF NMR experiments in combination with isotropic restraints obtained from MAS-ssNMR. This background should equip the reader with a critical overview of the geometric principles underpinning CSA and DC and their utilization in the determination of membrane protein structure and architecture within fully hydrated lipid bilayers. Note that a mere structural refinement disregards dynamics (motions), which are encoded in the NMR measurements. The former not only assumes the existence of a single structure, but it also overestimates the magnitude of anisotropic parameters, which may cause structural distortions. For processes such as signal transduction, enzymatic activity, ion transport, etc, motions often represent the main driving force for biological function but are neglected in classical structural determination. The dynamic refinement using RAOD-MD emerges as a faithful representation of the conformational ensemble that proteins explore in the timescale of NMR (Lindorff-Larsen *et al* 2005, De Simone *et al* 2009, De Simone *et al* 2011) and provides essential information on biological function, with an obvious advantage over other high-resolution techniques such as x-ray crystallography and cryoEM.

## Acknowledgments

This work was partially supported by the National Institutes of Health (GM64742 and HL144130) and the American Heart Association (19POST34420009 to DKW).

## References

Achuthan S, Asbury T, Hu J, Bertram R, Cross T A and Quine J R 2008 Continuity conditions and torsion angles from ssNMR orientational restraints *J. Magn. Reson.* **191** 24–30

Ahuja S *et al* 2013 A model of the membrane-bound cytochrome $b_5$-cytochrome P450 complex from NMR and mutagenesis data *J. Biol. Chem.* **288** 22080–95

Aisenbrey C, Prongidi-Fix L, Chenal A, Gillet D and Bechinger B 2009 Side chain resonances in static oriented proton-decoupled $^{15}$N solid-state NMR spectra of membrane proteins *J. Am. Chem. Soc.* **131** 6340–1

Asbury T, Quine J R, Achuthan S, Hu J, Chapman M S, Cross T A and Bertram R 2006 pipath: an optimized algorithm for generating α-helical structures from PISEMA data *J. Magn. Reson.* **183** 87–95

Awosanya E O, Lapin J and Nevzorov A A 2020 NMR 'Crystallography' for uniformly ($^{13}$C, $^{15}$N)-labeled oriented membrane proteins *Angew. Chem. Int. Ed.* **59** 3554–7

Bechinger B and Opella S J 1991 Flat-coil probe for NMR spectroscopy of oriented membrane samples *J. Magn. Reson. (1969)* **95** 585–8

Bermejo G A, Clore G M and Schwieters C D 2012 Smooth statistical torsion angle potential derived from a large conformational database via adaptive kernel density estimation improves the quality of NMR protein structures *Protein Sci.* **21** 1824–36

Bertram R, Asbury T, Fabiola F, Quine J R, Cross T A and Chapman M S 2003 Atomic refinement with correlated solid-state NMR restraints *J. Magn. Reson.* **163** 300–9

Bertram R, Quine J R, Chapman M S and Cross T A 2000 Atomic refinement using orientational restraints from solid-state NMR *J. Magn. Reson.* **147** 9–16

Buck-Koehntop B A, Mascioni A, Buffy J J and Veglia G 2005 Structure, dynamics, and membrane topology of stannin: a mediator of neuronal cell apoptosis induced by trimethyltin chloride *J. Mol. Biol.* **354** 652–65

Cady S D, Schmidt-Rohr K, Wang J, Soto C S, Degrado W F and Hong M 2010 Structure of the amantadine binding site of influenza M2 proton channels in lipid bilayers *Nature* **463** 689–92

Chen K and Tjandra N 2012 The use of residual dipolar coupling in studying proteins by NMR ed G Zhu *NMR of Proteins and Small Biomolecules.* (Berlin, Heidelberg: Springer)

Cheng X, Jo S, Marassi F M and Im W 2013 NMR-based simulation studies of Pf1 coat protein in explicit membranes *Biophys. J.* **105** 691–8

Chipot C *et al* 2018 Perturbations of native membrane protein structure in alkyl phosphocholine detergents: a critical assessment of NMR and biophysical studies *Chem. Rev.* **118** 3559–607

Cornilescu G, Delaglio F and Bax A 1999 Protein backbone angle restraints from searching a database for chemical shift and sequence homology *J. Biomol. NMR* **13** 289–302

De Angelis A A, Howell S C, Nevzorov A A and Opella S J 2006 Structure determination of a membrane protein with two trans-membrane helices in aligned phospholipid bicelles by solid-state NMR spectroscopy *J. Am. Chem. Soc.* **128** 12256–67

De Simone A, Gustavsson M, Montalvao R W, Shi L, Veglia G and Vendruscolo M 2013 Structures of the excited states of phospholamban and shifts in their populations upon phosphorylation *Biochemistry* **52** 6684–94

De Simone A, Montalvao R W and Vendruscolo M 2011 Determination of conformational equilibria in proteins using residual dipolar couplings *J. Chem. Theory Comput.* **7** 4189–95

De Simone A, Mote K R and Veglia G 2014 Structural dynamics and conformational equilibria of SERCA regulatory proteins in membranes by solid-state NMR restrained simulations *Biophys. J.* **106** 2566–76

De Simone A, Richter B, Salvatella X and Vendruscolo M 2009 Toward an accurate determination of free energy landscapes in solution states of proteins *J. Am. Chem. Soc.* **131** 3810–1

Denny J K, Wang J, Cross T A and Quine J R 2001 PISEMA powder patterns and PISA wheels *J. Magn. Reson.* **152** 217–26

Dürr U H N, Gildenberg M and Ramamoorthy A 2012 The magic of bicelles lights up membrane protein structure *Chem. Rev.* **112** 6054–74

Dvinskikh S, Dürr U, Yamamoto K and Ramamoorthy A 2006a A high-resolution solid-state NMR approach for the structural studies of bicelles *J. Am. Chem. Soc.* **128** 6326–7

Dvinskikh S V, Yamamoto K and Ramamoorthy A 2006b Separated local field NMR spectroscopy by windowless isotropic mixing *Chem. Phys. Lett.* **419** 168–73

Esteban-Martín S, Strandberg E, Fuertes G, Ulrich A S and Salgado J 2009 Influence of whole-body dynamics on $^{15}$N PISEMA NMR spectra of membrane proteins: a theoretical analysis *Biophys. J.* **96** 3233–41

Gopinath T, Mote K R and Veglia G 2011 Proton evolved local field solid-state nuclear magnetic resonance using Hadamard encoding: theory and application to membrane proteins *J. Chem. Phys.* **135** 074503

Gopinath T and Veglia G 2009 Sensitivity enhancement in static solid-state NMR experiments via single and multiple-quantum dipolar coherences *J. Am. Chem. Soc.* **131** 5754–6

Gopinath T, Verardi R, Traaseth N J and Veglia G 2010 Sensitivity enhancement of separated local field experiments: application to membrane proteins *J. Phys. Chem.* B **114** 5089–95

Haeberlen U 1976 *High Resolution NMR in Solids Selective Averaging* (Amsterdam: Elsevier)

Harris R K, Becker E D, Cabral De Menezes S M, Granger P, Hoffman R E and Zilm K W 2008 Further conventions for NMR shielding and chemical shifts (IUPAC Recommendations 2008) *Pure Appl. Chem.* **80** 59–84

Hester R K, Ackerman J L, Neff B L and Waugh J S 1976 Separated local field spectra in NMR: determination of structure of solids *Phys. Rev. Lett.* **36** 1081–3

Holt A, Rougier L, Réat V, Jolibois F, Saurel O, Czaplicki J, Antoinette Killian J and Milon A 2010 Order parameters of a transmembrane helix in a fluid bilayer: case study of a WALP peptide *Biophys. J.* **98** 1864–72

Hu W, Lazo N D and Cross T A 1995 Tryptophan dynamics and structural refinement in a lipid bilayer environment: solid state NMR of the gramicidin channel *Biochemistry* **34** 14138–46

Hu W, Lee K C and Cross T A 1993 Tryptophans in membrane proteins: indole ring orientations and functional implications in the gramicidin channel *Biochemistry* **32** 7035–47

Humphrey W, Dalke A and Schulten K 1996 VMD: visual molecular dynamics *J. Mol. Graph.* **14** 33–8

Ketchem R, Hu W and Cross T 1993 High-resolution conformation of gramicidin A in a lipid bilayer by solid-state NMR *Science* **261** 1457–60

Ketchem R R, Hu W, Tian F and Cross T A 1994 Structure and dynamics from solid-state NMR spectroscopy *Structure* **2** 699–701

Kim S, Quine J R and Cross T A 2001 Complete cross-validation and *R*-factor calculation of a solid-state NMR derived structure *J. Am. Chem. Soc.* **123** 7292–8

Lapin J and Nevzorov A A 2019 Validation of protein backbone structures calculated from NMR angular restraints using Rosetta *J. Biomol. NMR* **73** 229–44

Lee W, Tonelli M and Markley J L 2015 NMRFAM-SPARKY: enhanced software for biomolecular NMR spectroscopy *Bioinformatics* **31** 1325–7

Lewis B A, Harbison G S, Herzfeld J and Griffin R G 1985 NMR structural analysis of a membrane protein: bacteriorhodopsin peptide backbone orientation and motion *Biochemistry* **24** 4671–9

Lindorff-Larsen K, Best R B, Depristo M A, Dobson C M and Vendruscolo M 2005 Simultaneous determination of protein structure and dynamics *Nature* **433** 128–32

Mahalakshmi R and Marassi F M 2008 Orientation of the *Escherichia coli* outer membrane protein OmpX in phospholipid bilayer membranes determined by solid-state NMR *Biochemistry* **47** 6531–8

Marassi F M and Opella S J 2000 A solid-state NMR index of helical membrane protein structure and topology *J. Magn. Reson.* **144** 150–5

Marassi F M and Opella S J 2003 Simultaneous assignment and structure determination of a membrane protein from NMR orientational restraints *Protein Sci.* **12** 403–11

Mascioni A, Eggimann B L and Veglia G 2004 Determination of helical membrane protein topology using residual dipolar couplings and exhaustive search algorithm: application to phospholamban *Chem. Phys. Lipids* **132** 133–44

Mascioni A, Karim C, Barany G, Thomas D D and Veglia G 2002 Structure and orientation of sarcolipin in lipid environments *Biochemistry* **41** 475–82

Mascioni A and Veglia G 2003 Theoretical analysis of residual dipolar coupling patterns in regular secondary structures of proteins *J. Am. Chem. Soc.* **125** 12520–6

Michalek M, Salnikov E S and Bechinger B 2013 Structure and topology of the Huntingtin 1–17 membrane anchor by a combined solution and solid-state NMR approach *Biophys. J.* **105** 699–710

Mote K R, Gopinath T, Traaseth N J, Kitchen J, Gor'kov P L, Brey W W and Veglia G 2011 Multidimensional oriented solid-state NMR experiments enable the sequential assignment of uniformly $^{15}$N labeled integral membrane proteins in magnetically aligned lipid bilayers *J. Biomol. NMR* **51** 339–46

Mouritsen O G and Bloom M 1984 Mattress model of lipid-protein interactions in membranes *Biophys. J.* **46** 141–53

Murray D T, Hung I and Cross T A 2014 Assignment of oriented sample NMR resonances from a three transmembrane helix protein *J. Magn. Reson.* **240** 34–44

Nevzorov A A, Moltke S, Heyn M P and Brown M F 1999 Solid-state NMR line shapes of uniaxially oriented immobile systems *J. Am. Chem. Soc.* **121** 7636–43
Nevzorov A A and Opella S J 2003 Structural fitting of PISEMA spectra of aligned proteins *J. Magn. Reson.* **160** 33–9
Nevzorov A A and Opella S J 2007 Selective averaging for high-resolution solid-state NMR spectroscopy of aligned samples *J. Magn. Reson.* **185** 59–70
Opella S J 2013 Structure determination of membrane proteins in their native phospholipid bilayer environment by rotationally aligned solid-state NMR spectroscopy *Acc. Chem. Res.* **46** 2145–53
Page R C, Kim S and Cross T A 2008 Transmembrane helix uniformity examined by spectral mapping of torsion angles *Structure* **16** 787–97
Park S H, Berkamp S, Cook G A, Chan M K, Viadiu H and Opella S J 2011 Nanodiscs versus macrodiscs for NMR of membrane proteins *Biochemistry* **50** 8983–5
Park S H *et al* 2012 Structure of the chemokine receptor CXCR1 in phospholipid bilayers *Nature* **491** 779–83
Park S H, De Angelis A A, Nevzorov A A, Wu C H and Opella S J 2006 Three-dimensional structure of the transmembrane domain of Vpu from HIV-1 in aligned phospholipid bicelles *Biophys. J.* **91** 3032–42
Park S H, Marassi F M, Black D and Opella S J 2010 Structure and dynamics of the membrane-bound form of Pf1 coat protein: implications of structural rearrangement for virus assembly *Biophys. J.* **99** 1465–74
Perrin B S JR, Pastor R W and Cotten M 2014 Combining NMR spectroscopic measurements and molecular dynamics simulations to determine the orientation of amphipathic peptides in lipid bilayers *Advances in Biological Solid-State NMR: Proteins and Membrane-Active Peptides* ed A Naito and F Separovic (Cambridge: Royal Society of Chemistry)
Prosser R S, Hunt S A, Dinatale J A and Vold R R 1996 Magnetically aligned membrane model systems with positive order parameter: switching the sign of $S_{zz}$ with paramagnetic ions *J. Am. Chem. Soc.* **118** 269–70
Prosser R S, Volkov V B and Shiyanovskaya I V 1998 Novel chelate-induced magnetic alignment of biological membranes *Biophys. J.* **75** 2163–9
Quine J R, Achuthan S, Asbury T, Bertram R, Chapman M S, Hu J and Cross T A 2006 Intensity and mosaic spread analysis from PISEMA tensors in solid-state NMR *J. Magn. Reson.* **179** 190–8
Radoicic J, Park S H and Opella S J 2018 Macrodiscs comprising SMALPs for oriented sample solid-state NMR spectroscopy of membrane proteins *Biophys. J.* **115** P22–5
Ramamoorthy A, Wu C H and Opella S J 1997 Magnitudes and orientations of the principal elements of the $^{1}$H chemical shift, $^{1}$H–$^{15}$N dipolar coupling, and $^{15}$N chemical shift interaction tensors in $^{15}N_{\varepsilon1}$-tryptophan and $^{15}N_{\pi}$-histidine side chains *J. Am. Chem. Soc.* **119** 10479–86
Ramamoorthy A, Wu C H and Opella S J 1999 Experimental aspects of multidimensional solid-state NMR correlation spectroscopy *J. Magn. Reson.* **140** 131–40
Russell R W, Fritz M P, Kraus J, Quinn C M, Polenova T and Gronenborn A M 2019 Accuracy and precision of protein structures determined by magic angle spinning NMR spectroscopy: for some 'with a little help from a friend' *J. Biomol. NMR* **73** 333–46
Saitô H, Ando I and Ramamoorthy A 2010 Chemical shift tensor—the heart of NMR: insights into biological aspects of proteins *Prog. Nucl. Magn. Reson. Spectrosc.* **57** 181–228

Sanders C R, Hare B J, Howard K P and Prestegard J H 1994 Magnetically-oriented phospholipid micelles as a tool for the study of membrane-associated molecules *Prog. Nucl. Magn. Reson. Spectrosc* **26** 421–44

Sanz-Hernández M, Vostrikov V V, Veglia G and De Simone A 2016 Accurate determination of conformational transitions in oligomeric membrane proteins *Sci. Rep.* **6** 23063

Schwieters C D, Bermejo G A and Clore G M 2018 Xplor-NIH for molecular structure determination from NMR and other data sources *Protein Sci.* **27** 26–40

Schwieters C D, Kuszewski J J and Marius Clore G 2006 Using Xplor-NIH for NMR molecular structure determination *Prog. Nucl. Magn. Reson. Spectrosc.* **48** 47–62

Schwieters C D, Kuszewski J J, Tjandra N and Clore G M 2003 The Xplor-NIH NMR molecular structure determination package *J. Magn. Reson.* **160** 65–73

Segrest J P, De Loof H, Dohlman J G, Brouillette C G and Anantharamaiah G M 1990 Amphipathic helix motif: classes and properties *Proteins* **8** 103–17

Senes A, Chadi D C, Law P B, Walters R F S, Nanda V and Degrado W F 2007 $E_z$, a depth-dependent potential for assessing the energies of insertion of amino acid side-chains into membranes: derivation and applications to determining the orientation of transmembrane and interfacial helices *J. Mol. Biol.* **366** 436–48

Separovic F, Gehrmann J, Milne T, Cornell B A, Lin S Y and Smith R 1994 Sodium ion binding in the gramicidin A channel. Solid-state NMR studies of the tryptophan residues *Biophys. J.* **67** 1495–500

Separovic F, Pax R and Cornell B 1993 NMR order parameter analysis of a peptide plane aligned in a lyotropic liquid crystal *Mol. Phys.* **78** 357–69

Sharma M, Yi M, Dong H, Qin H, Peterson E, Busath D D, Zhou H X and Cross T A 2010 Insight into the mechanism of the influenza a proton channel from a structure in a lipid bilayer *Science* **330** 509–12

Shi L, Cembran A, Gao J and Veglia G 2009a Tilt and azimuthal angles of a transmembrane peptide: a comparison between molecular dynamics calculations and solid-state NMR data of sarcolipin in lipid membranes *Biophys. J.* **96** 3648–62

Shi L, Traaseth N J, Verardi R, Cembran A, Gao J and Veglia G 2009b A refinement protocol to determine structure, topology, and depth of insertion of membrane proteins using hybrid solution and solid-state NMR restraints *J. Biomol. NMR* **44** 195–205

Shi L, Traaseth N J, Verardi R, Gustavsson M, Gao J and Veglia G 2011 Paramagnetic-based NMR restraints lift residual dipolar coupling degeneracy in multidomain detergent-solubilized membrane proteins *J. Am. Chemi. Soc.* **133** 2232–41

Straus S K, Scott W R P and Watts A 2003 Assessing the effects of time and spatial averaging in $^{15}N$ chemical shift/$^{15}N$–$^{1}H$ dipolar correlation solid state NMR experiments *J. Biomol. NMR* **26** 283–95

Sun P *et al* 2018 Crystal structure of the bacterial acetate transporter SatP reveals that it forms a hexameric channel *J. Biol. Chem.* **293** 19492–500

Tesch D M, Pourmoazzen Z, Awosanya E O and Nevzorov A A 2018 Uniaxial diffusional narrowing of NMR lineshapes for membrane proteins reconstituted in magnetically aligned bicelles and macrodiscs *Appl. Magn. Reson.* **49** 1335–53

Tian Y, Lu G J, Marassi F M and Opella S J 2014a Structure of the membrane protein MerF, a bacterial mercury transporter, improved by the inclusion of chemical shift anisotropy constraints *J. Biomol. NMR* **60** 67–71

Tian Y, Schwieters C D, Opella S J and Marassi F M 2012 AssignFit: a program for simultaneous assignment and structure refinement from solid-state NMR spectra *J. Magn. Reson.* **214** 42–50

Tian Y, Schwieters C D, Opella S J and Marassi F M 2014b A practical implicit solvent potential for NMR structure calculation *J. Magn. Reson.* **243** 54–64

Tian Y, Schwieters C D, Opella S J and Marassi F M 2015 A practical implicit membrane potential for NMR structure calculations of membrane proteins *Biophys. J.* **109** 574–85

Tian Y, Schwieters C D, Opella S J and Marassi F M 2017 High quality NMR structures: a new force field with implicit water and membrane solvation for Xplor-NIH *J. Biomol. NMR* **67** 35–49

Tjandra N, Omichinski J G, Gronenborn A M, Marius Clore G and Bax A 1997 Use of dipolar $^{1}H–^{15}N$ and $^{1}H–^{13}C$ couplings in the structure determination of magnetically oriented macromolecules in solution *Nat. Struct. Biol.* **4** 732–8

Traaseth N J, Shi L, Verardi R, Mullen D G, Barany G and Veglia G 2009 Structure and topology of monomeric phospholamban in lipid membranes determined by a hybrid solution and solid-state NMR approach *Proc. Natl. Acad. Sci. USA* **106** 10165–70

Veglia G, Traaseth N J, Shi L, Verardi R, Gopinath T and Gustavsson M 2012 1.11 The hybrid solution/solid-state NMR method for membrane protein structure determination *Comprehensive Biophysics* ed E Egelman (Amsterdam: Elsevier)

Verardi R, Shi L, Traaseth N J, Walsh N and Veglia G 2011 Structural topology of phospholamban pentamer in lipid bilayers by a hybrid solution and solid-state NMR method *Proc. Natl Acad. Sci. USA* **108** 9101–6

Vostrikov V V, Mote K R, Verardi R and Veglia G 2013 Structural dynamics and topology of phosphorylated phospholamban homopentamer reveal its role in the regulation of calcium transport *Structure* **21** 2119–30

Wang J *et al* 2000 Imaging membrane protein helical wheels *J. Magn. Reson.* **144** 162–7

Wang S, Gopinath T and Veglia G 2019 Improving the quality of oriented membrane protein spectra using heat-compensated separated local field experiments *J. Biomol. NMR* **73** 617–24

Weber D K and Veglia G 2019 A theoretical assessment of the structure determination of multi-span membrane proteins by oriented sample solid-state NMR spectroscopy *Aust. J. Chem.* **73** 246–6

Weber D K, Wang S, Markley J L, Veglia G and Lee W 2020 PISA-SPARKY: an interactive SPARKY plugin to analyze oriented solid-state NMR spectra of helical membrane proteins *Bioinformatics* **36** 2915–6

Wu C H and Opella S J 2008 Shiftless nuclear magnetic resonance spectroscopy *J. Chem. Phy.* **128** 052312

Wu C H, Ramamoorthy A, Gierasch L M and Opella S J 1995 Simultaneous characterization of the amide $^{1}H$ Chemical Shift, $^{1}H–^{15}N$ dipolar, and $^{15}N$ chemical shift interaction tensors in a peptide bond by three-dimensional solid-state NMR spectroscopy *J. Am. Chem. Soc.* **117** 6148–9

Wu C H, Ramamoorthy A and Opella S J 1994 High-resolution heteronuclear dipolar solid-state NMR spectroscopy *J. Magn. Reson. Ser.* A **109** 270–2

Yan S, Suiter C L, Hou G, Zhang H and Polenova T 2013 Probing structure and dynamics of protein assemblies by magic angle spinning NMR spectroscopy *Acc. Chem. Res.* **46** 2047–58

Yao L, Grishaev A, Cornilescu G and Bax A 2010 Site-specific backbone amide $^{15}N$ chemical shift anisotropy tensors in a small protein from liquid crystal and cross-correlated relaxation measurements *J. Am. Chem. Soc.* **132** 4295–309

Yin Y and Nevzorov A A 2011 Structure determination in 'shiftless' solid state NMR of oriented protein samples *J. Magn. Reson.* **212** 64–73

**IOP** Publishing

Solid-State NMR
Applications in biomembrane structure
**Frances Separovic and Marc-Antoine Sani**

# Chapter 13

# Beyond structure: understanding dynamics in membrane protein complexes

**Nikita Malik and Lynmarie Thompson**

## Overview

Solid-state NMR is a powerful technique to study membrane proteins, especially because it can measure the structure and dynamics of functional states of these proteins in their native lipid bilayer environments. The mechanisms of many proteins involve dynamic changes between two or more conformations and NMR offers an array of tools to measure dynamics in membrane proteins at various timescales. This chapter describes some of these methods for measurement of both site-specific as well as bulk dynamics, including the recent shift towards fast MAS and proton-detected experiments. To illustrate the range of experiments available to address diverse proteins with different functions, we describe approaches to probe dynamics in three structurally different systems, in each case focusing on how measurements of dynamics advance understanding of the mechanism of these proteins. *Anabaena* sensory rhodopsin provides an example of an alpha helical transmembrane protein, and outer membrane proteins represent transmembrane beta barrels, both of which have been well-characterized by NMR. Finally, bacterial chemoreceptor protein complexes serve as an example that illustrates measurements of functionally important dynamics on a system with sparse NMR data. Creative strategies and methodological advances are making it possible to extend solid-state NMR studies to biologically important systems with challenging NMR properties and contribute key understanding of the underlying mechanisms.

## 13.1 Introduction

Membrane proteins are an important and diverse class of proteins that perform critical biological functions in all organisms, from viruses to mammals. These proteins are characterized by their association with lipid membranes and are structurally different from globular proteins. Most membrane proteins have a

doi:10.1088/978-0-7503-2532-5ch13  

highly hydrophobic region that contacts the lipid membrane, making these proteins insoluble in aqueous solutions. Some of the most fundamental biological processes, such as communication, sensing, transport of ions and metabolites, and bioenergetics require membrane proteins. Such a wide range of functions is possible due to the different conformations that these proteins can adopt at different stages of their functional cycle (Baker and Baldus 2014). Additionally, membrane proteins exhibit different properties depending on the membrane composition, such as cholesterol content and the lipid head groups present (Mandala *et al* 2018). The structural plasticity of membrane proteins combined with their association with the dynamic membrane environment makes it difficult to investigate these proteins in their native environment.

Most high-resolution structures of membrane proteins have been obtained using x-ray crystallography, but there has been a steady increase in structure determination of these proteins using cryo-electron microscopy. For crystallographic studies, the major challenge is the need to solubilize the protein from its native lipid environment and then crystallize it within a detergent environment. Maintaining native structure in detergents during purification and crystallization can be difficult, as many membrane proteins show reduced stability in detergents (Stangl *et al* 2012, Yang *et al* 2014). Furthermore, studies have shown that the membrane-mimics can have a huge impact on the biological function and structural properties of some membrane proteins (Zhou and Cross 2013). Advancements in cryo-EM have eased some constraints on sample preparation by allowing the use of native-like lipid environments and not requiring crystallization. However, both x-ray and cryo-EM can be difficult when a protein is dynamic. For instance, the intrinsic dynamics in G protein coupled receptors (GPCRs) hampered the crystallization efforts for a long time, and modifications like mutagenesis and binding of antibodies that suppressed protein motion were required to produce high quality protein crystals (Chen *et al* 2012, Rasmussen *et al* 2007, Sarkar *et al* 2008). Furthermore, it can be difficult to observe dynamic regions of proteins with either x-ray crystallography or cryo-EM. Finally, these methods provide static snapshots of the protein under study at cryogenic temperature, and do not offer a comprehensive view of the structure/dynamics at physiological conditions. NMR is an indispensable tool to gain high resolution insights into dynamic protein regions.

Dynamics are not only a challenge to structural biology, but also often an essential aspect of the mechanisms of proteins. While the three-dimensional structure of a protein provides key insights into its organization, it is well known that the internal dynamics are crucial for biological function: enzymatic activation, allosteric regulation, and ligand binding often depend on the plasticity of proteins (Henzler-Wildman and Kern 2007, Kovermann *et al* 2016, Lisi and Loria 2016). NMR provides powerful approaches for measuring protein dynamics at a wide range of timescales (Palmer 2004, Palmer and Massi 2006). Solution NMR methods have been applied to small membrane proteins in detergent micelles (Kang and Li 2011); these non-native detergent micelle environments are chosen in order to achieve the rapid tumbling required for solution NMR. But their utility for membrane proteins is limited, because they can perturb the structure and dynamics.

Some NMR studies employ alternatives to detergent micelles, including detergent-lipid bicelles, amphipols, and nanodiscs (made with membrane scaffold protein, polymers, or peptides), each suitable to specific systems or studies (Klöpfer and Hagn 2019). Despite these improved membrane-mimetics, a fundamental limitation of solution NMR is that rapid tumbling cannot be achieved for large proteins and multi-protein complexes. Studies of larger systems such as membrane proteins in their native environment (lipid bilayer vesicles) or large membrane proteins/complexes are limited by slow tumbling, making solid-state NMR the preferred method to characterize dynamics of these proteins (Lamley *et al* 2015, Lewandowski 2013, Schanda and Ernst 2016).

Solid-state nuclear magnetic resonance (ssNMR) offers opportunities to probe dynamic conformational ensembles of membrane proteins at the atomic level. Because neither rapid isotropic tumbling or 3D crystalline order are required for ssNMR, it is an excellent tool to probe membrane proteins in their native membrane environment. Without the size restriction of solution NMR experiments, larger protein-lipid complexes can be studied, provided sufficient spectral resolution is achieved. ssNMR approaches have been applied to whole cells, cellular membranes, vesicles that mimic biological membranes, and bicelles to study both rigid and dynamic membrane proteins. Seven helix receptors in native membranes (Etzkorn *et al* 2007), $H^+$-ATP synthase subunit c (Kobayashi *et al* 2006), DsbB (disulphide bond forming enzyme) (Li *et al* 2007), and LH2 (light harvesting protein complex) (Gammeren *et al* 2005) are a few of the many examples of membrane proteins with near complete resonance assignments. Advances in methodology development and flexibility in the range of sample preparation has resulted in structure elucidation of a number of membrane proteins, including influenza A M2 proton channel in DOPC/DOPE (Sharma *et al* 2010), microcrystalline DsbB-DsbA complex with native lipids (Tang *et al* 2011), *Anabaena* sensory rhodopsin (ASR) in DMPC/DMPA lipids (Wang *et al* 2013a), and human phospholamban (PLN) in DOPC/DOPE lipids (Verardi *et al* 2011). The mechanism of a membrane protein cannot be deciphered by its structure alone, but also requires understanding of how the protein moves. Here, we discuss some of the solid-state NMR strategies that can be used to study functionally important membrane protein dynamics. Although NMR of oriented samples has provided many important insights into membrane protein structure and dynamics (Hansen *et al* 2015), this review focuses on magic angle spinning (MAS) methods applied to a range of membrane protein systems (with and without sequence-specific assignments) and recent promising advances that should expand the applicability of such approaches to provide insights into additional important membrane proteins and their complexes.

## 13.2 Membrane protein dynamics

A comprehensive description of the mechanism of a membrane protein cannot be based only on its structure, as many proteins undergo dynamic changes between two or more conformations to perform essential biological functions. For example, the continuous exchange between the two states of the multidrug resistance transporter

EmrE (figure 13.1(a)) allows the transport of its substrates across the membrane (figure 13.1(b, c)) (Morrison *et al* 2011). Changes in the membrane protein conformation can be prompted by a variety of factors including ligand binding, ion conduction, change in lipid environment, and interaction with receptors. For instance, conformational dynamics of the potassium channel KcsA has been reported in response to changes in potassium ion concentration as well as due to toxin binding (Bhate *et al* 2010, Lange *et al* 2006). These functionally important cases are examples of the wide range of motions experienced by membrane proteins. Some motions cause a global change in the structure such as reorientation of domains relative to each other, or refolding of the membrane protein to transition

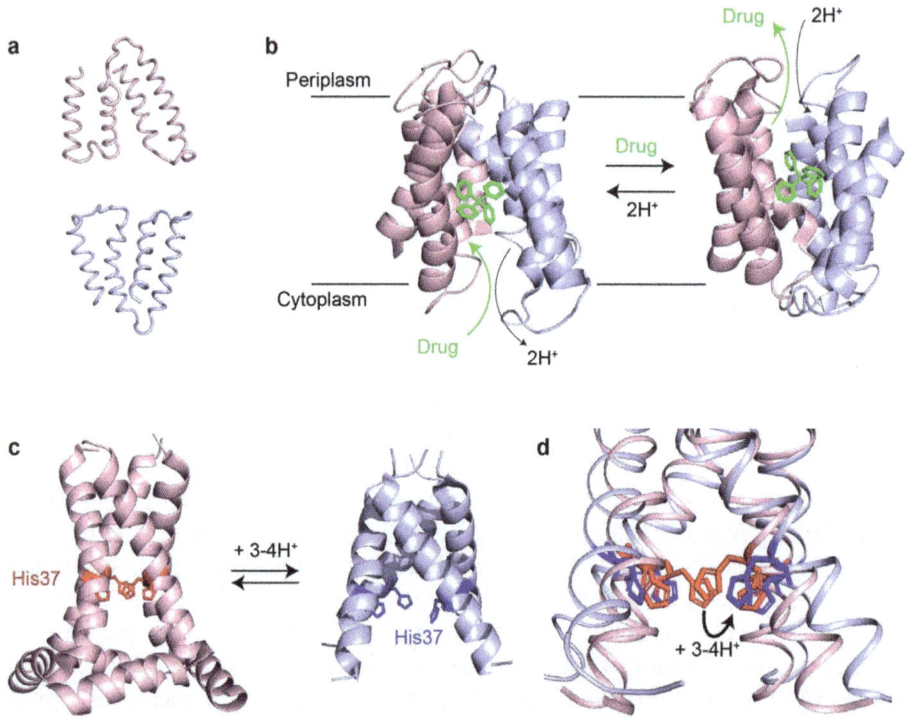


**Figure 13.1.** Structural changes in membrane proteins are critical for function. Active transporter EmrE functions through a global reorientation converting between inward and outward facing conformations (Morrison *et al* 2011), and the proton conduction mechanism through influenza M2 protein requires a local change in sidechain orientation of the His37 residue (Hu *et al* 2010). (a). The EmrE dimer (PDB: 3B5D) is composed of asymmetric antiparallel subunits (pale pink and pale blue), with equal populations of both monomer orientations present in the native inner membrane in *E coli.* (b). Antiport to achieve drug efflux occurs via the single site alternating access model, where both the substrates share a single binding site and conformational change occurs only when the substrate (two protons or one polyaromatic cation) is bound (PDB: 3B5D). (c). M2 channel at lower pH (pale blue, PDB: 3C9J) has a His37 conformation (dark blue, sticks) that keeps the sidechains far from each other forming a spacious pore, and at higher pH (pale pink, PDB: 2L0J) the neutral His37 ring (red, sticks) stacks in a manner that creates an electron dense region that repels water. This helps in proton shuttling, with the histidine ring reorientations facilitating the transfer of protons. (d). Change in the His37 ring positions in the structures at high and low pH.

from the aqueous to membrane-bound state (Bowie 2005, Huysmans *et al* 2010), while other motions cause more local changes like methyl rotations, torsional fluctuations, and aromatic ring flips (figure 13.1(d)). Protein dynamics, which include a wide range of conformational changes occurring at different timescales, are complex processes to measure and characterize. Additionally, in some cases the functionally relevant conformational state may be short-lived and sparsely populated, so detecting and studying these minor forms represents additional challenges. Thus, characterizing protein dynamics calls for experiments that are sensitive to different types of motions at varied timescales.

ssNMR can provide global and residue-level dynamic information across a range of timescales, providing a valuable means of characterizing the structural plasticity displayed by membrane proteins. Motions at different timescales have distinctive effects on NMR spectra. For example, depending on the timescale of a structural fluctuation that changes the environment of a nucleus, there may be apparent line narrowing (fast exchange), exchange broadening (intermediate exchange), or a broad unresolved resonance (slow exchange). Protein dynamics are also reflected in spectral intensities. For example, motional averaging of dipolar couplings reduces overall signal intensities in a cross polarization (CP) experiment, and reduces signal intensities of specific residues in a double-quantum spectrum (Schneider *et al* 2010). CP spectra, which rely on dipolar couplings to transfer magnetization from protons to carbon or nitrogen, are heavily biased by motions. Thus, systems that are expected to yield rapid polarization build-up can show significantly reduced CP if the protein region is experiencing motion. For example, the increase in signal intensity in $^{13}$C-CP experiments upon addition of EGF indicated ligand-induced stabilization of the epidermal growth factor receptor (EGFR). Additional DNP-supported ssNMR experiments confirmed the reduction in local dynamics after ligand binding by demonstrating reduced line widths in frozen spectra of EGFR labeled with specific residues (M, F, T, and L) upon adding EGF (Kaplan *et al* 2016). Some proteins have sufficient dynamics to have CP-invisible regions. These often can be detected by scalar coupling-based experiments like Insensitive Nuclei Enhanced by Polarization Transfer (INEPT). The Baldus group identified a high degree of disorder in the cytoplasmic domain of phospholamban (PLN) by employing INEPT-based experiments (Andronesi *et al* 2005). Further multidimensional experiments enabled them to assign resonances for two regions of PLN, the moderately stable transmembrane α-helix detected in dipolar transfer-based experiments and the unstructured cytoplasmic domain detected in INEPT transfer-based experiments. While interpreting the data from CP and INEPT blocks, it should be noted that the CP signal strength depends on the contact time and the strength of dipolar coupling, and the INEPT signal strength depends on the $T_2$ relaxation rate.

For measuring internal motions on a timescale above a few nanoseconds, ssNMR may be more informative and sample a wider range since solution NMR measurements of internal motions are restricted to timescales shorter than the molecular tumbling (Schanda and Ernst 2016). ssNMR can be used to directly probe both amplitudes and timescales of motion. Dipolar order parameters (the ratio of the observed to the expected dipolar coupling) report on amplitudes of motion on the submicrosecond timescale and have found widespread use (Helmus *et al* 2010, Lorieau *et al* 2008). Spin relaxation is sensitive to both timescale and amplitude of motions and $^{15}$N spins can be used as probes to study internal dynamics of solid proteins (Cole and Torchia 1991). Longitudinal relaxation $R_1$ is sensitive to picosecond-nanosecond timescale motions, and the transverse relaxation rates $R_{1\rho}$ and $R_2$ can be used to probe slower, nanosecond to microsecond motions. Methods to measure $^{15}$N $R_1$ and $R_{1\rho}$ relaxation rates to study site-specific dynamics have been developed (Giraud *et al* 2005, 2004, Krushelnitsky *et al* 2010) and applied to a variety of proteins. For example, measurement of $^{15}$N $R_1$ and $R_{1\rho}$ for all resolved amide signals for *K. pneumoniae* outer membrane protein (KpOmpA) reconstituted in *E.coli* polar extract liposomes revealed that residues showing higher $^{15}$N $R_{1\rho}$ indicating slow internal motions were located near strand extremities (Saurel *et al* 2017). Motions in the intermediate timescale can be monitored by separated local field experiments like dipolar-chemical-shift (DIPSHIFT) that enable determination of sub-microsecond motions by quantitatively fitting dipolar dephasing trajectories using specific models of molecular motions (deAzevedo *et al* 2008). Hong *et al* (2012) applied this technique on the transmembrane domain of influenza proton channel M2 in lipid bilayers to measure order parameters for specific labeled sidechains. They demonstrated that lowering the pH increases dynamics of histidine-37, allowing the imidazole ring to flip, which is necessary for proton transport (Hu *et al* 2010). For more complex systems, it should be noted that important functional information about a protein can also be deduced without complete resonance assignments. A $^{13}$C MAS study on G-protein-coupled neuropeptide Y Type 2 receptor in DMPC membranes assessed the dynamics of the protein in the absence of any site-specific information (Schmidt *et al* 2014). The comparison between direct polarization, cross polarization (CP) and INEPT-based experiments suggested that at least part of the receptor is mobile in membranes. They further quantitatively measured the motionally averaged $^1$H, $^{13}$C dipolar coupling and calculated order parameters of CH, $CH_2$, and $CH_3$ groups. Surprisingly, the order parameters showed consistently low values ($S_{CH}$ = 0.55, $S_{CH2}$ = 0.33, and $S_{CH3}$ = 0.17) corresponding to amplitudes of approximately 50° for the backbone and 50–60° for the sidechains, indicating significant structural fluctuations in the protein, which may be important during conformational change upon receptor activation (Schmidt *et al* 2014).

Recent developments in ssNMR instrumentation that allow access to very fast magic angle spinning (MAS) regimes and higher field strengths have significantly improved the sensitivity and resolution of spectra of biomolecular systems, with additional benefits for measurements of dynamics. MAS rates exceeding 100 kHz allow $^1$H detection that has dramatically enhanced the sensitivity over traditional $^{13}$C detected experiments (Xue *et al* 2017). For example, experiments at high field strength (23.4 T) and 100 kHz MAS on 0.5 mg of a fully protonated heptahelical membrane protein proteorhodopsin in a 0.7 mm rotor resulted in $^1$H line widths of $< 0.2$ ppm and sensitivity similar to that of ~2 mg protein in a 1.3 mm rotor (60 kHz MAS) at the same field strength (Lalli *et al* 2017). Very fast MAS also improves approaches for sequence-specific assignments, measurement of order parameters, and long distance measurements (De Paëpe 2012, Han *et al* 2013, Hou *et al* 2013, 2011). $^{1}$H-detected MAS has been used to map the rigid and dynamic domains of the cardiac membrane protein phospholamban (PLN) reconstituted in lipid bilayers and enabled the detection of weakly populated (<5%) excited states of PLN (Gopinath *et al* 2017). Higher MAS also achieves more complete averaging of inter-atomic interactions which simplifies interpretation of relaxation parameters (Lewandowski 2013). Weingarth and co-workers obtained highly resolved $^{1}$H–$^{15}$N spectra of KcsA, with inverse fractional deuteration of the protein in lipid bilayers, that enabled site-specific measurements of $^{15}$N $R_{1\rho}$ (Medeiros-Silva *et al* 2016). The measurements highlighted ns to μs motional enhancements in a highly conserved protein segment involved in ion conduction. Such advancements should make it easier to study more complex membrane proteins and streamline the methodology for measurement of dynamics in these proteins.

## 13.3 NMR studies of dynamics of specific membrane protein systems

Unlike solution state NMR, there is no standard method or protocol that is extensively used for measuring motions across many systems. Instead ssNMR users choose amongst a variety of approaches, depending on the system under study that determines both the quality of the spectra and the questions of interest. Here we describe the measurement and interpretation of dynamics in three types of membrane proteins to illustrate the potential of different NMR methods to address a variety of questions and systems. The *Anabaena* sensory rhodopsin provides an example of an alpha helical bundle with near complete backbone assignments, and thus a variety of site-specific dynamic studies can be performed on the protein. The outer membrane protein beta barrels have detailed structural information available, which makes them good targets to measure similar site-specific dynamics using new methods with fast MAS and $^{1}$H detection. We conclude with the multi-protein complex formed by bacterial chemoreceptor proteins as an example of a challenging system with sparse NMR data available, to describe strategies for measuring functionally important dynamics that may be applicable to other similarly complex systems.

### 13.3.1 Anabaena sensory rhodopsin

Rhodopsins are a large family of photochemically active membrane proteins with retinal as the chromophore, present both in microbes and eukaryotes. The *Anabaena* sensory rhodopsin (ASR) is a unique microbial rhodopsin found in the freshwater cyanobacterium *Anabaena* sp. PCC7120 (Jung *et al* 2003). This bacterial rhodopsin interacts with a 14 kDa cytoplasmic soluble transducer (ASRT) (Wang *et al* 2011a) to regulate genes related to photosynthesis in *Anabaena*. Specifically, the photo-activation of ASR releases ASRT, which can bind to DNA and regulate gene expression (Jung 2007). The high-resolution crystal structure of ASR revealed an organization that is characteristic of rhodopsin, consisting of seven transmembrane helices with retinal covalently bound to Lys210 on the seventh helix (G) through a Schiff base (Vogeley *et al* 2004). Additionally, one of the extracellular loops

(B-C loop) was not observed due to disorder. ASR can switch between two stable conformations corresponding to the all-*trans* and 13-*cis* conformations of retinal, depending on the wavelength of illumination. The electron density indicated a mixture of retinal isomers after the white light illumination and two alternate conformations of the Schiff base-forming Lys210 and Ser86 that are connected to retinal via the hydrogen bonded network. The light-modulated interaction of ASR with the transducer highlights its ability to undergo dynamic changes to perform its biological function, making it an excellent system to study using NMR.

ASR is dimeric in crystals but forms stable symmetric trimers in lipids; thus studies of ASR in lipids can monitor structure and dynamics of a more native state. Early MAS ssNMR studies on uniformly $^{13}C$, $^{15}N$-labeled ASR reconstituted in DMPC/DMPA (9:1) lipids revealed that the ASR conformation in lipids differs from that in the crystal structure (Shi *et al* 2011, Wang *et al* 2011b). The key deviations from the crystal structure include the ordered beta hairpin in the B-C loop on the extracellular side, and an uninterrupted helix G. The retinal chromophore was found to be in an all-*trans* conformation in the dark, based on the chemical shifts of C13 and C20 carbons of retinal, and the $^{15}N$ Schiff base. The chemical shifts of ASR in liposomes are very similar to those of the protein bound to *E. coli* lipid membrane (Ward *et al* 2015), indicating that the structure does not change significantly with lipid composition. NMR detection of hydrogen–deuterium (H/D) exchange has been successfully utilized to understand the conformational changes upon activation of ASR, by probing the difference in exchange between samples in dark and light (Wang *et al* 2011b). The protein undergoes conformational changes in the solvent-accessible surfaces upon illumination and shows localized increases in the exchange in the transmembrane regions (red sites in figure 13.2(a)). Notably, half of the cytoplasmic side of the seventh helix (helix G), which has little exchange in the dark, becomes almost completely exposed upon illumination, and the cytoplasmic ends of the second and third helices (helix B and C) also show more exchange (figure 13.2(a)), which indicates their displacement when activated may allow water to penetrate to the core of ASR. The large light-induced conformational change in helix G seen by H/D-exchange with NMR is consistent with previous FTIR measurements (Kawanabe *et al* 2008), and is proposed to occur upon transition from the all-*trans* to 13-*cis* retinal form that would affect the binding affinity of ASR to its soluble transducer.

An extensive effort for resonance assignment of ASR using samples grown on glycerol labeled at positions 1, 3 or 2 allowed a near-complete backbone assignment (Wang *et al* 2013b), which has enabled measurement of site-specific dynamics in the protein. Ladizhansky and co-workers investigated site-specific dynamics of ASR by measuring dipolar order parameters and $^{15}N$ $R_{1\rho}$ relaxation rates (Good *et al* 2014). Four dipolar couplings ($^{15}N$–$^{1}H_N$, $^{13}C_\alpha$–$^{1}H_\alpha$, $^{13}C'$–$^{15}N$, and $^{13}C_\alpha$–$^{15}N$) per residue were measured to characterize the amplitude of submicrosecond timescale motions that average these interactions in ASR, using 3D dipolar shift correlation (DIPSHIFT) experiments. Each coupling was determined in a separate 3D experiment with a series of 2D NCA or NCO experiment recorded as a function of dipolar

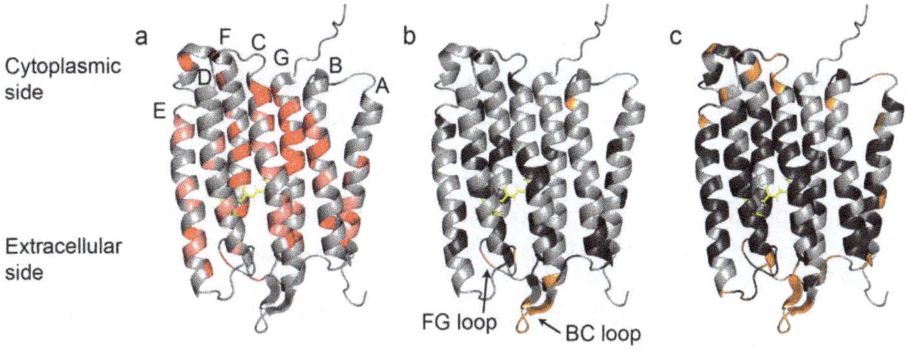


**Figure 13.2.** Dynamics in ASR. (a). Light-induced conformational change revealed by H/D exchange increases solvent exposure of the cytoplasmic side of helices G, C, and B (Wang *et al* 2011b). light red indicates residues with significantly greater deuterium uptake when exchanged under illumination vs dark conditions (at least 50% reduction in cross peak intensities), and dark red indicates residues that undergo complete exchange in the illuminated state. b. The BC and FG loops (highlighted in orange) show nanosecond timescale motions ($^{15}$N $R_{1\rho} > 10^{-1}$ s); dark grey represents all the residues for which $R_{1\rho}$ was measured (Good *et al* 2014). c. ASR residues at the exposed end of the helices and loops (orange) undergo nanosecond timescale motions (elevated $^{15}$N $R_1$ and field dependence of $R_1$) and the residues at the middle of the TM helices (dark grey) experience picosecond timescale motions. The average $R_1$ for helices was in the range 0.001–0.015 $s^{-1}$, and residues highlighted in orange have $R_1 > 0.019$ $s^{-1}$ (Good *et al* 2017). All dynamics are represented on the ASR monomer from the trimer structure (PDB: 2M3G).

evolution time, and different recoupling sequences (ZF-TEDOR to measure $^{13}$C′–$^{15}$N and $^{13}C_{\alpha}$–$^{15}$N, and TMREV to measure $^{15}$N–$^{1}H_N$ and $^{13}C_{\alpha}$–$^{1}H_{\alpha}$). Interestingly, they did not see a large difference in order parameters between the transmembrane and the solvent-exposed residues, indicating that the domains undergo motions of similar amplitudes. To probe the timescale of motions, $^{15}$N transverse relaxation rates ($R_1$ and $R_{1\rho}$) were measured. Though such experiments are performed routinely in solution NMR, they are difficult to execute on immobilized samples, as the relaxation in solids is dominated by coherent effects on the decay of transverse magnetization, due to incomplete averaging of dipolar couplings. This can be overcome by sample deuteration that dilutes the proton bath and reduces interproton couplings (Krushelnitsky *et al* 2010). Alternatively, fast magic angle spinning (>20 kHz for $^{15}$N, and >60 kHz for $^{13}$C) significantly reduces the proton-driven spin diffusion contributions to longitudinal relaxation rates (Lewandowski *et al* 2010). Fast MAS also makes it possible to accurately measure transverse relaxation rates: values of $^{15}$N $R_{1\rho}$ that match the predicted incoherent $R_2$ can be obtained on a protonated sample by using fast MAS rates of >45 kHz in combination with a moderate $B_1$ spinlock and without heteronuclear decoupling (Lewandowski *et al* 2011). $R_{1\rho}$ is enhanced by motions in the nanosecond to microsecond timescales. The measurements of $^{15}$N transverse spin–spin relaxation rates ($R_{1\rho}$) in ASR were performed on fully protonated samples at fast MAS (50 kHz) at 800 MHz. The measurements showed elevated relaxation rates for sites (orange in figure 13.2(b)) in the extracellular loops between the second and third helix and the sixth and seventh helix (B–C and F–G loops). These results indicate slow motions, estimated to be on the ~70 nanosecond timescale for the B–C loop and ~190 ns for the F–G loop (Good *et al* 2014). Interestingly, the slow motion of the F–G loop in ASR indicates that it may serve as a hinge point for the movement of helix G, which was proposed in the H/D exchange study described above. The hairpin in the B–C loop is a structurally conserved motif in many bacterial rhodopsins, and its mobility may be required for structural changes occurring in the retinal pocket during the photocycle. This should be especially important in ASR, as the protein is expected to undergo large conformational changes between the ASRT-bound and unbound states.

Recently, this study was further expanded to include the measurement of $^{15}N$ longitudinal relaxation $R_1$ at two temperatures and two different fields (600 and 800 MHz), with samples subjected to MAS rates of 19 kHz (Good *et al* 2017). The study could also resolve a higher number of cross peaks than the $R_{1\rho}$ study, thus providing insights into dynamics at additional residues. The $^{15}N$ $R_1$ relaxation rates report on the motions occurring on the picosecond to nanosecond timescale, and all of the data ($R_1$, $R_{1\rho}$, and order parameters) were analyzed with multiple models to deduce motions occurring throughout ASR. For non-proline residues, calculated curves based on a Simple Model Free approach predict a field dependence for nanosecond motions, with faster relaxation at lower field, but no field dependence for fast picosecond motions. The measured $R_1$ values (figure 13.2(c)) were lower for the center of the TM region (dark gray sites) than for the exposed regions (orange sites). The central portions of the helices showed no field dependence, which indicates the dynamics at the center of the helices are dominated by fast motions in the picosecond timescale. The exposed end of helices and loops exhibited higher $R_1$ rates, and their field dependence indicates that these residues undergo nanosecond timescale motions. Interestingly, elevated $R_{1\rho}$ rates were also observed for these regions (data only for BC and FG loops) in the previous study. For the analysis of temperature dependence, it was noted that correlation times will decrease at higher temperatures, with opposite effects on $R_1$ for regions with motions on different timescales. Shorter correlation time will decrease $R_1$ if picosecond motions dominate, but will increase $R_1$ if nanosecond motions dominate, for nanosecond correlation times beyond the $T_1$ minimum. For either nanosecond or picosecond timescale motions, shorter correlation times will decrease $R_{1\rho}$. Measurements at higher temperature showed a small decrease in $R_1$ for the middle portions of the TM helices, consistent with picosecond motions in this region. In contrast, $R_1$ showed a significant increase at higher temperature for cytoplasmic ends of helices A, B, C, D, and F and the loop regions, suggesting contributions from slower nanosecond timescale motions. As expected, at higher temperature there was also a decrease in $R_{1\rho}$ rates, especially in the BC and FG loops. In summary, analysis of both the $R_1$ and $R_{1\rho}$ rates indicate that faster picosecond motions are dominant in the middle portion of the transmembrane helices, and slower nanosecond timescale motions increase towards the cytoplasmic side of the helices as well as in the interhelical loops. Motions on the cytoplasmic side will affect the interaction of ASR with its transducer and thus may be functionally important.

### 13.3.2 Outer membrane proteins

Outer membrane proteins (OMPs) are involved in various biological activities, like cell shape maintenance, cell–cell recognition, protein translocation, and ion transport (Koebnik *et al* 2000). They are antiparallel β-barrel proteins situated in the outer membranes of bacteria and eukaryotic mitochondria and remain attractive targets for NMR studies due to the large chemical shift dispersion of the β-sheet residues. High resolution structures of many OMPs like OmpA (Arora *et al* 2001), PagP (Hwang *et al* 2002), and OmpG (Liang and Tamm 2007) in micelles have been determined using solution NMR. Recently, MAS ssNMR using a combination of $^{13}$C and $^{1}$H-detected experiments has been used to determine high resolution structures of some OMPs in lipid bilayers, such as OmpG in bilayers of *E. coli* lipid extract (Retel *et al* 2017), and *Yersinia pestis* Ail (Adhesion invasion locus) in phospholipid liposomes (Yao *et al* 2017).

*K. pneumoniae* is a human respiratory pathogen and KpOmpA binds to human bronchial epithelial cells through the Toll like receptor (TLR)2, resulting in the activation of the immune system (Pichavant *et al* 2003). The solution NMR structure of transmembrane β-barrel KpOmpA revealed 8 β-strands, 4 extracellular loops (L1–4) and 3 periplasmic turns (T1–3) (Renault *et al* 2009). KpOmpA in lipid bilayers was first investigated by traditional $^{13}$C-detected MAS NMR, in a study that illustrates one approach for obtaining insight into protein dynamics when the spectral resolution is insufficient for sequence assignments (Iordanov *et al* 2012). The authors identified two distinct clusters of backbone $C_\beta$–$C_\alpha$ correlations for Ser and Thr, one corresponding to the β-barrel regions and the other to the loops, based on assignments from solution NMR (Renault *et al* 2009) and consistent with the secondary structure dependence of these chemical shifts. Measurements of $T_1$ and CP buildup curves (yielding $T_{CP}$ and $T_{1\rho}$), along with comparison of 1D CP and $^{1}$H–$^{13}$C INEPT experiments, showed that the Ser/Thr residues in the β-barrel region (purple in figure 13.3(a)) were rigid (longer $T_1$ relaxation times, shorter $T_{CP}$, and absence of the Thr $C_\beta$ signal from the INEPT spectrum), while the residues in the loop region (red in figure 13.3(a)) were more dynamic (Iordanov *et al* 2012). These results agreed with the observation of dynamics in the solution NMR study (Renault *et al* 2009).

Recently developed $^{1}$H-detected experiments have made it possible to measure dynamics of specific residues in OmpA to gain insight into its mechanism. $^{1}$H-detected MAS measurements of protein dynamics have been developed primarily on microcrystalline samples (Gauto *et al* 2017, Knight *et al* 2012, Schanda *et al* 2010), and have been recently applied to non-crystalline preparations (Dannatt *et al* 2016, Smith *et al* 2016) including membrane proteins in near-native environments. Protein dynamics in *Klebsiella pneumoniae* OmpA (KpOmpA) have been measured using $^{1}$H-detected NMR at 23.4 T and 60 kHz MAS (Saurel *et al* 2017). Saurel and co-workers assigned 49 OmpA residues, using pulse sequences that contain multiple cross polarization steps and thus become filters to select the rigid regions (Saurel *et al* 2017). As expected, the assigned residues were from the β-barrel of the protein, and the intensities of the peaks became weaker at the edge of the β-barrel, which was the

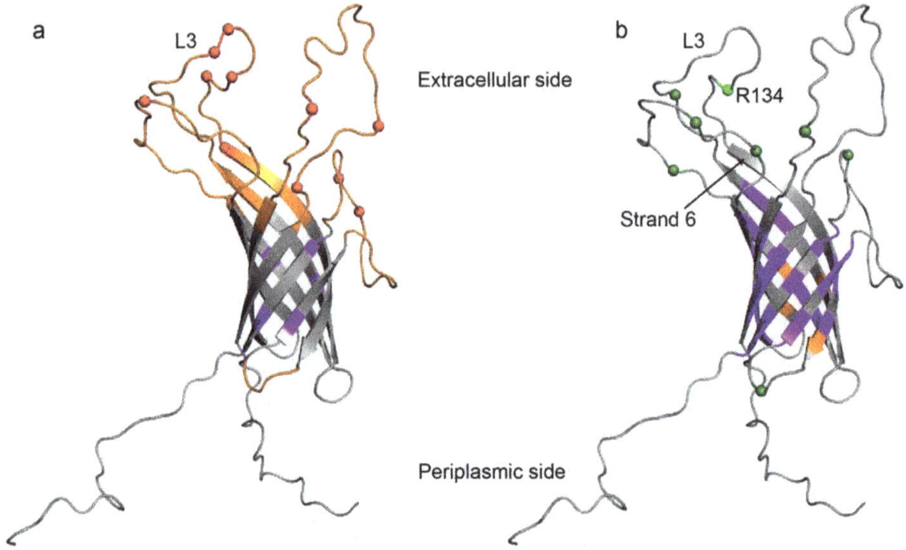


**Figure 13.3.** Dynamics of KpOmpA in lipid bilayers. (a). Ser/Thr residues distributed throughout the OmpA sequence were analyzed in two groups based on their $C_\alpha$ and $C_\beta$ chemical shifts (Iordanov *et al* 2012). Those in the barrel with a beta strand chemical shift (purple) are rigid based on their absence in INEPT spectra, and their long $T_1$ and short $T_{CP}$ values. Those in the loops with a random coil shift (red, $C_\alpha$ atoms shown as spheres) are flexible, with shorter $T_1$ and longer $T_{CP}$ values in CP spectra, and are also observed in INEPT spectra. The results are consistent with solution NMR measurements that showed residues throughout the four loops (orange) to be dynamic, based on $^1$H–$^{15}$N NOE and conformational exchange measurements. (b). Residue-specific mobility measured by $^1$H detected fast MAS methods. The residues in β-barrel are rigid within strands 1–4 and 7–8 (purple) based on the $^{15}$N $R_{1\rho}$ and GAF model analysis; a few residues (orange) exhibit higher $^{15}$N $R_{1\rho}$ indicative of slow motions. Although trypsin sites are found throughout the loops (dark green, $C_\alpha$ atoms shown as spheres), proteolysis occurs only at one site in Loop 3 (R134, bright green $C_\alpha$ atom shown as sphere), indicating this loop exhibits large structural fluctuations. Solution NMR structure of KpOmpA used for representing changes (PDB: 2K0L).

first marker of increased local flexibility in that region. No loop residues were detected by this approach. They measured residue-specific amide $^1$H–$^{15}$N dipolar couplings by REDOR to gain insight into the amplitudes of the dynamics. The observed high values for the order parameter (average value 0.88), indicate that the protein does not undergo fast axial diffusion in this concentrated sample preparation (high protein:lipid ratio). They used a Gaussian axial fluctuation (GAF) model to show that both the motional amplitude reflected by the measured order parameters and the timescale reflected by relatively constant values of $^{15}$N $R_1$ and $R_{1\rho}$ are consistent with a uniform rocking of the β-barrels with a small amplitude (10°) occurring on the hundreds of nanoseconds timescale. Further analysis of the $^{15}$N $R_{1\rho}$ revealed that the β-barrel is rigid within strands 1–4 and 7–8 (sites depicted in purple in figure 13.3(b)) with an average rate of 12 s$^{-1}$. The few residues that exhibit rates higher than 20 s$^{-1}$ (G67, A98, T112, and W151; orange in figure 13.3(b)) are mostly located close to the end of the strands, and the increased $^{15}$N $R_{1\rho}$ indicates they are subject to slow internal motions. The authors suggest that the absence of peaks from the extracellular side of strand 6 (first nine residues following loop 3, figure 13.3(b)) was due to an even larger increase in the $^{15}$N $R_{1\rho}$ and that this strand experiences transient unfolding. Since no residues were assigned in the loops, proteolysis experiments were employed as a complementary probe. Surprisingly, trypsin cleavage occurred at only one (R134, highlighted bright green in figure 13.3(b) of 8 trypsin sites (dark green in figure 13.3(b)) found throughout the loops, specifically in the C-terminal half of loop 3 (joining strands 5–6), which indicates fluctuations sufficient for trypsin access and is consistent with transient unfolding of the adjacent residues in strand 6. This disorder in loop 3, which was identified as a key binding epitope of KpOmpA, may facilitate binding of KpOmpA to different cell surface receptors that will potentially dictate the adherence of the pathogen to the surface of the human cells (Saurel *et al* 2017).

$^1$H-detected dynamics studies are not limited to site-specific measurements. Meier's group assessed the connection between lipid bilayer fluidity and the *bulk* dynamics of OmpX protein embedded in lipid bilayer nanodiscs with a $^1$H-detected study at 60 and 110 kHz MAS (Lakomek *et al* 2017). Samples in lipid nanodiscs were examined above and below the lipid phase transition temperature in both fully protonated and deuterated/back-exchanged form. Highest signal intensities were observed at temperatures below the lipid phase transition; however, better spectral resolution was obtained above the transition temperature. Estimation of the bulk $^1$H and $^{15}$N $R_2'$ coherence decay rates show that both increase with the increase in temperature. This would decrease the signal intensity, and also indicates that line widths increase below the lipid phase transition due to inhomogeneous line broadening rather than increased transverse relaxation rates. The study indicates that lipid dynamics influence the overall motion of the membrane protein under study and highlights the need for careful interpretation of protein dynamics measured *in vitro* when deducing their role in protein function.

### 13.3.3 Bacterial transmembrane chemoreceptors

Bacterial chemotaxis receptors are transmembrane proteins that bind to specific attractants in the environment and transmit this information across the membrane to direct the swimming behavior of the bacterium. The receptor trimers of dimers are coupled to cytoplasmic proteins CheA (kinase) and CheW (adaptor) to form large hexagonal signaling arrays that extend ~200 nm in the membrane and have an overall molecular weight >100 MDa. Although high-resolution structures of the entire array are not available, models have been created by docking crystallographic and NMR structures of protein fragments into electron density maps from electron cryotomography of receptors in whole cells (Briegel *et al* 2012, Cassidy *et al* 2020, 2015, Liu *et al* 2012). The transmembrane receptor binds to the ligand in the periplasm and relays the signal through the HAMP domain to the cytoplasmic domain (figure 13.4(a)). It is proposed that changes in dynamics in the cytoplasmic domain of the receptor are important for signaling by the complex (Swain *et al* 2009, Zhou *et al* 2009), but the mechanism of signal propagation by the receptor to control activity of the kinase (CheA) remains poorly understood. Solution NMR of binary

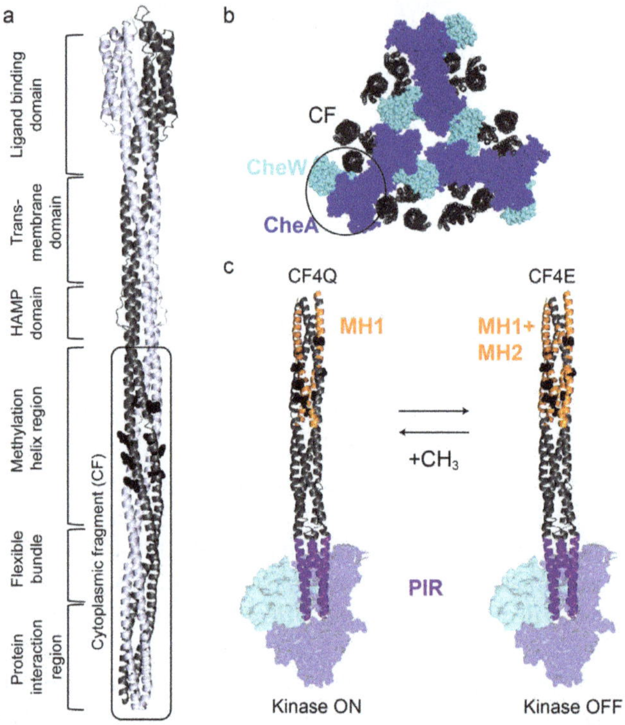


**Figure 13.4.** Dynamics of *E. coli* Asp receptor. (a). Model of full-length Asp receptor homodimer, highlighting the cytoplasmic fragment used to assemble functional complexes for MAS NMR studies (black rectangle). Monomers are shown in different colors (grey and pale blue) for clarity, and methylation sites are shown as black spheres. (b). Bottom-up view of the chemoreceptor array baseplate shows CF trimer of dimers (dark grey helices) arranged in a hexagon linked by a ring of alternating kinase CheA (blue) and coupling protein CheW (cyan). The NMR sample consists of multiple hexagons that form an array. (c). Simplified view of complex of single CF dimer with CheA and CheW (region circled in figure 13.4(b)), with colors highlighting CF dynamics measured by NMR and HDX-MS. INEPT-detectable regions with ns or faster timescale motions are colored orange: methylation helix 1 (MH1) is dynamic in the methylated receptor CF4Q; both methylation helix 1 and 2 are dynamic in the unmethylated receptor CF4E (Kashefi *et al* 2019, Malik *et al* 2020). REDOR demonstrates that a small fraction of the receptor (~75 residues) has rigid-limit carbon–nitrogen dipolar couplings (Kashefi *et al* 2019), and this is likely to include the protein interaction region (purple) based on HDX-MS results that indicate the rest of the CF is partially disordered (Li *et al* 2019).

complexes of protein fragments have suggested which residues are involved in protein–protein interactions, but the minimum functional unit of a pair of receptor cytoplasmic fragment trimers of dimers, bridged by a CheA dimer and two CheW monomers is too large to investigate with solution NMR methods. Our laboratory employs and develops methods for preparing functional complexes that form native-like arrays, as depicted in figure 13.4(b) (Briegel *et al* 2014, Fowler *et al* 2010, Haglin *et al* 2017, Shrout *et al* 2003), for biophysical measurements to compare structure and dynamics of the kinase-on and kinase-off signaling states. Although site-specific residue assignments are not yet available for this complex system, the following illustrates how measurements of dynamics using ssNMR and complementary methods has provided insights into the signaling mechanism.

Comparison of CP and INEPT spectra of functional complexes indicate that a significant fraction of the receptor cytoplasmic fragment (CF) is dynamic on the ns or shorter timescale (INEPT-detectable) and these dynamics change with signaling state (Kashefi and Thompson 2017, Malik *et al* 2020). Methylation, which mediates adaptation to an ongoing stimulus, occurs on two helices that form a 4-helix bundle within the dimeric receptor. HC and HN INEPT spectra of U–$^{13}$C,$^{15}$N–CF in functional complexes were assigned according to residue type, as individual residues of the same type were not resolved. In the absence of site-specific assignments, resonances of unique and rare amino acids in the CF and biochemical approaches were used to establish that methylation helix 1 is observable by INEPT, demonstrating asymmetric dynamics in the methylation region in the kinase-ON state (methylated receptor mimic CF4Q, figure 13.4(c)). This high flexibility/disorder of methylation helix 1, which is the segment connecting the cytoplasmic domain to the rest of the receptor, constrains possible mechanisms of signal propagation. Specifically, such a flexible segment cannot propagate a conformational change to control kinase activity of CheA bound at the cytoplasmic tip (Kashefi and Thompson 2017). Instead, based on hydrogen exchange mass spectrometry results we have proposed that most of the cytoplasmic domain is partially disordered and signaling involves stabilization of this domain that in turn modulates its interaction with CheA to control kinase activity (Li *et al* 2019). A flexible connecting segment is compatible with such a modulated disorder signaling mechanism. Furthermore, the receptor flexibility changes with signaling state: complexes with a kinase-off mutant receptor fragment show a ~1.4-fold increase in INEPT intensity (Kashefi and Thompson 2017) and complexes with the unmethylated receptor fragment CF4E show a ~2-fold increase in INEPT intensity (Malik *et al* 2020). Measurement of the $^{1}$H $T_2'$ of the methylated versus unmethylated states accounted for some (1.2-fold) increase in intensity, suggesting an increase in methylation helix 1 flexibility. Furthermore, analysis of the resonances of the 2D INEPT spectra of the unmethylated receptor complexes revealed additional regions with ns-timescale dynamics. This study identified methylation helix 2 segments that become dynamic in the unmethylated state of the receptor (CF4E, figure 13.4(c)), based on the peak analysis of the INEPT spectrum (Malik *et al* 2020), in agreement with EPR and HDX-MS studies (Bartelli and Hazelbauer 2016, Li *et al* 2019). The increased disorder of both of the methylation helices in the unmethylated state is likely to disfavor binding interactions with CheA and turn off the kinase. Furthermore, increased dynamics in the methylation region may also increase the accessibility of the methylation sites to the methyltransferase, accounting for the increased rate of methylation of the kinase-off state.

Both CP-buildup curves and REDOR measurements indicate that extensive regions of the receptor fragments in functional complexes exhibit intermediate timescale dynamics (Kashefi *et al* 2019, Malik *et al* 2020). Comparison of the CP-build up curves under frozen and unfrozen conditions indicate that a significant fraction (~one quarter to one third) of the cytoplasmic fragment is not observable in either dipolar or J-based NMR spectra (Kashefi *et al* 2019, Malik *et al* 2020). REDOR experiments were used to investigate the dynamics of the CP-observable regions of the receptor fragment in functional complexes, by measuring $^{13}C\{^{15}N\}$ REDOR dephasing on uniformly $^{13}C$, $^{15}N$-labeled CF to estimate the fraction of the receptor with dynamics on the ms-timescale that average the ~1 kHz C-N dipolar coupling (Kashefi *et al* 2019). A REDOR filter was also used to remove signal from rigid backbone carbons and retain signals from backbone carbons with ms-timescale dynamics that would not be observed in dipolar-based experiments like NCA/NCO. REDOR curves for both $C_\alpha$ and CO in frozen samples matched the prediction for the rigid-limit dipolar couplings, demonstrating that the spectra can be properly analyzed to correct for overlap with sidechain peaks. For unfrozen samples, the CF complexes in the kinase-ON state displayed 60%–65% of the expected REDOR dephasing, which suggests that 35%–40% of the CF-backbone experiences mobility that averages $^{13}C$–$^{15}N$ dipolar coupling to near zero (Kashefi *et al* 2019). This allowed us to estimate that about 75 residues with rigid-limit carbon–nitrogen dipolar couplings will be detected in NCA and NCO spectra, and assignments of these spectra are in progress to identify the rigid residues of the cytoplasmic fragment. Based on our previous HDX-MS study demonstrating that only the protein interaction region of the CF is well-ordered (Li *et al* 2019), we anticipate that NCA and NCO spectra will include this 56-residue region where CF interacts with CheA and CheW (figure 13.4(c)).

These ssNMR studies of the Asp receptor cytoplasmic fragment within functional complexes demonstrate that measurements of dynamics can provide valuable mechanistic insights even without site-specific assignments and can also guide the design of NMR experiments to further investigate the system and measure site-specific motions and structure. Indeed, it is important to begin an NMR study of a protein with an assessment of its dynamics, and this can increase the feasibility of NMR for systems that are challenging due to their large size or partial disorder. For complex systems it is also important to apply complementary methods to aid in the interpretation of results, such as HDX (Li *et al* 2019), EPR spectroscopy (Bartelli and Hazelbauer 2016, 2015) and molecular dynamics (Park *et al* 2011, Samanta *et al* 2015).

## 13.4 Conclusions

Membrane proteins are involved in a variety of biological processes that often involve dynamics, which may occur at very different timescales. MAS ssNMR has emerged as a powerful tool to study dynamics in membrane proteins. We have described different approaches applicable to a variety of systems to measure dynamics for a range of timescales and emphasized their ability to provide insights into the molecular mechanisms of the membrane protein. The different methods can be applied to proteins with existing structural information as well as to systems where little or no structural information is available. The availability of higher magnetic fields and recent advances in fast MAS and $^{1}H$-detected methods have facilitated obtaining residue-specific information on various proteins. These methods should make it possible to extend measurements of residue-specific dynamics such as those discussed for *Anabaena* sensory rhodopsin and outer membrane proteins to more complex and challenging membrane proteins. For these and other systems we also describe strategies to measure dynamics that do not require assignments, including measuring bulk properties for the entire protein or for regions that can be distinguished via their NMR properties, such as chemical shifts or relaxation. Such approaches, which are described for bacterial chemoreceptors, are especially valuable for the many biologically important systems with challenging NMR properties including extensive dynamics and/or disorder that make it difficult to obtain complete resonance assignments. For this latter case, we also highlight how a combination of MAS NMR with other biophysical tools can yield insights into functionally important dynamics and help to decipher the protein mechanism. Importantly, the new techniques and creative strategies make it possible to use MAS ssNMR to study more membrane proteins and their complexes in non-crystalline, native-like environments and measure dynamics of their functional states. Thus, NMR measurements will continue to contribute key understanding of the mechanisms underlying biological activity of membrane proteins.

## Acknowledgement

Research on bacterial chemotaxis receptors in our laboratory is supported by National Institutes of Health Grant R01-GM120195. The authors thank Katherine Wahlbeck, Thomas Trans and Jessica Allen for their critical reviews of this chapter and all of our colleagues who have contributed to our research.

## References

Andronesi O C, Becker S, Seidel K, Heise H, Young H S and Baldus M 2005 Determination of membrane protein structure and dynamics by magic-angle-spinning solid-state NMR spectroscopy *J. Am. Chem. Soc.* **127** 12965–74

Arora A, Abildgaard F, Bushweller J H and Tamm L K 2001 Structure of outer membrane protein A transmembrane domain by NMR spectroscopy *Nat. Struct. Biol.* **8** 334–8

Baker L A and Baldus M 2014 Characterization of membrane protein function by solid-state NMR spectroscopy *Curr. Opin. Struct. Biol.* **27** 48–55

Bartelli N L and Hazelbauer G L 2015 Differential backbone dynamics of companion helices in the extended helical coiled-coil domain of a bacterial chemoreceptor *Protein Sci.* **24** 1764–76

Bartelli N L and Hazelbauer G L 2016 Bacterial chemoreceptor dynamics: helical stability in the cytoplasmic domain varies with functional segment and adaptational modification *J. Mol. Biol.* **428** 3789–804

Bhate M P, Wylie B J, Tian L and McDermott A E 2010 Conformational dynamics in the selectivity filter of KcsA in response to potassium ion concentration *J. Mol. Biol.* **401** 155–66

Bowie J U 2005 Solving the membrane protein folding problem *Nature* **438** 581–9

Briegel A, Li X, Bilwes A M, Hughes K T, Jensen G J and Crane B R 2012 Bacterial chemoreceptor arrays are hexagonally packed trimers of receptor dimers networked by rings of kinase and coupling proteins *Proc. Natl Acad. Sci.* **109** 3766 LP–71

Briegel A *et al* 2014 New insights into bacterial chemoreceptor array structure and assembly from electron cryotomography *Biochemistry* **53** 1575–85 5000614

Cassidy C K, Himes B A, Alvarez F J, Ma J, Zhao G, Perilla J R, Schulten K and Zhang P 2015 CryoEM and computer simulations reveal a novel kinase conformational switch in bacterial chemotaxis signaling *Elife* **4** e08419

Cassidy C K, Himes B A, Sun D, Ma J, Zhao G, Parkinson J S, Stansfeld P J, Luthey-Schulten Z and Zhang P 2020 Structure and dynamics of the *E. coli* chemotaxis core signaling complex by cryo-electron tomography and molecular simulations *Commun. Biol.* **3** 24

Chen K-Y M, Zhou F, Fryszczyn B G and Barth P 2012 Naturally evolved G protein-coupled receptors adopt metastable conformations *Proc. Natl Acad. Sci.* **109** 13284–9

Cole H B R and Torchia D A 1991 An NMR study of the backbone dynamics of staphylococcal nuclease in the crystalline state *Chem. Phys.* **158** 271–81

Dannatt H R W, Felletti M, Jehle S, Wang Y, Emsley L, Dixon N E, Lesage A and Pintacuda G 2016 Weak and transient protein interactions determined by solid-state NMR *Angew. Chemie Int. Ed.* **55** 6638–41

De Paëpe G 2012 Dipolar recoupling in magic angle spinning solid-state nuclear magnetic resonance *Annu. Rev. Phys. Chem.* **63** 661–84

deAzevedo E R, Saalwachter K, Pascui O, de Souza A A, Bonagamba T J and Reichert D 2008 Intermediate motions as studied by solid-state separated local field NMR experiments *J. Chem. Phys.* **128** 104505

Etzkorn M, Martell S, Andronesi O C, Seidel K, Engelhard M and Baldus M 2007 Secondary structure, dynamics, and topology of a seven-helix receptor in native membranes, studied by solid-state NMR spectroscopy *Angew. Chemie Int. Ed.* **46** 459–62

Fowler D J, Weis R M and Thompson L K 2010 Kinase-active signaling complexes of bacterial chemoreceptors do not contain proposed receptor–receptor contacts observed in crystal structures *Biochemistry* **49** 1425–34

Gammeren A J, van, Hulsbergen F B, Hollander J G and de Groot H J M 2005 Residual backbone and side-chain $^{13}$C and $^{15}$N resonance assignments of the intrinsic transmembrane light-harvesting 2 protein complex by solid-state magic angle spinning NMR spectroscopy *J. Biomol. NMR* **31** 279–93

Gauto D F, Hessel A, Rovó P, Kurauskas V, Linser R and Schanda P 2017 Protein conformational dynamics studied by $^{15}$N and $^{1}$H R1ρ relaxation dispersion: application to wild-type and G53A ubiquitin crystals *Solid State Nucl. Magn. Reson.* **87** 86–95

Giraud N, Blackledge M, Goldman M, Böckmann A, Lesage A, Penin F and Emsley L 2005 Quantitative analysis of backbone dynamics in a crystalline protein from nitrogen-15 spin–lattice relaxation *J. Am. Chem. Soc.* **127** 18190–201

Giraud N, Böckmann A, Lesage A, Penin F, Blackledge M and Emsley L 2004 Site-specific backbone dynamics from a crystalline protein by solid-state NMR spectroscopy *J. Am. Chem. Soc.* **126** 11422–3

Good D, Pham C, Jagas J, Lewandowski J R and Ladizhansky V 2017 Solid-State NMR provides evidence for small-amplitude slow domain motions in a multispanning transmembrane α-helical protein *J. Am. Chem. Soc.* **139** 9246–58

Good D B, Wang S, Ward M E, Struppe J, Brown L S, Lewandowski J R and Ladizhansky V 2014 Conformational dynamics of a seven transmembrane helical protein anabaena sensory rhodopsin probed by solid-state NMR *J. Am. Chem. Soc.* **136** 2833–42

Gopinath T, Nelson S E D and Veglia G 2017 $^{1}$H-detected MAS solid-state NMR experiments enable the simultaneous mapping of rigid and dynamic domains of membrane proteins *J. Magn. Reson.* **285** 101–7

Haglin E R, Yang W, Briegel A and Thompson L K 2017 His-tag-mediated dimerization of chemoreceptors leads to assembly of functional nanoarrays *Biochemistry* **56** 5874–85

Han Y *et al* 2013 Magic angle spinning NMR reveals sequence-dependent structural plasticity, dynamics, and the spacer peptide 1 conformation in HIV-1 capsid protein assemblies *J. Am. Chem. Soc.* **135** 17793–803

Hansen S K, Bertelsen K, Paaske B, Nielsen N C and Vosegaard T 2015 Solid-state NMR methods for oriented membrane proteins *Prog. Nucl. Magn. Reson. Spectrosc.* **88–89** 48–85

Helmus J J, Surewicz K, Surewicz W K and Jaroniec C P 2010 Conformational flexibility of Y145Stop human prion protein amyloid fibrils probed by solid-state nuclear magnetic resonance spectroscopy *J. Am. Chem. Soc.* **132** 2393–403

Henzler-Wildman K and Kern D 2007 Dynamic personalities of proteins *Nature* **450** 964–72

Hong M, Zhang Y and Hu F 2012 Membrane protein structure and dynamics from NMR spectroscopy *Annu. Rev. Phys. Chem.* **63** 1–24

Hou G, Paramasivam S, Yan S, Polenova T and Vega A J 2013 Multidimensional magic angle spinning NMR spectroscopy for site-resolved measurement of proton chemical shift anisotropy in biological solids *J. Am. Chem. Soc.* **135** 1358–68

Hou G, Yan S, Sun S, Han Y, Byeon I-J L, Ahn J, Concel J, Samoson A, Gronenborn A M and Polenova T 2011 Spin diffusion driven by R-symmetry sequences: applications to homonuclear correlation spectroscopy in MAS NMR of biological and organic solids *J. Am. Chem. Soc.* **133** 3943–53

Hu F, Luo W and Hong M 2010 Mechanisms of proton conduction and gating in influenza M2 proton channels from solid-state NMR *Science* **330** 505 LP–8

Huysmans G H M, Baldwin S A, Brockwell D J and Radford S E 2010 The transition state for folding of an outer membrane protein *Proc. Natl Acad. Sci.* **107** 4099–104

Hwang P M, Choy W-Y, Lo E I, Chen L, Forman-Kay J D, Raetz C R H, Privé G G, Bishop R E and Kay L E 2002 Solution structure and dynamics of the outer membrane enzyme PagP by NMR *Proc. Natl Acad. Sci. USA* **99** 13560–5

Iordanov I, Renault M, Réat V, Bosshart P D, Engel A, Saurel O and Milon A 2012 Dynamics of Klebsiella pneumoniae OmpA transmembrane domain: the four extracellular loops display restricted motion behavior in micelles and in lipid bilayers *Biochim. Biophys. Acta – Biomembr.* **1818** 2344–53

Jung K-H 2007 The distinct signaling mechanisms of microbial sensory rhodopsins in Archaea, Eubacteria and Eukarya *Photochem. Photobiol.* **83** 63–9

Jung K-H, Trivedi V D and Spudich J L 2003 Demonstration of a sensory rhodopsin in eubacteria *Mol. Microbiol.* **47** 1513–22

Kang C and Li Q 2011 Solution NMR study of integral membrane proteins *Curr. Opin. Chem. Biol.* **15** 560–9

Kaplan M *et al* 2016 EGFR dynamics change during activation in native membranes as revealed by NMR *Cell* **167** 1241–51.e11

Kashefi M, Malik N, Struppe J O and Thompson L K 2019 Carbon–nitrogen REDOR to identify MS-timescale mobility in proteins *J. Magn. Reson.* **305** 5–15

Kashefi M and Thompson L K 2017 Signaling-related mobility changes in bacterial chemotaxis receptors revealed by solid-state NMR *J. Phys. Chem.* B **121** 8693–705

Kawanabe A, Furutani Y, Yoon S R, Jung K-H and Kandori H 2008 FTIR study of the L intermediate of anabaena sensory rhodopsin: structural changes in the cytoplasmic region *Biochemistry* **47** 10033–40

Klöpfer K and Hagn F 2019 Beyond detergent micelles: the advantages and applications of non-micellar and lipid-based membrane mimetics for solution-state NMR *Prog. Nucl. Magn. Reson. Spectrosc.* **114–115** 271–83

Knight M J, Pell A J, Bertini I, Felli I C, Gonnelli L, Pierattelli R, Herrmann T, Emsley L and Pintacuda G 2012 Structure and backbone dynamics of a microcrystalline metalloprotein by solid-state NMR *Proc. Natl Acad. Sci.* **109** 11095–100

Kobayashi M, Matsuki Y, Yumen I, Fujiwara T and Akutsu H 2006 Signal assignment and secondary structure analysis of a uniformly [$^{13}$C, $^{15}$N]-labeled membrane protein, $H^+$-ATP synthase subunit c, by magic-angle spinning solid-state NMR *J. Biomol. NMR* **36** 279–93

Koebnik R, Locher K P and Van Gelder P 2000 Structure and function of bacterial outer membrane proteins: barrels in a nutshell *Mol. Microbiol.* **37** 239–53

Kovermann M, Rogne P and Wolf-Watz M 2016 Protein dynamics and function from solution state NMR spectroscopy *Q. Rev. Biophys.* **49** e6

Krushelnitsky A, Zinkevich T, Reichert D, Chevelkov V and Reif B 2010 Microsecond time scale mobility in a solid protein as studied by the $^{15}$N $R_{1\rho}$ site-specific NMR relaxation rates *J. Am. Chem. Soc.* **132** 11850–3

Lakomek N-A, Frey L, Bibow S, Böckmann A, Riek R and Meier B H 2017 Proton-detected NMR spectroscopy of nanodisc-embedded membrane proteins: MAS solid-state vs solution-state methods *J. Phys. Chem.* B **121** 7671–80

Lalli D, Idso M N, Andreas L B, Hussain S, Baxter N, Han S, Chmelka B F and Pintacuda G 2017 Proton-based structural analysis of a heptahelical transmembrane protein in lipid bilayers *J. Am. Chem. Soc.* **139** 13006–12

Lamley J M, Öster C, Stevens R A and Lewandowski J R 2015 Intermolecular interactions and protein dynamics by solid-state NMR spectroscopy *Angew. Chemie Int. Ed.* **54** 15374–8

Lange A, Giller K, Hornig S, Martin-Eauclaire M F, Pongs O, Becker S and Baldus M 2006 Toxin-induced conformational changes in a potassium channel revealed by solid-state NMR *Nature* **440** 959–62

Lewandowski J R 2013 Advances in solid-state relaxation methodology for probing site-specific protein dynamics *Acc. Chem. Res.* **46** 2018–27

Lewandowski J R, Sass H J, Grzesiek S, Blackledge M and Emsley L 2011 Site-specific measurement of slow motions in proteins *J. Am. Chem. Soc.* **133** 16762–5

Lewandowski J R, Sein J, Sass H J, Grzesiek S, Blackledge M and Emsley L 2010 Measurement of site-specific $^{13}$C spin–lattice relaxation in a crystalline protein *J. Am. Chem. Soc.* **132** 8252–4

Li X, Eyles S J and Thompson L K 2019 Hydrogen exchange of chemoreceptors in functional complexes suggests protein stabilization mediates long-range allosteric coupling *J. Biol. Chem.* **294** 16062–79

Li Y, Berthold D A, Frericks H L, Gennis R B and Rienstra C M 2007 Partial $^{13}$C and $^{15}$N chemical-shift assignments of the disulfide-bond-forming enzyme DsbB by 3D magic-angle spinning NMR spectroscopy *Chem. Bio. Chem.* **8** 434–42

Liang B and Tamm L K 2007 Structure of outer membrane protein G by solution NMR spectroscopy *Proc. Natl Acad. Sci.* **104** 16140–5

Lisi G P and Loria J P 2016 Solution NMR spectroscopy for the study of enzyme allostery *Chem. Rev.* **116** 6323–69

Liu J, Hu B, Morado D R, Jani S, Manson M D and Margolin W 2012 Molecular architecture of chemoreceptor arrays revealed by cryoelectron tomography of *Escherichia coli* minicells *Proc. Natl Acad. Sci.* **109** E1481 LP–8

Lorieau J L, Day L A and McDermott A E 2008 Conformational dynamics of an intact virus: order parameters for the coat protein of Pf1 bacteriophage *Proc. Natl Acad. Sci.* **105** 10366–71

Malik N, Wahlbeck K A and Thompson L K 2020 Strategies for identifying dynamic regions in protein complexes: flexibility changes accompany methylation in chemotaxis receptor signaling states *Biochim. Biophys. Acta Biomembr.* **1862** 183312

Mandala V S, Williams J K and Hong M 2018 Structure and dynamics of membrane proteins from solid-state NMR *Annu. Rev. Biophys.* **47** 201–22

Medeiros-Silva J, Mance D, Daniëls M, Jekhmane S, Houben K, Baldus M and Weingarth M 2016 $^1$H-detected solid-state NMR studies of water-inaccessible proteins *in vitro* and *in situ* *Angew. Chemie Int. Ed.* **55** 13606–10

Morrison E A, DeKoster G T, Dutta S, Vafabakhsh R, Clarkson M W, Bahl A, Kern D, Ha T and Henzler-Wildman K A 2011 Antiparallel EmrE exports drugs by exchanging between asymmetric structures *Nature* **481** 45–50

Palmer A G 2004 NMR characterization of the dynamics of biomacromolecules *Chem. Rev.* **104** 3623–40

Palmer A G and Massi F 2006 Characterization of the dynamics of biomacromolecules using rotating-frame spin relaxation NMR spectroscopy *Chem. Rev.* **106** 1700–19

Park H, Im W and Seok C 2011 Transmembrane signaling of chemotaxis receptor tar: insights from molecular dynamics simulation studies *Biophys. J.* **100** 2955–63

Pichavant M, Delneste Y, Jeannin P, Fourneau C, Brichet A, Tonnel A-B and Gosset P 2003 Outer membrane protein a from klebsiella pneumoniae activates bronchial epithelial cells: implication in neutrophil recruitment *J. Immunol.* **171** 6697–705

Rasmussen S G F *et al* 2007 Crystal structure of the human $\beta_2$ adrenergic G-protein-coupled receptor *Nature* **450** 383–7

Renault M, Saurel O, Czaplicki J, Demange P, Gervais V, Löhr F, Réat V, Piotto M and Milon A 2009 Solution state NMR structure and dynamics of KpOmpA, a 210 residue transmembrane domain possessing a high potential for immunological applications *J. Mol. Biol.* **385** 117–30

Retel J S *et al* 2017 Structure of outer membrane protein G in lipid bilayers *Nat. Commun.* **8** 2073

Samanta D, Borbat P P, Dzikovski B, Freed J H and Crane B R 2015 Bacterial chemoreceptor dynamics correlate with activity state and are coupled over long distances *Proc. Natl Acad. Sci.* **112** 2455 LP–60

Sarkar C A, Dodevski I, Kenig M, Dudli S, Mohr A, Hermans E and Plückthun A 2008 Directed evolution of a G protein-coupled receptor for expression, stability, and binding selectivity *Proc. Natl Acad. Sci.* **105** 14808–13

Saurel O, Iordanov I, Nars G, Demange P, Le Marchand T, Andreas L B, Pintacuda G and Milon A 2017 Local and global dynamics in Klebsiella pneumoniae outer membrane protein a in lipid bilayers probed at atomic resolution *J. Am. Chem. Soc.* **139** 1590–7

Schanda P and Ernst M 2016 Studying dynamics by magic-angle spinning solid-state NMR spectroscopy: principles and applications to biomolecules *Prog. Nucl. Magn. Reson. Spectrosc.* **96** 1–46

Schanda P, Meier B H and Ernst M 2010 Quantitative analysis of protein backbone dynamics in microcrystalline ubiquitin by solid-state NMR spectroscopy *J. Am. Chem. Soc.* **132** 15957–67

Schmidt P, Thomas L, Müller P, Scheidt H A and Huster D 2014 The G-protein-coupled neuropeptide Y receptor type 2 is highly dynamic in lipid membranes as revealed by solid-state NMR spectroscopy *Chem.—A Eur. J.* **20** 4986–92

Schneider R, Seidel K, Etzkorn M, Lange A, Becker S and Baldus M 2010 Probing molecular motion by double-quantum ($^{13}$C,$^{13}$C) solid-state NMR spectroscopy: application to ubiquitin *J. Am. Chem. Soc.* **132** 223–33

Sharma M, Yi M, Dong H, Qin H, Peterson E, Busath D D, Zhou H-X and Cross T A 2010 Insight into the mechanism of the influenza A proton channel from a structure in a lipid bilayer *Science* **330** 509–12 1191750

Shi L, Kawamura I, Jung K-H, Brown L S and Ladizhansky V 2011 Conformation of a seven-helical transmembrane photosensor in the lipid environment *Angew. Chemie Int. Ed.* **50** 1302–5

Shrout A L, Montefusco D J and Weis R M 2003 Template-directed assembly of receptor signaling complexes *Biochemistry* **42** 13379–85

Smith A A, Testori E, Cadalbert R, Meier B H and Ernst M 2016 Characterization of fibril dynamics on three timescales by solid-state NMR *J. Biomol. NMR* **65** 171–91

Stangl M, Veerappan A, Kroeger A, Vogel P and Schneider D 2012 Detergent properties influence the stability of the glycophorin A transmembrane helix dimer in lysophosphatidylcholine micelles *Biophys. J.* **103** 2455–64

Swain K E, Gonzalez M A and Falke J J 2009 Engineered socket study of signaling through a four-helix bundle: evidence for a Yin–Yang mechanism in the kinase control module of the aspartate receptor *Biochemistry* **48** 9266–77

Tang M, Sperling L J, Berthold D A, Schwieters C D, Nesbitt A E, Nieuwkoop A J, Gennis R B and Rienstra C M 2011 High-resolution membrane protein structure by joint calculations with solid-state NMR and x-ray experimental data *J. Biomol. NMR* **51** 227–33

Verardi R, Shi L, Traaseth N J, Walsh N and Veglia G 2011 Structural topology of phospholamban pentamer in lipid bilayers by a hybrid solution and solid-state NMR method *Proc. Natl Acad. Sci. USA* **108** 9101–6

Vogeley L, Sineshchekov O A, Trivedi V D, Sasaki J, Spudich J L and Luecke H 2004 Anabaena sensory rhodopsin: a photochromic color sensor at 2.0 *Science* **306** 1390–3

Wang S, Kim S Y, Jung K-H, Ladizhansky V and Brown L S 2011a A eukaryotic-like interaction of soluble cyanobacterial sensory rhodopsin transducer with DNA *J. Mol. Biol.* **411** 449–62

Wang S, Munro R A, Shi L, Kawamura I, Okitsu T, Wada A, Kim S-Y, Jung K-H, Brown L S and Ladizhansky V 2013a Solid-state NMR spectroscopy structure determination of a lipid-embedded heptahelical membrane protein *Nat. Methods* **10** 1007–12

Wang S, Shi L, Kawamura I, Brown L S and Ladizhansky V 2011b Site-specific solid-state NMR detection of hydrogen-deuterium exchange reveals conformational changes in a 7-helical transmembrane protein *Biophys. J.* **101** L23–5

Wang S, Shi L, Okitsu T, Wada A, Brown L S and Ladizhansky V 2013b Solid-state NMR $^{13}$C and $^{15}$N resonance assignments of a seven-transmembrane helical protein anabaena sensory rhodopsin *Biomol. NMR Assign.* **7** 253–6

Ward M E, Wang S, Munro R, Ritz E, Hung I, Gor'kov P L, Jiang Y, Liang H, Brown L S and Ladizhansky V 2015 In situ structural studies of Anabaena sensory rhodopsin in the *E. coli* membrane *Biophys. J.* **108** 1683–96

Xue K, Sarkar R, Motz C, Asami S, Camargo D C R, Decker V, Wegner S, Tosner Z and Reif B 2017 Limits of resolution and sensitivity of proton detected MAS solid-state NMR experiments at 111 kHz in deuterated and protonated proteins *Sci. Rep.* **7** 7444

Yang Z *et al* 2014 Membrane protein stability can be compromised by detergent interactions with the extramembranous soluble domains *Protein Sci.* **23** 769–89

Yao Y, Dutta S K, Park S H, Rai R, Fujimoto L M, Bobkov A A, Opella S J and Marassi F M 2017 High resolution solid-state NMR spectroscopy of the *Yersinia pestis* outer membrane protein Ail in lipid membranes *J. Biomol. NMR* **67** 179–90

Zhou H-X and Cross T A 2013 Influences of membrane mimetic environments on membrane protein structures *Annu. Rev. Biophys.* **42** 361–92

Zhou Q, Ames P and Parkinson J S 2009 Mutational analyses of HAMP helices suggest a dynamic bundle model of input-output signalling in chemoreceptors *Mol. Microbiol.* **73** 801–14

**IOP** Publishing

Solid-State NMR
Applications in biomembrane structure
**Frances Separovic and Marc-Antoine Sani**

# Chapter 14

# Solid-state NMR methods for investigations of membrane protein structure and dynamics

**Peng Xiao and Vladimir Ladizhansky**

## Abstract

Membrane proteins play critical roles in many cellular processes. They are found in cell membranes, and their structure, dynamics and function depend on the lipid environment. Solid-state NMR is uniquely suited for the studies of membrane proteins in lipid bilayers. In this chapter, we review recent developments in solid-state NMR approaches for structure determination and for characterization of dynamics of membrane proteins, including multidimensional techniques for spectroscopic assignments and dipolar recoupling techniques for structural measurements, fast magic angle spinning NMR for improved sensitivity via proton detection and novel approaches to probing internal protein dynamics.

## 14.1 Introduction

Membrane proteins carry out critical cellular functions such as signal transduction, transport of molecules across cell membranes, ATP synthesis, cellular respiration and photosynthesis. They account for roughly one third of the human genome. Misfolding and misfunction of membrane proteins are directly implicated in various human diseases, making them important drug targets.

Membrane proteins have evolved to function within cell membranes. A full understanding of the mechanism of function requires detailed structural and dynamic information on the protein and its lipid environment. Among high-resolution structure determination methods, solid-state NMR (ssNMR) offers the unique prospect to characterize membrane proteins in their native-like lipid environment. Compared to other high-resolution techniques of x-ray crystallography, solution NMR and electron microscopy, the applicability of ssNMR is not restricted by molecular weight (from either end) and is not limited by the requirements of solubility or crystallinity. Applications of ssNMR to membrane proteins

doi:10.1088/978-0-7503-2532-5ch14  

are compatible with a wide range of lipid compositions and biochemical and reconstitution conditions, thereby offering both flexibility in closely mimicking cellular membrane environments and in choosing optimal sample preparation conditions. Moreover, recent developments in Dynamic Nuclear Polarization (DNP) NMR (Hall *et al* 1997, Lilly Thankamony *et al* 2017) have opened unique opportunities for analyzing membrane protein structures in intact cell membranes (Narasimhan *et al* 2020).

With the great potential of ssNMR there come many challenges. The main source of these challenges arises from the presence of anisotropic interactions which result from the restricted mobility of proteins within lipid bilayers. These anisotropic interactions lead to strong line broadening and low sensitivity and need to be removed for the analysis of molecular structure. Uniform alignment of protein molecules in the magnetic field of an NMR spectrometer is one of the two approaches used to achieve high resolution. It results in narrow NMR spectral lines and allows extracting from the spectra both the orientation of helices and backbone structure (Murray *et al* 2013, Opella 2015, Opella and Marassi 2017).

Magic Angle Spinning (MAS) (Andrew *et al* 1958, Lowe 1959) ssNMR is another approach which is used to reduce the anisotropic line broadening by spinning the sample about an axis pointing at the 'magic angle' of 54.7° with respect to the external magnetic field. Over the past two decades MAS ssNMR has emerged as a universally applicable method for studies of biomolecules. In this chapter, we review MAS ssNMR techniques for the characterization of structure and dynamics of membrane proteins and highlight advances enabled by recent developments in high magnetic field instrumentation and fast MAS.

## 14.2 Structure determination

### 14.2.1 Sample preparation

Preparation of functional and structurally homogeneous protein samples is the first step towards obtaining high quality spectra for protein NMR spectroscopy. Typically, ssNMR experiments require milligram quantities of a protein isotopically enriched with $^{13}$C and $^{15}$N. For proton-detected NMR and for the studies of protein dynamics, deuteration is often required.

Most ssNMR protein samples are produced in *Escherichia coli* (*E. coli*) bacterial expression systems which are chosen for their low cost, rapid growth rates, high expression yields, as well as for flexibility in the choice of isotopic labeling schemes and the availability of deuteration protocols. Uniform $^{13}$C and $^{15}$N enrichment ([$^{15}$N, $^{13}$C]) is the most straightforward labeling strategy, in which the protein is expressed on a minimal medium containing uniformly $^{13}$C-labeled glucose as the sole carbon source and $^{15}$N-labeled ammonium sulfate or ammonium chloride as the sole nitrogen source. Multidimensional ssNMR experiments conducted on uniformly [$^{15}$N, $^{13}$C]-labeled protein samples allow monitoring of protein structure, interactions and dynamics site-specifically. For large membrane proteins, a certain degree of simplification is generally required for improving spectral resolution, validating resonance assignments and streamlining structural measurements. Spectral resolution and crowding can be improved by spin dilution through a number of simplifying labeling schemes, such as sparse $^{13}$C labeling and selective amino acid labeling or unlabeling. A common alternate 'checkerboard' carbon labeling scheme is achieved by replacing uniformly $^{13}$C-labeled glucose with glycerols labeled at positions 1 and 3 ([1,3–$^{13}$C]), or at position 2 ([2–$^{13}$C]) as the sole carbon source (Castellani *et al* 2002, Hong and Jakes 1999, Lemaster and Kushlan 2001). Alternate labeling patterns effectively suppress most of the one-bond $^{13}$C–$^{13}$C dipolar and J-couplings, thereby reducing the linewidth and spectral overlap. This dilution of $^{13}$C spins also reduces dipolar truncation effects and facilitates long-range distance measurements which are critical for structure determination, as discussed in section 14.2.4. Additional carbon dilution approaches which use singly labeled [1–$^{13}$C]-glucose or [2–$^{13}$C]-glucose as carbon sources, introduce $^{13}$C only at methyl sites or backbone $C_\alpha$ sites, respectively, allowing for a higher degree of spectral simplification and linewidth reduction (Loquet *et al* 2010, 2011, Lundström *et al* 2007). However, this labeling scheme has the drawback that only half the sites can be labeled, which leads to reduced sensitivity.

Selective incorporation of isotopically labeled amino acids (forward labeling) in the otherwise unlabeled protein can also facilitate spectral simplification by reducing the number of NMR-visible residue types (Eddy *et al* 2015, Lange *et al* 2010, Sinnige *et al* 2015). Similarly, amino acid-specific unlabeling (reverse labeling) is also useful for lifting ambiguity in resonance assignments (Etzkorn *et al* 2007, Shi *et al* 2009a, 2009b, Yang *et al* 2011). Forward labeling is achieved by supplementing excess quantities of isotopically labeled amino acids of certain types to an unlabeled medium during expression; conversely, reverse labeling can be achieved using uniformly [$^{15}$N, $^{13}$C]-labeled expression medium supplemented with unlabeled amino acids. It should be noted that the metabolic pathways of some amino acids are often crossed, leading to inter-conversion between amino acids known as scrambling (Lacabanne *et al* 2018, Waugh 1996). In the case of metabolic scrambling, special auxotrophic strains have been developed to reduce or eliminate these effects (Iwasaki *et al* 2012, Lin *et al* 2015, 2011).

Yeast-based expression systems are valuable alternative expression platforms for the production of eukaryotic membrane proteins that do not express in *E. coli*, e.g., G-protein-coupled receptors (GPCRs) (Byrne 2015, Hartmann *et al* 2016). In particular, protocols for the methylotrophic yeast host *Pichia pastoris*, using $^{13}$C-labeled methanol as a primary carbon source, have been extensively investigated (Clark *et al* 2018, Fan *et al* 2015, Morgan *et al* 2000, Pickford and O'Leary 2004, Rodriguez and Krishna 2001, Wood and Komives 1999). Successful production of functional uniformly [$^{15}$N, $^{13}$C] labeled membrane protein samples with high levels of incorporation of isotopic labels has been demonstrated for a seven-helical photoreceptor *Leptosphaeria* rhodopsin (Fan *et al* 2011) and for human Aquaporins 1 and 2, for which ssNMR spectra of excellent resolution have been obtained (Emami *et al* 2013, Fan *et al* 2011, Munro *et al* 2020). Sparse $^{13}$C labeling approach has also been extended to *P. pastoris*. It is achieved by supplementing the natural abundance methanol with specifically $^{13}$C-labeled glycerols in the growth medium (Liu *et al* 2016). Such labeling results in improved carbon linewidths, and also facilitates observation of long-range correlations in the carbon–carbon spectra, but at the expense of the strong scrambling effects and an overall reduced labeling efficiency. Perdeuteration combined with specific $^{13}C$ labeling of methyl groups has also been demonstrated for this host (Clark *et al* 2015).

Higher eukaryotic expression systems, such as transfected mammalian and baculovirus-infected insect cells, have been under development for the production of isotopically labeled eukaryotic proteins (Dutta *et al* 2012, Egorova-Zachernyuk *et al* 2011, Franke *et al* 2018, Fuccio *et al* 2016, Kofuku *et al* 2018). Protocols are also available for *in vitro* cell free expression systems (Focke *et al* 2016, Klammt *et al* 2012, Linser *et al* 2014, Sobhanifar *et al* 2010) which have been used to produce membrane proteins for ssNMR studies (Abdine *et al* 2010, Kang *et al* 2014).

SsNMR offers the critical advantage of studying membrane proteins in the physiologically relevant environment of a lipid bilayer, or even in cell membranes. Isotopically labeled proteins can be reconstituted into lipid vesicles (liposomes) as model membranes, in which a variety of biochemical conditions, including the lipid composition, temperature and bilayer fluidity, protein-to-lipid ratio, pH and salinity, can be tuned to achieve optimal NMR spectral resolution and sensitivity, while at the same time maintaining the correct fold and functional activity. Other membrane model systems are also available for ssNMR, including bicelles (Dürr *et al* 2013, 2012), lipid nanodiscs (Bayburt *et al* 2002, Bayburt and Sligar 2003) and macrodiscs (Park *et al* 2011). Bicelles are a popular choice for Oriented Sample (OS) ssNMR experiments as they can be macroscopically aligned in a magnetic field (Marcotte and Auger 2005, Prosser *et al* 2006). Membrane proteins can also be reconstituted into nanodiscs of well-defined sizes, which are controlled by varying the length of the membrane scaffold protein (MSP) (Ding *et al* 2015, Hagn *et al* 2018, Kijac *et al* 2007, Mörs *et al* 2013). A larger version of nanodiscs, macrodiscs, in which MSP is replaced by a synthetic polymer styrene maleic acid (SMA), are in some cases better suited for ssNMR studies (Park *et al* 2020, Watt and Rienstra 2014). Lastly, intact cell membranes or even whole cells represent the most attractive membrane systems (Brown and Ladizhansky 2015, Kaplan *et al* 2016, 2015, Narasimhan *et al* 2020, 2019, Renault *et al* 2012a). 'In-membrane' or 'in-cell' ssNMR faces several challenges ranging from chemical heterogeneity of the samples to extreme spectral crowding and low sensitivity. Sample preparation protocols aimed to address these problems have been proposed, including the use of *E. coli* strains that are void of the most ubiquitous non-essential membrane proteins (Renault *et al* 2012b); isotopic labeling that is specifically targeted to the protein of interest (Jacso *et al* 2012); partial sample purification, e.g., separation of inner and outer membranes (Marani *et al* 2006), as well as the use of nickel affinity column (Everberg *et al* 2006) to concentrate NMR samples by isolating cell membranes that contain the protein of interest (Ward *et al* 2015b).

### 14.2.2 Dipolar recoupling techniques

Multidimensional correlation spectroscopy is the key element in ssNMR studies of macromolecular structure as it provides the site-specific resolution necessary for the determination of three-dimensional structures of biological macromolecules. Interatomic correlations in NMR spectra are established through polarization transfers between interacting nuclear spins and are used for obtaining resonance assignments and structural restraints.

In immobilized proteins, the main transfer methods utilize through-space dipole–dipole interactions. Through-bond J-couplings are also employed, especially for studies of dynamic molecular fragments such as flexible loops (Chen *et al* 2007a, 2007b, Tian *et al* 2009) The dipolar and chemical shift anisotropy (CSA) interactions are averaged out under MAS and can be selectively reintroduced through the application of phase- and/or amplitude modulated radio-frequency (RF) fields. Many methods, termed recoupling techniques, have been developed over the years for the reintroduction of homonuclear and heteronuclear dipolar interactions.

Homonuclear recoupling methods range from frequency-selective methods which reintroduce interactions within a small frequency range, such as rotational resonance ($R^2$) (Raleigh *et al* 1988) which requires matching a small multiple (1 or 2) of MAS frequency to the difference in the isotropic chemical shifts between two interacting spins, to broadband methods which recouple interactions between all carbon spins within a broad chemical shift range (Carravetta *et al* 2000, Gregory *et al* 1995, Hohwy *et al* 1999, 1998, Lee *et al* 1995, Sun *et al* 1995, Tycko and Dabbagh 1990). The choice of homonuclear recoupling methods depends on the MAS rate regime. Under moderate MAS conditions (<20 kHz), homonuclear polarization transfers between carbon atoms in multidimensional NMR experiments are commonly accomplished using Proton-Driven Spin Diffusion (PDSD) (Bloembergen 1949, Szeverenyi *et al* 1982) and Dipolar Assisted Rotational Resonance (DARR) (Morcombe *et al* 2004, Takegoshi *et al* 2001). These second-order based recoupling methods are robust and easy to set up, and readily transfer polarization between carbons separated by 3–4 bonds. At higher frequencies where both PDSD and DARR become much less efficient, other broadband second-order recoupling methods are available, including PAR (Proton-Assisted Recoupling) (De Paëpe *et al* 2008, Lewandowski *et al* 2007), PARIS-xy (Phase Alternated Recoupling Irradiation) (Weingarth *et al* 2010, 2009), SHANGHAI (Second-order Hamiltonian among Analogous Nuclei Generated by Hetero-nuclear Assistance Irradiation) (Hu *et al* 2011), COmbined $R2_n^\nu$-Driven (CORD) (Hou *et al* 2013) and MIRROR-SD (Mixed Rotational and ROtary Resonance) (Scholz *et al* 2008).

Some first-order recoupling sequences require application of RF fields at multiples of MAS frequencies, e.g., C7 (Lee *et al* 1995) and POST-C7 (Hohwy *et al* 1998), and become impractical at high MAS frequencies due to power limitations. In contrast, some techniques such as RFDR (Radio Frequency-Driven Recoupling) (Bennett *et al* 1998, 1992), HORROR (HOmonucleaR ROtary Resonance) (Nielsen *et al* 1994) and its adiabatic version DREAM (Dipolar Recoupling Enhanced by Amplitude Modulation) (Verel *et al* 2001), perform well and do not require proton decoupling at high spinning frequencies (Bayro *et al* 2008, Ernst *et al* 2004, 2003, Huang *et al* 2011). There are also a few methods optimized to perform over a broad range of spinning conditions and without proton decoupling, such as CORD-RFDR (Lu *et al* 2015) and SPC-*n* recoupling sequences (Courtney and Rienstra 2016).

Among heteronuclear dipolar recoupling methods, cross polarization (CP) (Pines *et al* 1972) is commonly used in multidimensional experiments to transfer polarization between protons and carbon or nitrogen spins, as well as between carbons and nitrogens (Schaefer *et al* 1979). CP is realised through the application of simultaneous spin-lock pulses satisfying the Hartmann–Hahn matching condition (Hartmann and Hahn 1962) to the two types of nuclear spins that are being recoupled. Proton-enhanced CP (Pines *et al* 1972) of carbon or nitrogen spins is a standard procedure in the majority of ssNMR experiments, and it is used to excite signals from rigid regions of proteins. CP enhances the signal from low-$\gamma$ nuclei and also allows for faster acquisition and reduced experimental time. CP between carbon and nitrogen spins serves to transfer polarizations in many heteronuclear multidimensional experiments (Schaefer *et al* 1979). To compensate for the inhomogeneity of the RF field, variable amplitude CP techniques are routinely used nowadays (Hediger *et al* 1995, 1994).

Heteronuclear dipolar interactions can also be recoupled using rotor-synchronized pulse sequences such as Rotational Echo DOuble Resonance (REDOR) (Gullion and Schaefer 1989), and Transferred Echo DOuble Resonance (TEDOR) (Hing *et al* 1993, 1992) techniques. In samples with extensive incorporation of $^{13}C$ and $^{15}N$ labels, frequency-selective REDOR (Jaroniec *et al* 2001) and band-selective and zero-filtered versions of TEDOR (Jaroniec *et al* 2002a) have been developed for quantitative measurements of weak $^{13}C$–$^{15}N$ dipolar couplings and for obtaining structural restraints. REDOR methods have also been used in deuterated proteins to accurately measure $^{1}H$–$^{15}N$ dipolar order parameters for studies of protein dynamics (Schanda *et al* 2011, 2010). More detailed discussion of recoupling techniques can be found in references (Bennett *et al* 1994, De Paëpe 2012, Dusold and Sebald 2000, Ladizhansky 2009).

### 14.2.3 Resonance assignments

Knowledge of chemical shifts is a prerequisite for site-specific analysis of protein structure, dynamics and interactions. Chemical shifts are also sensitive reporters on the local electrostatic environment and serve as fingerprints of protein conformation. Assignment of chemical shifts, i.e., mapping resonances to specific atoms in the sequence, is an essential step in the structural and dynamic analyses of a protein.

Most membrane protein studies to date have relied on carbon detection. While initial assignments obtained for small microcrystalline proteins mainly used two-dimensional (2D) chemical shift correlation spectroscopy (Franks *et al* 2005, Marulanda *et al* 2005, McDermott *et al* 2000, Pauli *et al* 2001), these methods are generally insufficient for application in membrane proteins. Modern assignment strategies combine 2D spectroscopy with better resolved three-dimensional (3D) spectra that correlate chemical shifts of backbone carbon and nitrogen atoms and side chains and allow identification of the amino acid and linking them into contiguous fragments.

A minimal suite of assignment experiments includes three types of 3D correlations: the NCACX, the NCOCX and the CANCO/CONCA (figure 14.1(A)).

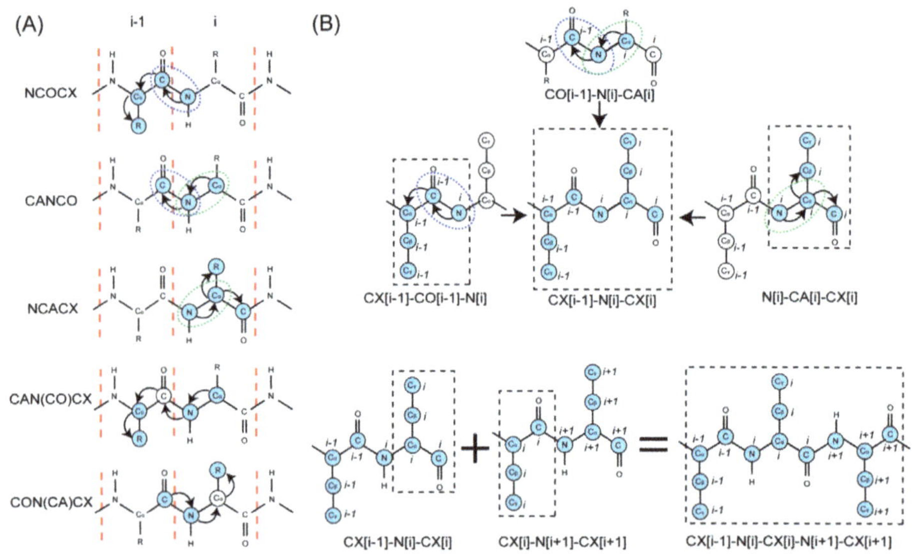


**Figure 14.1.** Carbon-detected ssNMR assignment strategies. (A) Schematic transfer steps for the NCOCX, CANCO and NCACX experiments. CAN(CO)CX and CON(CA)CX are extensions of CANCO for removing assignment ambiguities. The initial $^{1}H/^{15}N$ and $^{1}H/^{13}C$ CP steps are omitted for clarity. (B) Top panel illustrates how the extended spin systems CX[i-1]-N[i]-CX[i] are obtained by matching NCOCX with CANCO, and by matching NCACX with CANCO. Blue dashed lines highlight the CO[i-1]-N[i] spin pairs found in both CANCO and NCOCX. Green dashed lines highlight the N[i]-CA[i] spin pairs found in both CANCO and NCACX. Lower panel shows further linkage of the extended spin systems CX[i-1]-N[i]-CX[i]-N[i+1]-CX[i+1] by matching the CX[i] shifts (highlighted in dashed boxes) from the CX[i-1]-N[i]-CX[i] system and the CX[i]-N[i+1]-CX[i+1] system.

The CANCO experiment correlates the three backbone atoms CA[i], N[i] and CO[i-1] within two adjacent residues. The NCACX experiment correlates chemical shifts of the backbone atoms N[i], CA[i] to the backbone CO and side chain atoms CX[i] (CX = $C_\beta$, $C_\delta$, $C_\gamma$ etc) of the same residue. Side chains of the amino acids have unique chemical shift patterns: observation of these side chain resonances is critical as it allows the identification of the amino acid type. The NCOCX experiment establishes an inter-residue correlation between the N[i] shifts and the carbon shifts CO[i-1] and CX[i-1] of the preceding residue; based on the unique side chain chemical shift patterns, many amino acid types can also be identified in the NCOCX experiment.

In all these experiments, spin polarization is selectively transferred between $^{15}N$ and either CA or CO via SPECtrally Induced Filtering In Combination with CP (SPECIFIC CP) (Baldus *et al* 1998), whereas the transfers of polarization between carbon atoms are typically accomplished using through-space homonuclear recoupling techniques such as DARR in which the range of transfer can be tuned through the length of the DARR mixing time. Additionally, high-sensitivity 2D carbon–carbon correlation spectra can also be utilized to assist side-chain assignments.

The three experiments, CANCO, NCACX, and NCOCX, are used to establish sequential links between residues. Each CO[i-1]-N[i]-CA[i] peak in the CANCO spectrum is linked to the N[i]-CO[i-1]-CX[i-1] peaks in the NCOCX spectrum by matching N[i] and CO[i-1] shifts. Similarly, the same peaks are linked to the N[i]-CA [i]-CX[i] peaks in the NCACX spectrum by matching N[i] and CA[i] shifts. The resulting CX[i-1]-N[i]-CX[i] systems contain information for two sequential amino acids (figure 14.1(B) upper panel). They can be further linked into longer contiguous fragments containing three sequential residues CX[i-1]-N[i]-CX[i]-N[i+1]-CX[i+1] by matching the side chain CX[i] shifts (figure 14.1(B) lower panel). Once three or more sequential residues are identified, their chemical shifts can be unambiguously mapped onto the amino acid sequence.

Additional experiments can be employed when NCACX and NCOCX experiments do not provide sufficient resolution or sensitivity. Broadband recoupling methods that are commonly used in both experiments disperse polarization evenly between all the carbon atoms in the side chains, and often result in overall low peak intensities. Additionally, the spectral overlaps, e.g., between $C_\beta$ chemical shift regions and $C_\gamma/C_\delta$ shifts in some amino acids, may also hinder an unambiguous identification of spin systems. The sensitivity can be improved by directing polarization transfers for obtaining short range one-bond correlations between carbon spins through the use of band-selective recoupling. For example, band-selective transfers are readily accomplished in the NCACB experiment by using the DREAM scheme (Verel *et al* 2001) which mostly results in one-bond $C_\alpha/C_\beta$ transfers, thereby improving the transfer efficiency in the $C_\beta$ spectral region.

More complex 3D experiments can be used to resolve assignment ambiguities. The CONCA experiment can be extended to CAN(CO)CX and CON(CA)CX (Li *et al* 2008, Shi *et al* 2009a) correlations (figure 14.1(A)). A suite of experiments containing NCACB, N(CO)CACB and CAN(CO)CA has been used by Böckmann, Meier and coworkers (Schuetz *et al* 2010) for more reliable linking of spin systems using better dispersed $C_\beta$ shifts instead of the congested CO shifts, albeit at the expense of sensitivity due to the additional transfer steps. Similarly, Lange and co-workers have proposed band-selective homonuclear cross-polarization (BSH-CP) as a more efficient CO/CA transfer method and developed a set of complementary experiments using additional NCOCA/NCOCACB and NCACO correlations (Shi *et al* 2014).

Four-dimensional (4D) correlation experiments have also been implemented for assigning chemical shifts in large proteins. First demonstrated on a small protein $\beta$1 immunoglobulin binding domain of protein G (GB1) (Franks *et al* 2007), the 4D CONCACX, CANCOCX and its variation 4D CANCOCA experiments were used for assignments in the disulfide bond generating membrane protein DsbB (Li *et al* 2008), proteorhodopsin (Shi *et al* 2009b) and KcsA (Wylie *et al* 2014).

The high-dimensionality correlation schemes involve many polarization transfer steps and inevitably suffer from low sensitivity. Sergeyev, McDermott and co-workers have shown that these limitations can be overcome by using DNP to create the highly polarized state of nuclear spins. DNP signal enhancements of 35–60 obtained on the Pf1 bacteriophage coat protein were sufficient to overcome the sensitivity limitations and to record an experiment that includes correlations between side chain atoms of neighboring amino acids (Sergeyev *et al* 2017).

Lastly, spectral simplification using amino acid specific labeling/unlabeling or sparse $^{13}C$ labeling schemes can greatly assist assignments of membrane proteins. Unlabeling of the amino acid types that are generally abundant in transmembrane (TM) helices, such as Val, Leu, Ile and Phe, removes many overlapped signals and simplifies assignments (Etzkorn *et al* 2010, 2007, Schneider *et al* 2008, Shi *et al* 2009a). Using sparse $^{13}C$ labeling, Wang *et al* have shown that 2D and 3D experiments recorded on alternately labeled membrane protein samples significantly improve spectral resolution, resulting in additional well-resolved resonances and more complete assignments (figure 14.2(A&B)) (Wang *et al* 2013b). The improvement is particularly noticeable for the aromatic residues (figure 14.2(C&D)) which are usually difficult to identify and assign due to poor resolution and signal-to-noise

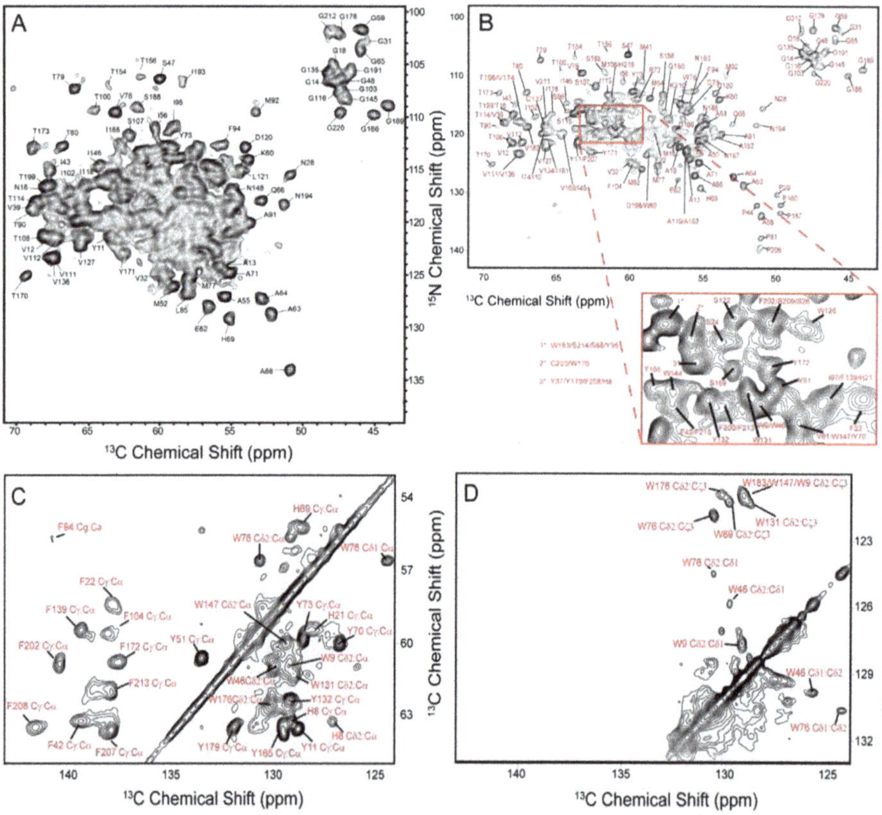


**Figure 14.2.** Sparse $^{13}C$ labeling and spectral simplification. (A–B) A comparison of 2D NCA spectra obtained from U [$^{13}C$, $^{15}N$]-labeled ASR sample in DMPC/DMPA lipids (A) and from [2–$^{13}C$]-labeled ASR sample in the same lipids. (C–D) Selected regions from the 2D $^{13}C$–$^{13}C$ correlation spectrum obtained from the [2–$^{13}C$]-labeled ASR sample. Aromatic-aliphatic correlation region is shown in (C), and aromatic-aromatic correlation region is in (D). (B–D) are reproduced with permission from Springer Nature: Wang *et al* (2013b).

of their side chain signals (Hiller *et al* 2008). More detailed discussions of the ssNMR assignment methodologies can be found in a recent comprehensive review article (Higman 2018).

### 14.2.4 NMR structural restraints

In NMR-based analysis, a structural model of a protein is derived from a large number of structural restraints. The primary source of structural information is the inter-atomic distances, which are obtained mainly from the application of dipolar recoupling techniques. Dipolar recoupling methods have been successfully integrated into multidimensional experiments, in which the distances between spatially close atoms are extracted from the intensities of the corresponding cross-peaks. Long-range distances between atoms (>3 Å) separated by several residues (five or more in α-helical structures) are of primary importance for structure determination, as they constrain the global fold of a protein. However, the major obstacle in obtaining long-range distance restraints from ssNMR spectra is the dipolar truncation effect, a phenomenon in which the polarization transfers between distant, weakly coupled atoms are suppressed by the strong dipolar interactions, e.g., by those between neighbouring carbon atoms (Bayro *et al* 2009a). This effect is present in many homonuclear and heteronuclear dipolar recoupling experiments which produce non-commuting pairwise effective Hamiltonians (Bayro *et al* 2009a, Hodgkinson and Emsley 1999, Hohwy *et al* 1999, Ladizhansky *et al* 2003b).

The dipolar truncation effect in homonuclear $^{13}C$–$^{13}C$ distance measurements can be minimized through the combined use of the spin dilution labeling schemes which remove the strongest one-bond dipolar couplings, recoupling methods that are less sensitive to dipolar truncation, and/or by using frequency- or band-selective recoupling schemes which reintroduce only the weak structurally constraining interactions. Multiple carbon–carbon restraints can, in principle, be simultaneously collected in a single 2D correlation experiment using these recoupling techniques.

Among homonuclear dipolar recoupling experiments, proton-assisted second-order recoupling such as DARR, PDSD, PAR, MIRROR, etc, have been shown to be less sensitive to dipolar truncation (Grommek *et al* 2006, Scholz *et al* 2007). These pulse sequences generate effective Hamiltonians that depend on the cross terms between heteronuclear and homonuclear dipolar couplings (Grommek *et al* 2006), and can achieve efficient long-range polarization transfer over a wide range of isotropic chemical shifts. Their performance for collecting structural restraints is further enhanced by employing sparse labeling strategies. Using [1,3–$^{13}C$] and [2–$^{13}C$] labeled glycerols is the most common and economic strategy as the alternate labeling patterns effectively eliminate most of the strong one-bond $^{13}C$–$^{13}C$ dipolar couplings. At moderate MAS frequencies, PDSD and DARR applied to sparsely labeled samples has been shown to lead to long-range structural restraints. Their utility for structure determination has been demonstrated on microcrystalline proteins (Castellani *et al* 2002, Franks *et al* 2008, Zech *et al* 2005) and later extended to membrane proteins: *Anabaena* sensory rhodopsin (ASR) (Wang *et al* 2013b), DsbB (Tang *et al* 2013) and outer membrane protein G (OmpG)

(Retel *et al* 2017). These methods were also used to solve structure of uniformly $^{13}C$-labeled membrane protein, *Yersinia enterocolitica* adhesin A (YadA) (Shahid *et al* 2012).

At faster MAS rates exceeding ~20 kHz, PDSD and DARR become less efficient. As an alternative, PAR was demonstrated to be able to generate transfers over long range distances up to 6–7 Å at 65 kHz in uniformly labeled microcrystalline Crh (De Paëpe *et al* 2008) and GB1 (Lewandowski *et al* 2009), and was used to collect long-range restraints in membrane proteins YadA (Shahid *et al* 2012) and in influenza A M2 proton channel (Andreas *et al* 2015). PARIS-xy and SHANGHAI also work efficiently over a wide range of spinning rates from 30–60 kHz, and they have been shown to provide efficient long-range transfer over 3.5–4 Å distances in amyloid fibrils (Shen *et al* 2013, Weingarth *et al* 2011). CORD has shown high transfer efficiency at 40 kHz for both aliphatic-aliphatic and aliphatic-carbonyl carbon regions (Hou *et al* 2013), and was recently used for obtaining restraints in the structure calculation of the influenza B M2 proton channel (Mandala *et al* 2020).

Frequency-selective or band-selective recoupling methods represent another avenue for collecting long-range restraints without suffering from dipolar truncation (Costa *et al* 1997b, Levitt *et al* 1998, Raleigh *et al* 1988, Ramachandran *et al* 2003, Verhoeven *et al* 2004). Here, the dipolar interactions are reintroduced only between nuclear spins within a specific range of isotropic chemical shifts, while all other dipolar interactions between spins outside that range remain averaged out by MAS. Such methods have been used in both uniformly and sparsely $^{13}C$-labeled macromolecules. Among the frequency-selective techniques, $R^2$ and related methods have been most widely utilized for homonuclear carbon–carbon distance measurements, including in membrane proteins. Early applications of $R^2$ were used to determine conformation of retinal in bacteriorhodopsin (bR) (Creuzet *et al* 1991) and for the structure determination of melittin (Lam *et al* 2001). Band-selective versions of $R^2$ are useful for simultaneously recoupling a large number of remote spin pairs, e.g., between the carbonyls and the aliphatic side chain spins. $R^2$ Width ($R^2W$) (Costa *et al* 2003, Ramachandran *et al* 2003) and the band-selective version of $R^2$, Homogeneously Broadened $R^2$ ($HBR^2$) (Janik *et al* 2007, Peng *et al* 2008), have been used in structure determination and refinement of GB1 (Peng *et al* 2008, van der Wel *et al* 2009), amyloid fibrils (van der Wel *et al* 2010), and of membrane proteins ASR (Wang *et al* 2013a) and human Aquaporin-1 (hAQP1) (Dingwell *et al* 2019). Other band-selective recoupling schemes such as the narrowband RFDR (Goobes *et al* 2000, Goobes and Vega 2002), BAnd-SElective (BASE) RFDR (Bayro *et al* 2009b) and SEASHORE (Shift-Evolution-Assisted Selective HOmonuclear REcoupling) (Paravastu and Tycko 2006) have also been proposed for structure determination.

Frequency-selective (fs) REDOR and band-selective and zero-filtered versions of TEDOR are specifically optimized for the simultaneous collection of multiple $^{15}N$–$^{13}C$ distance restraints in proteins with extensive incorporation of $^{13}C$ and $^{15}N$ labels. These methods have been used for the determination of the 3D structure of a tripeptide MLF (Rienstra *et al* 2002), probing protein structure and inter-protein interfaces of microcrystalline proteins and amyloid fibrils (Helmus *et al* 2011, 2010, 2008, Jaroniec *et al* 2002b, Nieuwkoop *et al* 2009, Nieuwkoop and Rienstra 2010, Shahid *et al* 2012), and structure determination of membrane proteins (Cady *et al* 2010, Hu *et al* 2010, Maciejko *et al* 2015, Shahid *et al* 2012, Tang *et al* 2013). $^{13}C$–$^{2}H$ fsREDOR was used for quantitative characterization of the amantadine binding sites in the influenza A M2 proton channel by Hong and co-workers (Cady *et al* 2010). The range of distance restraints that can be probed by fsREDOR can be further extended by measuring stronger dipolar interactions, e.g., between carbon spins and spins with high gyromagnetic ratios, such as fluorine or proton. Proton detected $^{13}C$-dephased $^{1}H$ fsREDOR was recently reported by the Rienstra lab (Ghosh and Rienstra 2017), whereas $^{13}C$–$^{19}F$ fsREDOR was used to determine interactions between M2 and fluorinated analog of cholesterol (Elkins *et al* 2017) and for structure determination of the influenza B M2 channel (Mandala *et al* 2020). TEDOR measurements enhanced by DNP were used to determine the oligomeric interface of proteorhodopsin (Maciejko *et al* 2015).

NHHC and CHHC ($^{15}N$- and/or $^{13}C$-detected proton diffusion) experiments (Lange *et al* 2002) combine proton–proton spin diffusion with nitrogen and/or carbon detection, and encode information on inter-proton proximities. Although less sensitive compared to DARR, these approaches are not subject to dipolar truncation effects and have been used to measure proton–proton distances for structure determination in a small peptide (Lange *et al* 2005), microcrystalline protein Crh (Loquet *et al* 2008) and GB1 (Nieuwkoop and Rienstra 2010), amyloid fibrils (Loquet *et al* 2018), and membrane proteins (Shahid *et al* 2012, Wang *et al* 2013a).

The majority of distance restraints in ssNMR are obtained through the application of nuclear spin–spin dipolar recoupling techniques, and their detectable range is typically limited to 5–6 Å. Paramagnetic Relaxation Enhancement (PRE) methods (Bertini *et al* 2001, Clore and Iwahara 2009, Jaroniec 2015, 2012) utilize the much stronger electron–nuclear couplings and are suitable for probing longer range distances up to 15–20 Å. PRE measurements require a paramagnetic tag that can be introduced into a protein through chemical modification of the reactive cysteine side chains, which are either endogenously present or introduced by site-directed mutagenesis. S-(1-oxyl-2,2,5,5-tetramethyl-2,5-dihydro-1H-pyrrol-3-yl)methyl methanesulfonothioate (MTSL) is a commonly used nitroxide radical for PRE measurements. The unpaired electron of MTSL enhances the transverse relaxation rates of the nuclear spins within approximately 15 Å radius, causing line-broadening and significant attenuation of their NMR signals, and thereby providing qualitative structural restraints. In membrane proteins, PRE measurements have been successfully used for the characterization of the oligomerization interface of ASR (Wang *et al* 2012) and assisted in refining its structure (Wang *et al* 2013a), as illustrated in figure 14.3. Importantly, the same MTSL-labeled mutants can be used for Double Electron-Electron Resonance (DEER) measurements, enabling additional long-range restraints on the oligomeric arrangement and simultaneously improving structural resolution (Milikisiyants *et al* 2017).

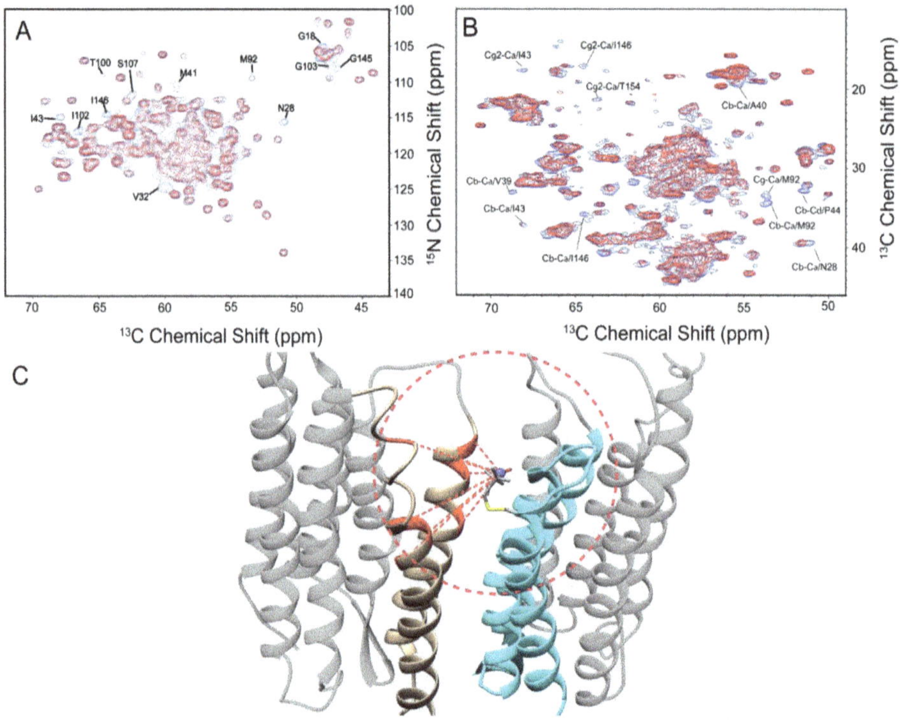


**Figure 14.3.** PRE detection of the oligomeric interface in the param S26C mutant of ASR in DMPC/DMPA lipids. (A–B) 2D NCA correlation spectra (A) and 2D $^{13}C$–$^{13}C$ correlation spectra (B) showing comparison of paramagnetically labeled sample (red) and its diamagnetic control (blue). Attenuated peaks in the red spectra indicate residues in the vicinity of the paramagnetic center (~15 Å radius shown as red circle). (C) Graphical representation of the PRE restraints at the oligomeric interface. Helices of two interacting monomers are shown in different colors. Paramagnetic center is shown in blue, and residues with attenuated signals are in red. Only intermonomer PRE effects are shown for clarity. (A–B) are reproduced with permission from Wang *et al* (2012). Copyright 2012, American Chemical Society.

In structure calculation and refinement, the internuclear distance restraints are complemented by experimental torsion angle restraints. The chemical shift index (CSI) (Wishart and Sykes 1994) analysis represents a simple and efficient method to quickly evaluate secondary structure by comparing the site-specific chemical shifts of $C_\alpha$ and $C_\beta$ atoms with their random coil values. The TALOS+ (Torsion Angle Likelihood Obtained from Shift and sequence similarity) (Cornilescu *et al* 1999, Shen *et al* 2009) program represents an empirical way for estimating torsion angles from backbone chemical shifts and has been commonly employed in structure determination. The torsion angles predicted by TALOS+ add valuable restraints that significantly improve the resolution of structural models derived from both solution and ssNMR, and have been used for structure determination of several membrane proteins (Mandala *et al* 2020, Retel *et al* 2017, Shahid *et al* 2012, Tang *et al* 2013, Wang *et al* 2013a). Additionally, more precise torsion angles can be probed using recoupling methods that correlate relative orientations of dipolar

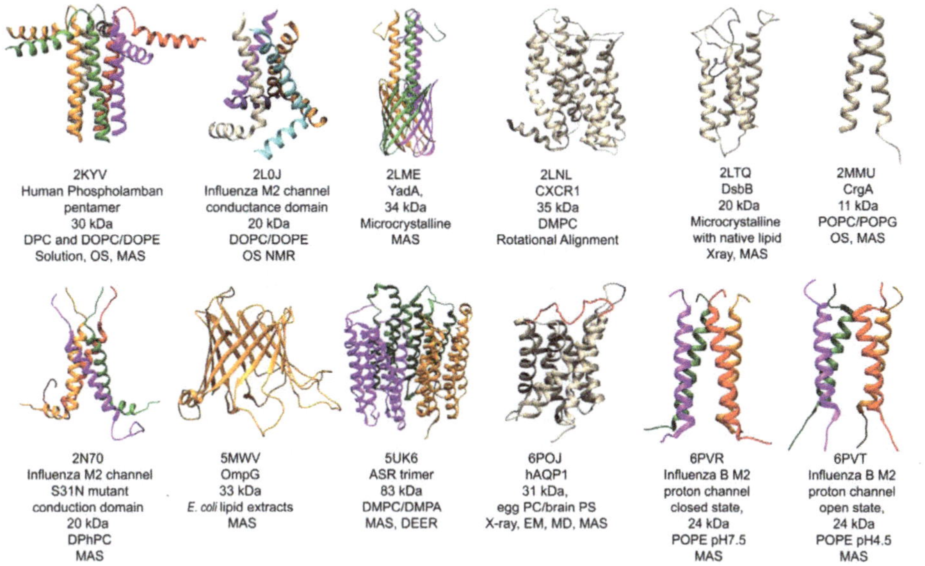


**Figure 14.4.** Selected high-resolution membrane protein structures determined or refined using ssNMR restraints: Human phospholamban (Verardi *et al* 2011); Influenza A M2 channel (Andreas *et al* 2015, Sharma *et al* 2010), YadA (Shahid *et al* 2012), CXCR1 (Park *et al* 2012), DsbB (Tang *et al* 2013), CrgA (Das *et al* 2015), ASR trimer (Milikisiyants *et al* 2017), OmpG (Retel *et al* 2017), hAQP1 (Dingwell *et al* 2019), and Influenza B M2 channel (Mandala *et al* 2020). Protein names, molecular weights, lipid systems, type of ssNMR methods and PDB codes are listed for each structure. Individual subunits are differentiated by different colors. Refined region in hAQP1 from NMR restraints is shown in red.

and/or chemical shift tensors, e.g., HNCH (Rienstra *et al* 2002), HN[i+1]CH (Jaroniec *et al* 2004), NCCN (Costa *et al* 1997a, Ladizhansky *et al* 2003a), and Vector Angle (VEAN) restraints (Franks *et al* 2008). Such methods can, in principle, provide more accurate torsional restraints, albeit they require high sensitivity which is not often available in applications to membrane proteins. In figure 14.4 we show examples of high-resolution structures of membrane protein determined by ssNMR.

## 14.3 Proton detection in membrane proteins

Most ssNMR experiments rely on the detection of low gamma nuclei, $^{15}N$ or, more commonly, $^{13}C$. The detection of protons exploits the advantages of their high natural abundance and high gyromagnetic ratio, offers higher sensitivity with reduced experimental time, and provides an additional chemical shift dimension in complex multidimensional correlation experiments (see chapter 15).

In solids, proton detection is complicated by lack of resolution due to the significant $^{1}H$ line broadening. The strong proton–proton dipolar interactions need to be sufficiently suppressed in order to enable site-specific proton detection. Two approaches have emerged for solving this problem, sometimes combined together. They have been tested on microcrystalline proteins and later extended to membrane embedded proteins. The first approach employs proton dilution by extensive deuteration. A high level of deuteration can be achieved by recombinantly expressing a protein in a perdeuterated growth medium. Protons can then be reintroduced at the exchangeable sites for detection through deuterium/hydrogen (D/H) back-exchange, with desired degree of protonation level achieved by incubating the sample in a buffer with a controlled $H_2O/D_2O$ ratio. Early work by the Reif lab and the Zilm lab have demonstrated that 100 % back-exchange of amide protons can effectively reduce proton linewidths to ~0.2 ppm in microcrystalline proteins, α-spectrin SH3 domain (Chevelkov *et al* 2003) and ubiquitin (Paulson *et al* 2003), even at moderate MAS frequencies of ~10–20 kHz. Additional improvement of proton linewidths down to ~0.05 ppm was achieved in SH3 by further diluting protons to 10% reprotonation (Chevelkov *et al* 2006). The trade-off between the resolution gain (approximately three-fold improvement in proton linewidth by going from 100% to 10% back-exchanged protonation levels) and the sensitivity losses due to the reduced number of protons has prompted Oschkinat and coworkers to examine the optimal proton back-exchange level: 30%–40% back-exchange reprotonation level was found to be optimal in SH3 for overall sensitivity and resolution at spinning rates of ~24 kHz (Akbey *et al* 2010). These requirements could be further relaxed at faster MAS frequencies: similar linewidths were observed in the SH3 domain with 100% back-exchanged amide protons at the MAS rate of 60 kHz (Lewandowski *et al* 2011a).

Application of perdeuteration and back-exchange in α-helical membrane proteins is challenging. Many multispan membrane proteins forming α-helical bundles have a hydrophobic core which is not readily accessible to polar solvent because of the tight interhelical packing and lipid interactions. The exchangeability of protons is limited to the solvent-exposed sites, rendering the TM core region mostly invisible for $^1$H NMR detection (Bolton *et al* 2018, Linser *et al* 2011, Ward *et al* 2011). The problem is further complicated by lack of generally applicable *in vitro* unfolding and refolding protocols for polytopic α-helical membrane proteins (Harris and Booth 2012, Popot 2014, Stanley and Fleming 2008).

This challenge may be addressed by using partial isotopic labeling schemes. Protons can be selectively introduced into the aliphatic regions from partially protonated sources during expression to achieve homogeneous distribution in both solvent-accessible and solvent-inaccessible regions. Some examples include: Reduced Adjoining Protonation (RAP) which uses deuterated glucose in partially deuterated buffer containing 5%–15% $H_2O$ (Asami *et al* 2012, 2010); Fractional Deuteration (FD) which uses protonated glucose in fully deuterated buffer (Mance *et al* 2015); and the Proton Clouds method which uses deuterated glucose supplemented with uniformly $^1$H,$^{13}$C,$^{15}$N-labeled amino acids in fully deuterated buffer (Sinnige *et al* 2014).

The use of highly deuterated medium often results in reduced protein yields and lengthy expression time in bacterial cells and may not even be compatible with certain eukaryotic expression systems, e.g., mammalian cells. Inverse Fractional Deuteration (iFD) represents a suitable labeling scheme for addressing this issue, as was recently demonstrated for the *E. coli* system using bacterial potassium channel KcsA (Medeiros-Silva *et al* 2016). KcsA was expressed using perdeuterated

$^{13}C$-labeled glucose as the sole carbon source in an otherwise fully protonated medium. The resolution of proton signals in the TM region was further improved by removing the signals from solvent-exposed residues through H/D exchange. It is worth noting that random incorporation of deuterons in these partially isotopic labeling schemes inevitably leads to mixtures of isotopomers and the heterogeneity induced line broadening may compromise sensitivity and resolution, most notably in the methyl groups (Paepe *et al* 2017).

The second approach does not require extensive deuteration but employs ultrafast MAS to reduce the $^{1}H$–$^{1}H$ homonuclear dipolar couplings. It has been shown that the observed proton linewidth is inversely proportional to the MAS rate (Chevelkov *et al* 2006, Nishiyama *et al* 2011, Penzel *et al* 2019, Reif and Griffin 2003, Sternberg *et al* 2018). Ultrafast MAS opens up many promising opportunities to exploit proton detection in membrane proteins. At MAS rate of 60 kHz, proton linewidths of ~0.25–0.4 ppm are achieved without deuteration (Lewandowski *et al* 2011a, Marchetti *et al* 2012, Vasa *et al* 2016), which may not be sufficient for full assignment of resonances. Further proton line narrowing down to ~0.15–0.2 ppm can be achieved on fully protonated membrane protein samples at spinning rates of 100–111 kHz. This has facilitated spectroscopic assignments and distance restraint measurements in relatively large proteins (Jain *et al* 2017, Lalli *et al* 2017, Lecoq *et al* 2019, Stanek *et al* 2016). The utility of ultrafast MAS for protein structure determination has also been demonstrated on fully protonated GB1 and *Acinetobacter* phage 205 coat protein (Andreas *et al* 2016). In the case of a heptahelical transmembrane protein proteorhodopsin (PR), proton resolution obtained at 100 kHz MAS on a fully protonated sample was comparable to that obtained on a perdeuterated and 100 % back-exchanged sample measured at 60 kHz MAS (figure 14.5 right panel). Importantly, the spectrum from the fully protonated sample contained cross-peaks corresponding to the hydrophobic TM region which was not accessible to solvent in the perdeuterated case (figure 14.5 left panel) (Lalli *et al* 2017).

Resonance assignment strategies for proton-detected ssNMR experiments exploit both dipolar and J-couplings between nuclear spins and rely on sequences of polarization transfer steps that are similar to solution-state assignment methods. At fast spinning rates above 60 kHz, sufficiently long $^{13}C$ coherence lifetimes enable efficient homonuclear scalar-based transfers between carbon atoms (Barbet-Massin *et al* 2014, Bertini *et al* 2011). Amide protons are detected in perdeuterated samples and spin systems can be sequentially linked through the shared N–H pairs in a series of 3D spectra, with the third dimension encoding chemical shifts of the aliphatic carbons and carbonyl atoms ($C_\alpha$, $C_\beta$ and CO) of either the same or the preceding residue (Barbet-Massin *et al* 2014, Fricke *et al* 2017, Knight *et al* 2011, Zhou *et al* 2012). A basic set of experiments includes (H)CONH and (H)CANH chemical shift correlations which establish inter- and intra-residue connectivities, respectively. They are supplemented by extended sets of additional inter-carbon connectivities, (H)CO(CA)NH, (H)(CO)CA(CO)NH, (H)(CA)CB(CA)NH and (H)(CA)CB(CACO)NH, in which $C_\alpha$ and $C_\beta$ chemical shifts are used to identify the amino acid types (figure 14.6(A)). The joint analysis of all sets links resonances into long contiguous fragments that can be uniquely matched onto the protein sequence.

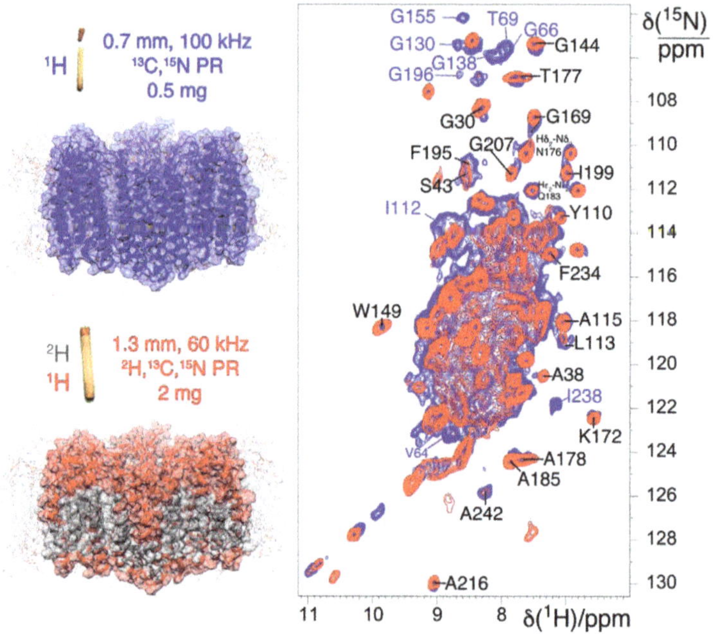


**Figure 14.5.** Proton detection experiments on a seven-helical membrane protein PR in DMPC/DMPA lipids. Left: PR oligomers are shown (PDB: 2L6X) with key sample preparation and experimental conditions indicated above. Detectable region in the fully protonated sample is shown in blue, and for the perdeuterated sample back-exchanged sample in red. Right: an overlay of the 2D $^1$H–$^{15}$N HSQC spectra obtained from fully protonated PR at 100 kHz MAS (blue), and from perdeuterated back-exchanged PR sample at 60 kHz MAS (red). The TM core region remains invisible in the deuterated sample but is detectable and results in additional peaks in the fully protonated sample (blue spectrum). Reproduced with permission from Lalli *et al* (2017) copyright American Chemical Society.

In fully protonated samples at ultrafast MAS rates of 100 kHz or faster, the detection is no longer limited to amide protons. Alpha protons can be additionally exploited to provide complementary chemical shift correlation sets (Stanek *et al* 2016). (H)N(CO)CAH and (H)NCAH experiments record inter- and intra-residue correlations, respectively, whereas (H)CO(N)CAH and (H)COCAH, provide sequential matching and an additional (H)(CA)CBCAH can be used for identifying amino acid type through $C_\beta$ resonances (figure 14.6(B)). In these kind of experiments, the residual proton signals from water and lipids can be suppressed using MISSISSIPPI (Multiple Intense Solvent Suppression Intended for Sensitive Spectroscopic Investigation of Protonated Proteins, Instantly) scheme (Zhou *et al* 2012, Zhou and Rienstra 2008).

## 14.4 Dynamics

### 14.4.1 Protein dynamics

Protein motions play an important role in various biomolecular processes such as ligand binding, protein folding, signaling and enzymatic catalysis. These motions

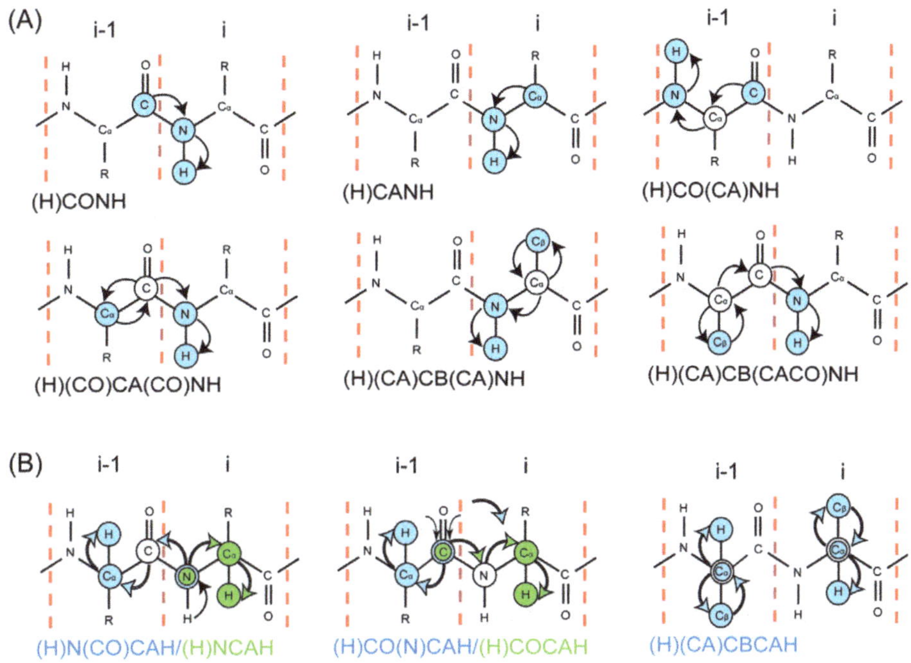


**Figure 14.6.** Assignment strategies in $^1$H-detected ssNMR. The initial $^1H/^{15}N$ and $^1H/^{13}C$ CP steps are omitted for clarity. (A) Experiments using $^1H_N$ detection are applicable in both fully protonated and perdeuterated and back-exchanged samples: (H)CONH, (H)CANH, (H)CO(CA)NH, (H)(CO)CA(CO)NH, (H)(CA)CB(CA) NH and (H)(CA)CB(CACO)NH. (B) Experiments using $^1H_\alpha$ detection are applicable in fully protonated samples: (H)N(CO)CAH, (H)NCAH, (H)CO(N)CAH, (H)COCAH, and (H)(CA)CBCAH.

can be of local, collective and global nature and involve dynamic processes which occur on a wide range of timescales and amplitudes. While globular proteins undergo rapid molecular tumbling in solution, with typical tumbling rates falling in the range from a few nanoseconds to few tens of nanoseconds (Jarymowycz and Stone 2006), membrane-embedded proteins reside in a more viscous and constrained environment of a cell membrane. They do not undergo global rotational diffusion in a bilayer but may experience translational and uniaxial rotational diffusion. The overall membrane protein mobility is affected by protein-lipid and protein-protein interactions, as well as by the hydration level and physical state of the lipid bilayer (Alenghat and Golan 2013, Smith 2012). Annular lipids mediate interactions between membrane proteins and the bilayer, and its fluidity can influence both the uniaxial rotational and lateral diffusion. Lipid cofactors are more tightly bound to TM helices, subunits or oligomeric interfaces, and may serve as stabilizing agents for the protein. Restriction of the protein mobility may further arise from the formation of protein complexes and oligomerization.

All these factors restrain the global dynamics of membrane proteins as well as their internal motions. The relatively rigid TM regions are mostly confined within the hydrophobic portion of a bilayer. They are stabilized by strong hydrogen bonds and interact with both lipids and other proteins. These interactions restrict the fast local motions, which involve individual atoms or small group of atoms, as well as collective motions of larger groups of atoms in the TM core, e.g., those of individual helices and protein domains. By contrast, the loop regions experience a very different polar environment and are generally more mobile. Nevertheless, their mobility may be also affected by interactions with ligands, soluble proteins (Gilman 1987, Palczewski 2006, Shimada *et al* 2019, Smith 2010, Spudich *et al* 2000) and other extracellular materials which may serve as structurally stabilizing factors.

These complex dynamic motions are not readily accessible to solution NMR and other methods. SsNMR is uniquely suited for characterizing membrane protein dynamics site-specifically and on a wide range of timescales with the potential to contribute to greater understanding of both membrane protein energetics and function.

### 14.4.2 NMR methods for studying protein dynamics

Many NMR approaches have been developed for probing protein dynamics at atomic resolution. In particular, nuclear spin relaxation, which arises from the stochastic fluctuations of local magnetic fields, is exceptionally sensitive to motions and can be used to probe both timescales and amplitudes of internal protein dynamics.

NMR can be used to monitor several relaxation rates that report on motions on a wide range of timescales. The longitudinal (spin-lattice) relaxation rate $R_1$ is sensitive to fast picosecond to nanosecond motions, whereas the transverse (spin-spin) relaxation rate $R_2$ and spin-lattice relaxation rate in the rotating frame $R_{1\rho}$ are sensitive reporters on the nanosecond to microsecond timescale motions. In solution NMR, slower processes occurring on the microsecond to millisecond timescales, such as chemical exchange, can be probed by ZZ exchange, Carr–Purcell–Meiboom–Gill (CPMG) $R_2$ relaxation dispersion or $R_{1\rho}$ dispersion experiments (Baldwin and Kay 2009, Korzhnev and Kay 2008, Mittermaier and Kay 2006, Palmer *et al* 2001). Some of these methods can also be adapted for ssNMR applications. Slow dynamics in the solid state can also be investigated using CODEX (Centerband-Only Detection of EXchange) experiments (DeAzevedo *et al* 2000a, 2000b, Reichert *et al* 2003). In solution NMR, order parameters can be extracted from $R_1$ and $R_{1\rho}$ measurements as well as from heteronuclear NOE measurements to yield estimates for the amplitude of motions. In solids, the order parameters can be directly estimated from measurement of effective dipolar interactions.

Great progress has been made recently with the development of ssNMR techniques for relaxation measurements. In comparison to solution NMR, ssNMR relaxation can be used to probe internal motions over a wider range of timescales. For proteins in solution, the internal motions that occur on a similar timescale of the overall molecular tumbling cannot be readily extracted from the relaxation rates due to the inference with tumbling motions. This makes the nanosecond-microsecond motions hardly accessible to solution NMR measurements (Palmer 2004, Palmer and Massi 2006). In contrast, the lack of overall tumbling in solids unmasks the nanosecond-microsecond internal motions. In principle, ssNMR relaxation measurements can provide information on the full range of timescales.

The lack of tumbling in solids leads to substantial residual anisotropic interactions even under MAS. The major complication for both $R_1$ and $R_{1\rho}/R_2$ measurements arises from the coherent effects associated with the incomplete averaging of dipolar couplings between low gamma spins ($^{13}C$, $^{15}N$) and protons as well as between protons. In particular, these interactions lead to strong PDSD effects under moderate MAS frequencies. If PDSD occurs on a timescale which is comparable to or faster than the $T_1$ relaxation time, it results in averaging of the relaxation effects across several residues and hinders site-specific analysis. Furthermore, in $R_2/R_{1\rho}$ measurements, the coherent contributions from dipolar couplings dominate the relaxation pathways even under strong heteronuclear decoupling conditions, making the dynamics-related stochastic contributions difficult to detect.

Recent advances in ssNMR instrumentation and sample preparation techniques have partially alleviated these limitations. For site-specific $R_1$ measurements, the PDSD induced rate-averaging effects can be sufficiently suppressed by employing fast MAS, sometimes in combination with proton spin dilution. Due to the relatively weak $^{15}N$–$^{15}N$ couplings, spinning rates on the order of 20 kHz or faster are required for site-specific $^{15}N$ $R_1$ measurements in fully protonated samples (Giraud *et al* 2007, Krushelnitsky *et al* 2006). This modest experimental requirement makes $^{15}N$ $R_1$ measurement the most commonly used experiment for studying site-specific backbone dynamics (Chevelkov *et al* 2009, 2008, 2007, Giraud *et al* 2005, 2004, Zinkevich *et al* 2013), including in membrane embedded proteins such as potassium channel KcsA (Ader *et al* 2010), ASR (Good *et al* 2017) and outer membrane protein A of *Klebsiella pneumoniae* (KpOmpA) (Saurel *et al* 2017).

The stronger $^{13}C$–$^{13}C$ and $^{13}C$–$^{1}H$ couplings make experiments for enabling site-specific $^{13}C$ $R_1$ measurements more demanding. Lewandowski *et al* have shown that ultrafast MAS rates in excess of 60 kHz are required to effectively suppress PDSD for reliable site-specific $R_1$ measurements of carbonyl carbons in fully protonated proteins (Lewandowski *et al* 2010b). The suppression of PDSD solely by ultrafast MAS is less effective for the aliphatic and aromatic carbons due to their much stronger couplings to protons. Further sample modification, such as proton and carbon dilution, in combination with fast MAS (>50 kHz) are necessary for site-specific measurements of the $R_1$ rates of aliphatic carbons (Asami *et al* 2015).

Extensive deuteration combined with heteronuclear decoupling is a viable way to reduce the coherent contributions to $^{15}N$ $R_{1\rho}$ at spinning rates below 40 kHz (Krushelnitsky *et al* 2010, Ma *et al* 2014, Quinn and Mcdermott 2012, Zinkevich *et al* 2013), thereby allowing accurate quantification of motions on the intermediate timescale. An alternative approach to $^{15}N$ $R_{1\rho}$ measurements in fully protonated proteins was recently demonstrated by Lewandowski and Emsley who have shown that coherent contributions to $^{15}N$ $R_{1\rho}$ can be effectively suppressed under fast MAS (>50 kHz) and moderate lock-field strength above 10 kHz (Lewandowski *et al* 2011b). Under these conditions, the stochastic contributions due to nanosecond to microsecond motions begin to dominate the relaxation pathways. As discussed above, using fast MAS conditions for $^{15}N$ $R_{1\rho}$ measurements is advantageous in the context of applications to membrane proteins, as such measurements can be performed in fully protonated samples without the need for deuteration. In conjunction with $R_1$ and order parameter measurements, $^{15}$N $R_{1\rho}$ relaxation methods have been extensively used for quantifying the timescales and amplitudes of picosecond–microsecond motions, including in membrane proteins (Good *et al* 2017, 2014, Saurel *et al* 2017).

$^{15}$N $R_{1\rho}$ relaxation dispersion experiments ($R_{1\rho}$ rate measured as a function of the lock-field amplitude) have also been used in ssNMR for probing slow conformational exchange. The $R_{1\rho}$ dispersion due to the fluctuation of isotropic chemical shifts between exchange states can be observed at low lock-field amplitudes and is used in solution NMR to probe slow exchange kinetics on the microsecond–millisecond timescale. Under MAS conditions, the $R_{1\rho}$ dispersion profile contains two maxima related to the coherent effects due to the rotary resonance recoupling ($R^3$) (Oas *et al* 1988). Stochastic conformational fluctuations involving reorientation of anisotropic CSA and/or dipolar interactions on a microsecond–millisecond time scale result in additional broadening of the $R^3$ conditions (Ma *et al* 2014, Quinn and McDermott 2009), and can be used to quantify conformational dynamics. This method, termed NEar-Rotary-Resonance Relaxation Dispersion (NERRD) (Kurauskas *et al* 2017), was recently used to characterize the angular amplitude of backbone motions on a microsecond timescale in several microcrystalline proteins (Krushelnitsky *et al* 2018, Kurauskas *et al* 2017, Ma *et al* 2014, Marion *et al* 2019).

In ssNMR, the amplitudes of motions occurring on the submicrosecond timescale can be estimated from the partial averaging of anisotropic interactions. Motional amplitudes are quantified by the dipolar order parameters $S^2$ which are defined as the square of the ratios of the experimentally observed dipolar couplings, typically between directly bonded nuclei, and their rigid limits. The latter only depends on the inter-nuclear distance and can be calculated based on the known bond length. The site-specific measurement of the apparent dipolar coupling strengths is usually performed using Dipolar Chemical Shift (DIPSHIFT) correlation spectroscopy (Cobo *et al* 2012, Hong *et al* 1997), with dipolar couplings reintroduced using recoupling sequences such as Lee–Goldburg CP (Rossum *et al* 2000), transferred MREV (Hohwy *et al* 2000), REDOR (Schanda *et al* 2010), R-symmetry type sequences (Hou *et al* 2014, 2011) or other heteronuclear recoupling methods. The most commonly measured are the backbone dipolar order parameters, $S^2_{CH}$, $S^2_{NH}$ and $S^2_{NC}$, and sometimes amplitudes of side chain motions can be quantified as well (Chevelkov *et al* 2009, Good *et al* 2017, Helmus *et al* 2010, Lorieau and McDermott 2006, Lorieau *et al* 2008, Quinn *et al* 2018, Schmidt *et al* 2014).

By combing $R_1$, $R_{1\rho}$ relaxation measurements with dipolar order parameter measurements, the site-specific details of protein dynamics, such as timescales and amplitudes of motions, can be modelled. Simple model free and extended model free analyses (Bremi and Brüschweiler 1997, Brüschweiler and Wright 1994, Lipari and Szabo 1982a, 1982b) have been used in peptides, small microcrystalline proteins and amyloid fibrils (Chevelkov *et al* 2009, 2007, Knight *et al* 2012, Krushelnitsky *et al* 2018, Lakomek *et al* 2017, Lamley *et al* 2015, Lewandowski *et al* 2010a, Lienin *et al* 1998,

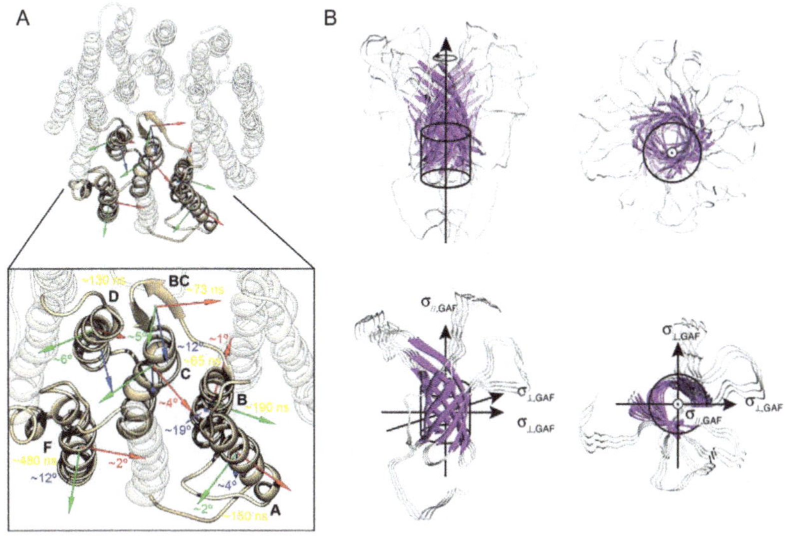


**Figure 14.7.** Internal membrane protein dynamics modelled from ssNMR relaxation and order parameter measurements. (A) Collective motions of transmembrane α-helices B, C, D, F and of the B-C loop in ASR. Timescales are indicated in yellow, rotational amplitudes of rotations and their axes are shown in the same colors. Reproduced with permission from Good *et al* (2017). (B) Collective dynamics of the β-barrel membrane protein KpOmpA. Motions are characterized as a superposition of the rotation around the bilayer normal, and the overall rocking motion of the entire barrel with the amplitudes $\sigma_{\|} = 6°$ and $\sigma_{\perp} = 10°$. Reproduced with permission from Saurel *et al* (2017). Copyright 2017, American Chemical Society.

Mollica *et al* 2012, Schanda *et al* 2010, Smith *et al* 2016, Zinkevich *et al* 2013). These relaxation-based approaches can also provide detailed insights into membrane protein dynamics, with rich information content comparable to that available from solution NMR relaxation measurements. However, such measurements require high signal-to-noise and are time consuming. To the best of our knowledge, there are only a few relaxation studies of membrane proteins published to date, which used fast MAS methods, specifically those of an α-helical ASR (Good *et al* 2017, 2014) and of a β-barrel protein (KpOmpA) (Saurel *et al* 2017). Two types of small amplitude motions were found in ASR. In addition to local picosecond motions commonly present in proteins, there are also slower collective 'rocking' motions of the TM helices on the time scales of tens to hundreds of nanoseconds, as well as larger collective motions of the cytoplasmic ends of helices and the inter-helical loops (figure 14.7(A)) (Good *et al* 2017). For KpOmpA, relaxation data suggest that along with rotational diffusion, there is a small amplitude collective rocking motion of the entire β-barrel on a timescale of hundreds of nanoseconds and slower local motions for residues close to the loops (figure 14.7(B)) (Saurel *et al* 2017).

### 14.4.3 Opportunity for observing mobile regions in membrane proteins

TM domains of membrane proteins are constrained within the lipid bilayer. Insufficient motional averaging of anisotropic interactions in rigid TM regions allows for through-space dipolar based recoupling methods to be used for signal excitation and polarization transfers. By contrast, extra-membranous loops and solvent exposed terminal regions often experience local molecular motions of large amplitude. Such motions may lead to greatly reduced dipolar interactions and longer transverse coherence lifetimes, thereby resulting in narrow linewidths (Andronesi *et al* 2005, de Vlugt *et al* 2020, Etzkorn *et al* 2007, Schrottke *et al* 2017, Siemer *et al* 2006, Ward *et al* 2015a, Yang *et al* 2011, Zhong *et al* 2007). The mobile regions of membrane proteins are often 'invisible' to the through-space ssNMR techniques due to the motional averaging of dipolar interactions, but can be selectively excited using the through-bond INEPT (Insensitive Nuclei Enhanced by Polarization Transfer) (Burum and Ernst 1980) method. The INEPT method can be further combined with the through-bond polarization transfer schemes such as TOBSY (ThrOugh-Bond correlation SpectroscopY) (Baldus and Meier 1996) for multidimensional spectroscopy. These approaches have been used for probing mobile regions in fully protonated, uniformly $^{13}$C-labeled membrane protein samples, and yielded excellent resolution in the proton dimension (ca. 0.1–0.2 ppm) even at moderate MAS frequencies and without the need for deuteration (Andronesi *et al* 2005, de Vlugt *et al* 2020, Ward *et al* 2015a, Yang *et al* 2011, Zhong *et al* 2007). Nevertheless, heteronuclear $^{1}$H/$^{13}$C/$^{15}$N spectroscopy is problematic because localized motions are insufficient to yield long $T_2$ lifetimes comparable to those observed in globular proteins in solution. Accordingly, the relaxation-related signal losses result in poor polarization transfer efficiency, particularly for the $^{15}$N–$^{13}$C through-bond polarization transfers which exploit weak $^{15}$N–$^{13}$C J-couplings. Such experiments would benefit from the implementation of proton detection as demonstrated for the non-annular lipids bound to a seven-helical membrane protein ASR where sensitivity improvement of $\sim$10 was observed in comparison to carbon detection (Ward *et al* 2015a), the conformationally excited state of the cytoplasmic domain of phospholamban (PLN) (Gopinath *et al* 2017), and the intrinsically disordered domains of amyloid fibrils (Falk and Siemer 2016).

## 14.5 Conclusions and perspective

In this chapter, we have discussed briefly the existing and emerging MAS ssNMR methodologies and illustrated their application to membrane proteins. Understanding and exploration of fundamental NMR interactions has led to the development of many robust pulse sequences that now form the basis for structural analysis of macromolecular structure. The intrinsically low sensitivity of ssNMR measurements remains the key limitation. Towards resolving this challenge, novel methods such as the development of high field magnets and cryogenically cooled MAS ssNMR probes (Hassan *et al* 2020), proton detection methods and DNP techniques, will continue to be at the forefront of future developments.

## Acknowledgements

This research was supported by the Natural Sciences and Engineering Research Council of Canada, RGPIN-2020-04489 to VL.

## References

Abdine A, Verhoeven M A, Park K H, Ghazi A, Guittet E, Berrier C, Van Heijenoort C and Warschawski D E 2010 Structural study of the membrane protein MscL using cell-free expression and solid-state NMR *J. Magn. Reson.* **204** 155–9

Ader C, Pongs O, Becker S and Baldus M 2010 Protein dynamics detected in a membrane-embedded potassium channel using two-dimensional solid-state NMR spectroscopy *BBA—Biomembr.* **1798** 286–90

Akbey Ü, Lange S, Franks W T, Linser R, Rehbein K, Diehl A, Rossum B-J, van, Reif B and Oschkinat H 2010 Optimum levels of exchangeable protons in perdeuterated proteins for proton detection in MAS solid-state NMR spectroscopy *J. Biomol. NMR* **46** 67–73

Alenghat F J and Golan D E 2013 Membrane protein dynamics and functional implications in mammalian cells *Curr. Top. Membr.* **72** 89–120

Andreas L B *et al* 2016 Structure of fully protonated proteins by proton-detected magic-angle spinning NMR *Proc. Natl Acad. Sci. USA* **113** 9187–92

Andreas L B, Reese M, Eddy M T, Gelev V, Ni Q Z, Miller E A, Emsley L, Pintacuda G, Chou J J and Griffin R G 2015 Structure and mechanism of the influenza A M2 18–60 dimer of dimers *J. Am. Chem. Soc.* **137** 14877–86

Andrew E R, Bradbury A and Eades R G 1958 Nuclear magnetic resonance spectra from a crystal rotated at high speed *Nature* **182** 1659

Andronesi O C, Becker S, Seidel K, Heise H, Young H S and Baldus M 2005 Determination of membrane protein structure and dynamics by magic-angle-spinning solid-state NMR spectroscopy *J. Am. Chem. Soc.* **127** 574–81

Asami S, Porter J R, Lange O F and Reif B 2015 Access to Cα backbone dynamics of biological solids by $^{13}$C $T_1$ relaxation and molecular dynamics simulation *J. Am. Chem. Soc.* **137** 1094–100

Asami S, Schmieder P, Reif B and Mu H 2010 High resolution $^{1}$H-detected solid-state NMR spectroscopy of protein aliphatic resonances: access to tertiary structure information *J. Am. Chem. Soc.* **132** 15133–5

Asami S, Szekely K, Schanda P, Meier B H and Reif B 2012 Optimal degree of protonation for $^{1}$H detection of aliphatic sites in randomly deuterated proteins as a function of the MAS frequency *J. Biomol. NMR* **54** 155–68

Baldus M and Meier B H 1996 Total correlation spectroscopy in the solid state. The use of scalar couplings to determine the through-bond connectivity *J. Magn. Reson. Ser.* A **121** 65–9

Baldus M, Petkova A T, Herzfeld J and Griffin R G 1998 Cross polarization in the tilted frame: assignment and spectral simplification in heteronuclear spin systems *Mol. Phys.* **95** 8976

Baldwin A J and Kay L E 2009 NMR spectroscopy brings invisible protein states into focus *Nat. Chem. Biol.* **5** 808–14

Barbet-Massin E *et al* 2014 Rapid proton-detected NMR assignment for proteins with fast magic angle spinning *J. Am. Chem. Soc.* **136** 12489–97

Bayburt T H, Grinkova Y V and Sligar S G 2002 Self-assembly of discoidal phospholipid bilayer nanoparticles with membrane scaffold proteins *Nano Lett.* **2** 853–6

Bayburt T H and Sligar S G 2003 Self-assembly of single integral membrane proteins into soluble nanoscale phospholipid bilayers *Protein Sci.* **12** 2476–81

Bayro M J, Ramachandran R, Caporini M A, Eddy M T, Bayro, Marvin J, Ramachandran R, Caporini M A, Eddy M T and Griffin R G 2008 Radio frequency-driven recoupling at high magic-angle spinning frequencies: homonuclear recoupling sans heteronuclear decoupling *J. Chem. Phys.* **128** 052321

Bayro M J, Huber M, Ramachandran R, Davenport T C, Meier B H, Griffin R G, Meier B H, Ernst M and Griffin R G 2009a Dipolar truncation in magic-angle spinning NMR recoupling experiments *J. Chem. Phys.* **130** 114506

Bayro M J, Maly T, Birkett N R, Dobson C M and Griffin R G 2009b Long-range correlations between aliphatic $^{13}$C nuclei in protein MAS NMR spectroscopy *Angew. Chem. Int. Ed.* **48** 5708–10

Bennett A E, Griffin R G, Ok J H and Vega S 1992 Chemical shift correlation spectroscopy in rotating solids: radio frequency-driven dipolar recoupling and longitudinal exchange *J. Chem. Phys.* **96** 8624

Bennett A E, Grillin R G and Vega S 1994 Recoupling of homo- and heteronuclear dipolar interactions in rotating solids *Solid-State NMR IV Methods and Applications of Solid-State NMR* (Berlin: Springer) pp 1–77

Bennett A E, Rienstra C M, Griffiths J M, Zhen W, Lansbury P T Jr and Griffin R G 1998 Homonuclear radio frequency-driven recoupling in rotating solids *J. Chem. Phys.* **108** 9463

Bertini I, Emsley L, Felli I C, Laage S, Lesage A, Lewandowski J R, Marchetti A, Pierattelli R and Pintacuda G 2011 High-resolution and sensitivity through-bond correlations in ultra-fast magic angle spinning (MAS) solid-state NMR *Chem. Sci.* **2** 345–8

Bertini I, Luchinat C and Piccioli M 2001 Paramagnetic probes in metalloproteins *Methods Enzym.* **339** 314–40

Bloembergen N 1949 On the interaction of nuclear spins in a crystalline lattice *Physica* **15** 386–426

Bolton D, Brown L S and Ladizhansky V 2018 Partial solid-state NMR $^{1}$H, $^{13}$C, $^{15}$N resonance assignments of a perdeuterated back-exchanged seven-transmembrane helical protein *Anabaena* sensory rhodopsin *Biomol. NMR Assign.* **12** 237–42

Bremi T and Brüschweiler R 1997 Locally anisotropic internal polypeptide backbone dynamics by NMR relaxation *J. Am. Chem. Soc.* **7863** 6672–3

Brown L S and Ladizhansky V 2015 Membrane proteins in their native habitat as seen by solid-state NMR spectroscopy *Protein Sci.* **24** 1333–46

Brüschweiler R and Wright P E 1994 NMR order parameters of biomolecules: a new analytical representation and application to the gaussian axial fluctuation model *J. Am. Chem. Soc.* **116** 8426–7

Burum D P and Ernst R R 1980 Net polarization transfer via a J-ordered state for signal enhancement of low-sensitivity nuclei *J. Magn. Reson.* **39** 163–8

Byrne B 2015 Pichia pastoris as an expression host for membrane protein structural biology *Curr. Opin. Struct. Biol.* **32** 9–17

Cady S D, Schmidt-Rohr K, Wang J, Soto C S, Degrado W F and Hong M 2010 Structure of the amantadine binding site of influenza M2 proton channels in lipid bilayers *Nature* **463** 689–92

Carravetta M, Eden M, Zhao X, Brinkmann A and Levitt M H 2000 Symmetry principles for the design of radio frequency pulse sequences in the nuclear magnetic resonance of rotating solids *Chem. Phys. Lett.* **321** 205–15

Castellani F, Rossum B, Van, Diehl A, Schubert M, Rehbein K and Oschkinat H 2002 Structure of a protein determined by solid-state magic-angle-spinning NMR spectroscopy *Nature* **420** 23–6

Chen L, Kaiser J M, Lai J, Polenova T, Yang J, Rienstra C M and Mueller L J 2007a J-based 2D homonuclear and heteronuclear correlation in solid-state proteins *Magn. Reson. Chem.* **45** S84–92

Chen L, Kaiser J M, Polenova T, Yang J, Rienstra C M and Mueller L J 2007b Backbone assignments in solid-state proteins using J-based 3D heteronuclear correlation spectroscopy *J. Am. Chem. Soc.* **129** 10650–1

Chevelkov V, Diehl A and Reif B 2008 Measurement of $^{15}$N-$T_1$ relaxation rates in a perdeuterated protein by magic angle spinning solid-state nuclear magnetic resonance spectroscopy *J. Chem. Phys.* **128** 052316

Chevelkov V, Fink U and Reif B 2009 Accurate determination of order parameters from $^{1}$H, $^{15}$N dipolar couplings in MAS solid-state NMR experiments *J. Am. Chem. Soc.* **131** 14018–22

Chevelkov V, Rehbein K, Diehl A and Reif B 2006 Ultrahigh resolution in proton solid-state NMR spectroscopy at high levels of deuteration *Angew. Chem. Int. Ed.* **45** 3878–81

Chevelkov V, van Rossum B J, Castellani F, Rehbein K, Diehl A, Hohwy M, Steuernagel S, Engelke F, Oschkinat H and Reif B 2003 $^{1}$H detection in MAS solid-state NMR spectroscopy of biomacromolecules employing pulsed field gradients for residual solvent suppression *J. Am. Chem. Soc.* **125** 7788–9

Chevelkov V, Zhuravleva A V, Xue Y, Reif B and Skrynnikov N R 2007 Combined analysis of $^{15}$N relaxation data from solid- and solution-state NMR spectroscopy *J. Am. Chem. Soc.* **129** 12594–5

Clark L, Dikiy I, Rosenbaum D M and Gardner K H 2018 On the use of *Pichia pastoris* for isotopic labeling of human GPCRs for NMR studies *J. Biomol. NMR* **71** 203–11

Clark L, Zahm J A, Ali R, Kukula M, Bian L, Patrie S M, Gardner K H, Rosen M K and Rosenbaum D M 2015 Methyl labeling and TROSY NMR spectroscopy of proteins expressed in the eukaryote Pichia pastoris *J. Biomol. NMR* **62** 239–45

Clore G M and Iwahara J 2009 Theory, practice, and applications of paramagnetic relaxation enhancement for the characterization of transient low-population states of biological macromolecules and their complexes *Chem. Rev.* **109** 4108–39

Cobo M F, Achilles A, Reichert D, DeAzevedo E R and Saalwächter K 2012 Recoupled separated-local-field experiments and applications to study intermediate-regime molecular motions *J. Magn. Reson. J.* **221** 85–96

Cornilescu G, Delaglio F and Bax A 1999 Protein backbone angle restraints from searching a database for chemical shift and sequence homology *J. Biomol. NMR* **13** 289–302

Costa P R, Gross J D, Hong M and Griffin R G 1997a Solid-state NMR measurement of Psi in peptides: a NCCN 2Q-heteronuclear local field experiment *Chem. Phys. Lett.* **280** 95–103

Costa P R, Sun B Q and Griffin R G 1997b Rotational resonance tickling: accurate internuclear distance measurement in solids *J. Am. Chem. Soc.* **119** 10821–30

Costa P R, Sun B and Griffin R G 2003 Rotational resonance NMR: separation of dipolar coupling and zero quantum relaxation *J. Magn. Reson.* **164** 92–103

Courtney J M and Rienstra C M 2016 Efficient dipolar double quantum filtering under magic angle spinning without a $^{1}$H decoupling field *J. Magn. Reson.* **269** 152–6

Creuzet F, McDermott A, Gebhard R, van der Hoef K, Spijker-Assink M B, Herzfeld J, Lugtenburg J, Levitt M H and Griffin R G 1991 Determination of membrane protein structure by rotational resonance NMR: bacteriorhodopsin *Science* **251** 783–6

Das N, Dai J, Hung I, Rajagopalan M, Zhou H and Cross T A 2015 Structure of CrgA, a cell division structural and regulatory protein from *Mycobacterium tuberculosis*, in lipid bilayers *Proc. Natl Acad. Sci. USA* **112** E119–26

De Paëpe G 2012 Dipolar recoupling in magic angle spinning solid-state nuclear magnetic resonance *Annu. Rev. Phys. Chem.* **63** 661–84

De Paëpe G, Lewandowski J R, Loquet A, Böckmann A and Griffin R G 2008 Proton assisted recoupling and protein structure determination *J. Chem. Phys.* **129** 245101

DeAzevedo E R, Hu W-G, Bonagamba T J and Schmidt-Rohr K 2000a Principles of centerband-only detection of exchange in solid-state nuclear magnetic resonance, and extension to four-time centerband-only detection of exchange *J. Chem. Phys.* **112** 8988

DeAzevedo E R, Kennedy S B and Hong M 2000b Determination of slow motions in extensively isotopically labeled proteins by magic-angle-spinning $^{13}$C-detected $^{15}$N exchange NMR *Chem. Phys. Lett.* **321** 43–8

de Vlugt J E, Xiao P, Munro R, Charchoglyan A, Brewer D, Al-Abdul-Wahid M S, Brown L S and Ladizhansky V 2020 Identifying lipids tightly bound to an integral membrane protein *BBA—Biomembr.* **1862** 183345

Ding Y, Fujimoto L M, Yao Y and Marassi F M 2015 Solid-state NMR of the *Yersinia pestis* outer membrane protein Ail in lipid bilayer nanodiscs sedimented by ultracentrifugation *J. Biomol. NMR* **61** 275–86

Dingwell D A, Brown L S and Ladizhansky V 2019 Structure of the functionally important extracellular loop C of human aquaporin 1 obtained by solid-state NMR under nearly physiological conditions *J. Phys. Chem.* B **123** 7700–10

Dürr U H N, Gildenberg M and Ramamoorthy A 2012 The magic of bicelles lights up membrane protein structure *Chem. Rev.* **112** 6054–74

Dürr U H N, Soong R and Ramamoorthy A 2013 When detergent meets bilayer: birth and coming of age of lipid bicelles *Prog. Nucl. Magn. Reson. Spectrosc.* **69** 1–22

Dusold S and Sebald A 2000 Dipolar recoupling under magic-angle spinning conditions *Annu. Rep. Nmr Spectrosc.* **41** 185–264

Dutta A, Saxena K, Schwalbe H and Klein-Seetharaman J 2012 Isotope labeling in mammalian cells *Methods Mol. Biol.* **831** 55–69

Eddy M T, Su Y, Silvers R, Andreas L, Clark L, Wagner G, Pintacuda G, Emsley L and Griffin R G 2015 Lipid bilayer-bound conformation of an integral membrane beta barrel protein by multidimensional MAS NMR *J. Biomol. NMR* **61** 299–310

Egorova-Zachernyuk T A, Bosman G J and Degrip W J 2011 Uniform stable-isotope labeling in mammalian cells: formulation of a cost-effective culture medium *Appl. Microbiol. Biotechnol.* **89** 397–406

Elkins M R, Williams J K, Gelenter M D, Dai P, Kwon B, Sergeyev I V, Pentelute B L and Hong M 2017 Cholesterol-binding site of the influenza M2 protein in lipid bilayers from solid-state NMR *Proc. Natl Acad. Sci. USA* **114** 12946–51

Emami S, Fan Y, Munro R, Ladizhansky V and Brown L S 2013 Yeast-expressed human membrane protein aquaporin-1 yields excellent resolution of solid-state MAS NMR spectra *J. Biomol. NMR* **55** 147–55

Ernst M, Detken A, Böckmann A and Meier B H 2003 NMR spectra of a microcrystalline protein at 30 kHz MAS *J. Am. Chem. Soc.* **125** 15807–10

Ernst M, Meier M A, Tuherm T, Samoson A and Meier B H 2004 Low-power high-resolution solid-state NMR of peptides and proteins *J. Am. Chem. Soc.* **126** 4764–5

Etzkorn M, Martell S, Andronesi O C, Seidel K, Engelhard M and Baldus M 2007 Secondary structure, dynamics, and topology of a seven-helix receptor in native membranes, studied by solid-state NMR spectroscopy *Angew. Chem. Int. Ed.* **46** 459–62

Etzkorn M, Seidel K, Li L, Martell S, Geyer M, Engelhard M and Baldus M 2010 Complex formation and light activation in membrane-embedded sensory rhodopsin II as seen by solid-state NMR spectroscopy *Structure* **18** 293–300

Everberg H, Clough J, Henderson P, Jergil B, Tjerneld F and Ramirez I B 2006 Isolation of Escherichia coli inner membranes by metal affinity two-phase partitioning *J. Chromatogr.* A **1118** 244–52

Falk A S and Siemer A B 2016 Dynamic domains of amyloid fibrils can be site-specifically assigned with proton detected 3D NMR spectroscopy *J. Biomol. NMR* **66** 159–62

Fan Y, Emami S, Munro R, Ladizhansky V and Brown L S 2015 Isotope labeling of eukaryotic membrane proteins in yeast for solid-state NMR *Methods Enzymol.* **565** 193–212

Fan Y, Shi L, Ladizhansky V, Brown L S and Page S D S 2011 Uniform isotope labeling of a eukaryotic seven-transmembrane helical protein in yeast enables high-resolution solid-state NMR studies in the lipid environment *J. Biomol. NMR* **49** 151–61

Focke P J, Hein C, Ho B, Matulef K, Bernhard F, Do V and Valiyaveetil F I 2016 Combining *in vitro* folding with cell free protein synthesis for membrane protein expression *Biochemistry* **55** 4212–9

Franke B, Opitz C, Isogai S, Grahl A, Delgado L, Gossert A D and Grzesiek S 2018 Production of isotope-labeled proteins in insect cells for NMR *J. Biomol. NMR* **71** 173–84

Franks W T, Kloepper K D, Wylie B J and Rienstra C M 2007 Four-dimensional heteronuclear correlation experiments for chemical shift assignment of solid proteins *J. Biomol. NMR* **39** 107–31

Franks W T, Wylie B J, Schmidt H L, Nieuwkoop A J, Mayrhofer R M, Shah G J, Graesser D T and Rienstra C M 2008 Dipole tensor-based atomic-resolution structure determination of a nanocrystalline protein by solid-state NMR *Proc. Natl Acad. Sci. USA* **105** 4621–6

Franks W T, Zhou D H, Wylie B J, Money B G, Graesser D T, Frericks H L, Sahota G and Rienstra C M 2005 Magic-angle spinning solid-state NMR spectroscopy of the $\beta_1$ immunoglobulin binding domain of protein G (GB1): $^{15}$N and $^{13}$C chemical shift assignments and conformational analysis *J. Am. Chem. Soc.* **127** 12291–305

Fricke P, Chevelkov V, Zinke M, Giller K, Becker S and Lange A 2017 Backbone assignment of perdeuterated proteins by solid-state NMR using proton detection and ultrafast magic-angle spinning *Nat. Protoc.* **12** 764–82

Fuccio C, Luchinat E, Barbieri L, Neri S and Fragai M 2016 Algal autolysate medium to label proteins for NMR in mammalian cells *J. Biomol. NMR* **64** 275–80

Ghosh M and Rienstra C M 2017 $^1$H-detected REDOR with fast magic-angle spinning of a deuterated protein *J. Phys. Chem.* B **121** 8503–11

Gilman A G 1987 G proteins: transducers of receptor-generated signals *Annu. Rev. Biochem.* **56** 615–49

Giraud N, Blackledge M, Bockmann A and Emsley L 2007 The influence of nitrogen-15 proton-driven spin diffusion on the measurement of nitrogen-15 longitudinal relaxation times *J. Magn. Reson.* **184** 51–61

Giraud N, Blackledge M, Goldman M, Bockmann A, Lesage A, Penin F and Emsley L 2005 Quantitative analysis of backbone dynamics in a crystalline protein from nitrogen-15 spin-lattice relaxation *J. Am. Chem. Soc.* **127** 18190–201

Giraud N, Bockmann A, Lesage A, Penin F, Blackledge M and Emsley L 2004 Site-specific backbone dynamics from a crystalline protein by solid-state NMR spectroscopy *J. Am. Chem. Soc.* **126** 11422–3

Goobes G, Boender G J and Vega S 2000 Spinning-frequency-dependent narrowband RF-driven dipolar recoupling *J. Magn. Reson.* **146** 204–19

Goobes G and Vega S 2002 Improved narrowband dipolar recoupling for homonuclear distance measurements in rotating solids *J. Magn. Reson.* **154** 236–51

Good D, Pham C, Jagas J, Lewandowski J R and Ladizhansky V 2017 Solid-State NMR provides evidence for small-amplitude slow domain motions in a multispanning transmembrane α-helical protein *J. Am. Chem. Soc.* **139** 9246–58

Good D B, Wang S, Ward M E, Struppe J, Brown L S, Lewandowski J R and Ladizhansky V 2014 Conformational dynamics of a seven transmembrane helical protein *Anabaena* Sensory Rhodopsin probed by solid-state NMR *J. Am. Chem. Soc.* **136** 2833–42

Gopinath T, Nelson S E D, Soller K J and Veglia G 2017 Probing the conformationally excited states of membrane proteins via $^{1}$H-detected MAS solid-state NMR spectroscopy *J. Phys. Chem.* B **121** 4456–65

Gregory D M, Mitchell D J, Stringer J A, Kiihne S, Shiels J C, Callahan J, Mehta M A and Drobny G P 1995 Windowless dipolar recoupling—the detection of weak dipolar couplings between spin-1/2 nuclei with large chemical-shift anisotropies *Chem. Phys. Lett.* **246** 654–63

Grommek A, Meier B H and Ernst M 2006 Distance information from proton-driven spin diffusion under MAS *Chem. Phys. Lett.* **427** 404–9

Gullion T and Schaefer J 1989 Rotational-echo double-resonance NMR *J. Magn. Reson.* **81** 196–200

Hagn F, Nasr M L and Wagner G 2018 Assembly of phospholipid nanodiscs of controlled size for structural studies of membrane proteins by NMR *Nat. Protoc.* **13** 79–98

Hall D A, Maus D C, Gerfen G J, Inati S J, Becerra L R, Dahlquist F W and Griffin R G 1997 Polarization-enhanced NMR spectroscopy of biomolecules in frozen solution *Science* **276** 930–2

Harris N J and Booth P J 2012 Folding and stability of membrane transport proteins *in vitro Biochim. Biophys. Acta* **1818** 1055–66

Hartmann L, Kugler V and Wagner R 2016 Expression of eukaryotic membrane proteins in *Pichia pastoris Methods Mol. Biol.* **1432** 143–62

Hartmann S R and Hahn E L 1962 Nuclear double resonance in the rotating frame *Phys. Rev.* **128** 2042

Hassan A *et al* 2020 Sensitivity boosts by the CPMAS CryoProbe for challenging biological assemblies *J. Magn. Reson.* **311** 106680

Hediger S, Meier B H and Ernst R R 1995 Adiabatic passage Hartmann-Hahn cross-polarization in NMR under magic angle sample spinning *Chem. Phys. Lett.* **240** 449–56

Hediger S, Meier B H, Kurur N D, Bodenhausen G and Ernst R R 1994 NMR cross polarization by adiabatic passage through the *Chem. Phys. Lett.* **223** 283–8

Helmus J J, Nadaud P S, Höfer N and Jaroniec C P 2008 Determination of methyl $^{13}$C–$^{15}$N dipolar couplings in peptides and proteins by three-dimensional and four-dimensional magic-angle spinning solid-state NMR spectroscopy *J. Chem. Phys.* **128** 052314

Helmus J J, Surewicz K, Apostol M I, Surewicz W K and Jaroniec C P 2011 Intermolecular alignment in Y145Stop human prion protein amyloid fibrils probed by solid-state NMR spectroscopy *J. Am. Chem. Soc.* **133** 13934–7

Helmus J J, Surewicz K, Surewicz W K and Jaroniec C P 2010 Conformational flexibility of Y145Stop human prion protein amyloid fibrils probed by solid-state nuclear magnetic resonance spectroscopy *J. Am. Chem. Soc.* **132** 2393–403

Higman V A 2018 Solid-state MAS NMR resonance assignment methods for proteins *Prog. Nucl. Magn. Reson. Spectrosc.* **106–107** 37–65

Hiller M, Higman V A, Jehle S, van Rossum B J, Kuhlbrandt W and Oschkinat H 2008 [2,3-$^{13}$C]-labeling of aromatic residues—getting a head start in the magic-angle-spinning NMR assignment of membrane proteins *J. Am. Chem. Soc.* **130** 408–9

Hing A W, Vega S and Schaefer J 1992 Transferred-echo double-resonance NMR *J. Magn. Reson.* **96** 205–9

Hing A W, Vega S and Schaefer J 1993 Measurement of heteronuclear dipolar coupling by transferred-echo double-resonance NMR *J. Magn. Reson. Ser.* A **103** 151–62

Hodgkinson P and Emsley L 1999 The accuracy of distance measurements in solid-state NMR *J. Magn. Reson.* **139** 46–59

Hohwy M, Jakobsen H J, Edén M, Levitt M H and Nielsen N C 1998 Broadband dipolar recoupling in the nuclear magnetic resonance of rotating solids: A compensated C7 pulse sequence *J. Chem. Phys.* **108** 2686

Hohwy M, Jaroniec C P, Reif B, Rienstra C M and Griffin R G 2000 Local structure and relaxation in solid-state NMR: Accurate measurement of amide N–H bond lengths and H–N–H bond angles *J. Am. Chem. Soc.* **122** 3218–9

Hohwy M, Rienstra C M, Jaroniec C P and Griffin R G 1999 Fivefold symmetric homonuclear dipolar recoupling in rotating solids: Application to double quantum spectroscopy *J. Chem. Phys.* **110** 7983

Hong M, Gross J D, Rienstra C M, Griffin R G, Kumashiro K K and Schmidt-Rohr K 1997 Coupling amplification in 2D MAS NMR and its application to torsion angle determination in peptides *J. Magn. Reson.* **129** 85–92

Hong M and Jakes K 1999 Selective and extensive $^{13}$C labeling of a membrane protein for solid-state NMR investigations *J. Biomol. NMR* **14** 71–4

Hou G, Byeon I L, Ahn J, Gronenborn A M and Polenova T 2011 $^{1}$H–$^{13}$C/$^{1}$H–$^{15}$N heteronuclear dipolar recoupling by R-symmetry sequences under fast magic angle spinning for dynamics analysis of biological and organic solids *J. Am. Chem. Soc.* **133** 18646–55

Hou G, Lu X, Vega A J and Polenova T 2014 Accurate measurement of heteronuclear dipolar couplings by phase-alternating R-symmetry (PARS) sequences in magic angle spinning NMR spectroscopy *J. Chem. Phys.* **141** 104202

Hou G, Yan S, Trébosc J, Amoureux J-P and Polenova T 2013 Broadband homonuclear correlation spectroscopy driven by combined $R2_n^{\nu}$ sequences under fast magic angle spinning for NMR structural analysis of organic and biological solids *J. Magn. Reson.* **232** 18–30

Hu B, Lafon O, Trébosc J, Chen Q and Amoureux J 2011 Broad-band homo-nuclear correlations assisted by $^{1}$H irradiation for bio-molecules in very high magnetic field at fast and ultra-fast MAS frequencies *J. Magn. Reson.* **212** 320–9

Hu F, Luo W and Hong M 2010 Mechanisms of proton conduction and gating in influenza M2 proton channels from solid-state NMR *Science* **330** 505–9

Huang K Y, Siemer A B and McDermott A E 2011 Homonuclear mixing sequences for perdeuterated proteins *J. Magn. Reson.* **208** 122–7

Iwasaki T, Fukazawa R, Miyajima-Nakano Y, Baldansuren A, Matsushita S, Lin M T, Gennis R B, Hasegawa K, Kumasaka T and Dikanov S A 2012 Dissection of hydrogen bond interaction network around an iron–sulfur cluster by site-specific isotope labeling of hyper-thermophilic archaeal rieske-type ferredoxin *J. Am. Chem. Soc.* **134** 19731–8

Jacso T, Franks W T, Rose H, Fink U, Broecker J, Keller S, Oschkinat H and Reif B 2012 Characterization of membrane proteins in isolated native cellular membranes by dynamic nuclear polarization solid-state NMR spectroscopy without purification and reconstitution *Angew. Chem. Int. Ed.* **51** 432–5

Jain M G *et al* 2017 Selective $^{1}H$–$^{1}H$ distance restraints in fully protonated proteins by very fast magic-angle spinning solid-state NMR *J. Phys. Chem. Lett.* **8** 2399–405

Janik R, Peng X and Ladizhansky V 2007 $^{13}C$–$^{13}C$ distance measurements in U-$^{13}C$,$^{15}N$-labeled peptides using rotational resonance width experiment with a homogeneously broadened matching condition *J. Magn. Reson.* **188** 129–40

Jaroniec C P 2012 Solid-state nuclear magnetic resonance structural studies of proteins using paramagnetic probes *Solid State Nucl. Magn. Reson.* **43–44** 1–13

Jaroniec C P 2015 Structural studies of proteins by paramagnetic solid-state NMR spectroscopy *J. Magn. Reson.* **253** 50–9

Jaroniec C P, Filip C and Griffin R G 2002a 3D TEDOR NMR experiments for the simultaneous measurement of multiple carbon-nitrogen distances in uniformly $^{13}C$,$^{15}N$-labeled solids *J. Am. Chem. Soc.* **124** 10728–42

Jaroniec C P, MacPhee C E, Astrof N S, Dobson C M and Griffin R G 2002b Molecular conformation of a peptide fragment of transthyretin in an amyloid fibril *Proc. Natl Acad. Sci. USA* **99** 16748–53

Jaroniec C P, MacPhee C E, Bajaj V S, McMahon M T, Dobson C M and Griffin R G 2004 High-resolution molecular structure of a peptide in an amyloid fibril determined by magic angle spinning NMR spectroscopy *Proc. Natl Acad. Sci. USA* **101** 711–6

Jaroniec C P, Tounge B A, Herzfeld J and Griffin R G 2001 Frequency selective heteronuclear dipolar recoupling in rotating solids: accurate $^{13}C$–$^{15}N$ distance measurements in uniformly $^{13}C$,$^{15}N$-labeled peptides *J. Am. Chem. Soc.* **123** 3507–19

Jarymowycz V A and Stone M J 2006 Fast time scale dynamics of protein backbones: NMR relaxation methods, applications, and functional consequences *Chem. Rev.* **106** 1624–71

Kang S J, Todokoro Y, Yumen I, Shen B, Iwasaki I, Suzuki T, Miyagi A, Yoshida M, Fujiwara T and Akutsu H 2014 Active-site structure of the thermophilic Foc-subunit ring in membranes elucidated by solid-state NMR *Biophys. J.* **106** 390–8

Kaplan M *et al* 2015 Probing a cell-embedded megadalton protein complex by DNP-supported solid-state NMR *Nat. Methods* **12** 649–52

Kaplan M, Pinto C, Houben K and Baldus M 2016 Nuclear magnetic resonance (NMR) applied to membrane-protein complexes *Q. Rev. Biophys.* **49** e15

Kijac A Z, Li Y, Sligar S G and Rienstra C M 2007 Magic-angle spinning solid-state NMR spectroscopy of nanodisc-embedded human CYP3A4 *Biochemistry* **46** 13696–703

Klammt C *et al* 2012 Facile backbone structure determination of human membrane proteins by NMR spectroscopy *Nat. Methods* **9** 834–9

Knight M J *et al* 2011 Fast resonance assignment and fold determination of human superoxide dismutase by high-resolution proton-detected solid-state MAS NMR spectroscopy *Angew. Chem. Int. Ed.* **50** 11697–701

Knight M J, Pell A J, Bertini I, Felli I C, Gonnelli L, Pierattelli R, Herrmann T, Emsley L and Pintacuda G 2012 Structure and backbone dynamics of a microcrystalline metalloprotein by solid-state NMR *Proc. Natl Acad. Sci. USA* **109** 11095–100

Kofuku Y *et al* 2018 Deuteration and selective labeling of alanine methyl groups of $\beta_2$—adrenergic receptor expressed in a baculovirus-insect cell expression system *J. Biomol. NMR* **71** 185–92

Korzhnev D M and Kay L E 2008 Probing invisible, low-populated states of protein molecules by relaxation dispersion NMR spectroscopy: an application to protein folding *Acc. Chem. Res.* **41** 442–51

Krushelnitsky A, Brauniger T and Reichert D 2006 N-15 spin diffusion rate in solid-state NMR of totally enriched proteins: the magic angle spinning frequency effect *J. Magn. Reson.* **182** 339–42

Krushelnitsky A, Gauto D, Rodriguez D C, Schanda P and Saalwächter K 2018 Microsecond motions probed by near-rotary-resonance $R_{1\rho}$ $^{15}$N MAS NMR experiments: the model case of protein overall-rocking in crystals *J. Biomol. NMR* **71** 53–67

Krushelnitsky A, Zinkevich T, Reichert D, Chevelkov V and Reif B 2010 Microsecond time scale mobility in a solid protein as studied by the $^{15}$N $R_{1\rho}$ site-specific NMR relaxation rates *J. Am. Chem. Soc.* **132** 11850–3

Kurauskas V *et al* 2017 Slow conformational exchange and overall rocking motion in ubiquitin protein crystals *Nat. Commun.* **8** 145

Lacabanne D, Meier B H and Böckmann A 2018 Selective labeling and unlabeling strategies in protein solid-state NMR spectroscopy *J. Biomol. NMR* **71** 141–50

Ladizhansky V 2009 Homonuclear dipolar recoupling techniques for structure determination in uniformly $^{13}$C-labeled proteins *Solid State Nucl. Magn. Reson.* **36** 119–28

Ladizhansky V, Jaroniec C P, Diehl A, Oschkinat H and Griffin R G 2003a Measurement of multiple psi torsion angles in uniformly $^{13}$C,$^{15}$N-labeled alpha-spectrin SH3 domain using 3D $^{15}$N–$^{13}$C–$^{13}$C–$^{15}$N MAS dipolar-chemical shift correlation spectroscopy *J. Am. Chem. Soc.* **125** 6827–33

Ladizhansky V, Vinogradov E, Rossum B-J, van, Groot H J M and de, Vega S 2003b Multiple-spin effects in fast magic angle spinning Lee –Goldburg cross-polarization experiments in uniformly labeled compounds *J. Chem. Phys.* **118** 5547

Lakomek N, Penzel S, Lends A, Cadalbert R, Ernst M and Meier B H 2017 Microsecond dynamics in ubiquitin probed by solid-state N NMR spectroscopy $R_{1\rho}$ relaxation experiments under fast MAS (60–110 kHz) *Chem. Eur. J.* **23** 9425–33

Lalli D, Idso M N, Andreas L B, Hussain S, Baxter N, Han S, Chmelka B F and Pintacuda G 2017 Proton-based structural analysis of a heptahelical transmembrane protein in lipid bilayers *J. Am. Chem. Soc.* **139** 13006–12

Lam Y, Wassall S R, Morton C J, Smith R and Separovic F 2001 Solid-state NMR structure determination of melittin in a lipid environment *Biophys. J.* **81** 2752–61

Lamley J M, Lougher M J, Sass J, Rogowski M, Grzesiekb S and Lewandowski J R 2015 Unraveling the complexity of protein backbone dynamics with combined $^{13}$C and $^{15}$N solid-state NMR relaxation measurements *Phys. Chem. Chem. Phys.* **17** 21997–2008

Lange A, Becker S, Seidel K, Giller K, Pongs O and Baldus M 2005 A concept for rapid protein-structure determination by solid-state NMR spectroscopy *Angew. Chem. Int. Ed.* **44** 2089–92

Lange A, Luca S and Baldus M 2002 Structural constraints from proton-mediated rare-spin correlation spectroscopy in rotating solids *J. Am. Chem. Soc.* **124** 9704–5

Lange V, Becker-Baldus J, Kunert B, van Rossum B J, Casagrande F, Engel A, Roske Y, Scheffel F M, Schneider E and Oschkinat H 2010 A MAS NMR study of the bacterial ABC transporter ArtMP *Chem. Bio. Chem.* **11** 547–55

Lecoq L, Schledorn M, Wang S and Smith-penzel S 2019 100 kHz MAS proton-detected NMR spectroscopy of hepatitis B virus capsids *Front Mol Biosci.* **6** 58

Lee Y K, Kurur N D, Helmle M, Johannessen O G, Nielsen N C and Levitt M H 1995 Efficient dipolar recoupling in the NMR of rotating solids. A sevenfold symmetric radiofrequency pulse sequence *Chem. Phys. Lett.* **242** 304–9

Lemaster D M and Kushlan D M 2001 Dynamical mapping of *E.coli* thioredoxin via $^{13}$C NMR relaxation analysis *J. Am. Chem. Soc.* **7863** 9255–64

Levitt M H, Raleigh D P, Creuzet F and Griffin R G 1998 Theory and simulations of homonuclear spin pair systems in rotating solids *J. Chem. Phys.* **92** 6347

Lewandowski J R, De Paëpe G, Eddy M T, Struppe J, Maas W and Griffin R G 2009 Proton assisted recoupling at high spinning frequencies *J. Phys. Chem.* B **113** 9062–9

Lewandowski J R, De Paëpe G and Griffin R G 2007 Proton assisted insensitive nuclei cross polarization *J. Am. Chem. Soc.* **129** 728–9

Lewandowski J R, Dumez J-N, Akbey Ü, Lange S, Emsley L and Oschkinat H 2011a Enhanced resolution and coherence lifetimes in the solid-state NMR spectroscopy of perdeuterated proteins under ultrafast magic-angle spinning *J. Phys. Chem. Lett.* **2** 2205–11

Lewandowski J R, Sass H J, Grzesiek S, Blackledge M and Emsley L 2011b Site-specific measurement of slow motions in proteins *J. Am. Chem. Soc.* **133** 16762–5

Lewandowski J R, Sein J, Blackledge M and Emsley L 2010a Anisotropic collective motion contributes to nuclear spin relaxation in crystalline proteins *J. Am. Chem. Soc.* **132** 1246–8

Lewandowski J R, Sein J, Sass H J, Grzesiek S, Blackledge M and Emsley L 2010b Measurement of site-specific $^{13}$C spin-lattice relaxation in a crystalline protein *J. Am. Chem. Soc.* **132** 8252–4

Li Y, Berthold D A, Gennis R B and Rienstra C M 2008 Chemical shift assignment of the transmembrane helices of DsbB, a 20-kDa integral membrane enzyme, by 3D magic-angle spinning NMR spectroscopy *Protein Sci.* **17** 199–204

Lienin S F, Bremi T, Brutscher B, Brüschweiler R and Ernst R R 1998 Anisotropic intramolecular backbone dynamics of ubiquitin characterized by NMR relaxation and MD computer simulation *J. Am. Chem. Soc.* **120** 9870–9

Lilly Thankamony S A, Wittmann J J, Kaushik M and Corzilius B 2017 Dynamic nuclear polarization for sensitivity enhancement in modern solid-state NMR *Prog. Nucl. Magn. Reson. Spectrosc.* **103** 120–95

Lin M T, Fukazawa R, Miyajima-nakano Y, Matsushita S, Choi S K, Iwasaki T and Gennis R B 2015 *Escherichia coli* auxotroph host strains for amino acid-selective isotope labeling of recombinant proteins *Methods Enzymol.* **565** 45

Lin M T, Sperling L J, Frericks Schmidt H L, Tang M, Samoilova R I, Kumasaka T, Iwasaki T, Dikanov S A, Rienstra C M and Gennis R B 2011 A rapid and robust method for selective isotope labeling of proteins *Methods* **55** 370–8

Linser R *et al* 2011 Proton-detected solid-state NMR spectroscopy of fibrillar and membrane proteins *Angew. Chem. Int. Ed.* **50** 4508–12

Linser R, Gelev V, Hagn F, Arthanari H, Hyberts S G and Wagner G 2014 Selective methyl labeling of eukaryotic membrane proteins using cell-free expression *J. Am. Chem. Soc.* **136** 11308–10

Lipari G and Szabo A 1982a Model-free approach to the interpretation of nuclear magnetic resonance relaxation in macromolecules. 1. Theory and range of validity *J. Am. Chem. Soc.* **104** 4546–59

Lipari G and Szabo A 1982b Model-free approach to the interpretation of nuclear magnetic resonance relaxation in macromolecules. 2. Analysis of experimental results *J. Am. Chem. Soc.* 4559–70

Liu J, Liu C, Fan Y, Munro R A, Ladizhansky V, Brown L S and Wang S 2016 Sparse $^{13}$C labelling for solid-state NMR studies of *P. pastoris* expressed eukaryotic seven-transmembrane proteins *J. Biomol. NMR* **65** 7–13

Loquet A, Bardiaux B, Gardiennet C, Blanchet C, Baldus M, Nilges M, Malliavin T and Böckmann A 2008 3D structure determination of the Crh protein from highly ambiguous solid-state NMR restraints *J. Am. Chem. Soc.* **130** 3579–89

Loquet A, El N, Stanek J, Berbon M, Bardiaux B, Pintacuda G and Habenstein B 2018 3D structure determination of amyloid fibrils using solid-state NMR spectroscopy *Methods* **138–139** 26–38

Loquet A, Giller K, Becker S and Lange A 2010 Supramolecular interactions probed by $^{13}$C–$^{13}$C solid-state NMR *J. Am. Chem. Soc.* **132** 15164–6

Loquet A, Lv G, Giller K, Becker S and Lange A 2011 $^{13}$C spin dilution for simplified and complete solid-state NMR resonance assignment of insoluble biological assemblies *J. Am. Chem. Soc.* **133** 4722–5

Lorieau J L, Day L A and McDermott A E 2008 Conformational dynamics of an intact virus: order parameters for the coat protein of Pf1 bacteriophage *Proc. Natl Acad. Sci. USA* **105** 10366–71

Lorieau J and McDermott A E 2006 Order parameters based on $^{13}$C$^{1}$H,$^{13}$C$^{1}$H2 and $^{13}$C$^{1}$H3 heteronuclear dipolar powder patterns: a comparison of MAS-based solid-state NMR sequences *Magn. Reson. Chem.* **44** 334–47

Lowe I J 1959 Free induction decays of rotating solids *Phys. Rev. Lett.* **2** 285–7

Lu X, Guo C, Hou G and Polenova T 2015 Combined zero-quantum and spin-diffusion mixing for efficient homonuclear correlation spectroscopy under fast MAS: broadband recoupling and detection of long-range correlations *J. Biomol. NMR* **61** 7–20

Lundström P, Teilum K, Carstensen T, Bezsonova I, Wiesner S, Hansen D F, Religa T L, Akke M and Kay L E 2007 Fractional $^{13}$C enrichment of isolated carbons using [1-$^{13}$C]- or [2-$^{13}$C]-glucose facilitates the accurate measurement of dynamics at backbone Ca and side-chain methyl positions in proteins *J. Biomol. NMR* **38** 199–212

Ma P, Haller J D, Zajakala J, Macek P, Sivertsen A C, Willbold D, Boisbouvier J and Schanda P 2014 Probing transient conformational states of proteins by solid-state $R_{1\rho}$ relaxation-dispersion NMR spectroscopy *Angew. Chem. Int. Ed.* **53** 4312–7

Maciejko J, Mehler M, Kaur J, Lieblein T, Morgner N, Ouari O, Tordo P, Becker-Baldus J and Glaubitz C 2015 Visualizing specific cross-protomer interactions in the homo-oligomeric membrane protein proteorhodopsin by dynamic- nuclear-polarization-enhanced solid-state NMR *J. Am. Chem. Soc.* **137** 9032–43

Mance D, Sinnige T, Kaplan M, Narasimhan S, Dani M, Houben K, Baldus M and Weingarth M 2015 An efficient labelling approach to harness backbone and side-chain protons in $^{1}$H-detected solid-state NMR spectroscopy *Angew. Chem. Int. Ed.* **54** 15799–803

Mandala V S, Loftis A R, Shcherbakov A A, Pentelute B L and Hong M 2020 Atomic structures of closed and open influenza mechanism *Nat. Struct. Mol. Biol.* **27** 160–7

Marani P, Wagner S, Baars L, Genevaux P, de Gier J W, Nilsson I, Casadio R and von Heijne G 2006 New *Escherichia coli* outer membrane proteins identified through prediction and experimental verification *Protein Sci.* **15** 884–9

Marchetti A *et al* 2012 Backbone assignment of fully protonated solid proteins by $^1$H detection and ultrafast magic-angle-spinning NMR spectroscopy *Angew. Chem. Int. Ed.* **51** 10756–9

Marcotte I and Auger M 2005 Bicelles as model membranes for solid- and solution-state NMR studies of membrane peptides and proteins *Concepts Magn. Reson. Part* A **24A** 17–37

Marion D, Gauto D F, Ayala I, Giandoreggio-barranco K G-B and Schanda P 2019 Microsecond protein dynamics from combined Bloch-McConnell and near-rotary-resonance $R_{1\rho}$ relaxation-dispersion MAS NMR *Chem. Phy. Chem.* **20** 276–84

Marulanda D, Tasayco M L, Cataldi M, Arriaran V and Polenova T 2005 Resonance assignments and secondary structure analysis of *E. coli* thioredoxin by magic angle spinning solid-state NMR spectroscopy *J. Phys. Chem.* B **109** 18135–45

McDermott A, Polenova T, Bockmann A, Zilm K W, Paulsen E K, Martin R W and Montelione G T 2000 Partial NMR assignments for uniformly ($^{13}$C,$^{15}$N)-enriched BPTI in the solid state *J. Biomol. NMR* **16** 209–19

Medeiros-Silva J, Mance D, Dani M, Jekhmane S, Houben K, Baldus M and Weingarth M 2016 $^1$H-detected solid-state NMR studies of water-inaccessible proteins *in vitro* and *in situ* Angewandte *Angew. Chem.Int. Ed.* **55** 13606–10

Milikisiyants S, Wang S, Munro R A, Donohue M, Ward M E, Bolton D, Brown L S, Smirnova T I, Ladizhansky V and Smirnov A I 2017 Oligomeric structure of *Anabaena* sensory rhodopsin in a lipid bilayer environment by combining solid-state NMR and long-range DEER constraints *J. Mol. Biol.* **429** 1903–20

Mittermaier A and Kay L E 2006 New tools provide new insights in NMR studies of protein dynamics *Science* **312** 224–9

Mollica L, Baias M, Lewandowski R, Wylie B J, Sperling L J, Rienstra C M, Emsley L and Blackledge M 2012 Atomic-resolution structural dynamics in crystalline proteins from NMR and molecular simulation *J. Phys. Chem. Lett.* **3** 3657–62

Morcombe C R, Gaponenko V, Byrd R A and Zilm K W 2004 Diluting abundant spins by isotope edited radio frequency field assisted diffusion *J. Am. Chem. Soc.* **126** 7196–7

Morgan W D, Kragt A and Feeney J 2000 Expression of deuterium-isotope-labelled protein in the yeast pichia pastoris for NMR studies *J. Biomol. NMR* **17** 337–47

Mörs K, Roos C, Scholz F, Wachtveitl J, Dötsch V, Bernhard F and Glaubitz C 2013 Modified lipid and protein dynamics in nanodiscs Karsten *Biochim. Biophys. Acta* **1828** 1222–9

Munro R, de Vlugt J, Ladizhansky V and Brown L S 2020 Improved protocol for the production of the low-expression eukaryotic membrane protein human aquaporin 2 in *Pichia pastoris* for solid-state NMR *Biomolecules* **10** 434

Murray D T, Das N and Cross T A 2013 Solid state NMR strategy for characterizing native membrane protein structures *Acc. Chem. Res.* **46** 2172–81

Narasimhan S, Folkers G E and Baldus M 2020 When small becomes too big: expanding the use of in-cell solid-state NMR spectroscopy *Chem. plus. chem.* **85** 760–8

Narasimhan S, Scherpe S, Paioni A L, Van Der Zwan J, Folkers G E, Ovaa H and Baldus M 2019 DNP-supported solid-state NMR spectroscopy of proteins inside mammalian cells *Angew. Chem. Int. Ed.* **58** 12969–73

Nielsen N C, Bildso H, Jakobsen H J, Levitt M H, Nielsen N C, Bilds H and Jakobsen H J 1994 Double-quantum homonuclear rotary resonance: Efficient dipolar recovery in magicangle spinning nuclear magnetic resonance *J. Chem. Phys.* **101** 1805

Nieuwkoop A J and Rienstra C M 2010 Supramolecular protein structure determination by site-specific long-range intermolecular solid state NMR spectroscopy *J. Am. Chem. Soc.* **132** 7570–1

Nieuwkoop A J, Wylie B J, Franks W T, Shah G J and Rienstra C M 2009 Atomic resolution protein structure determination by three-dimensional transferred echo double resonance solid-state nuclear magnetic resonance spectroscopy resonance spectroscopy *J. Chem. Phys.* **131** 095101

Nishiyama Y, Endo Y, Nemoto T, Utsumi H, Yamauchi K, Hioka K and Asakura T 2011 Very fast magic angle spinning $^{1}H$–$^{14}N$ 2D solid-state NMR: Sub-micro-liter sample data collection in a few minutes *J. Magn. Reson.* **208** 44–8

Oas T G, Griffin R G and Levitt M H 1988 Rotary resonance recoupling of dipolar interactions in solid-state nuclear magnetic resonance spectroscopy *J. Chem. Phys.* **89** 692

Opella S J 2015 Solid-state NMR and membrane proteins *J. Magn. Reson.* **253** 129–37

Opella S J and Marassi F M 2017 Applications of NMR to membrane proteins *Arch. Biochem. Biophys.* **628** 92–101

Paepe D C, Stanek J, Jaudzems K, Tars K, Andreas L B and Pintacuda G 2017 Is protein deuteration beneficial for proton detected solid-state NMR at and above 100 kHz magic-angle spinning? *Solid State Nucl. Magn. Reson.* **87** 126–36

Palczewski K 2006 G protein–coupled receptor rhodopsin *Annu. Rev. Biochem.* **75** 743–67

Palmer A G III 2004 NMR characterization of the dynamics of biomacromolecules *Chem. Rev.* **105** 3623–40

Palmer A G III, Kroenke C D and Loria J P 2001 Nuclear magnetic resonance methods for quantifying microsecond-to-millisecond motions in biological macromolecules *Methods in Enzymol* **339** 204–38

Palmer A G III and Massi F 2006 Characterization of the dynamics of biomacromolecules using rotating-frame spin relaxation NMR spectroscopy *Chem. Rev.* **106** 1700–19

Paravastu A K and Tycko R 2006 Frequency-selective homonuclear dipolar recoupling in solid state NMR *J. Chem. Phys.* **124** 194303

Park S H, Berkamp S, Cook G A, Chan M K, Viadiu H and Opella S J 2011 Nanodiscs versus macrodiscs for NMR of membrane proteins *Biochemistry* **50** 8983–5

Park S H *et al* 2012 Structure of the chemokine receptor CXCR1 in phospholipid bilayers *Nature* **491** 779–83

Park S H, Wu J, Yao Y, Singh C, Tian Y, Marassi F M and Opella S J 2020 Membrane proteins in magnetically aligned phospholipid polymer discs for solid-state NMR spectroscopy *BBA—Biomembr.* **1862** 183333

Pauli J, Baldus M, van Rossum B, de Groot H, Oschkinat H, Rossum B V, Groot H D and Oschkinat H 2001 Backbone and side-chain $^{13}C$ and $^{15}N$ signal assignments of the alpha-spectrin SH3 domain by magic angle spinning solid-state NMR at 17.6 Tesla *Chem. Bio. Chem.* **2** 272–81

Paulson E K, Morcombe C R, Gaponenko V, Dancheck B, Byrd R A and Zilm K W 2003 Sensitive high resolution inverse detection NMR spectroscopy of proteins in the solid state *J. Am. Chem. Soc.* **125** 15831–6

Peng X, Libich D, Janik R, Harauz G and Ladizhansky V 2008 Dipolar chemical shift correlation spectroscopy for homonuclear carbon distance measurements in proteins in the solid state: application to structure determination and refinement *J. Am. Chem. Soc.* **130** 359–69

Penzel S, Oss A, Liis M, Ago O, Anja S, Matthias B, Beat E and Ernst M 2019 Spinning faster: protein NMR at MAS frequencies up to 126 kHz *J. Biomol. NMR* **73** 19–29

Pickford A R and O'Leary J M 2004 Isotopic labeling of recombinant proteins from the methylotrophic yeast Pichia pastoris *Methods Mol. Biol.* **278** 17–33

Pines A, Gibby M G and Waugh J S 1972 Proton-enhanced nuclear induction spectroscopy. a method for high resolution NMR of dilute spins in solids *J. Chem. Phys.* **56** 1776

Popot J-L 2014 Folding membrane proteins *in vitro*: a table and some comments *Arch. Biochem. Biophys.* **564** 314–26

Prosser R S, Evanics F, Kitevski J L and Al-Abdul-Wahid M S 2006 Current applications of bicelles in NMR studies of membrane-associated amphiphiles and proteins *Biochemistry* **45** 8453–65

Quinn C M and McDermott A E 2009 Monitoring conformational dynamics with solid-state $R_{1\rho}$ experiments *J. Biomol. NMR* **45** 5–8

Quinn C M and Mcdermott A E 2012 Quantifying conformational dynamics using solid-state $R_{1\rho}$ experiments *J. Magn. Reson. J.* **222** 1–7

Quinn C M, Wang M, Fritz M P, Runge B, Ahn J, Xu C, Perilla J R, Gronenborn A M and Polenova T 2018 Dynamic regulation of HIV-1 capsid interaction with the restriction factor TRIM5 α identified by magic-angle spinning NMR and molecular dynamics simulations *Proc. Natl Acad. Sci. USA* **115** 11519–24

Raleigh D P, Levitt M H and Griffin R G 1988 Rotational resonance in solid state NMR *Chem. Phys. Lett.* **146** 71–6

Ramachandran R, Ladizhansky V, Bajaj V S and Griffin R G 2003 $^{13}C$–$^{13}C$ rotational resonance width distance measurements in uniformly $^{13}C$-labeled peptides *J. Am. Chem. Soc.* **125** 15623–9

Reichert D, Pascui O, Bonagamba T J, DeAzevedo E R and Schmidt A 2003 Scaling-down the CSA recoupling in S-CODEX 1D-MAS exchange experiments *Chem. Phys. Lett.* **380** 583–8

Reif B and Griffin R G 2003 $^{1}H$ detected $^{1}H$,$^{15}N$ correlation spectroscopy in rotating solids *J. Magn. Reson.* **160** 78–83

Renault M *et al* 2012a Solid-state NMR spectroscopy on cellular preparations enhanced by dynamic nuclear polarization *Angew. Chem. Int. Ed.* **51** 2998–3001

Renault M, Tommassen-van Boxtel R, Bos M P, Post J A, Tommassen J and Baldus M 2012b Cellular solid-state nuclear magnetic resonance spectroscopy *Proc. Natl Acad. Sci. USA* **109** 4863–8

Retel J S *et al* 2017 Structure of outer membrane protein G in lipid bilayers *Nat. Commun.* **8** 2073

Rienstra C M, Tucker-Kellogg L, Jaroniec C P, Hohwy M, Reif B, McMahon M T, Tidor B, Lozano-Perez T and Griffin R G 2002 De novo determination of peptide structure with solid-state magic-angle spinning NMR spectroscopy *Proc. Natl Acad. Sci. USA* **99** 10260–5

Rodriguez E and Krishna N R 2001 An economical method for $^{15}N/^{13}C$ isotopic labeling of proteins expressed in *Pichia pastoris J. Biochem.* **130** 19–22

Rossum B-J, van, de Groot C P, Ladizhansky V, Vega S and de Groot H J M 2000 A method for measuring heteronuclear ($^{1}H$–$^{13}C$) distances in high speed MAS NMR *J. Am. Chem. Soc.* **122** 3465–72

Saurel O, Iordanov I, Nars G, Demange P, Le Marchand T, Andreas L B, Pintacuda G and Milon A 2017 Local and global dynamics in *Klebsiella pneumoniae* outer membrane protein a in lipid bilayers probed at atomic resolution *J. Am. Chem. Soc.* **139** 1590–7

Schaefer J, Mckay R and Stejskal E 1979 Double-cross-polarization NMR of solids *J. Magn. Reson.* **34** 443–7

Schanda P, Meier B H and Ernst M 2011 Accurate measurement of one-bond H–X heteronuclear dipolar couplings in MAS solid-state NMR *J. Magn. Reson.* **210** 246–59

Schanda P, Meier B H and Ernst M 2010 Quantitative analysis of protein backbone dynamics in microcrystalline ubiquitin by solid-state NMR spectroscopy *J. Am. Chem. Soc.* **132** 15957–67

Schmidt P, Thomas L, Müller P, Scheidt H A and Huster D 2014 The G-protein-coupled neuropeptide Y receptor type 2 is highly dynamic in lipid membranes as revealed by solid-state NMR spectroscopy *Chem. Eur. J.* **20** 4986–92

Schneider R, Ader C, Lange A, Giller K, Hornig S, Pongs O, Becker S and Baldus M 2008 Solid-state NMR spectroscopy applied to a chimeric potassium channel in lipid bilayers *J. Am. Chem. Soc.* **130** 7427–35

Scholz I, Huber M, Manolikas T, Meier B H and Ernst M 2008 MIRROR recoupling and its application to spin diffusion under fast magic-angle spinning *Chem. Phys. Lett.* **460** 278–83

Scholz I, Meier B H and Ernst M 2007 Operator-based triple-mode Floquet theory in solid-state NMR *J. Chem. Phys.* **127** 204504

Schrottke S, Kaiser A, Vortmeier G, Els-Heindl S, Worm D, Bossel M, Schmidtl P, Scheidt H A, Beck-Sickinger A G and Huster D 2017 Expression, functional characterization, and solid-state NMR investigation of the g protein-coupled GHS receptor in bilayer membranes *Sci. Rep.* **7** 46128

Schuetz A, Wasmer C, Habenstein B, Verel R, Greenwald J, Riek R, Böckmann A and Meier B H 2010 Protocols for the sequential solid-state NMR spectroscopic assignment of a uniformly labeled 25 kDa protein: HET-s(1-227) *Chem. Bio. Chem.* **11** 1543–51

Sergeyev I V, Itin B, Rogawski R, Day L A and McDermott A E 2017 Efficient assignment and NMR analysis of an intact virus using sequential side-chain correlations and DNP sensitization *Proc. Natl Acad. Sci. USA* **114** 5171–6

Shahid S A, Bardiaux B, Franks W T, Krabben L, Habeck M, van Rossum B-J and Linke D 2012 Membrane-protein structure determination by solid-state NMR spectroscopy of micro-crystals *Nat. Methods* **9** 1212–7

Sharma M, Yi M, Dong H, Qin H, Peterson E, Busath D D, Zhou H and Cross T A 2010 Insight into the mechanism of the influenza a proton channel from a structure in a lipid bilayer *Science* **330** 509–13

Shen M, Liu Q, Trébosc J, Lafon O, Masuda Y, Takegoshi K, Amoureux J, Hu B and Chen Q 2013 Exploring various modulation-sideband recoupling conditions of SHA+ sequence at fast MAS *Solid State Nucl. Magn. Reson.* **55–56** 42–7

Shen Y, Delaglio F, Cornilescu G and Bax A 2009 TALOS+: a hybrid method for predicting protein backbone torsion angles from NMR chemical shifts *J. Biomol. NMR* **44** 213–23

Shi C, Fasshuber H K, Chevelkov V, Xiang S, Habenstein B, Vasa S K, Becker S and Lange A 2014 BSH-CP based 3D solid-state NMR experiments for protein resonance assignment *J. Biomol. NMR* **59** 15–22

Shi L, Ahmed M A, Zhang W, Whited G, Brown L S and Ladizhansky V 2009a Three-dimensional solid-state NMR study of a seven-helical integral membrane proton pump-structural insights *J. Mol. Biol.* **386** 1078–93

Shi L, Lake E M R, Ahmed M A M, Brown L S and Ladizhansky V 2009b Solid-state NMR study of proteorhodopsin in the lipid environment: secondary structure and dynamics *Biochim. Biophys. Acta* **1788** 2563–74

Shimada I, Ueda T, Kofuku Y, Eddy M T and Wüthrich K 2019 GPCR drug discovery: integrating solution NMR data with crystal and cryo-EM structures *Nat. Rev. Drug Discov.* **18** 59

Siemer A B, Arnold A A, Ritter C, Westfeld T, Ernst M, Riek R and Meier B H 2006 Observation of highly flexible residues in amyloid fibrils of the HET-s prion *J. Am. Chem. Soc.* **128** 13224–8

Sinnige T, Daniels M, Baldus M and Weingarth M 2014 Proton clouds to measure long-range contacts between nonexchangeable side chain protons in solid-state NMR *J. Am. Chem. Soc.* **136** 4452–5

Sinnige T, Houben K, Pritisanac I, Renault M, Boelens R and Baldus M 2015 Insight into the conformational stability of membrane-embedded BamA using a combined solution and solid-state NMR approach *J. Biomol. NMR* **61** 321–32

Smith A A, Testori E, Cadalbert R, Meier B H and Ernst M 2016 Characterization of fibril dynamics on three timescales by solid-state NMR *J. Biomol. NMR* **65** 171–91

Smith A W 2012 Lipid–protein interactions in biological membranes: a dynamic perspective *Biochim. Biophys. Acta* **1818** 172–7

Smith S O 2010 Structure and activation of the visual pigment rhodopsin *Annu. Rev. Biophys.* **39** 309–28

Sobhanifar S, Reckel S, Junge F, Schwarz D, Kai L, Karbyshev M, Löhr F, Bernhard F and Dötsch V 2010 Cell-free expression and stable isotope labelling strategies for membrane proteins *J. Biomol. NMR* **46** 33–43

Spudich J L, Yang C, Jung K and Spudich E N 2000 Retinylidene Proteins: structures and functions from archaea to humans *Annu. Rev. Cell Dev. Biol.* **16** 365–92

Stanek J *et al* 2016 NMR spectroscopic assignment of backbone and side-chain protons in fully protonated proteins: microcrystals, sedimented assemblies,and amyloid fibrils *Angew. Chem. Int.Ed.* **55** 15504–9

Stanley A M and Fleming K G 2008 The process of folding proteins into membranes: challenges and progress *Arch. Biochem. Biophys.* **469** 46–66

Sternberg U, Witter R, Kuprov I, Lamley J M, Oss A, Lewandowski J R and Samoson A 2018 $^{1}$H line width dependence on MAS speed in solid state NMR—comparison of experiment and simulation *J. Magn. Reson.* **291** 32–9

Sun B Q, Costa P R, Kocisko D, Lansbury P T Jr and Griffin R G 1995 Internuclear distance measurements in solid state nuclear magnetic resonance: dipolar recoupling via rotor synchronized spin locking *J. Chem. Phys.* **102** 702

Szeverenyi N M, Sullivan M J and Maciel G E 1982 Observation of spin exchange by two-dimensional Fourier transform $^{13}$C cross polarization-magic-angle spinning *J. Magn. Reson.* **47** 462–75

Takegoshi K, Nakamura S and Terao T 2001 $^{13}$C–$^{1}$H dipolar-assisted rotational resonance in magic-angle spinning NMR *Chem. Phys. Lett.* **344** 631–7

Tang M, Nesbitt A E, Sperling L J, Berthold D A, Schwieters C D, Gennis R B and Rienstra C M 2013 Structure of the disulfide bond generating membrane protein DsbB in the lipid bilayer *J. Mol. Biol.* **425** 1670–82

Tian Y, Chen L, Niks D, Kaiser J M, Lai J, Rienstra C M, Dunnb M F and Mueller L J 2009 J-Based 3D sidechain correlation in solid-state proteins *Phys. Chem. Chem. Phys.* **11** 7078–86

Tycko R and Dabbagh G 1990 Measurement of nuclear magnetic dipole–dipole couplings in magic angle spinning NMR *Chem. Phys. Lett.* **173** 461–5 van der Wel P C A, Eddy M T, Ramachandran R and Griffin R G 2009 Targeted $^{13}C$–$^{13}C$ distance measurements in a microcrystalline protein via J-decoupled rotational resonance width measurements *Chem. Phy. Chem.* **10** 1656–63

van der Wel P C A, Lewandowski R and Griffin R G 2010 Structural characterization of GNNQQNY amyloid fibrils by magic angle spinning NMR *Biochemistry* **49** 9457–69

Vasa S K, Rovó P, Giller K, Becker S and Linser R 2016 Access to aliphatic protons as reporters in non-deuterated proteins by solid-state NMR *Phys. Chem. Chem. Phys.* **18** 8359–63

Verardi R, Shi L, Traaseth N J, Walsh N and Veglia G 2011 Structural topology of phospholamban pentamer in lipid bilayers by a hybrid solution and solid-state NMR method *Proc. Natl Acad. Sci. USA* **108** 9101–6

Verel R, Ernst M and Meier B H 2001 Adiabatic dipolar recoupling in solid-state NMR: The DREAM scheme *J. Magn. Reson.* **150** 81–99

Verhoeven A, Williamson P T F, Zimmermann H, Ernst M and Meier B H 2004 Rotational-resonance distance measurements in multi-spin systems *J. Magn. Reson.* **168** 314–26

Wang S, Munro R A, Kim S Y, Jung K, Brown L S and Ladizhansky V 2012 Paramagnetic relaxation enhancement reveals oligomerization interface of a membrane protein *J. Am. Chem. Soc.* **134** 16995–8

Wang S, Munro R A, Shi L, Kawamura I, Okitsu T, Wada A, Kim S Y, Jung K H, Brown L S and Ladizhansky V 2013a Solid-state NMR spectroscopy structure determination of a lipid-embedded heptahelical membrane protein *Nat. Methods* **10** 1007–12

Wang S, Shi L, Okitsu T, Wada A, Brown L S and Ladizhansky V 2013b Solid-state NMR $^{13}C$ and $^{15}N$ resonance assignments of a seven-transmembrane helical protein *Anabaena* Sensory Rhodopsin *Biomol. NMR Assign.* **7** 253–6

Ward M E, Ritz E, Ahmed M A, Bamm V V, Harauz G, Brown L S and Ladizhansky V 2015a Proton detection for signal enhancement in solid-state NMR experiments on mobile species in membrane proteins *J. Biomol. NMR* **63** 375–88

Ward M E, Shi L, Lake E, Krishnamurthy S, Hutchins H, Brown L S and Ladizhansky V 2011 Proton-detected solid-state NMR reveals intramembrane polar networks in a seven-helical transmembrane protein proteorhodopsin *J. Am. Chem. Soc.* **133** 17434–43

Ward M E, Wang S, Munro R, Ritz E, Hung I, Gor P L, Jiang Y, Liang H, Brown L S and Ladizhansky V 2015b In situ structural studies of *Anabaena* sensory rhodopsin in the *E. Coli* membrane *Biophys. J.* **108** 1683–96

Watt E D and Rienstra C M 2014 Recent advances in solid-state nuclear magnetic resonance techniques to quantify biomolecular dynamics *Anal. Chem.* **86** 58–64

Waugh D S 1996 Genetic tools for selective labeling of proteins with alpha-$^{15}N$-amino acids *J. Biomol. NMR* **8** 184–92

Weingarth M, Bodenhausen G and Tekely P 2010 Broadband magnetization transfer using moderate radio-frequency fields for NMR with very high static fields and spinning speeds *Chem. Phys. Lett.* **488** 10–6

Weingarth M, Demco D E, Bodenhausen G and Tekely P 2009 Improved magnetization transfer in solid-state NMR with fast magic angle spinning *Chem. Phys. Lett.* **469** 342–8

Weingarth M, Masuda Y, Takegoshi K, Bodenhausen G and Tekely P 2011 Sensitive $^{13}C$–$^{13}C$ correlation spectra of amyloid fibrils at very high spinning frequencies and magnetic fields *J. Biomol. NMR* **50** 129–36

Wishart D S and Sykes B D 1994 The $^{13}$C chemical-shift index: a simple method for the identification of protein secondary structure using $^{13}$C chemical-shift data *J. Biomol. NMR* **4** 171–80

Wood M J and Komives E A 1999 Production of large quantities of isotopically labeled protein in Pichia pastoris by fermentation *J. Biomol. NMR* **13** 149–59

Wylie B J, Bhate M P and McDermott A E 2014 Transmembrane allosteric coupling of the gates in a potassium channel *Proc. Natl Acad. Sci. USA* **111** 185–90

Yang J, Aslimovska L and Glaubitz C 2011 Molecular dynamics of proteorhodopsin in lipid bilayers by solid-state NMR *J. Am. Chem. Soc.* **133** 4874–81

Zech S G, Wand A J and McDermott A E 2005 Protein structure determination by high-resolution solid-state NMR spectroscopy: application to microcrystalline ubiquitin *J. Am. Chem. Soc.* **127** 8618–26

Zhong L, Bamm V V, Ahmed M A, Harauz G and Ladizhansky V 2007 Solid-state NMR spectroscopy of 18.5 kDa myelin basic protein reconstituted with lipid vesicles: spectroscopic characterisation and spectral assignments of solvent-exposed protein fragments *Biochim. Biophys. Acta* **1768** 3193–205

Zhou D H, Nieuwkoop A J, Berthold D A, Comellas G, Sperling L J, Tang M, Shah G J, Brea E J, Lemkau L R and Rienstra C M 2012 Solid-state NMR analysis of membrane proteins and protein aggregates by proton detected spectroscopy *J. Biomol. NMR* **54** 291–305

Zhou D H and Rienstra C M 2008 High-performance solvent suppression for proton detected solid-state NMR *J. Magn. Reson.* **192** 167–72

Zinkevich T, Chevelkov V, Reif B, Saalwächter K and Krushelnitsky A 2013 Internal protein dynamics on ps to μs timescales as studied by multi-frequency 15N solid-state NMR relaxation *J. Biomol. NMR* **57** 219–35

# Chapter 15

# Proton-detected solid-state NMR and its applications to membrane proteins

**Julia Kotschy and Rasmus Linser**

## Abstract

Proton detection has revolutionized the field of solid-state NMR recently. Here we comprehensively explain the general aspects of this methodology for protein targets, including its peculiar strengths for resonance assignment, structure calculation, and the determination of their dynamics. The wealth of possibilities make proton-detected solid-state NMR spectroscopy a well-suited tool for challenging proteins, including membrane proteins. With small sample amounts required, NMR characterization is facilitated particularly for those proteins that are difficult to express with high yields, with molecular weights extending the traditional limitations and being continuously expanded in current research. The chapter closes with a review of membrane protein studies that have been performed in the context of proton detection to date.

## 15.1 Introduction

In biological solution NMR spectroscopy, protons have always been the most important nuclei. This is true not only for signal detection but also for spin–spin correlations in multi-dimensional experiments and for their role as reporters of intermolecular interactions, as witnessed from chemical shift perturbations at ligand binding sites or in protein–protein interaction interfaces. Historically, this convention has evolved for practical reasons: Protons represent a large share of most molecules' content of nuclei, both in organic chemistry and biology. And, for most other elements, bestowing target molecules with sufficient levels of NMR-active isotopes (for instance $^{13}C$) has long been a challenge. However, the gyromagnetic ratio of protons is also the highest of the whole periodic table (apart from radioactive tritium), which generally makes it the most sensitive nucleus to work with. This applies to the occupancy differences among energy levels according to the

doi:10.1088/978-0-7503-2532-5ch15

Boltzmann distribution, to any currents induced in the detection coil upon signal acquisition, as well as to spin–spin interactions exploited for example in through-space correlations.

Solid-state NMR (ssNMR), by contrast, has traditionally been applied using $^{13}C$-detected methods, which is due to the spectroscopic hurdles—in the absence of fast tumbling motion—that are associated with the high gyromagnetic ratio $\gamma$ of protons. Apart from the advantages for sensitivity, a large $\gamma$ also incurs a large dipole–dipole interaction among protons and between protons and other nuclei, which is often associated with very short coherence lifetimes and as such impedingly broad signal linewidths. Even though heteronuclear detection, a viable alternative, has paved the way for successful ssNMR over decades (see other chapters), not only has the sensitivity and information content associated with protons been partly sacrificed this way, but also other drawbacks, like the need for high-power proton decoupling sequences, heating effects, and the demand for short acquisition times in both direct and indirect dimensions have been a necessary evil in ssNMR. Only with the onset of very fast magic-angle spinning (MAS) and use of largely deuterated samples, could the strong dipole–dipole interactions associated with protons be more effectively tamed, and protons have shifted now into the focus for ssNMR spectroscopy. For reasons given in more detail below, this development has been going hand in hand with the fabrication of smaller and smaller rotors for MAS (see figure 15.1).

## 15.2 Fast MAS and deuteration

The homonuclear dipole–dipole (DD) coupling can be described with the dipolar Hamiltonian as

$$\hat{H}_{IS}^{DD} = \mu_0 \frac{\gamma_I \gamma_S \hbar}{4\pi r_{IS}^3} \frac{3\cos^2\beta - 1}{2}(3I_zS_z - \mathbf{I}\cdot\mathbf{S}) \quad (15.1)$$

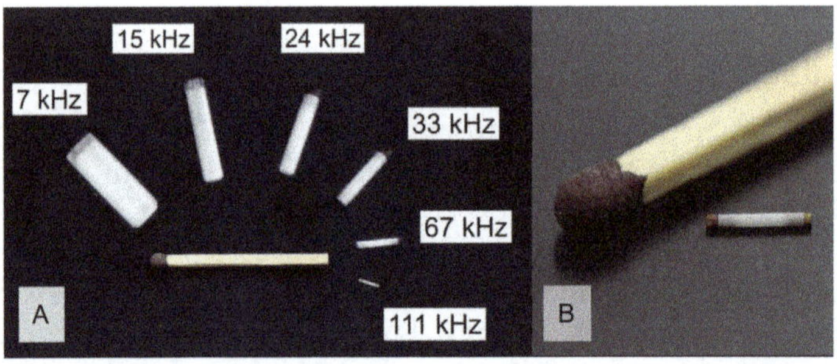


**Figure 15.1.** A subset of the commercially available magic-angle spinning rotors and their maximum MAS rotation frequencies: **(A)** For faster MAS frequencies, the rotor diameter needs to shrink, which is associated with less material in the coil. Luckily, with smaller diameters, coils tend to have better performance, and proton detection and long coherence lifetimes may be additional benefits of faster spinning frequencies. **(B)** Fast-MAS rotors are usually composed of a hollow tube bestowed with a winged top cap and a cylindrical bottom cap.

where $r_{IS}$ is the internuclear distance, $\gamma_I$ and $\gamma_S$ are the gyromagnetic ratios of nuclei I and S, $\hbar$ is Planck's constant divided by $2\pi$, $\mu_0$ is the permeability of free space, and $\beta$ is the angle between the internuclear vector and the direction of the external magnetic field $B_0$. Apart from these numerical factors, the last term describes the difference of the product of the $z$-components of spins I and S, $I_z$ and $S_z$, and the scalar product of both spin vectors **I** and **S**.

We will ignore the spin part of the interaction on the right and rather focus on the constants on the left and the spatial part in the middle: on one hand, a large size of the dipole–dipole interaction, in particular for proton–proton dipolar couplings but also for proton–heteronuclear couplings, can be readily deduced from the dependence of the DD coupling on $\gamma$ of both interaction partners. On the other, the angle $\beta$ obviously plays an important role in modulating the size of the interaction (see below). Conversely, equation (15.1) also allows derivation of potential remedies against the detrimental consequences of strong DD couplings:

i. Partially replacing protons with deuterons will replace a large share of the $^1$H–$^1$H interactions with $^1$H–$^2$H interactions, where $^2$H has a $\gamma$ approximately six times smaller. On the other hand, the remaining $^1$H–$^1$H interactions tend to have a much larger internuclear distance $r$ (and thus a DD coupling that is much smaller). Proton dilution has been successfully applied in various ways, dependent on the information content sought as well as the MAS regime available (see below). With MAS frequencies around 10–20 kHz, proton–proton dipolar couplings could, for the first time, be effectively overcome to such an extent that proton line shapes upon direct detection were made up of homogeneous contributions from shimming, motion-induced relaxation, and differences in magnetic susceptibility (Chevelkov *et al* 2006). Consequently, in ssNMR of well-ordered nano- or micro-crystalline preparations, protons behaved as in the solution state (Linser *et al* 2008, Linser *et al* 2010b, Schanda *et al* 2009), overcoming many of the drawbacks previously associated with protons, like the mentioned heating effects (Linser *et al* 2007). The samples used comprised not only a complete substitution of non-exchangeable (aliphatic/aromatic) protons, but also a partial reduction (90%–70%) of exchangeable protons (i.e., at amide sites) (Akbey *et al* 2009, Linser *et al* 2011b). Figure 15.2(A) compares an $^1$H–$^{15}$N correlation obtained under such conditions with other conditions as described below for the SH3 domain of chicken α-spectrin. For spinning frequencies at around 60 kHz MAS, exchangeable (amide) sites in deuterated samples have usually been back-protonated to 100 % (figure 15.2(B)), leading to (almost) the same proton linewidths as for example in figure 15.2(A) (Lewandowski *et al* 2011a, Knight *et al* 2012, Schanda *et al* 2010, Lewandowski *et al* 2011b). Fully protonated samples with no proton dilution/deuteration have been used with increasing feasibility for faster MAS (figure 15.2(C)) (Zhou *et al* 2007a, Xiang *et al* 2016, Andreas *et al* 2016), although direct proton–proton interactions are not yet fully suppressed even at MAS frequencies up to 150 kHz (Xue *et al* 2017).

Since the NMR linewidth also depends on heterogeneous contributions (i.e., sample inhomogeneity in cases when non-crystalline samples are used),

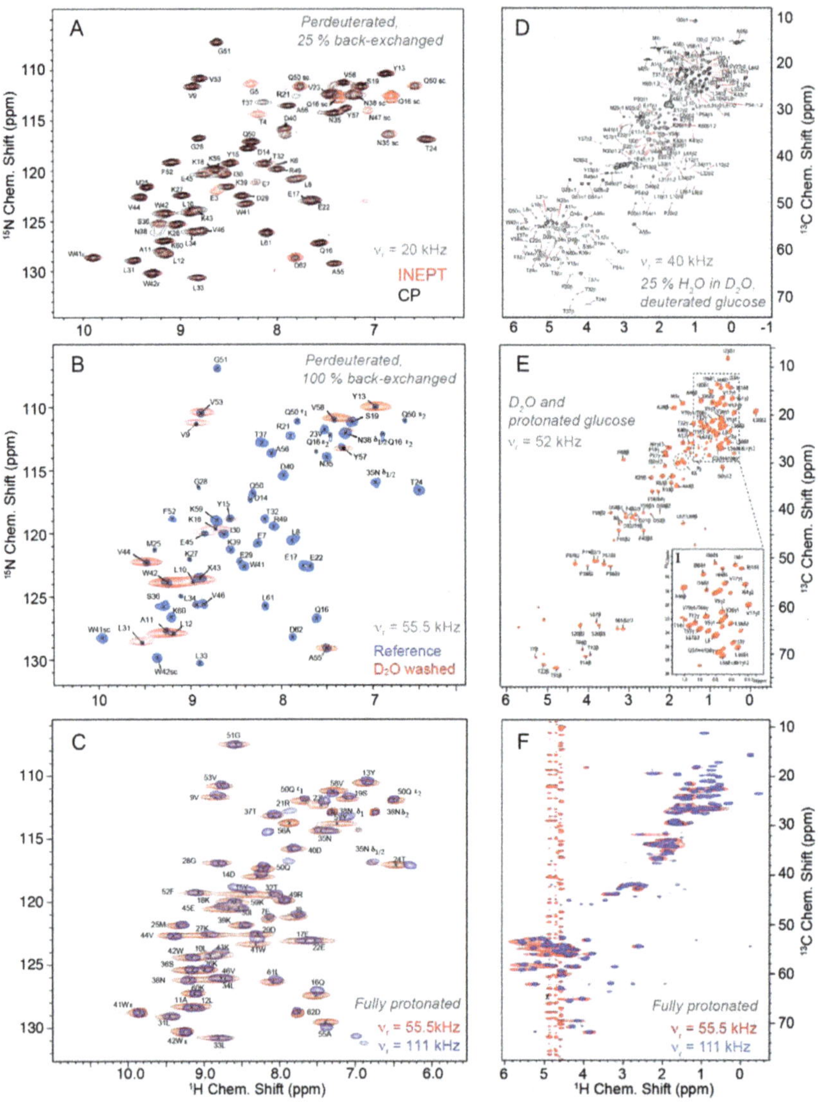


**Figure 15.2.** Commensuration of proton content and MAS frequencies for amide and aliphatic protons. (A)–(C) Comparison of $^{1}H$–$^{15}N$ spectra for: (A) 25 % proton back-exchange in otherwise deuterated background spun at 20 kHz, obtained with either scalar or dipolar transfers (adapted with permission from Linser *et al* (2010a), copyright 2010 American Chemical Society); (B) 100 % proton back-exchange in otherwise deuterated background, spun at 55 kHz (blue), overlaid with a spectrum of an H/D exchanged sample in red (adapted by permission from Grohe *et al* (2017), copyright 2017 Springer Nature); and (C) for a fully protonated sample, spun at either 55.5 (red) or 111 kHz (blue). (D)–(F) Comparison of aliphatic $^{1}H$–$^{13}C$ correlations, recorded (D) using random adjoint protonation from 25 % $H_2O$/75 % $D_2O$ and deuterated glucose, spun at 40 kHz (adapted with permission from Asami and Reif (2013), copyright 2013 American Chemical Society); (E), with fractional deuteration at 52 kHz (adapted from Mance *et al* (2015), copyright 2015 the authors); or (F) with a fully protonated sample, spun at 111 kHz (adapted from Vasa *et al* 2019), copyright 2019 John Wiley and Sons Ltd). Apart from (E), recorded on ubiquitin, all spectra were recorded on the SH3 domain of chicken α-spectrin.

full suppression of the homogeneous contributions to the proton linewidth by deuteration or faster MAS may not always be useful. This is the case for heterogeneous preparations as are often found, for example, for fibrillar proteins. In such instances, the drawbacks of sample deuteration (see below) or faster spinning (in particular, the smaller rotors needed) can be avoided, and fully protonated samples have been used successfully already at 60 kHz MAS (Xiang *et al* 2016, Xiang *et al* 2017).

ii. A second way to ameliorate the effective couplings is derived from the spatial term. Even though the angle $\beta$ can obviously not be adjusted to an ideal value for all $^1H$–$^1H$ spin pairs at the same time, MAS can average the effective inter-spin vector to the angle of the rotor axis with respect to $B_0$, leading to an effective value of $\beta = 54.7°$, where the spatial term becomes zero to first-order approximation. Whereas the $^1H$–$^1H$ dipolar interactions of deuterated and proton-back-exchanged samples can be effectively averaged out by MAS frequencies of 10–60 kHz, dependent on the actual proton dilution strategy employed, maximal proton–proton dipolar couplings for geminal protons at aliphatic sites amount to approximately 110 kHz, setting a high benchmark for MAS frequencies to become sufficiently effective (Böckmann *et al* 2015, Xue *et al* 2017). At 110 kHz MAS, in particular for amide protons, homogeneous linewidths on the order of 50–80 Hz are obtained, a regime that can definitely be called 'useful' (figure 15.2(C)) and allows for manifold aspects of NMR-based structural biology studies even for complex proteins (Vasa *et al* 2019).

Increasing MAS rates require smaller and smaller rotor diameters (figure 15.1) due to increasing centrifugal forces and velocities with increasing distance from the rotor axis. This entails dangers due to rotor cracking and turbulences for air flows faster than the speed of sound. As such, construction of very fast MAS rotors and probes has been a substantial technological effort, which concatenates requirements for high mechanostability (due to the high centrifugal forces) and precision upon manufacturing of the parts involved. Even though smaller rotors allow less material (and hence contain fewer nuclei that can be detected) in the coil, the better relative performance of smaller-diameter coils as well as the high $\gamma$ of protons, which can now be detected (with potentially sharp linewidths, sample homogeneity permitting), have made proton-detected, fast-MAS an extremely powerful tool (Vasa *et al* 2018a).

Use of deuterated proteins with their exchangeable protons partially back-exchanged has been the main preparative technique for proton-detected ssNMR so far, affording high-intensity and sharp amide proton signals for backbone assignment, for dynamics measurements (Schanda and Ernst, 2016, Lewandowski, 2013, Rovó *et al* 2019) and for structure calculation (Zhou *et al* 2007b, Knight *et al* 2012, Linser *et al* 2014, Shi *et al* 2015, Grohe *et al* 2019). Some major drawbacks, however, accompany this strategy: (i) perdeuteration, including $D_2O$-based growth media, is a hurdle for recombinant expression of many target proteins, (ii) amide site back-exchange is hampered in the case of large hydrophobic protein cores whenever unfolding and refolding procedures for quantitative exchange are unavailable, and (iii) the information content associated with non-exchangeable protons is not yet accessible. To overcome these issues, alternative ways for partial deuteration in non-exchangeable sites have been developed in addition: instead of involving $D_2O$ and deuterated carbon sources, either $D_2O/H_2O$ mixtures have been used, a labelling that has been called Random Adjoint Protonation (Asami *et al* 2010, Asami and Reif, 2013) (RAP, see figure 15.2(D)), or only one of these two potentially proton-carrying media ingredients (water and glucose/glycerol) is employed in a deuterated fashion, whereas the other carries protons (see figure 15.2(E)). The latter labeling strategy has been termed Fractional Deuteration or inverse Fractional Deuteration (iFD), dependent on which of the proton sources is employed (Mance *et al* 2015, Medeiros-Silva *et al* 2016). All of these ways introduce a certain, amino acid type-dependent, stochastic degree of protonation in non-exchangeable positions, which can then be used for structure calculation, chemical shift perturbation and relaxation experiments. In any case, a protonation degree <100 % is aimed at (implying a reduced number of spins detected) to enable sharper linewidths. Alternatively, Stereo-Array Isotope Labelling (SAIL) (Gauto *et al* 2019) or precursors for methyl groups in otherwise deuterated background (Agarwal *et al* 2006, Huber *et al* 2012) can be used to implement selective and thus more quantitative proton spins.

## 15.3 Spectroscopic possibilities of proton-detected ssNMR

Given the possibility of proton detection with adequate sensitivity and line shapes, the information content of protons can be exploited in manifold ways (Vasa *et al* 2018a), which is applicable to any kind of target from inorganic materials to membrane proteins and membrane constituents themselves. The following will comment on the possibilities for ssNMR with respect to *signal assignments* (in particular for protein resonances), *water accessibilities*, *intermolecular interactions*, assessment of *molecular structures* and elucidation of internal molecule *dynamics*.

### 15.3.1 General trends in proton-detected ssNMR

Proton-detected ssNMR enables correlations between multiple spins, each of whose chemical shifts can be encoded in the transfer pathway. The characterization of the proton shifts themselves in one dimension, combined with long coherence lifetimes in indirect evolution periods as well as during magnetization transfers, facilitates experiments of three and more dimensions with relatively few magnetization transfer steps. In contrast to solution NMR of large proteins, the high transfer efficiencies of the individual magnetization transfers are not affected by the target molecular weight. As such, whereas carbon-detected methods have often relied on 2D experiments (Andronesi *et al* 2008), proton-detected ssNMR regularly employs 3D schemes and currently witnesses the trend toward four dimensions (Xiang *et al* 2016, Vasa *et al* 2018b, Klein *et al* 2020 submitted, Fraga *et al* 2017). Section 15.3.2 shows multiple assignment experiments in three dimensions, but assignment experiments with as few as four transfer steps can be implemented using sensitive 4D schemes, like the hCACONH and hCOCANH shown also in section 15.3.2 (Xiang *et al* 2016). Similarly useful, through-space experiments for structure calculation can be easily implemented as 4D approaches, like the hNH-RFDR-hNH experiment (commensurate to an HSQC-NOESY-HSQC) or derivatives as shown in section 15.3.4 (Linser *et al* 2011a, Huber *et al* 2012). Such 'higher-dimensionality' experiments would suffer from very long experimental times due to the many increments in the various indirect dimensions that need to be acquired. For this reason, non-uniform sampling strategies, reviewed for example by Mobli and coworkers (Miljenović *et al* 2017), have been adapted from solution NMR techniques, which can nowadays be implemented with a low level of effort required on the user side. Upon reconstruction of the data, the 4D spectra can be used as if they had been acquired using uniform sampling and just like the 3D data sets, simply with an additional 'scrollbar' to define an additional shift value at which the 2D planes to be depicted will be extracted. Figure 15.3 shows a 3D, 4D, and 5D version of the same proton-detected HNCACONH experiment (on differently complex samples), demonstrating the increasing resolution upon increase of dimensionality (Xiang *et al* 2015, Vasa *et al* 2018b, Klein *et al* 2020 submitted). Whereas the 3D HncacoNH strips encoding inter-residual proton shifts were obtained from a 7 kDa protein (the SH3 domain of chicken α-spectrin) (Xiang *et al* 2015), the 4D HNcacoNH walk was obtained from a 29 kDa protein (human carbonic anhydrase hCAII) (Vasa *et al* 2018b), and the 5D HNCAcoNH walk was obtained on the 72 kDa tryptophan synthase (just counting the asymmetric unit) (Klein *et al* 2020 submitted).

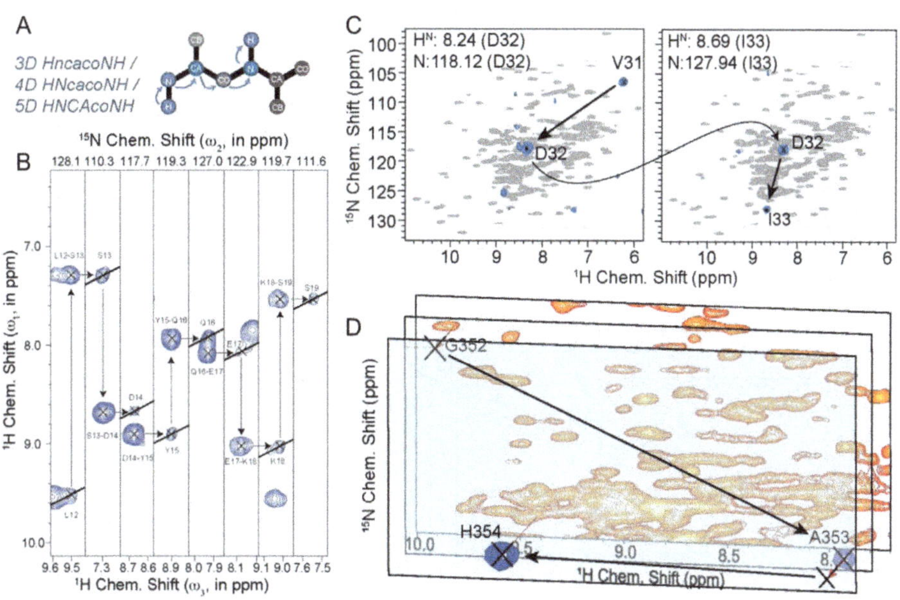


**Figure 15.3.** Implementation of additional dimensions in the same experiment: (A) The HNCACONH experiment (Xiang *et al* 2015) as a potential 3D, 4D, or 5D experiment; (B) implementation as a 3D HncacoNH, shown for SH3 (adapted by permission from Xiang *et al* (2015), copyright 2015 Springer Nature); (C) implementation as a 4D HNcacoNH, shown for hCAII, (adapted with permission from Vasa *et al* (2018b), copyright 2018 American Chemical Society); and (D) implementation as a 5D HNCAcoNH, shown for tryptophan synthase (Klein *et al* 2020 submitted).

Higher-dimensionality data is particularly useful when signal overlap is obtained due to a large number of amino acids (high-molecular-weight proteins) or when high heterogeneity of the sample leads to signal overlap, as for instance for fibrillar proteins (Xiang *et al* 2016, Xiang *et al* 2017). Since many proteins of potential interest for NMR-based structural biology are larger than 30 to 40 kDa, even 4D spectra often are sometimes not well-enough resolved. 5D data sets, as shown in section 15.3.2, can be pulse-programmed and recorded easily with state-of-the-art acquisition software. Their processing and handling in the course of the assignment process, however, requires slightly different procedures (Kosiński *et al* 2017). Processing of a full 5D data set would easily amount to terabytes of data, which exceeds both storage and random-access memories of current computer hardware. Instead, since a subset of the spectrum, in particular 3D data correlating three out of the five nuclei correlated in the 5D, has usually been recorded beforehand; 2D planes from the additional two shifts can be constructed specifically at the positions of the 3D source spectrum by sparse multi-dimensional Fourier transformation (Kazimierczuk *et al* 2009). Figure 15.4 compares the decrease of shift degeneracy upon increase in dimensionality for the same residue E135 in tryptophan synthase, whose connectivity is fully resolved only in the 5D spectrum (Klein *et al* 2020 submitted).

In addition to the processing/data handling strategies described above, higher-dimensionality ssNMR spectra can be recorded as projection spectroscopy (Klein *et al* 2018, Orton *et al* 2020). This involves simultaneous evolution of indirect evolution periods in a linear way, providing data that resemble projections of a

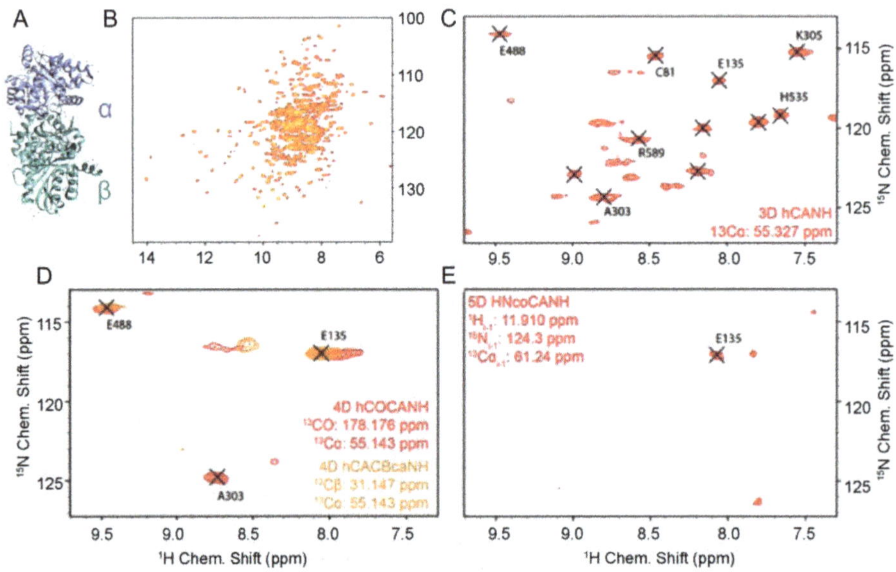


**Figure 15.4.** Disambiguation of sequential contacts by increasing dimensionality in tryptophan synthase: (A) Structure of the 72 kDa asymmetric subunit of tryptophan synthase, (B) 2D $^{1}H$–$^{15}N$ correlation, (C) 3D hCANH, (D) 4D hCACONH, and (E) 5D HNcoCANH. All spectra were recorded on deuterated, $^{13}C$-, $^{15}N$-labeled protein back-exchanged in 100 % $H_2O$ at 700 MHz proton Larmor frequency.

multi-dimensional spectrum in a (usually) two-dimensional space. Then, the multiple 2D projections can be read out and the underlying higher-dimensionality spectrum reconstructed in an automated fashion (APSY) (Hiller *et al* 2005b). Whereas for reasonable results with APSY, excessive peak overlap in 2D spectra should not be present, the approach is another elegant way of reducing higher-dimensionality data (demonstrated in solids with up to five and in solution with up to seven dimensions) to something that current hard- and software can handle.

### 15.3.2 Signal assignment in proton-detected ssNMR

Similar to solution NMR, in addition to exploitation of protons as a source of magnetization, their potential availability as viable detection nuclei strongly suggests their exploitation for detection (Vasa *et al* 2018a). Signal assignment, the initial hurdle in protein NMR with usually thousands of signals, hinges on sequential backbone correlations successively connecting chemical shifts of amino acids in the primary sequence. Both the amide proton NH as well as the alpha proton, Hα, have a relatively high chemical shift dispersion and can be used to constitute 'sequential walks'. In the underlying multi-dimensional experiments, a combination of chemical shifts, acquired in the various indirect dimensions of the experiment, is correlated with the detected proton signal (Zhou *et al* 2007a, Linser *et al* 2008). Figure 15.5 depicts a selection of proton-detected shift correlations for sequential assignment, often recorded in pairs of complementary information for making matches between successive residues.

Just like for carbon-detected ssNMR or protein NMR in solution, multiple complementary strategies (pairs of experiments) exist to achieve match making between strips of successive residues. Much like in solution NMR, the chemical shifts that are usually taken into account for match making are Cα (figure 15.5(B)), Cβ (figure 15.5(C)) and CO (figure 15.5(D)) (Linser *et al* 2008, Knight *et al* 2011, Barbet-Massin *et al* 2013, Penzel *et al* 2015). However, in proton-detected ssNMR, the successive (or previous) amide N or $H^N$ also can be used for this purpose in a straightforward (sensitive) manner and irrespective of target molecular weight (figure 15.5(E), which shows the NNH version of the HNH experiment in figure 15.3) (Xiang *et al* 2015, Andreas *et al* 2015b). Additionally, experiments correlating sets of sidechain shifts (in particular the group of aliphatic carbon shifts) with the backbone shifts have been developed (see below) (Linser 2011, Kulminskaya *et al* 2016, Vasa *et al* 2018b, Klein *et al* 2020 submitted). These are particularly useful for the identification of the amino acid type, which is crucial for assigning stretches of connected amino acids to the protein primary sequence.

Sidechain assignment represented a severe hurdle in the early years of proton-detected ssNMR spectroscopy. With aliphatic and aromatic protons being absent (leaving only carbon sidechain assignments as a possibility), both a sensitive detection nucleus as well as a source of polarization were absent. In addition, in the absence of strong proton dipolar couplings (due to fast MAS and deuterated carbons) proton-driven long-range magnetization transfers as in carbon-detected methodology could not be used, and one-bond mixing only would cause prohibitive

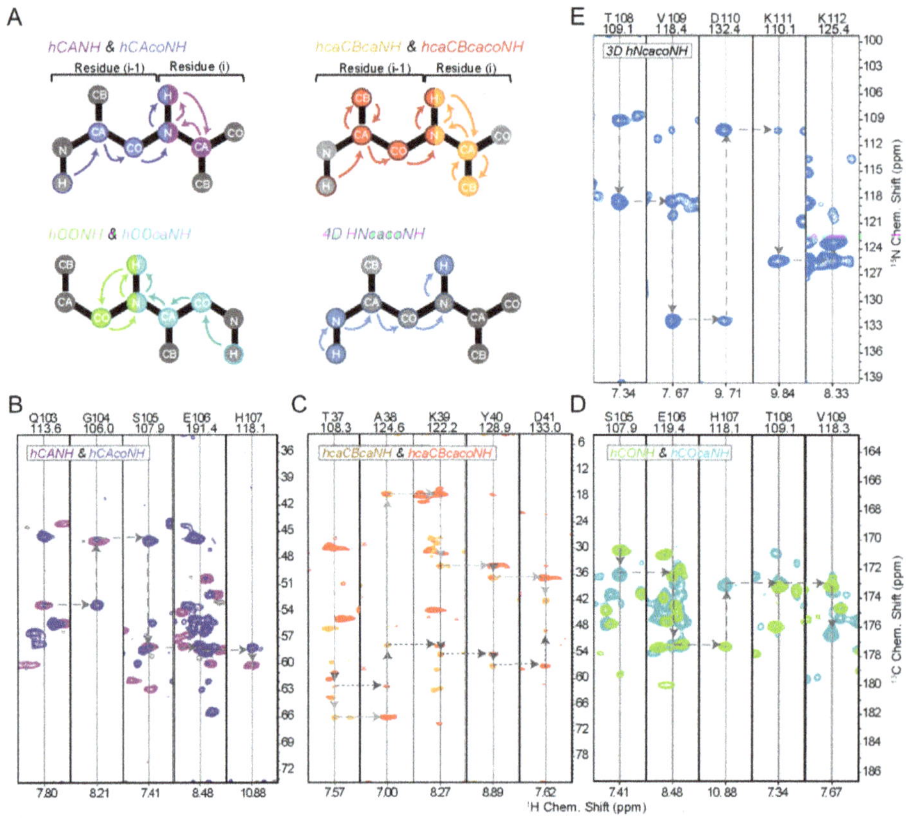

**Figure 15.5.** Amide proton-detected experiments for backbone and sidechain carbon assignments as schemes (A) and depicted as exemplary strips (B–E) from 29 kDa carbonic anhydrase II (hCAII) (Vasa *et al* 2018b). The first three experiments in (A) show pairs of 3D carbon-match-based experiments, demonstrated in (B) by an hCANH/hCAcoNH backbone walk, in (C) by an hcaCBcaNH/hcaCBcacoNH, and in (D) by an hCONH/hCOcaNH combination. The blue, fourth scheme in (A) (as well as the strips in (E) depict an amide-to-amide correlation, recorded here with encoding of $^{15}$N shifts for the interresidual amino acid. Figure adapted with permission from Vasa *et al* (2018b), copyright 2018 American Chemical Society.

levels of ambiguity. Instead, besides long-range heteronuclear transfers from the backbone, like long-range $^{1}$H–$^{13}$C (Linser 2012) or $^{15}$N–$^{13}$C TEDOR correlations (Nieuwkoop *et al* 2009), long-range carbon–carbon mixing schemes were implemented, like RFDR, TOBSY, and MOCCA, that could correlate the aliphatic carbon shifts either with residual sidechain protons (in methyl groups) or the backbone amides (Agarwal and Reif 2008, Linser 2011, Tang *et al* 2010, Kulminskaya *et al* 2015, Kulminskaya *et al* 2016). The resulting experiments have now enabled the correlation of sidechain carbons with their backbone amide shifts in up to four dimensions and allow unambiguous sidechain assignment in deuterated samples (Klein *et al* 2020 submitted). Figure 15.6(A) shows a CC correlation in deuterated hCAII as an example for homonuclear carbon–carbon mixing (using composite-pulse MOCCA) (Vasa *et al* 2018b) and its implementation into a 3D and

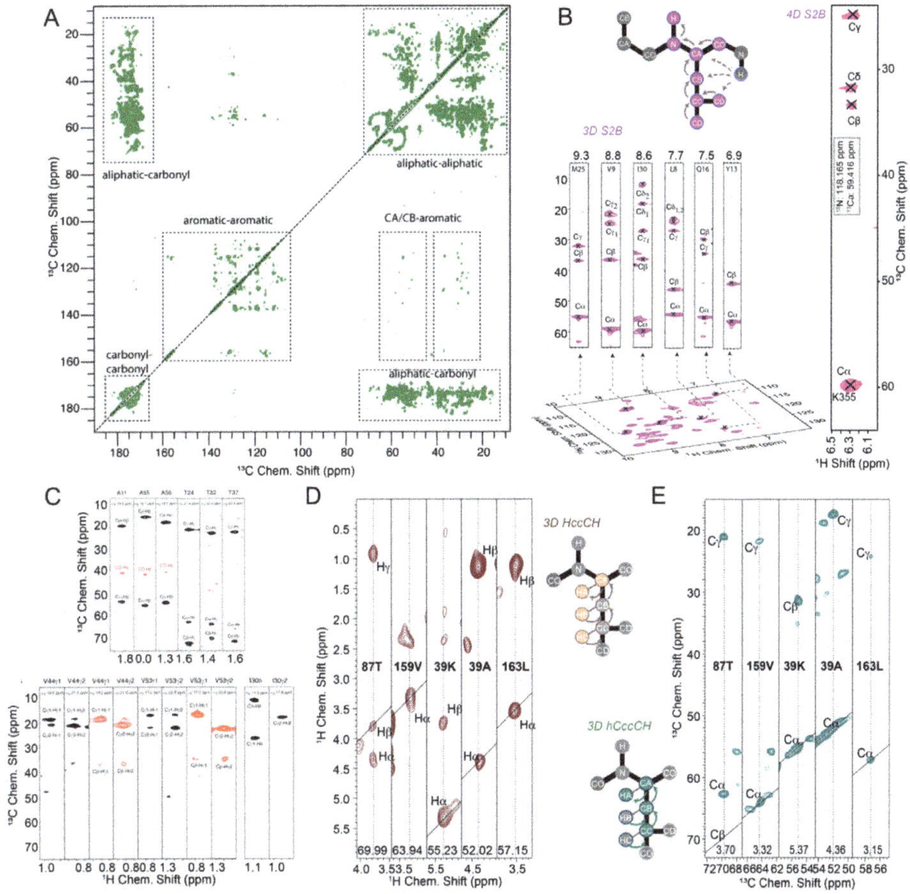


**Figure 15.6.** Sidechain assignment in proton-detected ssNMR: (A) Scalar carbon–carbon mixing under fast MAS using composite-pulse MOCCA, applied to hCAII at 55 kHz (adapted with permission from Vasa *et al* (2018b), copyright 2018 American Chemical Society); (B) proton-detected sidechain-to-backbone (S2B) experiment as 3D (hCccNH, applied to the SH3 domain, (adapted from Kulminskaya *et al* (2016) with permission from The Royal Society of Chemistry) or 4D experiments (hCccCNH, applied to tryptophan synthase) (Klein *et al* 2020 submitted), both at 55 kHz MAS; (C) Methyl-detected sidechain correlation, based on TOBSY scalar mixing and applied to the SH3 domain at 20 kHz MAS (adapted from Agarwal and Reif (2008), copyright 2008, with permission from Elsevier); (D) and (E) 3D HccCH and hCccCH, respectively, for fully protonated samples using Waltz-16 mixing and applied to hCAII (adapted from Vasa *et al* 2019), copyright (2019) John Wiley and Sons Ltd).

4D sidechain-to-backbone experiment (figure 15.6(B)) (Klein *et al* 2020 submitted). Whole-sidechain correlations using TOBSY (Linser 2011) were implemented as methyl proton-detected experiments first (figure 15.6(C)) (Agarwal and Reif 2008). For fully protonated samples at very fast MAS, carbon–carbon mixing has been achieved using isotropic mixing sequences like WALTZ-16 or DIPSY, and 3D correlations like hCCH and HcCH have enabled full assignment of both sidechain carbons and protons (figures 15.6(D) and 15.6(E)) (Andreas *et al* 2016, Vasa *et al* 2019). In the presence of aliphatic protons with useful line width, Hα detection also is as an attractive alternative for sequential chemical shift assignment (Stanek *et al* 2016). In a more sophisticated way, some HN and Hα detected experiments can even be combined and recorded at the same time (Stanek *et al* 2020). These latest efforts rely on time-shared heteronuclear acquisition periods after simultaneous $^{1}H$–$^{15}N$ and $^{1}H$–$^{13}C$ cross polarization (CP) (Linser *et al* 2011a) as well as the possibility to record and store multiple FIDs within a single scan (Sharma *et al* 2016).

Whereas for micro-crystalline preparations, higher-dimensionality approaches have been applied only for targets larger than 25 kDa, severe signal overlap in 2D and 3D space can also be encountered for smaller proteins in cases where the lines are very broad. This is demonstrated in figure 15.7(A) for Tau fibrils, which, however, can be assigned to a large extent by increasing the dimensionality of the assignment experiments to four (figure 15.7(B)) (Xiang *et al* 2016).

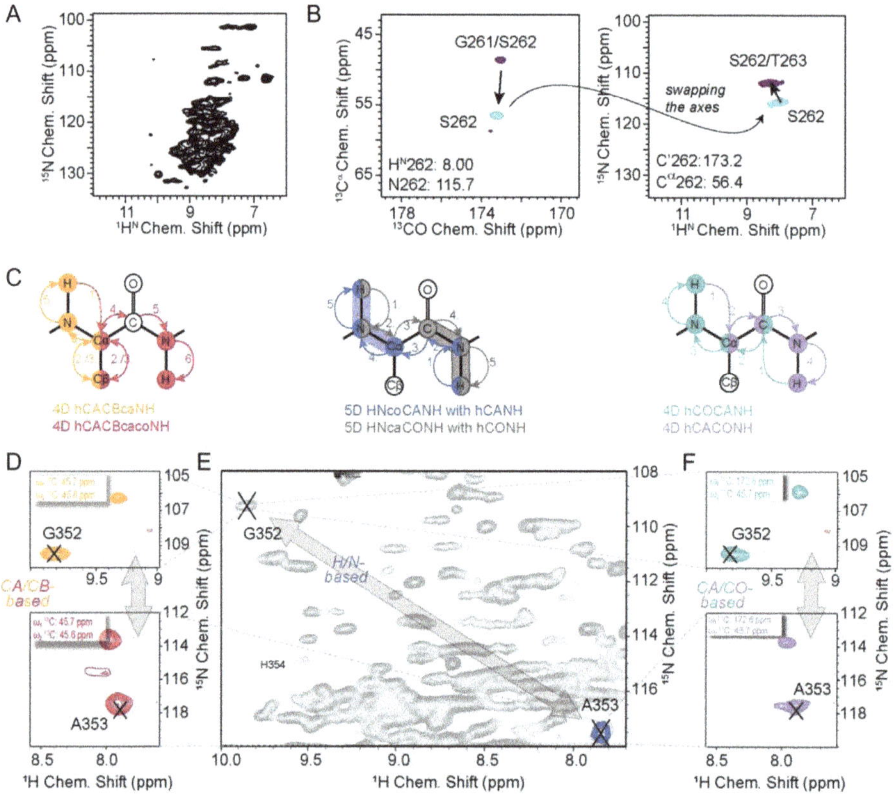


**Figure 15.7.** Higher-dimensionality approaches for heterogeneous preparations and facilitation of match making by complementary walks: (A) heterogeneous Tau fibrils (fully protonated, at 55 kHz MAS), and (B) 4D spectra hCACONH and hCOCANH for assignment (both adapted from Xiang *et al* 2013 with permission from The Royal Society of Chemistry, Creative Commons Attribution 3.0 Unported Licence (Xiang *et al* 2016)). (C) Schemes for complementary higher-dimensionality experiments. (D) Slices from a 4D hCACBcaNH/hCACBcaoNH pair from 72 kDa tryptophan synthase as deuterated and 100 % $H^{N}$ back-exchanged sample at 55 kHz MAS, in conjunction with (E) 5D HNcoCANH slices (extracted at $^{1}H/^{15}N/^{13}CA$ shifts of G352 and overlaid over the $^{1}H$–$^{15}N$ correlation), as well as (F) slices from a 4D hCACONH/hCOCANH pair (Klein *et al* 2020 submitted).

Despite the potency of each individual experiment, multiple such approaches need to be applied in parallel for targets of high molecular weight. This is shown in figure 15.7(C–F) for tryptophan synthase. Whereas the sequential amide experiment (implemented here as a 5D HNCAcoNH) allows matching of $^{1}H$–$^{15}N$ shift pairs, the 4D hCACONH/hCOCANH pair allows matching of CA and CO shifts, and the hCACBcaNH/hCACBcacoNH 4D pair in addition allows matching of adjacent CB shifts (Klein *et al* 2020 submitted). Aiding the backbone walks, sidechain-to-backbone experiments like the hCCNH (see above) (Linser 2011), acquired via MOCCA, RFDR, TOBSY or WALTZ for mixing of magnetization between sidechain carbons, can be employed as a complement to determine the amino acid type, which helps the mapping of matching strips to the known primary sequence of the target protein.

### 15.3.3 H/D exchange and interactions with small molecules

A facilitated detection of amide protons has enabled methodology for assessing the exchangeability of such protons. Using a mixture of $H_2O$ and $D_2O$, Lopez Del Amo *et al* 2010 could track the fate of an amide-bound proton or deuteron over time by selecting for a certain state at the amide nitrogen (proton-bound or deuteron-bound) and allowing variable times for exchange (Lopez Del Amo *et al* 2010). With this experiment, the fast fluctuation of H/D at amides (ms-s timescale) can be quantified, whereas the longer timescale of slowly exchanging amide sites (hrs-weeks), which exist at non-exposed sites and hydrogen-bonded amides, can be followed by real-time proton-detected ssNMR correlations after a buffer exchange with varied protonation ratio (Grohe *et al* 2017). Figure 15.8(A,B) and 8(C,D) depict spectra and rates for amides in the SH3 domain of chicken α-spectrin in the slow and the fast regimes, respectively.

Not only solvent interactions, but also the interaction with small molecules can be easily tracked by proton-detected ssNMR. The principle exploited is the same as for any other kind of NMR, where the association of a binding partner induces a change in the environment of a nucleus and thus a change in its chemical shift. An advantage of proton-detected experiments is the availability of a very sensitive reporter for the binding event, as well as the possibility to easily combine manifold added shift perturbations from different nuclei, as is represented in figure 15.8(E) for an inhibitor of the enzyme human carbonic anhydrase II (hCAII), where the value of $CSP = (\Delta H^2 + \Delta N^2/10 + \Delta C\alpha^2/4)^{0.5}$ is depicted (Vasa *et al* 2018b). The shift perturbations can be used to verify the binding site of the small molecule, as shown in figure 15.8(F), or can even be involved in structure calculation as pseudo-structural restraints for intermolecular interactions, as shown in figure 15.8(G–H) for another inhibitor of this enzyme (Vasa *et al* 2019).

### 15.3.4 Protein structure calculation in proton-detected ssNMR

Even though protein dynamics is becoming more and more the real 'selling point' of ssNMR, protein structure elucidation has always been an important feature. Protein

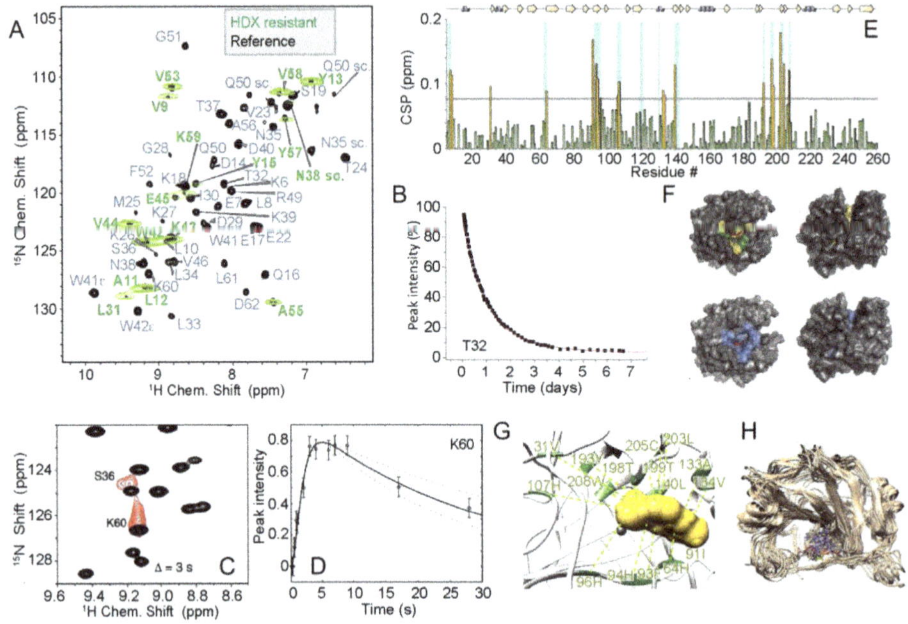


**Figure 15.8.** Exploitation of proton-detected ssNMR for quantification of H/D exchange and for localization of ligand binding: (A) Real-time HDX experiment (after overnight incubation with deuterated buffer), and (B) quantification *via* successive spectral acquisition (adapted by permission from Grohe *et al* (2017), copyright 2017 Springer Nature). (C) Steady-state HDX experiment according to Lopez-del Amo *et al* in which ms-s timescale exchange can be monitored, with fitted intensities over time in (D) (adapted by permission from Lopez Del Amo *et al* (2010), copyright 2010 Springer Nature). (E) Chemical-shift perturbation as a function of residue in hCAII with ligand dorzolamide, read out using hCANH spectra (adapted with permission from Vasa *et al* (2018b), copyright 2018 American Chemical Society). (F) Comparison of experimental (yellow) and expected areas (blue) for data shown in (E)) (adapted with permission from Vasa *et al* 2018b), copyright 2018 American Chemical Society). Green areas do not yield data due to slow exchange. (G) Close-up of inhibitor binding site with chemical-shift perturbations used as pseudo restraints for structure calculation for the hCAII: acetazolamide complex with the full complex structure in (H) (adapted from Vasa *et al* 2019), copyright 2019 John Wiley and Sons Ltd).

structure elucidation by NMR, in simple terms, relies on the acquisition of (hundreds to thousands of) experimental parameters ('restraints') by NMR experiments and successive 'simulated-annealing molecular dynamics' calculations. In this MD step, so-called pseudo-potential energy terms, which reflect the measured experimental properties, are employed in addition to *a priori* information like van-der-Waals and electrostatic potentials to guide the *in silico* folding of the protein primary sequence from an elongated linear chain into the correctly folded structure. In solution NMR, experimental restraints cover information on interatomic distances, bond angles, angles between inter-spin vector and the external magnetic field, and others. Solid-state NMR in general as well as proton-detected ssNMR make use of many of these concepts, too. Leaving aside the possibility of $^3J$-based dihedral angle determination in solution and dipolar vector angle determination at slow MAS, the implementation of dihedral-angle restraints is performed in most structure calculations on the basis of chemical shifts and comparison with data bank information using TALOS (Shen *et al* 2009), irrespective of the aggregation state or technical regime. As such, apart from the consideration of $H^N$ and Hα for secondary chemical shifts, proton-detected ssNMR does not offer much novelty here. For distance restraints, however, carbon-detected methodology and proton-detected ssNMR each differ from the solution NMR counterpart. Whereas traditionally, NOESY restraints are replaced in solids by information from long-range carbon–carbon correlations, for example from DARR (Takegoshi *et al* 2001), PDSD (Bloembergen 1949), PAR (De Paëpe *et al* 2008), PAIN (Lewandowski *et al* 2007) schemes and others, proton-detected ssNMR has used direct proton–proton mixing in the sense of RFDR (Bennett *et al* 1992), DREAM (Verel *et al* 2001) or more sophisticated variants like BASS (Duong *et al* 2018) transfer (Linser *et al* 2011a, Linser *et al* 2014, Shi *et al* 2015, Grohe *et al* 2019, Knight *et al* 2012, Huber *et al* 2012, Vasa *et al* 2019).

The main difference comes with the facilitated involvement of more nuclei than in CC correlations, which is crucial since in each peak of such correlations *two* residues have to be unambiguously identified. As an example, an H/N-RFDR-H/N 4D correlation (Linser *et al* 2014, Linser *et al* 2011a) is shown in figure 15.9(A) and a methyl–methyl 4D (Huber *et al* 2012) in figure 15.9(B), in which both the source and sink residue of the interresidual transfer are characterized with *two* nuclei each. The possibility of long-range dipolar transfers also enables facilitated carbon editing in such experiments without a relay transfer through nitrogen, as is shown in the 4D $H^N$,N-RFDR-$H^N$,Cα correlation on hydrophobin rodlets in figure 15.9(C). For all of these experiments, both $N^H/H^N$ and $C^{ali}/H^{ali}$ transfers can be affected at the same time when CP transfers are used that both match the Hartman–Hahn condition between H & N and H & C (Linser *et al* 2011a). As such, 'time-shared' correlations between different kinds of protons, including amides and aliphatic (or aromatic) protons, can be acquired at the same time, which brings a wealth of information in a single experiment (Linser *et al* 2011a). Figure 15.9(D) shows two different strips from 3D experiments in which both, $^{13}$C-bound protons and amide protons, are considered simultaneously for an hC/NH-RFDR-H through-space correlation. The implementation of time-shared acquisition of distance restraints upon evolution of different heteronuclear shift at the same time, can be implemented both upon characterization of the source and the sink moiety.

In any homonuclear through-space correlations, the diagonal signal is usually much stronger than the cross peaks of interest. In this case, a simple subtraction approach for data in which only the diagonal peak is present can help circumvent artifacts like baseline distortions and 'wiggles', which can be critical for truncated indirect dimensions. In the hNH-RFDR-hNH 4D experiment of figure 15.9(A), such a diagonal suppression technique has been applied (Linser *et al* 2014). Thus, the diagonal peak intensity of K26 is reduced to the same level as the cross peaks.

Protein NMR structure calculation from cross-peak intensity is compromised by multiple effects that are not a direct reporter of distances. Nevertheless, in direct proton–proton correlations in deuterated and (quantitatively) back-exchanged samples, inaccuracies from differential relaxation, source polarization and spin

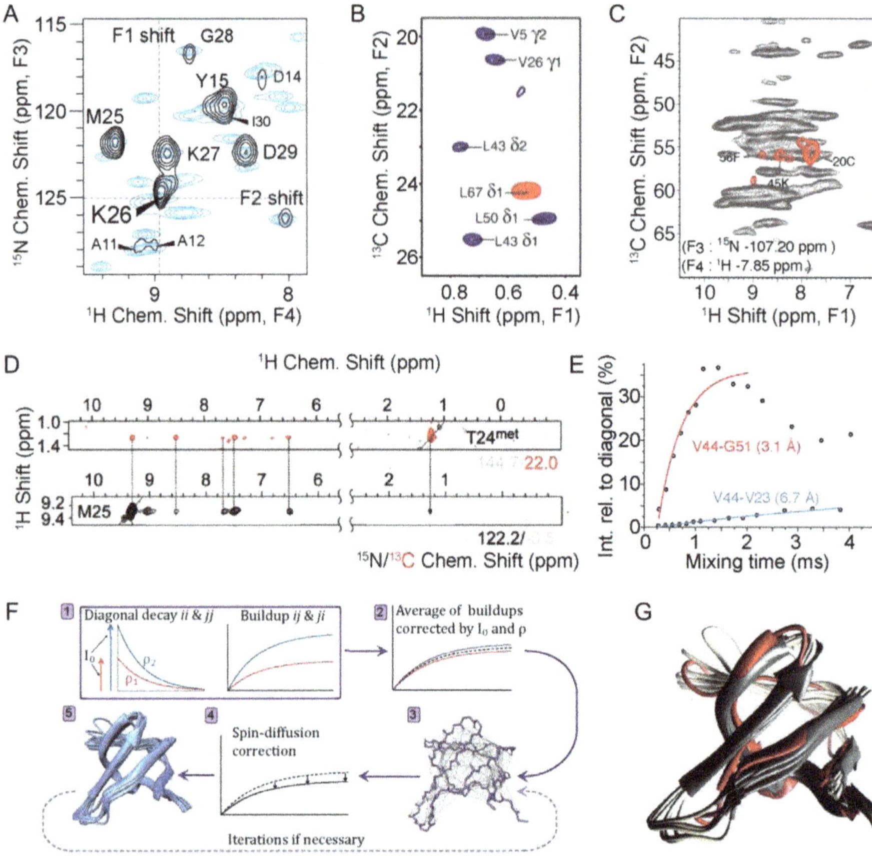


**Figure 15.9.** Proton-detected solid-state NMR protein structure determination: (A) 4D HNh-RFDR-hNH correlation (black) for amide-amide distance restraints, shown here with diagonal peak cancellation for K26, overlaid with an $^{1}H$–$^{15}N$ correlation (blue) (adapted from Linser *et al* (2011a), Linser *et al* (2014), copyright 2014 American Chemical Society). (B) 4D hCH-DREAM-hCH experiment for methyl–methyl distance restraints (reproduced with permission of Royal Society of Chemistry, from Huber *et al* (2012)). Using DREAM, the diagonal peak (red) has opposite sign with respect to direct cross peaks (blue). (C) 4D hCH-RFDR-hNH experiment (red) for additional disambiguation of restraints by involving Cα shifts (recorded on hydrophobin rodlets, overlaid with a $^{1}H$–$^{15}N$ correlation, black). (D) Time-shared acquisition of distance restraints, where contacts to methyls (upper strip) and to amides (lower strip) can be acquired simultaneously (adapted from Linser *et al* (2011a), copyright 2011 American Chemical Society). (E) Time-resolved buildup of RFDR intensities for two SH3 cross peaks, (F) the solids 'eRFDR' routine adapted from the eNOE (Vögeli *et al* 2009) concept, and (G) a backbone structure of the SH3 domain solely on the basis of amide-amide restraints with the x-ray structure (PDB 2NUZ) in red (using eRFDR), adapted from Grohe *et al* (2019) with permission from The Royal Society of Chemistry.

diffusion effects upon cross-peak buildup (figure 15.9(E)) have been addressed in more detail. Adapted from similar strategies in solution (eNOEs, introduced by Vögeli and Riek (Vögeli *et al* 2009, Vögeli *et al* 2015)), RFDR spectra can be similarly optimized (using the so-called 'eRFDR' framework), particularly by measuring buildups rather than single-time-point cross peak intensities, correcting for differential losses during heteronuclear transfers and chemical shift evolution, and by accounting for structural information and spin diffusion effects, back-calculated from preliminary structure calculation and treated in an iterative way (figure 15.9(F)) (Grohe *et al* 2019). Particularly for the short and mid-range restraints up to 5–6 Å, a high accuracy can be obtained, which enables a structure calculation with improved accuracy. Figure 15.9(G) shows a backbone fold from amide distances only, calculated under otherwise 'established' conditions using eRFDR restraints.

### 15.3.5 Relaxation and elucidation of protein dynamics

One of the major differences that the framework of proton detection and fast MAS brings about is how protein dynamics can be assessed. Whereas the quantification of chemical shift anisotropy and the strength of dipolar couplings within highly comparable groups, measured by recoupling sequences, has been the main tool for $^{13}$C-detected ssNMR (Hong *et al* 1997, van Rossum *et al* 2000, Tycko *et al* 1989), measuring $^{15}$N relaxation has become the main focus in the regime of fast MAS (Schanda and Ernst 2016). This is partly due to the increased difficulty of applying recoupling sequences at fast MAS with regard to the duty for the probe, even though feasible methodology has also been worked out (Hou *et al* 2012).

The dipolar coupling DD between $^1$H and $^{15}$N, for which a constant and known distance is usually assumed, is probed for example by recoupling *via* (standard or time-shifted) REDOR-type experiments (figure 15.10(A–B)) (Ma *et al* 2015, Schanda and Ernst 2016). Instead of REDOR, tailored CP experiments, like CP with phase inversion (CPPI), have been shown to be more robust with respect to proton–proton third-spin interactions (i.e., additional protons in the neighborhood), whenever complete sample deuteration is not desired (Asami and Reif 2017). The size of DD informs on the dipolar order parameter, which reflects motional averaging of the amide orientation by internal dynamics over a broad range of timescales. In addition, for MAS frequencies larger than approximately 40 kHz and in the absence of adjacent $^{13}$C coupling partners, $^{15}$N $R_{1\rho}$ (or also $^{13}$C $R_{1\rho}$) (Asami *et al* 2015) can be probed very reliably, which complements $^{15}$N $R_1$ measurements for specifically quantified intermediate-timescale (μs–ms) dynamics (Krushelnitsky *et al* 2010, Lewandowski *et al* 2011b, Knight *et al* 2012, Quinn, Mcdermott 2012, Haller and Schanda 2013). Particularly, in $^{15}$N-labeled, deuterated and proton-back-exchanged samples, $^{15}$N $R_{1\rho}$ measurements have been applied over a wide range of spin lock frequencies, which allows $^{15}$N Bloch–McConnel relaxation dispersion (BMRD) experiments in the solid state (Rovó and Linser 2018, Singh *et al* 2019, Öster *et al* 2019).

As in solution NMR, relaxation rates obtained under different spin lock field strengths are individually dependent on the underlying frequencies of protein motions, or in other words, on the exact form of the spectral density function. Under MAS, these frequencies in question are, however, modulated by the MAS spinning frequency, such that in particular for $R_{1\rho}$ an exploitable dependency on MAS and the spin lock field arises (Zinkevich *et al* 2013, Ma *et al* 2015, Rovó and

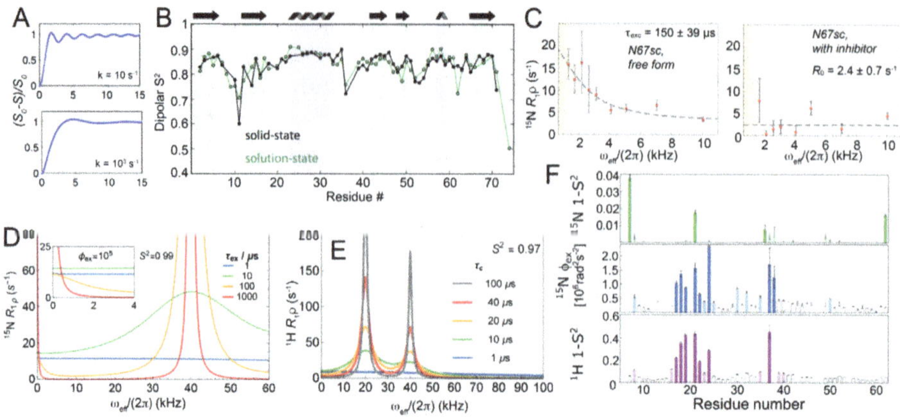

**Figure 15.10.** Assessment of protein dynamics in proton-detected ssNMR: (A) Simulated profiles expected for dipolar recoupling via REDOR (adapted from Schanda and Ernst (2016), copyright 2016, with permission from Elsevier), (B) dipolar order parameters of ubiquitin, compared with solution NMR data, adapted from Ma *et al* 2015 with permission from Springer Nature, (C) $^{15}$N $R_{1\rho}$ relaxation dispersion data of the N67 sidechain of hCAII with and without inhibitor, adapted from Singh *et al* (2019), copyright 2019 American Chemical Society, (D) $^{15}$N $R_{1\rho}$ relaxation rates in the BMRD (left and insert) and NERRD (center) regimes (adapted with permission from Rovó *et al* (2019), copyright 2019 American Chemical Society), (E) $^{1}$H $R_{1\rho}$, where homonuclear recoupling causes an additional peak at the half rotor resonance (HORROR) condition (adapted with permission from Rovó and Linser (2017), copyright 2017 American Chemical Society). Both analytical simulations, (D) and (E), assume 40 kHz rotor frequency; and (F) comparison of $^{15}$N NERRD, $^{15}$N BMRD, and $^{1}$H NERRD (from top down) as a qualitative reporter on dynamics. Due to the sensitivity of $^{1}$H NERRD to motion in the neighborhood, the proton dipolar-coupling network, its trends are similar to those in $^{15}$N BMRD data (adapted with permission from Rovó *et al* (2019), copyright 2019 American Chemical Society).

Linser 2017, Rovó *et al* 2019). Similarly to the $J(0)$ contribution for transverse relaxation in solution, $R_{1\rho}$ under MAS is strongly influenced by the contribution of conditions where the rotor frequency and the spin lock field strength are similar. The BMRD regime is similar to CPMG or $R_{1\rho}$ relaxation dispersion experiments in solution and is exemplified in figure 15.10(C) by two profiles (with and without inhibitor) from the hCAII active site. In addition, 'NEar-Rotary-Resonance relaxation Dispersion' (NERRD) measurements can be used under fast MAS for assessment of μs–ms timescale dynamics in solids (figure 15.10(D)) (Kurbanov *et al* 2011, Zinkevich *et al* 2013, Ma *et al* 2014, Rovó *et al* 2019). The NERRD effect arises due to recoupling of the anisotropic interactions, which are otherwise averaged out under MAS, when spin lock field strength and rotation frequency match. Whereas for rigid moieties this condition is a very narrow regime, it is broadened in the presence of internal dynamics in a quantifiable way. Both analytical equations (Kurbanov *et al* 2011, Rovó and Linser 2017, Zinkevich *et al* 2013) or numerically simulated data (Ma *et al* 2015, Kurauskas *et al* 2017) can be used for interpretation of the data in terms of motional timescales and order parameters. As opposed to BMRD, NERRD does not rely on low-intensity spin locks, which circumvents the deterioration of relaxation data due to the presence of strong direct spin interactions (Rovó and Linser 2018). Also, in contrast to BMRD, it probes changes in the anisotropic interactions (DD and CSA) rather than the isotropic chemical shift changes. This allows a local assessment of data and is in opposition to the influence of motional events in the vicinity, which would also modulate the shift of the nucleus in focus, sensed in BMRD measurements. Figure 15.10(D) shows the expected trends of BMRD and NERRD for a $^{15}N$ nucleus moving on different timescales in a qualitative way.

In addition to $^{15}N$ relaxation, proton-detected ssNMR spectroscopy also allows the measurement of $^{1}H$ relaxation (Rovó and Linser 2017, Gauto *et al* 2017, Rovó *et al* 2019). In proton NERRD experiments, the broadening of the half-rotary-resonance condition (the HORROR condition) can be probed, where only *homo*-nuclear dipolar couplings are reintroduced (figure 15.10(E)). In contrast to $^{15}N$ NERRD, these recoupled interactions are not local but depend on the relative position of the proton in focus with respect to other protons in its vicinity. As such, dynamic events can be monitored either by looking at the mobile residue itself *or* by specifically looking at protons in the vicinity of the hotspot that are not moving. This has the advantage that the presence of dynamics can be probed even if the dynamic residues themselves give rise to peaks that are exchange-broadened and thus undetectable. Whereas the actual rates of proton relaxation are only loosely associated with internal dynamics, the NERRD effect is indeed only observable in the presence of motion. Disadvantages of proton relaxation are that the direct spin interactions reintroduced at the HORROR conditions also depend on the local proton densities and are affected by spin diffusion effects. These effects render proton relaxation as a rather qualitative reporter on protein motion (Rovó and Linser 2017). A coherent picture can be obtained by combining the different types of data, as is shown for the SH3 domain at 35 °C in figure 15.10(F) (Rovó *et al* 2019).

## 15.4 Proton detection for membrane proteins

### 15.4.1 General considerations

Membrane proteins account for 50% of the mass of the cellular plasma membrane (Alberts *et al* 2002) and, in 2017, were targeted by ~67% of US FDA-approved drugs (Santos *et al* 2017). This is not surprising as membrane proteins play crucial roles in processes like cell adhesion, catalysis, membrane-associated transport, sensory perception and signal transduction (Cournia *et al* 2015). Nevertheless, only around 4% of all solved protein structures are structures of membrane proteins. This discrepancy reveals several inherent characteristics of membrane proteins (PDB, https://www.rcsb.org/, 05.02.20). Being involved in highly regulated signaling processes, the abundance of membrane proteins is often low and precisely regulated. As a consequence, over-expression can lead to cellular stress responses due for example to protein aggregation or toxic signaling effects decreasing the overall yield drastically. In addition, trans-membrane proteins especially have impaired solubility due to their amphipathic nature. Still, these proteins need to be solubilized and consequently are removed from their native environment, the cellular lipid bilayer, to allow for structure elucidation. This solubilization is performed upon addition of, e.g. detergent micelles. As their native environment changes, the stability of membrane proteins decreases and they can lose their native structure and/or activity. Therefore, the tools used for solubilization are crucial to produce protein samples representing structures relevant for protein function. Similarly, expression and purification parameters as well as the NMR sample requirements for structure and dynamics studies need to be adapted to the characteristics of each membrane protein. For NMR experiments, isotopic labelling as well as sample stability, homogeneity and amounts are critical factors. Here, proton-detected NMR reveals advantages over carbon-detected NMR because, for example, increased sensitivity of proton detection reduces the sample amount needed to perform similar experiments.

In this last section, we will present practical considerations to employ the methodology presented in the previous sections for membrane proteins, along with several examples in which these methods were successfully applied.

### 15.4.2 Preparative issues

*15.4.2.1 Deuterated expression and H/D exchange*

For ssNMR, deuterated membrane protein samples are often desired (see above), thus requiring optimized expression conditions in isotopically labelled media. As mentioned in the preceding sections, dipolar coupling networks have been a major restriction for high-resolution spectra of proteins in the past. Therefore, proton dilution via deuteration has been used extensively to improve peak resolution and detection sensitivity, as faster spinning via MAS was not yet available. For proton dilution, proteins are produced from cells which have been adapted to grow in deuterated media (see section 15.2), mostly using 100 % $D_2O$ with a deuterated carbon source like glucose or glycerol. To reintroduce protons after expression in fully deuterated media, H/D exchange is performed to repopulate the exchangeable (amide) sites. As described in section 15.2, quantitative back-exchange is a challenge for larger proteins due to the solvent-inaccessible hydrophobic core. For membrane proteins, this problem is aggravated due to the hydrophobic nature of partition of the residues on the protein surface (Schubeis *et al* 2018). As such, even when solubilization is achieved in the presence of aqueous ($H_2O$-based) buffers, a large fraction of the amino acid amide sites usually remains deuterated. Exemplarily, this is illustrated in figure 15.12(D) as only 90 out of 189 residues could be assigned and studied in terms of relaxation (Shi *et al* 2019). As most of the unassigned residues are in the highly hydrophobic region of the lipid bilayer, water-inaccessibility is most likely one of the main reasons for the absence of residue peaks. In favorable cases, proteins can be purified from inclusion bodies using denaturants like urea or guanidinium hydrochloride (where protonated buffers leverage a complete back-exchange) and refolded into a detergent-solubilized native state, for example by sudden dilution into a detergent-containing buffer without denaturant (Hiller *et al* 2008). Such proteins, e.g, many β-barrels like the outer membrane proteins (Omps) described below, are not only uncritical regarding this exchange but usually also deliver good protein yields.

Perdeuteration has been extensively used due to simple labelling schemes and is well suited to study the protein backbone, while most of the deuterons of the sidechains do not undergo H/D exchange. Instead, random adjoint protonation (RAP) (Asami *et al* 2010), fractional deuteration (Mance *et al* 2015) and inverse fractional deuteration (iFD) (Medeiros-Silva *et al* 2016) (see section 15.2) introduce protons in a stochastic way, which results in generation of a sidechain proton content of potential use for structure and interaction elucidation (Asami and Reif 2013, Vasa *et al* 2019). In the latter case, amide protons are also present despite an overall reduction of proton density. Thus, iFD can overcome potential problems associated with H/D exchange upon water inaccessibility or refolding. In figure 15.12(A), the partition of residues that are not accessible by water, as exemplified for a potassium channel, can be visually appreciated (Medeiros-Silva *et al* 2016). Which of these labelling schemes is most suitable depends on the spinning frequency employed as well as on the specific questions in focus.

For membrane proteins, the challenges associated with deuterated media, in particular with deuterated water, are critical not due only to back-exchange. In most cases unfolding/refolding of membrane proteins is difficult and cannot restore structure and activity, especially for larger, more complex membrane proteins (see below). Membrane proteins thus expressed and purified under native conditions generally have low expression yields, and their yields are even decreased in isotopically labelled media. As mentioned above, with MAS frequencies higher than 100 kHz, deuteration of samples is no longer obligatory to achieve highly resolved, proton-detected spectra (Vasa *et al* 2019, Andreas *et al* 2016). The differences in proton line widths for protonated compared to deuterated samples (see figure 15.2) is even less in cases with significant inhomogeneous contributions to the line width, which has been the case for most membrane proteins to date (Ward *et al* 2011).

*15.4.2.2 Solubilization, reconstitution and crystallization of membrane proteins*

Solubilization of membrane proteins can be achieved by different means, for example using detergents, i.e., in micelles or bicelles, via amphipols, lipid nanoparticles or lipid protein nanodiscs. Detergents are usually the method of choice due to easy, efficient, and standardized protocols as well as monetary considerations. In addition to forming micelles, with properties favorable for respective steps in the protocol with regard to the critical micelle concentration (CMC) as well as costs, the detergent of choice is usually a mild detergent with non-ionic and non-denaturing head groups. In contrast, harsher detergents with ionic or zwitterionic head groups may be more efficient during solubilization but can result in non-native protein structures (Seddon *et al* 2004). Although a large variety of detergents is available, they are often not favorable for structural and functional studies as detergents are chemically distinct from membrane lipids. Therefore, several alternative systems have been designed for solubilization as well as strategies during which the solubilized membrane protein is reconstituted into a native-like environment.

For solubilization, alternatives to detergents are amphipols as well as carrier systems which do not interact directly with the membrane protein but with a lipid bilayer surrounding the protein (Etzkorn *et al* 2013). Amphipols are amphipathic polymers which can bind and, therefore, solubilize proteins. In comparison to detergents, they do not form micelles but, similar to detergents, can introduce non-native conditions as they directly interact with the protein. Another approach is to use a mixture of lipids and detergents for solubilization, which forms bicelles of different sizes. These allow a close-to-native lipid environment directly at the protein while keeping the particles small, which is advantageous for solution NMR (short tumbling correlation time) and ssNMR (reasonable protein-to-lipid ratio if empty bicelles are removed). However, an impact of the detergent, which is still present in the final assembly, on protein properties cannot be excluded, and target membrane proteins still need to be compatible with the detergents during solubilization.

Solubilized membrane proteins can also be reconstituted into detergent-free, lipid-based carrier systems like, for example liposomes or nanodiscs. Liposomes consist of lipids only and can be loaded with proteins in very different ratios. In nanodiscs, dedicated membrane scaffold proteins (MSPs, usually apo-lipoproteins) compartmentalize and solubilize the belt of lipids around the membrane protein and, therefore, only indirectly interact with the proteins themselves (Hagn *et al* 2013). Reconstitution for both is typically performed by adding lipids (in the presence or absence of MSPs) to the membrane proteins, which previously were solubilized in detergent micelles. Next, the detergents are slowly removed by dialysis or detergent-binding beads, for which parameters like size, homogeneity and rigidity of the carrier system need to be optimized for the individual protein of interest (Mclean *et al* 2018, Hagn *et al* 2013). Most membrane protein preparations for ssNMR have employed liposomes that are pelleted into the NMR rotor during centrifugation (see below for examples), whereas (due to their solubility) nanodiscs are mostly a solution NMR tool (Hagn *et al* 2013). Nevertheless, nanodiscs also have been tested for ssNMR (Lakomek *et al* 2017, Bersch *et al* 2017). For membrane protein structure and dynamics studies, solubilization (and reconstitution) conditions need to keep the membrane protein in a state which is as native-like as possible. In addition, the spectral resolution of proteins in liposomes is directly affected by sample homogeneity, which can be negatively influenced by protein–protein contacts present at high protein:lipid ratios. However, only with sufficiently high protein:lipid ratios do ssNMR experiments deliver data with high information content, such that a compromise has to be found. Even though lipid reconstitution (e.g. in nanodics or liposomes) represents a more native-like environment than detergent or other synthetic constructs, a non-native lipid composition also may have its own influence on structure and dynamics. For this reason, a composition as close to the native environment should be sought.

Another, distinct approach is to study crystallized membrane proteins via ssNMR. Crystallizing proteins into 3D crystals adds another difficult step to sample preparation and, therefore, presents a potential obstacle. However, particularly narrow, homogeneous line widths in the presence of a lipid environment are also achieved with 2D-crystalline protein samples (Hiller *et al* 2005a). 2D crystallization, as witnessed by flat arrays of regular protein arrangements, has leveraged studies on the outer mitochondrial membrane protein G (OmpG), and the voltage-dependent anion channel (VDAC) has afforded very homogeneous, 2D-crystalline preparations, as described below.

### 15.4.3 Example studies using ssNMR for membrane proteins

As of 2020, many different membrane proteins have been subjected to proton-detected ssNMR spectroscopy. Even though we cannot describe all these studies here, at least several targets shall be discussed briefly in the following section.

Several members of the diverse group of outer membrane proteins have been studied by $^{1}H$-detected ssNMR. The outer membrane protein G (OmpG) is a 14-stranded β-barrel protein of around 300 amino acids and, therefore is one of the larger membrane proteins that have been studied by ssNMR so far. In 2011, Oschkinat, Reif and coworkers published $^{1}H$–$^{15}N$ correlation spectra of a perdeuterated, $^{13}C$-, $^{15}N$-labelled OmpG construct, which was reconstituted into perdeuterated *Escherichia coli* lipid extract bilayers and crystallized yielding highly homogeneous 2D crystals (see figure 15.11(A)) (Linser *et al* 2011b). The H/D back-exchange was performed in 70 % $D_2O$ and 30 % $H_2O$, while the protein was unfolded and refolded in the presence of urea. Proton line widths of 50–80 Hz were obtained in the ssNMR spectra. The experiments confirmed that the structure of OmpG in lipid bilayers shows high similarities to the structure of OmpG in micelles based on isotropic chemical shifts.

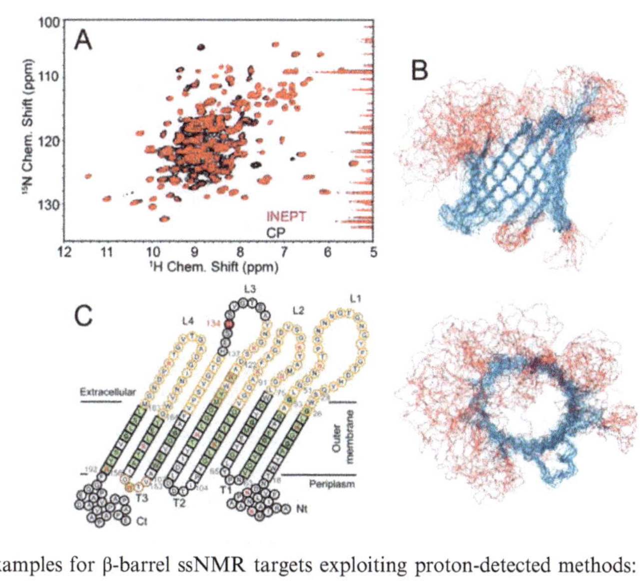


**Figure 15.11.** Examples for β-barrel ssNMR targets exploiting proton-detected methods: (A) comparison of OmpG $^{1}H$–$^{15}N$ correlation based on INEPT-based magnetization transfer (red) and dipolar transfer, in which only the residues with long $T_2$ relaxation times or rigid structural elements show, respectively (adapted from Linser *et al* 2011b), copyright (2011) John Wiley and Sons Ltd); (B) OmpG ssNMR structure (adapted from Retel *et al* (2017) with permission from Springer Nature). Sheet structures and flexible loops are depicted in blue and red, respectively. (C) ssNMR assignments in OmpA, with green colours representing the assigned rigid residues within the lipid bilayer (adapted with permission from Saurel *et al* (2017), copyright 2017 American Chemical Society).

In 2014, a collaboration with the Pintacuda lab resulted in an overall assignment of ~ 40 % of a crystallized sample of OmpG in *E. coli* lipid extracts using proton-detected 3D spectra at 60 kHz MAS (Barbet-Massin *et al* 2014). For 100 % back-exchanged, perdeuterated OmpG, line widths amounted to 130–180 Hz. More extensive assignment (of around 60 %) was achieved in this collaborative effort only in 2017, by combining $^{13}$C-detected, amino acid-specific labelling with proton-detected ssNMR upon perdeuteration (Retel *et al* 2017). In figure 15.11(B), the ssNMR structure obtained is shown. In the loop regions, decreased ssNMR signal intensities or absent resonance signals were explained by high structural flexibility. Retel *et al* also compared the ssNMR structure of OmpG with solution NMR and x-ray crystallography structures. For both ssNMR and solution NMR structures, OmpG has less extended β-sheets towards the extracellular membrane than in the x-ray structure and indicates rather higher structural mobility.

Belonging to the same protein family, site-specific dynamics of the outer membrane protein A (OmpA) of *Klebsiella pneumoniae*, reconstituted in *E. coli* polar extract liposomes, were investigated in 2017 using a 100 % proton back-exchanged, perdeuterated, $^{13}$C-, $^{15}$N-labelled sample (Saurel *et al* 2017). No L3 loop signals in the $^{1}$H–$^{15}$N correlation spectrum at 60 kHz MAS were observed (see figure 15.11(C)). Also based on $^{15}$N $R_1$ and $R_{1\rho}$ relaxation experiments as well as additional proteolysis experiments, the hypothesis of the L3 loop being sufficiently flexible to potentially mediate cell-cell adhesion interactions could be supported.

Outer membrane protein X (OmpX) was studied by Riek, Meier and coworkers in 2017, comparing ssNMR and solution NMR spectra of the protein reconstituted into nanodiscs consisting of 1,2-dimyristoyl-*sn*-glycero-3-phosphocholine/1,2-dimyristoyl-*sn*-glycero-3-phosphoglycerol (DMPC/DMPG) lipids. Here, fully protonated, $^{13}$C-, $^{15}$N-labelled OmpX resulted in a well-resolved $^{1}$H–$^{15}$N correlation spectrum with sharp proton line widths of 40–80 Hz at 110 kHz MAS compared to 30–40 Hz line widths observed in a solution NMR $^{1}$H–$^{15}$N TROSY spectrum.

Similar results were obtained for a $^{1}$H–$^{15}$N CP-HSQC spectrum of another outer membrane protein, the adhesion invasion locus (Ail) protein from *Yersinia pestis*, at 60 kHz MAS by the Opella and Marassi labs (Yao *et al* 2017). The Ail sample was labelled by fractional deuteration (see above) and reconstituted into nanodiscs consisting of a MSP1D1 ΔH5 protein belt filled with DMPC/DMPG lipids in a 3:1 ratio.

In 2017, Agarwal, Pintacuda and co-workers investigated the outer membrane protein AlkL from *Pseudomonas putida*, which is predicted to form a β-barrel involving eight β-strands and thought to navigate hydrophobic molecules through the outer membrane. Information on long-range contacts between amide, Cα and sidechain protons was obtained via the BASS-SD experiment at 111 kHz MAS on fully protonated, $^{13}$C-, $^{15}$N-labelled AlkL sample in deuterated DMPC liposomes (Jain *et al* 2017). In 2019, the Andreas lab investigated site-specific interaction of mobile molecules like water and lipids with surfaces of this protein (Najbauer *et al* 2019).

The voltage-dependent anion channel VDAC, also employed in the latter study, was investigated in detail by Griffin and coworkers. In 2015, Eddy *et al* compared a ssNMR $^{1}H$–$^{15}N$ correlation spectrum at 60 kHz MAS with a $^{1}H$–$^{15}N$ TROSY spectrum of VDAC in LDAO micelles (Eddy *et al* 2015). In fact, similar resolution was achieved. The ssNMR VDAC sample was perdeuterated, 100% proton back-exchanged and $^{13}C$-, $^{15}N$-labelled, reconstituted in DMPC lipid bilayers and 2D-crystallized. In addition, the β-barrel assembly machinery (BAM) complex, which assembles β-barrel outer membrane proteins, was investigated by Baldus and co-workers in 2019. In this case cellular, $^{1}H$-detected ssNMR spectroscopy was used on native membranes (Pinto *et al* 2019).

The family of rhodopsins is another group of membrane proteins studied by $^{1}H$-detected ssNMR spectroscopy. Proteorhodopsin (PR) has seven transmembrane helices and, upon light activation via the retinal cofactor, functions as a proton pump. The Ladizhansky lab in 2011 reported proton line widths of PR as low as 80–144 Hz for $^{1}H$–$^{15}N$ correlation spectra at 28 kHz MAS of perdeuterated and either 40% or 100% back-exchanged protein (Ward *et al* 2011). In 2017 fully protonated, $^{13}C$-, $^{15}N$-labelled PR yielded 190 Hz average proton line width at 100 kHz MAS, with the advantage that problems regarding exchangeable-proton back-exchange (see above) could be avoided (Lalli *et al* 2017). *Anabaena* sensory rhodopsin (ASR) belongs to the same family as PR and has been studied fully protonated, $^{13}C$-, $^{15}N$-labelled reconstituted in DMPC/DMPA liposomes of 9:1 ratio via ssNMR by Ladizhansky and coworkers in 2015. Proton-detected resonances corresponding to lipids and carbohydrates could be observed for a co-purified ASR sample from the expression organism *E. coli*. This can potentially contribute to study native-like interactions of ASR with lipids and carbohydrates.

Bacteriorhodopsin (bR) is another seven-helix integral membrane protein similar to the GPCR family, which can be obtained in a quasi-crystalline state in the purple membranes of *Halobacterium salinarum* without further purification (and without need of unfolding or refolding). A perdeuterated, $^{13}C$-, $^{15}N$-labelled sample, back-exchanged in the presence of 10% $H_2O$ (Reif and coworkers in 2011), yielded excellent, homogeneous line widths comparable to microcrystalline samples. As expected, however, only a subset of residues could be observed due to ineffective amide proton back-exchange (Linser *et al* 2011b). Watts and co-workers recorded $^{1}H$-detected HSQC spectra at 60 kHz MAS of a fully protonated, $^{13}C$-, $^{15}N$-labelled bR sample in native, purple membranes in 2015 (Dannatt *et al* 2015). Spectra corresponding to dipolar-based transfer experiments were characterized by proton line widths below 200 Hz, while the spectra for which scalar transfers were applied showed only very few residues due to flexibility on the chemical shift timescale.

The potassium channel KcsA has been characterized by the groups of Baldus and Weingarth in manifold aspects (see figure 15.12(A–B)). As such, it has also played a role in the course of methodological developments (for example for demonstration of fractional deuteration (Mance *et al* 2015) and inverse fractional deuteration (Medeiros-Silva *et al* 2016) labelling schemes, see above). In a biological sense, it was possible to investigate buried water close to the conduction selectivity filter for the closed-conductive KcsA state and, using the method for KcsA in *E. coli* polar lipid extract bilayers, obtain assignments and protein dynamics information of the transmembrane region. Subsequently, the filter region of KcsA was shown to be

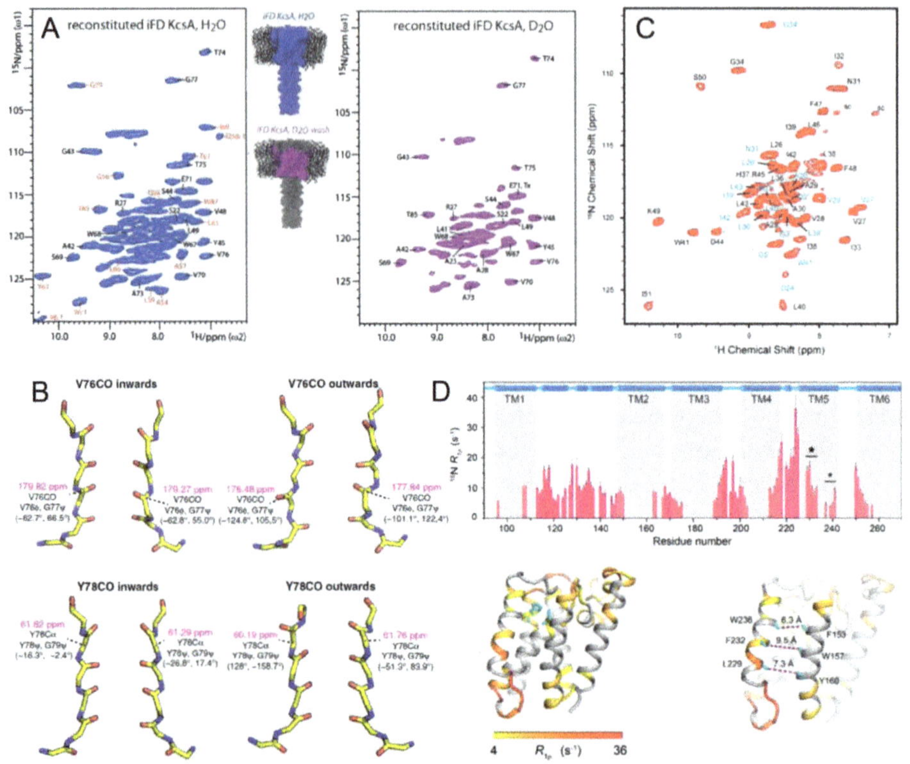


**Figure 15.12.** Examples of α-helical ssNMR targets: (A) $^{1}H$–$^{15}N$ correlation spectra of reconstituted KcsA potassium channel with fractional and inverse fractional deuteration (iFD) (peaks labelled in red and black, respectively) as well as before (peaks colored in blue) and after $D_2O$ wash (peaks colored in magenta) at 60 kHz MAS. Respective protein structures visualize the location of the detected protons in the corresponding coloring code (adapted with permission from Medeiros-Silva *et al* (2016), copyright 2016 John Wiley and Sons Ltd); (B) molecular-dynamics snapshots of inwards and outwards filter states for WT and E71I mutant of KcsA for V76CO and Y78CO with indicated chemical shifts (adapted from Jekhmane *et al* 2019 with permission from Springer Nature); and (C) $^{1}H$–$^{15}N$ correlation spectra of reconstituted $M2_{18-60}$ including backbone assignments (single cross-peaks labelled in blue, doubled cross-peaks in black) recorded at 60 kHz (adapted with permission from Andreas *et al* (2015a), copyright 2015 American Chemical Society); and (D) $^{15}N$ $R_{1\rho}$ relaxation rates from 3D hCANH-based relaxation experiments on reconstituted GlpG, plotted on the crystal structure (PDB 2IC8) (adapted with permission from Shi *et al* (2019), copyright 2019 American Chemical Society).

characterised by slow and rather collective molecular motion in the range of ns to μs based on $^{15}N$ $R_{1\rho}$ measurements. In collaboration with the Roux group in 2019, Weingarth and co-workers investigated KcsA mutants with regard to amino acid-specific functions in the filter region (see figure 15.12(C)). By mutating E71, the filter could be locked in a specific conformation distinct from wild-type KcsA. $^{1}H$–$^{15}N$ correlation spectra, as well as $R_{1\rho}$ relaxation and MD simulation data of KcsA embedded in *E. coli* polar lipids extract, revealed that the filter region of KcsA can adapt various conformational states depending on changes in hydrogen bonding and the water network (Jekhmane *et al* 2019).

A $^1H$–$^{15}N$ correlation spectrum of the perdeuterated Influenza A conductance domain from proton channel M2 in 1,2-diphytanoyl-*sn*-glycero-3-phosphocholine (DPhPC) liposomes at 60 kHz MAS was obtained by Pintacuda and coworkers in 2014. Assembled as tetramers and deeply embedded in the bilayer, the protein yielded well-resolved proton line widths of around 100 Hz (Barbet-Massin *et al* 2014). In 2015, the Griffin lab investigated the perdeuterated Influenza A $M2_{18-60}$ channel at 60 kHz MAS (see figure 15.12(C)), measuring inter- and intramolecular methyl–methyl and histidine–tryptophan distances to study the drug-resistant S31N mutant (Andreas *et al* 2015a). Andreas and colleagues also used proton-detected ssNMR to closely inspect the histidine sidechains in the pore of the channel (Movellan *et al* 2020).

A series of other membrane proteins has been studied by proton-detected ssNMR which does not directly fall into the classes of targets mentioned hitherto:

a. CzcD, a zinc diffusion facilitator protein, is a membrane protein facilitating cation diffusion. A perdeuterated $^{13}C$-, $^{15}N$-labelled sample of CzcD from *Cupriavidus metallidurans* was studied by $^1H$-detected ssNMR by Schanda and colleagues in 2017 (Bersch *et al* 2017). CzcD was solubilized by adding SMA polymer directly to the membrane fraction, preparing nanodiscs based on native lipids and yielding line widths of around 50–70 Hz at 55 kHz MAS.
b. Another (perdeuterated, $^{13}C$-, $^{15}N$-labelled) membrane protein, the disulfide bond formation protein B (DsbB) was studied by Rienstra and coworkers in 2012 (Zhou *et al* 2012). The data recorded at 38 kHz MAS complemented earlier $^{13}C$-detected experiments recorded in 2008 (Li *et al* 2008) for assigning the flexible loop regions and transmembrane sites close to the surface.
c. The 62 kDa YidC membrane protein was analyzed by the Baldus lab in 2018. The insertase from *E. coli* was prepared as native, cellular sample for ssNMR measurements at 55 kHz MAS as a perdeuterated and $^{13}C$-, $^{15}N$-labelled protein. This study, which combined ssNMR and electron cryotomography, yielded information on the overall fold of YidC under native conditions.
d. Finally, the first member of the family of rhomboid proteases, GlpG from *E. coli*, was studied by $^1H$-detected ssNMR by the Lange group in 2019 (Shi *et al* 2019). An N-terminally truncated construct of GlpG was expressed under perdeuterated, $^{13}C$-, $^{15}N$-labelled conditions and reconstituted in liposomes consisting of *E. coli* total lipid extract. In total, 90 residues could be assigned *de novo* and short-range contacts to water molecules revealed water accessibility at the active site of the protease. In addition, $R_{1\rho}$ relaxation dispersion measurements (see figure 15.12(D)) afforded dynamics of various regions of GlpG.

### 15.4.4 Perspectives

Solid-state NMR methods development, particularly in the direction of faster MAS and proton detection, has revolutionized the way ssNMR can be done nowadays.

The ameliorated demands for sample amounts required have been of major benefit in particular for membrane proteins. Whereas the signal-to-noise ratio per mg of protein is tremendous for deuterated samples spun for example at around 40–60 kHz, two significant bottlenecks remain in this MAS regime: perdeuterated expression and proton-back-exchange into the sample. Both difficulties are dramatically facilitated for the newly accessible spinning regime over 100 kHz, where fully protonated samples start to be really versatile. As such, manifold membrane protein studies will extremely benefit from these techniques. The decreased sensitivity and extended measurement times in comparison to deuterated and partially protonated samples in the MAS regime around 60 kHz—added to the overall costs of the hardware—however, retain solid-state NMR as an expensive undertaking generally, where the thin line of detailed, useful and, at the same time, affordable information needs to be achieved with great precision.

## References


Agarwal V, Diehl A, Skrynnikov N and Reif B 2006 High resolution $^{1}$H detected $^{1}$H,$^{13}$C correlation spectra in MAS solid-state NMR using deuterated proteins with selective $^{1}$H,$^{2}$H isotopic labeling of methyl groups *J. Am. Chem. Soc.* **128** 12620–1

Agarwal V and Reif B 2008 Residual methyl protonation in perdeuterated proteins for multi-dimensional correlation experiments in MAS solid-state NMR spectroscopy *J. Magn. Reson.* **194** 16–24

Akbey Ü, Lange S, Franks W T, Linser R, Rehbein K, Diehl A, Van Rossum B-J, Reif B and Oschkinat H 2009 Optimum levels of exchangeable protons in perdeuterated proteins for proton detection in MAS solid-state NMR spectroscopy *J. Biomol. NMR* **46** 67–73

Alberts B, Johnson A, Lewis J, Raff M, Roberts K and Walter P 2002 *Molecular Biology of the Cell* 4th edn (New York: Garland Science)

Andreas L B *et al* 2016 Structure of fully protonated proteins by proton-detected magic-angle spinning NMR *Proc. Natl Acad. Sci. USA* **113** 9187–92

Andreas L B, Reese M, Eddy M T, Gelev V, Ni Q Z, Miller E A, Emsley L, Pintacuda G, Chou J J and Griffin R G 2015a Structure and mechanism of the influenza A M218–60 dimer of dimers *J. Am. Chem. Soc.* **137** 14877–86

Andreas L B *et al* 2015b Protein residue linking in a single spectrum for magic-angle spinning NMR assignment *J. Biomol. NMR* **62** 253–61

Andronesi O C, Von Bergen M, Biernat J, Seidel K, Griesinger C, Mandelkow E and Baldus M 2008 Characterization of Alzheimer's-like paired helical filaments from the core domain of tau protein using solid-state NMR spectroscopy *J. Am. Chem. Soc.* **130** 5922–8

Asami S, Porter J R, Lange O F and Reif B 2015 Access to Cα backbone dynamics of biological solids by $^{13}$C $T_1$ relaxation and molecular dynamics simulation *J. Am. Chem. Soc.* **137** 1094–100

Asami S and Reif B 2013 Proton-detected solid-state NMR spectroscopy at aliphatic sites: application to crystalline systems *Acc. Chem. Res.* **46** 2089–97

Asami S and Reif B 2017 Comparative study of REDOR and CPPI derived order parameters by $^{1}$H-detected MAS NMR and MD simulations *J. Phys. Chem.* B **121** 8719–30

Asami S, Schmieder P and Reif B 2010 High resolution $^{1}$H-detected solid-state NMR spectroscopy of protein aliphatic resonances: access to tertiary structure information *J. Am. Chem. Soc.* **132** 15133–5

Barbet-Massin E *et al* 2013 Out-and-back $^{13}C$–$^{13}C$ scalar transfers in protein resonance assignment by proton-detected solid-state NMR under ultra-fast MAS *J. Biomol. NMR* **56** 379–86

Barbet-Massin E *et al* 2014 Rapid proton-detected NMR assignment for proteins with fast magic angle spinning *J. Am. Chem. Soc.* **136** 12489–97

Bennett A E, OK J H, Vega S and Griffin R G 1992 Chemical shift correlation spectroscopy in rotating solids: radio frequency-driven dipolar recoupling and longitudinal exchange *J. Chem. Phys.* **96** 8624–7

Bersch B, Dörr J M, Hessel A, Killian J A and Schanda P 2017 Proton-detected solid-state NMR spectroscopy of a zinc diffusion facilitator protein in native nanodiscs *Angew. Chem. Int. Ed.* **56** 2508–12

Bloembergen N 1949 On the interaction of nuclear spins in a crystalline lattice *Physica* **15** 386–426

Böckmann A, Ernst M and Meier B H 2015 Spinning proteins, the faster, the better? *J. Magn. Reson.* **253** 71–9

Chevelkov V, Rehbein K, Diel A and Reif B 2006 Ultra-high resolution in proton solid-state NMR spectroscopy at high levels of deuteration *Angew. Chem. Int. Ed.* **45** 3878–81

Cournia Z *et al* 2015 Membrane protein structure, function, and dynamics: a perspective from experiments and theory *J. Membr. Biol.* **248** 611–40

Dannatt H R W, Taylor G F, Varga K, Higman V A, Pfeil M-P, Asilmovska L, Judge P J and Watts A 2015 $^{13}C$- and $^{1}H$-detection under fast MAS for the study of poorly available proteins: application to sub-milligram quantities of a 7 trans-membrane protein *J. Biomol. NMR* **62** 17–23

De Paëpe G, Lewandowski J R, Loquet A, Böckmann A and Griffin R G 2008 Proton assisted recoupling and protein structure determination *J. Chem. Phys.* **129** 245101–1

Duong N T, Raran-Kurussi S, Nishiyama Y and Agarwal V 2018 Quantitative $^{1}H$–$^{1}H$ distances in protonated solids by frequency- selective recoupling at fast magic angle spinning NMR *J. Phys. Chem. Lett.* **9** 5948–54

Eddy M T, Su Y, Silvers R, Andreas L, Clark L, Wagner G, Pintacuda G, Emsley L and Griffin R G 2015 Lipid bilayer-bound conformation of an integral membrane beta barrel protein by multidimensional MAS NMR *J. Biomol. NMR* **61** 299–310

Etzkorn M, Raschle T, Hagn F, Gelev V, Rice A J, Walz T and Wagner G 2013 Cell-free expressed bacteriorhodopsin in different soluble membrane mimetics: biophysical properties and NMR accessibility *Structure* **5** 394–401

Fraga H *et al* 2017 Solid-state NMR H–N–(C)–H and H–N–C–C 3D/4D correlation experiments for resonance assignment of large proteins *Chem. Phys. Chem.* **18** 2697–703

Gauto D F, Hessel A, Rovó P, Kurauskas V, Linser R and Schanda P 2017 Protein conformational dynamics studied by $^{15}N$ and $^{1}H$ $R_{1\rho}$ relaxation dispersion: application to wild-type and G53A ubiquitin crystals *Solid State Nucl. Mag.* **87** 86–95

Gauto D F *et al* 2019 Aromatic ring dynamics, thermal activation, and transient conformations of a 468 kDa enzyme by specific $^{1}H$–$^{13}C$ labeling and fast magic-angle spinning NMR *J. Am. Chem. Soc.* **141** 11183–95

Grohe K, Movellan K T, Vasa S K, Giller K, Becker S and Linser R 2017 Non-equilibrium hydrogen exchange for determination of H-bond strength and water accessibility in solid proteins *J. Biomol. NMR* **68** 7–17

Grohe K, Nimerovsky E, Singh H, Vasa S K, SöLdner B, Vögeli B H, Rienstra C M and Linser R 2019 Exact distance measurements for structure and dynamics in solid proteins by fast magic angle spinning NMR *Chem. Commun.* **55** 7899–902

Hagn F, Etzkorn M, Raschle T and Wagner G 2013 Optimized phospholipid bilayer nanodiscs facilitate high-resolution structure determination of membrane proteins *J. Am. Chem. Soc.* **135** 1919–25

Haller J D and Schanda P 2013 Amplitudes and time scales of picosecond-to-microsecond motion in proteins studied by solid-state NMR: a critical evaluation of experimental approaches and application to crystalline ubiquitin *J. Biomol. NMR* **57** 263–80

Hiller M, Krabben L, Vinothkumar K R, Castellani F, Van Rossum B J, Kühlbrandt W and Oschkinat H 2005a Solid-state magic-angle spinning NMR of outer-membrane protein G from *Escherichia coli Chem. Bio. Chem.* **6** 1679–84

Hiller S, Fiorito F, Wüthrich K and Wider G 2005b Automated projection spectroscopy (APSY) *Proc. Natl Acad. Sci. USA* **102** 10876–81

Hiller S, Garces R G, Malia T J, Orekhov V Y, Colombini M and Wagner G 2008 Solution structure of the integral human membrane protein VDAC-1 in detergent micelles *Science* **321** 1206

Hong M, Gross J D, Rienstra C M, Griffin R G, Kumashiro K K and Schmidt-Rohr K 1997 Coupling amplification in 2D MAS NMR and its application to torsion angle determination in peptides *J. Magn. Reson.* **129** 85–92

Hou G, Byeon I-J L, Ahn J, Gronenborn A M and Polenova T 2012 Recoupling of chemical shift anisotropy by R-symmetry sequences in magic angle spinning NMR spectroscopy *J. Chem. Phys.* **137** 134201

Huber M, Böckmann A, Hiller S and Meier B H 2012 4D solid-state NMR for protein structure determination *Phys. Chem. Chem. Phys.* **14** 5239–46

Jain M G *et al* 2017 Selective $^{1}$H–$^{1}$H distance restraints in fully protonated proteins by very fast magic-angle spinning solid-state NMR *J. Phys. Chem. Lett.* **8** 2399–405

Jekhmane S, Medeiros-Silva J, LI J, Kümmerer F, Müller-Hermes C, Baldus M, Roux B and Weingarth M 2019 Shifts in the selectivity filter dynamics cause modal gating in K+ channels *Nat. Commun.* **10** 123

Kazimierczuk K, Zawadzka A and Koźmiński W 2009 Narrow peaks and high dimensionalities: exploiting the advantages of random sampling *J. Magn. Reson.* **197** 219–28

Klein A, Rovó P, Sakhrani V V, Skowronek P, Kukuk L, Bojarzyn T, Janning P, Vasa S K, Mueller L J and Linser R 2020 Partial resonance assignments in 70 kDa tryptophan synthase by five-dimensional solid-state NMR *submitted*

Klein A, Vasa S K and Linser R 2018 Automated projection spectroscopy in solid-state NMR *J. Biomol. NMR* **72** 163–70

Knight M J, Pell A J, Bertini I, Felli I C, Gonnelli L, Pierattelli R, Herrmann T, Emsley L and Pintacuda G 2012 Structure and backbone dynamics of a microcrystalline metalloprotein by solid-state NMR *Proc. Natl Acad. Sci. USA* **109** 11095–100

Knight M J *et al* 2011 Fast resonance assignment and fold determination of human superoxide dismutase by high-resolution proton-detected solid-state MAS NMR spectroscopy *Angew. Chem. Int. Ed.* **50** 11697–701

Kosiński K, Stanek J, Górka M J, Żerko S and Koźmiński W 2017 Reconstruction of non-uniformly sampled five-dimensional NMR spectra by signal separation algorithm *J. Biomol. NMR* **68** 129–38

Krushelnitsky A, Zinkevich T, Reichert D, Chevelkov V and Reif B 2010 Microsecond time scale mobility in a solid protein as studied by the $^{15}$N $R_{1\rho}$ site-specific NMR relaxation rates *J. Am. Chem. Soc.* **132** 11850–3

Kulminskaya N, Vasa S K, Giller K, Becker S, Kwan A, Sunde M and Linser R 2016 Access to side-chain carbon information in deuterated solids under fast mas through non-rotor-synchronized mixing *Chem. Commun.* **52** 268–71

Kulminskaya N, Vasa S K, Giller K, Becker S and Linser R 2015 Asynchronous through-bond homonuclear isotropic mixing: application to carbon-carbon transfer in perdeuterated proteins under MAS *J. Biomol. NMR* **63** 245–53

Kurauskas V *et al* 2017 Slow conformational exchange and overall rocking motion in ubiquitin protein crystals *Nat. Commun.* **8** 145

Kurbanov R, Zinkevich T and Krushelnitsky A 2011 The nuclear magnetic resonance relaxation data analysis in solids: general $R_1/R_{1\rho}$ equations and the model-free approach *J. Chem. Phys.* **135** 184104

Lakomek N-A, Frey L, Bibow S, Böckmann A, Riek R and Meier B H 2017 Proton-detected NMR spectroscopy of nanodisc-embedded membrane proteins: MAS solid-state vs solution-state methods *J. Phys. Chem.* B **121** 7671–80

Lalli D, Idso M N, Andreas L B, Hussain S, Baxter N, Han S, Chmelka B F and Pintacuda G 2017 Proton-based structural analysis of a heptahelical transmembrane protein in lipid bilayers *J. Am. Chem. Soc.* **139** 13006–12

Lewandowski J R 2013 Advances in solid-state relaxation methodology for probing site-specific protein dynamics *Acc. Chem. Res.* **46** 2018–27

Lewandowski J R, Depaepe G and Griffin R G 2007 Proton assisted insensitive nuclei cross polarization *J. Am. Chem. Soc.* **129** 728–9

Lewandowski J R, Dumez J-N, Akbey Ü, Lange S, Emsley L and Oschkinat H 2011a Enhanced resolution and coherence lifetimes in the solid-state NMR spectroscopy of perdeuterated proteins under ultrafast magic-angle spinning *J. Phys. Chem. Lett.* **2** 2205–11

Lewandowski J R, Sass H J, Grzesiek S, Blackledge M and Emsley L 2011b Site-specific measurement of slow motions in proteins *J. Am. Chem. Soc.* **133** 16762–5

Li Y, Berthold D A, Gennis R B and Rienstra C M 2008 Chemical shift assignment of the transmembrane helices of DsbB, a 20-kDa integral membrane enzyme, by 3D magic-angle spinning NMR spectroscopy *Protein Sci.* **17** 199–204

Linser R 2011 Side-chain to backbone correlations from solid-state NMR of perdeuterated proteins through combined excitation and long-range magnetization transfers *J. Biomol. NMR* **51** 221–6

Linser R 2012 Backbone assignment of perdeuterated proteins using long-range H/C-dipolar transfers *J. Biomol. NMR* **52** 151–8

Linser R, Bardiaux B, Higman V, Fink U and Reif B 2011a Structure calculation from unambiguous long-range amide and methyl $^1H$–$^1H$ distance restraints for a microcrystalline protein with MAS solid-state NMR spectroscopy *J. Am. Chem. Soc.* **133** 5905–12

Linser R, Bardiaux B, Hyberts S G, Kwan A H, Morris V K, Sunde M and Wagner G 2014 Solid-state NMR structure determination from diagonal-compensated, sparsely nonuniform-sampled 4D proton–proton restraints *J. Am. Chem. Soc.* **136** 11002–10

Linser R, Chevelkov V, Diehl A and Reif B 2007 Sensitivity enhancement using paramagnetic relaxation in MAS solid-state NMR of perdeuterated proteins *J. Magn. Reson.* **189** 209–16

Linser R *et al* 2011b Proton detected solid-state NMR of fibrillar and membrane proteins *Angew. Chem. Int. Ed.* **50** 4508–12

Linser R, Fink U and Reif B 2008 Proton-detected scalar coupling based assignment strategies in MAS solid-state NMR spectroscopy applied to perdeuterated proteins *J. Magn. Reson.* **193** 89–93

Linser R, Fink U and Reif B 2010a Assignment of dynamic regions in biological solids enabled by spin-state selective NMR experiments *J. Am. Chem. Soc.* **132** 8891–3

Linser R, Fink U and Reif B 2010b Narrow carbonyl resonances in proton-diluted proteins facilitate NMR assignments in the solid state *J. Biomol. NMR* **47** 1–6

Lopez Del Amo J-M, Fink U and Reif B 2010 Quantification of protein backbone hydrogen-deuterium exchange rates by MAS solid-state NMR spectroscopy *J. Biomol. NMR* **48** 203–12

Ma P, Haller J D, Zajakala J, Macek P, Sivertsen A C, Willbold D, Boisbouvier J and Schanda P 2014 Probing transient conformational states of proteins by solid-state $R_{1\rho}$ relaxation-dispersion NMR spectroscopy *Angew. Chem. Int. Ed.* **53** 4312–7

Ma P *et al* 2015 Observing the overall rocking motion of a protein in a crystal *Nat. Commun.* **6** 8361

Mance D, Sinnige T, Kaplan M, Narasimhan S, Daniëls M, Houben K, Baldus M and Weingarth M 2015 An efficient labelling approach to harness backbone and side-chain protons in $^1$H-detected solid-state NMR spectroscopy *Angew. Chem. Int. Ed.* **54** 15799–803

Mclean M A, Gregory M C and Sligar S G 2018 Nanodiscs: a controlled bilayer surface for the study of membrane proteins *Annu. Rev. Biophys.* **47** 107–24

Medeiros-Silva J, Mance D, Daniels M, Jekhmane S, Houben K, Baldus M and Weingarth M 2016 $^1$H-detected solid-state NMR studies of water-inaccessible proteins in vitro and in situ *Angew. Chem. Int. Ed.* **55** 13606–10

Miljenović T M, Jia X and Mobli M 2017 Nonuniform sampling in biomolecular NMR ed G A Webb *Modern Magnetic Resonance* (Cham: Springer)

Movellan K T, Wegstroth M, Overkamp K, Leonov A, Becker S and Andreas L B 2020 Imidazole–imidazole hydrogen bonding in the pH-sensing histidine side chains of influenza A M2 *J. Am. Chem. Soc.* **142** 2704–8

Najbauer E E, Movellan K T, Schubeis T, Schwarzer T, Castiglione K, Giller K, Pintacuda G, Becker S and Andreas L B 2019 Probing membrane protein insertion into lipid bilayers by solid-state NMR *Chem. Phys. Chem.* **20** 302–10

Nieuwkoop A J, Wylie B J, Franks W T, Shah G J and Rienstra C M 2009 Atomic resolution protein structure determination by three-dimensional transferred echo double resonance solid-state nuclear magnetic resonance spectroscopy *J. Chem. Phys.* **131** 95–101

Orton H W *et al* 2020 Protein NMR resonance assignment without spectral analysis: 5D SOlid-state Automated Projection SpectroscopY (SO-APSY) *Angew. Chem. Int. Ed.* **59** 2380–4

Öster C, Kosol S and Lewandowski J R 2019 Quantifying microsecond exchange in large protein complexes with accelerated relaxation dispersion experiments in the solid state *Sci. Rep.* **9** 11082

Penzel S, Smith A A, Agarwal V, Hunkeler A, Samoson A, Böckmann A, Ernst M and Meier B H 2015 Protein resonance assignment at MAS frequencies approaching 100 KHz: a quantitative comparison of J-coupling and dipolar-coupling-based transfer methods *J. Biomol. NMR* **63** 165–86

Pinto C, Mance D, Julien M, Daniels M, Weingarth M and Baldus M 2019 Studying assembly of the BAM complex in native membranes by cellular solid-state NMR spectroscopy *J. Struct. Biol.* **206** 1–11

Quinn C M and Mcdermott A E 2012 Quantifying conformational dynamics using solid-state R1q experiments *J. Magn. Reson.* **222** 1–7

Retel J S *et al* 2017 Structure of outer membrane protein G in lipid bilayers *Nat. Commun.* **8** 2073

Rovó P and Linser R 2017 Microsecond timescale proton rotating-frame relaxation under magic angle spinning *J. Phys. Chem.* B **121** 6117–30

Rovó P and Linser R 2018 Microsecond timescale protein dynamics: a combined solid-state NMR approach *Chem. Phys. Chem.* **19** 34–9

Rovó P, Smith C A, Gauto D, De Groot B L, Schanda P and Linser R 2019 Mechanistic insights into microsecond timescale motion of solid proteins using complementary $^{15}$N and $^{1}$H relaxation dispersion techniques *J. Am. Chem. Soc.* **141** 858–69

Santos R *et al* 2017 A comprehensive map of molecular drug targets *Nat. Rev. Drug Discov.* **16** 19–34

Saurel O, Iordanov I, Nars G, Demange P, Le Marchand T, Andreas L B, Pintacuda G and Milon A 2017 Local and global dynamics in *Klebsiella pneumoniae* outer membrane protein a in lipid bilayers probed at atomic resolution *J. Am. Chem. Soc.* **139** 1590–7

Schanda P and Ernst M 2016 Studying dynamics by magic-angle spinning solid-state NMR spectroscopy: principles and applications to biomolecules *Prog. Nucl. Magn. Reson. Spectrosc.* **96** 1–46

Schanda P, Huber M, Verel R, Ernst M and Meier B H 2009 Direct detection of $^{3h}J_{NC'}$ hydrogen-bond scalar couplings in proteins by solid-state NMR spectroscopy *Angew. Chem. Int. Ed.* **48** 9322–5

Schanda P, Meier B H and Ernst M 2010 Quantitative analysis of protein backbone dynamics in microcrystalline ubiquitin by solid-state NMR spectroscopy *J. Am. Chem. Soc.* **132** 15957–67

Schubeis T, Le Marchand T, Andreas L B and Pintacuda G 2018 $^{1}$H magic-angle spinning NMR evolves as a powerful new tool for membrane proteins *J. Magn. Reson.* **287** 140–52

Seddon A M, Curnow P and Booth P J 2004 Membrane proteins, lipids and detergents: not just a soap opera *BBA Biomembranes* **1666** 105–17

Sharma K, Madhu P K and Mote K R 2016 A suite of pulse sequences based on multiple sequential acquisitions at one and two radiofrequency channels for solid-state magic-angle spinning NMR studies of proteins *J. Biomol. NMR* **65** 127–41

Shen Y, Delaglio F, Cornilescu G and Bax A 2009 TALOS+: a hybrid method for predicting protein backbone torsion angles from NMR chemical shifts *J. Biomol. NMR* **44** 213–23

Shi C, Fricke P, Lin L, Chevelkov V, Wegstroth M, Giller K, Becker S, Thanbichler M and Lange A 2015 Atomic-resolution structure of cytoskeletal bactofilin by solid-state NMR *Sci. Adv.* **1** e1501087

Shi C, Öster C, Bohg C, Li L, Lange S, Chevelkov V and Lange A 2019 Structure and dynamics of the rhomboid protease GlpG in liposomes studied by solid-state NMR *J. Am. Chem. Soc.* **141** 17314–21

Singh H, Vasa S K, Jangra H, Rovó P, Päslack C, Das C K, Zipse H, Schäfer L V and Linser R 2019 Fast-microsecond dynamics of the protein-water network in the active site of human carbonic anhydrase II by solid-state NMR spectroscopy *J. Am. Chem. Soc.* **141** 19276–88

Stanek J *et al* 2016 NMR spectroscopic assignment of backbone and side-chain protons in fully protonated proteins: microcrystals, sedimented assemblies, and amyloid fibrils *Angew. Chem. Int. Ed.* **55** 15504–9

Stanek J, Schubeis T, Paluch P, Güntert P, Andreas L B and Pintacuda G 2020 Automated backbone NMR resonance assignment of large proteins using redundant linking from a single simultaneous acquisition *J. Am. Chem. Soc.* **142** 5793–9

Takegoshi K, Nakamura S and Terao T 2001 $^{13}$C–$^{1}$H dipolar-assisted rotational resonance in magic-angle spinning NMR *Chem. Phys. Lett.* **344** 631–7

Tang M, Comellas G, Mueller L J and Rienstra C M 2010 High resolution $^{13}$C-detected solid-state NMR spectroscopy of a deuterated protein *J. Biomol. NMR* **48** 103–11

Tycko R, Dabbagh G and Mirau P A 1989 Determination of chemical-shift-anisotropy lineshapes in a two-dimensional magic-angle-spinning NMR experiment *J. Magn. Reson.* **85** 265–74

Van Rossum B J, De Groot C P, Ladizhansky V, Vega S and De Groot H J M 2000 A method for measuring heteronuclear ($^{1}$H–$^{13}$C) distances in high speed MAS NMR *J. Am. Chem. Soc.* **122** 3465–72

Vasa S K, Rovó P and Linser R 2018a Protons as versatile reporters in solid-state NMR spectroscopy *Acc. Chem. Res.* **51** 1386–95

Vasa S K, Singh H, Grohe K and Linser R 2019 Assessment of a large enzyme-drug complex by proton-detected solid-state NMR without deuteration *Angew. Chem. Int. Ed.* **58** 5758–62

Vasa S K, Singh H, Rovó P and Linser R 2018b Dynamics and interactions of a 29-kDa human enzyme studied by solid-state NMR *J. Phys. Chem. Lett.* **9** 1307–11

Verel R, Ernst M and Meier B H 2001 Adiabatic dipolar recoupling in solid-state NMR: the DREAM scheme *J. Magn. Reson.* **150** 81–90

Vögeli B, Olsson S, Riek R and Güntert P 2015 Complementarity and congruence between exact NOEs and traditional NMR probes for spatial decoding of protein dynamics *J. Struct. Biol.* **191** 306–17

Vögeli B, Segawa T F, Leitz D, Sobol A, Choutko A, Trzesniak D, Van Gunsteren W and Riek R 2009 Exact distances and internal dynamics of perdeuterated ubiquitin from NOE buildups *J. Am. Chem. Soc.* **131** 17215–25

Ward M E, Shi L, Lake E, Krishnamurthy S, Hutchins H, Brown L S and Ladizhansky V 2011 Proton-detected solid-state NMR reveals intramembrane polar networks in a seven-helical transmembrane protein proteorhodopsin *J. Am. Chem. Soc.* **133** 17434–43

Xiang S, Biernat J, Mandelkow E, Becker S and Linser R 2016 Backbone assignment for minimal protein amounts of low structural homogeneity in the absence of deuteration *Chem. Commun.* **52** 4002–5

Xiang S, Grohe K, Rovó P, Vasa S, Giller K, Becker S and Linser R 2015 Sequential backbone assignment based on dipolar amide-to-amide correlation experiments *J. Biomol. NMR* **62** 303–11

Xiang S *et al* 2017 A two-component adhesive: tau fibrils arise from a combination of a well-defined motif and conformationally flexible interactions *J. Am. Chem. Soc.* **139** 2639–46

Xue K, Sarkar R, Motz C, Asami S, Camargo D C R, Decker V, Wegner S, Tosner Z and Reif B 2017 Limits of resolution and sensitivity of proton detected MAS solid-state NMR experiments at 111 kHz in deuterated and protonated *Proteins. Sci. Rep.* **7** 7444

Yao Y, Dutta S K, Park S H, Rai R, Fujimoto L M, Bobkov A A, Opella S J and Marassi F M 2017 High resolution solid-state NMR spectroscopy of the *Yersinia pestis* outer membrane protein Ail in lipid membranes *J. Biomol. NMR* **67** 179–90

Zhou D H, Nieuwkoop A J, Berthold D A, Comellas G, Sperling L J, Tang M, Shah G J, Brea E J, Lemkau L R and Rienstra C M 2012 Solid-state NMR analysis of membrane proteins and protein aggregates by proton detected spectroscopy *J. Biomol. NMR* **54** 291–305

Zhou D H, Shah G, Cormos M, Mullen C, Sandoz D and Rienstra C M 2007a Proton-detected solid-state NMR spectroscopy of fully protonated proteins at 40 kHz magic-angle spinning *J. Am. Chem. Soc.* **129** 11791–801

Zhou D H, Shea J J, Nieuwkoop A J, Franks W T, Wylie B J, Mullen C, Sandoz D and Rienstra C M 2007b Solid-state protein structure determination with proton-detected triple resonance 3D magic-angle spinning NMR spectroscopy *Angew. Chem. Int. Ed.* **46** 8380–3

Zinkevich T, Chevelkov V, Reif B, Saalwächter K and Krushelnitsky A 2013 Internal protein dynamics on ps to μs timescales as studied by multi-frequency $^{15}N$ solid-state NMR relaxation *J. Biomol. NMR* **57** 219–35

Lightning Source UK Ltd.
Milton Keynes UK
UKHW051607111121
393782UK00003B/113

9 780750 325301